Additional MATHEMATICS

Additional MATHEMATICS

Hugh Morrison • Alison Hughes • Anne Hunt • Mairead Tallon

Photo acknowledgements: pages 175 and 259 Image Select; page 276 The Wellcome Institute Library, London; page 350 The Royal Society; all others Mary Evans Picture Library.
Cover illustration: © 1948 M. C. Escher Foundation ® – Barn – Holland.

First published in 1994
by Hodder and Stoughton Limited, an Hachette UK Company
338 Euston Road, London NW1 3BH

Reprinted in 1994, 1995, 1996, 1998, 1999, 2000, 2001, 2002, 2003, 2004, 2005, 2006, 2007, 2008, 2009, 2010, 2012

Design by Eric Drewery

Typeset by Servis Filmsetting Ltd, Manchester in 10/12pt Times
Printed and bound by CPI Group (UK) Ltd, Croydon, CR0 4YY

A CIP catalogue record for this book is available from the British Library

ISBN: 978-0-719-55324-0

Contents

PURE MATHEMATICS

MECHANICS

Preface

Additional Mathematics provides an important bridge between GCSE mathematics and A/AS mathematics. The syllabus is followed by a wide spectrum of pupils including high attaining GSCE pupils wishing to broaden their mathematical experience and pupils taking the subject as a component of their sixth form curriculum.

For the former, the syllabus provides important information when making well-informed A and AS subject choices. Pupils gain valuable insights into the likely intellectual demands of advanced mathematics and have the opportunity to evaluate their capabilities in pure mathematics, mechanics and statistics. Additional Mathematics grades provide teachers with vital information with which to guide pupils planning their A level studies. Additional Mathematics equips those pupils following A and AS courses in the sciences (physics and biology in particular) with the necessary competencies to cope with the mathematical demands of these courses.

This book offers complete and comprehensive coverage of the pure mathematics, mechanics and statistics components of the Northern Ireland Schools Examinations and Assessment Council's Additional Mathematics syllabus, in a single volume. The book addresses topics which are central to all AO additional and AS mathematics syllabuses and pupils preparing for the University of Cambridge international examination in Additional Mathematics will find this book a valuable resource. Each chapter comprises detailed teaching notes, plentiful examples and carefully graded exercises. A wide range of examination questions are presented as worked examples in the text and also feature prominently in the exercises.

It is now recognised that pupils should perceive mathematics as the fruit of human endeavour and not as a series of timeless, abstract laws. To this end, each chapter begins with a profile of a mathematician closely identified with the ideas presented in the chapter.

Please note
The final answers to questions requiring the use of tables or calculators will normally be given correct to four significant figures.

Acknowledgements

The authors are indebted to the following for their permission to reproduce questions from past examination papers:

The Northern Ireland Schools Examinations and Assessment Council, the University of Cambridge Local Examinations Syndicate and the Oxford and Cambridge Schools Examination Board.

All answers to exercises are entirely the responsibility of the authors and not of the examination boards.

We would also like to express our gratitude to Claire Shannon and Alastair Edwards for their careful preparation of the manuscript and to Rose Wands of John Murray for her invaluable guidance and encouragement throughout.

CHAPTER 1

Some useful topics in algebra

The French mathematician **Pierre Fermat** (1601–1665) made a formidable contribution to mathematical thinking although he is best remembered for two popular mathematical ideas. He died believing that the formula $2^p + 1$ $(p = 2^n,$ $n = 1,2,3,\ldots)$ was the much sought-after law of prime numbers. The reader can easily confirm the 'law' for $n = 1$, 2 and 3. Furthermore, $n = 4$ gives 65 537 which is prime. Unfortunately, $n = 5$ generates 4 294 967 297 which is 6 700 417 times 641. Fermat acknowledged before his death that the proof of his 'law' had eluded him. However, he claimed to have successfully proved his *great* theorem that $x^n + y^n = z^n$ is impossible for positive integers x, y, z and $n > 2$. Three centuries of struggle by the world's finest mathematical minds to replicate the proof could have been avoided had Fermat not laid claim to the proof in the margin of his copy of a text by Diophantus. His reference to the proof of his great theorem reads '... I have assuredly found an admirable proof of this, but the margin is too narrow to contain it'. In his *Some Thoughts on Modern Mathematical Research* published in 1912, G. A. Miller refers to a reward of \$25 000 for the proof of Fermat's great theorem.

In view of the fact that the offered prize is about \$25 000 and that lack of marginal space in his copy of Diophantus was the reason given by Fermat for not communicating his proof, one might be tempted to wish that one could send credit for a dime back through the ages to Fermat and thus secure this coveted prize, if it actually existed. This might, however, result more seriously than one would at first suppose; for if Fermat had bought on credit a dime's worth of paper even during the year of his death, 1665, and if this bill had been drawing compound interest at the rate of 6% since that time, the bill would not amount to more than seven times as much as the prize.

In June 1993, the British mathematician Andrew Wiles presented a proof of the theorem based upon elliptical functions. However, at approximately 1000 pages in length, it is unlikely to be the method of proof which Fermat had in mind.

INTRODUCTION

This chapter addresses those algebraic topics of particular importance in Additional Mathematics, namely, the manipulation of algebraic fractions, the solution of three linear equations in three unknowns, and the solution of two simultaneous equations, one linear and one quadratic. A knowledge of basic algebra to Key Stage 4 is assumed.

ALGEBRAIC FRACTIONS

This sections details how algebraic fractions (such as

$$\frac{x+1}{4-x} \quad \frac{x^2-2x+1}{x^2-4} \quad \frac{x^2+x+2}{x-2}$$

etc.) are added, subtracted, multiplied and divided. The methods involved are best illustrated through a series of examples.

Addition and subtraction of algebraic fractions

EXAMPLE 1
Simplify

$$\frac{1}{x+2}+\frac{4}{x-3}$$

Solution 1
The first algebraic fraction is multiplied by 1 in the form $(x-3)/(x-3)$ and the second by 1 in the form $(x+2)/(x+2)$.

Then, since multiplying any expression by unity leaves it unchanged

$$\frac{1}{x+2}+\frac{4}{x-3}=\frac{x-3}{(x+2)(x-3)}+\frac{4(x+2)}{(x-3)(x+2)}$$

The algebraic fractions on the right-hand side of this equation may now be combined yielding

$$\frac{1}{x+2}+\frac{4}{x-3}=\frac{x-3+4(x+2)}{(x+2)(x-3)}$$

$$=\frac{5x+5}{(x+2)(x-3)}$$

EXAMPLE 2
Simplify

$$\frac{1}{2x+7}-\frac{1}{x-3}$$

Solution 2

Proceeding as in Example 1,

$$\frac{1}{2x+7} - \frac{1}{x-3} = \frac{x-3}{(2x+7)(x-3)} - \frac{2x+7}{(x-3)(2x+7)}$$

or $$\frac{1}{2x+7} - \frac{1}{x-3} = \frac{(x-3)-(2x+7)}{(2x+7)(x-3)}$$

Hence

$$\frac{1}{2x+7} - \frac{1}{x-3} = \frac{-x-10}{(2x+7)(x-3)}$$

EXAMPLE 3

Simplify

$$\frac{1}{(x+4)(x-7)} + \frac{1}{(x+2)(x+4)}$$

Solution 3

Here the common denominator will be $(x+4)(x-7)(x+2)$ and so

$$\frac{1}{(x+4)(x-7)} + \frac{1}{(x+2)(x+4)}$$

$$= \frac{(x+2)}{(x+4)(x-7)(x+2)} + \frac{(x-7)}{(x+2)(x+4)(x-7)}$$

It follows that

$$\frac{1}{(x+4)(x-7)} + \frac{1}{(x+2)(x+4)} = \frac{2x-5}{(x+4)(x-7)(x+2)}$$

EXAMPLE 4

Simplify

$$\frac{1}{3x^2+x-4} + \frac{1}{3+x-4x^2}$$

Solution 4

The first step is to factorise the quadratic expressions $3x^2+x-4$ and $3+x-4x^2$.

Using any standard procedure for factorising quadratics:

$$3x^2+x-4 = (3x+4)(x-1)$$

and $$3+x-4x^2 = (3+4x)(1-x)$$

The sum of fractions may therefore be rewritten:

$$\frac{1}{3x^2+x-4} + \frac{1}{3+x-4x^2} = \frac{1}{(3x+4)(x-1)} + \frac{1}{(3+4x)(1-x)}$$

Now $(x-1) = -(1-x)$ and so changing the sign of the numerator and denominator of the second fraction gives:

$$\frac{1}{3x^2+x-4}+\frac{1}{3+x-4x^2}=\frac{1}{(3x+4)(x-1)}-\frac{1}{(3+4x)(x-1)}$$

The right-hand side of this equation becomes

$$\frac{3+4x}{(3+4x)(3x+4)(x-1)}-\frac{(3x+4)}{(3x+4)(3+4x)(x-1)}$$

This sum may now be further simplified:

$$\frac{3+4x-(3x+4)}{(3+4x)(3x+4)(x-1)}$$

$$=\frac{3+4x-3x-4}{(3+4x)(3x+4)(x-1)}$$

$$=\frac{x-1}{(3+4x)(3x+4)(x-1)}$$

Finally, 'cancelling' the $(x-1)$ factors yields

$$\frac{1}{(3+4x)(3x+4)}$$

In summary:

$$\frac{1}{3x^2+x-4}+\frac{1}{3+x-4x^2}=\frac{1}{(3+4x)(3x+4)}$$

EXAMPLE 5

Simplify

$$1+\frac{3}{x+2}-\frac{2(11x+10)}{3x^2+4x-4}$$

Solution 5

Again the first step is to factorise the quadratic expression.

$$3x^2+4x-4=(3x-2)(x+2)$$

The task is therefore to simplify

$$1+\frac{3}{x+2}-\frac{2(11x+10)}{(3x-2)(x+2)}$$

Using the common denominator $(3x-2)(x+2)$ this expression may be rewritten as

$$\frac{(3x-2)(x+2)}{(3x-2)(x+2)}+\frac{3(3x-2)}{(3x-2)(x+2)}-\frac{2(11x+10)}{(3x-2)(x+2)}$$

$$= \frac{(3x-2)(x+2)+3(3x-2)-2(11x+10)}{(3x-2)(x+2)}$$

Simplifying the numerator by expanding the brackets as follows

$$\frac{3x^2+4x-4+9x-6-22x-20}{(3x-2)(x+2)}$$

eventually leads to

$$\frac{3x^2-9x-30}{(3x-2)(x+2)}$$

Factorising the numerator, the expression becomes

$$\frac{3(x-5)(x+2)}{(3x-2)(x+2)}$$

and 'cancelling' the $(x+2)$ factors it follows that

$$1+\frac{3}{x+2}-\frac{2(11x+10)}{3x^2+4x-4}=\frac{3(x-5)}{3x-2}$$

Multiplication and division of algebraic fractions

EXAMPLE 6

Simplify

$$\frac{Z^2-4Z+4}{Z^2+3Z}\times\frac{Z^2+6Z+9}{Z^2+Z-6}$$

Solution 6

Factorising the four quadratic expressions

$$\frac{(Z^2-4Z+4)(Z^2+6Z+9)}{(Z^2+3Z)(Z^2+Z-6)}=\frac{(Z-2)^2(Z+3)^2}{Z(Z+3)(Z+3)(Z-2)}$$

Cancelling $(Z-2)$ allows the right-hand side to be rewritten as

$$\frac{(Z-2)(Z+3)^2}{Z(Z+3)(Z+3)}$$

And cancelling the $(Z+3)^2$ in the numerator and denominator gives

$$\frac{Z^2-4Z+4}{Z^2+3Z}\times\frac{Z^2+6Z+9}{Z^2+Z-6}=\frac{Z-2}{Z}$$

EXAMPLE 7

Simplify

$$\frac{x^2-5x+6}{x^2-9x+18}\div\frac{x-2}{2x^2-12x}$$

Solution 7

Using the traditional 'turn upside-down and multiply' rule for dividing fractions

$$\frac{x^2-5x+6}{x^2-9x+18} \div \frac{x-2}{2x^2-12x} = \frac{x^2-5x+6}{x^2-9x+18} \times \frac{2x^2-12x}{x-2}$$

Now factorising where possible

$$\frac{x^2-5x+6}{x^2-9x+18} \div \frac{x-2}{2x^2-12x} = \frac{(x-3)(x-2)\,2x(x-6)}{(x-6)(x-3)(x-2)}$$

cancellation now yields:

$$\frac{x^2-5x+6}{x^2-9x+18} \div \frac{x-2}{2x^2-12x} = 2x$$

EXERCISE 1.1

Simplify the following algebraic expressions.

1. $\dfrac{3}{x+2} + \dfrac{5}{x-3}$
2. $\dfrac{4x}{2x-3} + \dfrac{3x}{2x+3}$
3. $x + \dfrac{x^2+xy}{x-y}$
4. $\dfrac{x}{4} + \dfrac{3x^2}{x+5}$
5. $\dfrac{4x}{x+1} + \dfrac{3x}{x-1} + \dfrac{2x^2}{x^2-1}$
6. $\dfrac{4x+3y}{x^2+2xy+y^2} + \dfrac{3x-2y}{x^2-y^2}$
7. $\dfrac{x}{x+y} + \dfrac{y}{x-y} + \dfrac{1}{x}$
8. $\dfrac{3x}{2x+2y} + \dfrac{5y}{3x-3y} + \dfrac{xy}{x^2-y^2}$
9. $\dfrac{1}{x^2-5x-14} + \dfrac{2}{x^2-4} + \dfrac{3}{x^2-9x+14}$
10. $x + \dfrac{x^2}{x+1} + \dfrac{x^2}{x-1}$

EXERCISE 1.2

Simplify the following algebraic expressions.

1. $\dfrac{4x^2+8x}{3x+9} \times \dfrac{6x+18}{5x^2+10x}$
2. $\dfrac{x^2+5x+6}{4x^2} \times \dfrac{6x}{x^2+x-2}$
3. $\dfrac{x-y}{2} \times \dfrac{4}{x^2-y^2} \times \dfrac{x+y}{1}$
4. $\dfrac{3x^2-6xy}{5y^2z} \times \dfrac{4z^2}{x^2-4y^2} \times \dfrac{10y}{9x}$
5. $\dfrac{x^2-y^2}{xy} \div \dfrac{x-y}{y}$
6. $\dfrac{x^2-9}{4} \div \dfrac{x+3}{12x}$

7. $\dfrac{(x+y)^2}{4x-4y} \div \dfrac{x-y}{6x+6y}$

8. $\dfrac{yx^2-x^3}{y+x} \div \dfrac{3x}{y^2-x^2}$

9. $\dfrac{x^2-16}{x^2-25} \div \dfrac{x+4}{x-5}$

THE SOLUTION OF TWO SIMULTANEOUS EQUATIONS, ONE LINEAR AND ONE QUADRATIC

In order to cope with problems involving the calculation of the area contained by the intersection of a line and a curve (see Chapter 6), the reader needs to be able to solve simultaneous equations where one equation represents a line and the other, a curve. Furthermore, while two lines intersect in a point, a line will generally cut a curve in two points. Line equations are of the form $y = mx + c$ (e.g. $y = 4x - 2$, $y = -2x + 7$, $x + 2y = 4$) while the equations of curves involve higher powers of x and/or y (e.g. $y = 4x^2 - 2$, $x = y^2 - 2$, $x^2 + y^2 = 4$). The examples which follow will help illustrate the **substitution method** for solving such simultaneous equations.

EXAMPLE 8

Solve the simultaneous equations.

$$5x + 9y = 1$$
$$x^2 - 9y = 13$$

Solution 8

The solution of this pair of equations is analogous to finding the points of intersection of a line and a curve. The line equation here is $5x + 9y = 1$. $5x + 9y = 1$ is therefore referred to as the 'linear' equation of the pair. The curve equation is $x^2 - 9y = 13$, the x^2 term's association with the quadratic equation resulting in this equation being classified 'quadratic'.

Now $5x + 9y = 1$ may be rearranged as $9y = 1 - 5x$.

Replacing the $9y$ term in the quadratic equation with $(1 - 5x)$ leads to:

$$x^2 - (1 - 5x) = 13$$
or $x^2 + 5x - 14 = 0$
Factorising, $(x + 7)(x - 2) = 0$
Hence $x = -7$ or $x = 2$.

The x coordinate of each point of intersection has been found i.e. the points of intersection are $(-7, ?)$ and $(2, ?)$. Since the points of intersection are, by their very nature, common to both the line and curve, they must satisfy the line equation in particular.

The y coordinates corresponding to $x = -7$ and $x = 2$ may hence be found as follows.

For $x = -7$: $5(-7) + 9y = 1 \Rightarrow y = 4$
for $x = 2$: $5(2) + 9y = 1 \Rightarrow y = -1$

The solutions are summarised as either $x = 2$ and $y = -1$ or $x = -7$ and $y = 4$.

EXAMPLE 9

Solve the simultaneous equations.

$$2x^2 - 3y^2 + 43 = 0$$
$$4x - 3y - 1 = 0$$

Solution 9

From the linear equation

$$4x = 1 + 3y \text{ or } x = \tfrac{1}{4}(1 + 3y).$$

Substituting for x in the quadratic:

$$2\{\tfrac{1}{4}(1 + 3y)\}^2 - 3y^2 + 43 = 0$$
or $$\tfrac{1}{8}(1 + 6y + 9y^2) - 3y^2 + 43 = 0$$
or $$1 + 6y + 9y^2 - 24y^2 + 344 = 0$$
or $$-15y^2 + 6y + 345 = 0$$

Hence

$$y = \frac{-6 \pm \sqrt{\{6^2 - 4(-15)(345)\}}}{-30}$$

$$y = \frac{-6 \pm 144}{-30}$$

$$y = -4.6 \text{ or } y = 5$$

Using the linear equation to obtain the x values appropriate to these values of y,

$y = -4.6 \Rightarrow 4x = 1 + 3(-4.6) \Rightarrow x = -3.2$
and for $y = 5 \Rightarrow 4x = 1 + 3(5) \Rightarrow x = 4$
Therefore $x = -3.2$ and $y = -4.6$ or $x = 4$ and $y = 5$.

EXERCISE 1.3

Solve the following simultaneous equations – one linear and one quadratic – giving your answers to three places of decimals where appropriate.

1. $y = x + 6$
$y = x^2$

2. $y = x + 35$
$y = 6x^2$

3. $y = -30$
$y = x^2 - 11x$

4. $y = x - 9$
$y = x^2 + 7x$

5. $2x + y = 6$
$x^2 - 2xy + 2y^2 = 4$

6. $3x + 4y = 10$
$2x^2 + 5xy = 3$

7. $4y + x = 5$
$x^2 + xy - 3y^2 = 5$

8. $2x + y = 3$
$x^2 - 2xy + y^2 = 8$

9. $x = 3(y - 2)$
$4y^2 = 3xy - 8$

10. $\frac{y}{3} = \frac{2x - 4}{5}$
$3(x + y) = xy - 4$

11. $2x^2 + y^2 = 9$
$\frac{x + 2}{4} = \frac{y - x}{3}$

12. $3y = 2(x + 1)$
$4x^2 - 3xy = 7$

13. $x^2 + 2y^2 = 18$
$\frac{4 + x}{3} = \frac{y + x}{2}$

14. $\frac{x}{3} = \frac{2y + 1}{2}$
$3(x - y) = xy + 1$

15. The line $y = 3x - 1$ intersects the curve $2x^2 + 2y^2 - x + y - 11 = 0$ at A and B. Calculate the coordinates of (i) A and B, (ii) the mid-point of AB.

(U.C.L.E.S. Additional Mathematics, November 1991)

16. Solve the simultaneous equations

$x^2 - xy + y^2 = 7$
$2x - y = 5$

(U.C.L.E.S. Additional Mathematics, June 1992)

17. Calculate the coordinates of the points of intersection of the curve $x^2 + y^2 = 8$ and the straight line $2x - y = 2$.

(U.C.L.E.S. Additional Mathematics, June 1991)

THE SOLUTION OF THREE SIMULTANEOUS EQUATIONS IN THREE UNKNOWNS

In solving real-life problems, from those in current electricity to economics and financial forecasting, mathematicians are frequently faced with the task of solving a large number of linear simultaneous equations. In order to illustrate that the familiar methods used to tackle the solution of two equations in two unknowns can be extended to the solution of a greater number of simultaneous equations, the focus now switches to three equations with three unknowns. A knowledge of the solution of linear simultaneous equations in two variables will be assumed in what follows.

It is instructive to illustrate how the methods used to solve simultaneous equations such as

$2x + 3y = -1$
$x + y = 0$

can be extended to accommodate the solution of a system of equations such as:

$$5x - y + 2z = 25 \quad (1)$$
$$3x + 2y - 3z = 16 \quad (2)$$
$$2x - y + z = 9 \quad (3)$$

The aim is to rewrite equations (1) to (3) so that the coefficients of x, y or z are equal (apart from a possible difference in sign). For example, if equation (1) were multiplied by 6, equation (2) by 10 and equation (3) by 15, then the term '$30x$' appears at the beginning of each of the new equations. Alternatively, equations (1) and (3) could be multiplied by 2 generating a '$2y$' term in all three equations (albeit $-2y$ in the case of equations (1) and (3)). Finally, multiplying equation (1) by 3, equation (2) by 2, and equation (3) by 6 produces three new equations each containing the term $6z$.

Mathematicians are adept at finding little 'tricks' which serve to minimise the more tedious aspects of problem solving. One such 'trick' in this particular case is to add the three equations and immediately find $10x = 50$ so $x = 5$.

In order to illustrate an effective means of solution for a system of equations such as that given above, the variable y shall now be eliminated from equations (1)–(3).

If equation (1) is multiplied by 2 and added to equation (2), y can be eliminated from these two equations. Thus

$$\begin{array}{rl} 10x - 2y + 4z &= 50 \\ 3x + 2y - 3z &= 16 \\ \hline 13x \qquad + z &= 66 \end{array} \quad (4)$$

Similarly y is eliminated from equations (2) and (3) by multiplying (3) by 2 and adding (2).

$$\begin{array}{lrl} \text{So} & 3x + 2y - 3z &= 16 \\ & 4x - 2y + 2z &= 18 \\ \hline \Rightarrow & 7x \qquad - z &= 34 \end{array} \quad (5)$$

Now equations (4) and (5) may be solved (i.e. eliminating z to find x) using the standard procedure for solving simple simultaneous linear equations in two variables.

$$\begin{array}{rl} 13x + z &= 66 \\ 7x - z &= 34 \\ \hline 20x \qquad &= 100 \end{array}$$
$$\Rightarrow x = 5$$

Substituting for x in (4) gives $13 \times 5 + z = 66 \Rightarrow z = 1$.

Substituting $x = 5$ and $z = 1$ in equation (1) yields:

$$5 \times 5 - y + 2 \times 1 = 25 \Rightarrow y = 2$$

The solution is therefore $x = 5$, $y = 2$, $z = 1$.

EXAMPLE 10

Three types of microcomputer, types A, B and C, are used in a university. In January the Physics Department purchased 3 microcomputers of type A, 4 of type B and 2 of type C at a total cost of £4600. At the same time the Chemistry Department purchased 2 microcomputers of type A, 3 of type B and 1 of type C at a total cost of £3000.

On 1st February the cost of type A microcomputers was increased by 20% and the cost of type C microcomputers was increased by £50. The cost of type B microcomputers remained unchanged.

In February the Mathematics Department purchased 5 microcomputers of type A, 10 of type B and 10 of type C at a total cost of £14 900.

(i) Calculate the cost of each type of microcomputer in January.
(ii) In February the Computer Science Department had a budget of £24 800. They bought 10 microcomputers of type A, 10 of type B and the remaining money from the budget was used to purchase microcomputers of type C. Calculate the number of type C microcomputers bought.

(N. Ireland Additional Mathematics, 1991)

Solution 10

Let £x, £y and £z be the January costs of microcomputers A, B and C respectively.

Now $3x + 4y + 2z = 4600$ (Physics Department)
$2x + 3y + z = 3000$ (Chemistry Department)

The February cost of computer A is £$1.20x$ and the February cost of computer C is £$(z + 50)$.

Therefore $5 \times 1.2x + 10y + 10(z + 50) = 14\,900$ (Mathematics Department).

Collecting the equations together and simplifying the third:

$$3x + 4y + 2z = 4600 \qquad (6)$$
$$2x + 3y + z = 3000 \qquad (7)$$
$$6x + 10y + 10z = 14\,400 \qquad (8)$$

x shall be eliminated in what follows:

$2 \times (6)$ gives $6x + 8y + 4z = 9200$

$3 \times (7)$ gives $6x + 9y + 3z = 9000$

$$\Rightarrow \quad -y + z = 200 \qquad (9)$$

$3 \times (7)$ gives $6x + 9y + 3z = 9000$

$1 \times (8)$ gives $6x + 10y + 10z = 14\,400$

$$\Rightarrow \quad -y - 7z = -5400 \qquad (10)$$

Solving (9) and (10) simultaneously

$$-y + z = 200$$
$$-y - 7z = -5400$$
Hence $8z = 5600 \Rightarrow z = 700$

Substituting $z = 700$ in equation (9) gives $-y + 700 = 200$
$\Rightarrow y = 500.$

Substituting $z = 700$ and $y = 500$ in (6) yields

$$3x + 4 \times 500 + 2 \times 700 = 4600 \Rightarrow x = 400.$$

Hence microcomputer A costs £400, B costs £500 and C costs £700 in January.

February prices are A = £400 × 1.20 or A = £480, B = £500 and C = £(700 + 50) or C = £750.

Suppose x computers of type C are purchased by the Computer Science department:

$$10 \times 480 + 10 \times 500 + x \times 750 = 24\,800 \Rightarrow x = 20.$$

Hence 20 type C microcomputers were purchased.

EXAMPLE 11

(a) Solve the simultaneous equations.

$$2x - y + 7z = -1$$
$$-3x + 2y - 2z = 12$$
$$5x + 3z = -7$$

(b) On flights between Belfast and London, an airline offers three types of fare – a key fare of £50, a normal fare of £70 and a standby fare of £40.

100 passengers travelled on each of two flights.

Compared with the first flight, the second flight had 10 more passengers travelling on key fares, 20 fewer passengers travelling on normal fares, and twice as many passengers travelling on standby fares.

The total income from fares on the first flight was £6300.

Calculate the numbers of passengers who paid key, normal and standby fares on each flight. Hence determine the total income from fares on the second flight.

(N. Ireland Additional Mathematics, 1988)

Solution 11

(a)
$$2x - y + 7z = -1 \quad (11)$$
$$-3x + 2y - 2z = 12 \quad (12)$$
$$5x + 3z = -7 \quad (13)$$

Eliminate y from (11) and (12)

$$\begin{array}{ll} 2 \times (11): & 4x - 2y + 14z = -2 \\ 1 \times (12): & -3x + 2y - 2z = 12 \\ \hline & x + 12z = 10 \end{array} \qquad (14)$$

Solving (13) and (14):

$$5x + 3z = -7$$
$$x + 12z = 10$$

These may be rewritten:

$$5x + 3z = -7$$
$$5x + 60z = 50$$

Subtracting: $-57z = -57 \Rightarrow z = 1$

Substituting $z = 1$ in (14) gives

$$x + 12 \times 1 = 10 \Rightarrow x = -2$$

Substituting $x = -2$ and $z = 1$ in (11) gives:

$$2 \times -2 - y + 7 \times 1 = -1 \Rightarrow y = 4$$

The complete solution is therefore $x = -2$, $y = 4$, $z = 1$.

b) Let k passengers fly on key fare, n fly normal fare and s fly standby fare.

For the first flight:

$$k + n + s = 100$$
$$50k + 70n + 40s = 6300$$

For the second flight:

$$(k + 10) + (n - 20) + 2s = 100$$

Summarising all the available information:

$$k + n + s = 100 \qquad (15)$$
$$k + n + 2s = 110 \qquad (16)$$
$$50k + 70n + 40s = 6300$$
or $$5k + 7n + 4s = 630 \qquad (17)$$

Subtracting (15) from (16) yields $s = 10$.

Substituting $s = 10$ in (16) and (17) leads to:

$$k + n = 90$$
$$5k + 7n = 590$$

or
$$\begin{array}{r} 5k + 5n = 450 \\ 5k + 7n = 590 \\ \hline 2n = 140 \Rightarrow n = 70 \end{array}$$

Also, substituting $s = 10$, $n = 70$ in (15) gives $k = 20$.

Therefore on the first flight 20 passengers paid key fares, 70 normal and 10 standby. On the second flight 30 flew key, 50 normal and 20 standby. The income from the second flight is given by:

$$£[30 \times 50 + 50 \times 70 + 20 \times 40] = £5800$$

The required income is therefore £5800.

EXERCISE 1.4

For questions 1 to 15 solve the set of simultaneous linear equations.

1. $x + y + z = 9$, $2x + y - z = 3$, $x - z = -2$

2. $x + y + z = 6$, $2x + 5y + 3z = 29$, $x - y + z = -4$

3. $x + 2y + z = 2$, $3x - y + 2z = -6$, $2x + y + 3z = -4$

4. $x + y + 2z = -4$, $4x + 2y + z = -15$, $2x + y - z = -6$

5. $x + y + z = 5$, $3x - 2y - z = 3$, $x + 2y + 3z = 2$

6. $x + 2y + z = 20$, $x - 2y + z = 0$, $4x - 3y + 2z = 15$

7. $2x + 3y + z = 13$, $3x + y + 4z = 28$, $x - y - z = 6$

8. $x + 2y + 3z = 17$, $2x + y - 3z = -11$, $3x - y + z = 4$

9. $x + y + 2z = 2$, $2x + y - 2z = 20$, $4x + 2y + z = 25$

10. $x + y + z = 3$, $2x - y + 3z = 7$, $3x + y + 2z = 8$

11. $x + y + z = 3$, $3x + 2y + z = 2$, $5x + 3y + 2z = 1$

12. $-x + y + 3z = -$, $2x - y - z = 11$, $x + 2y + 3z = -$

13. $2x + y + z = 6$, $x + 2y + z = 4$, $x + z = 6$

14. $x + 2y + z = 5$, $4x + 4y + z = 2$, $9x + 8y + z = -1$

15. $x + y + z = -2$, $x - 5y + z = -1$, $9x + 23y + z = 50$

16. (a) Why have the simultaneous equations

$$2x - 6y = 8$$
$$-x + 3y = 4$$

no solution?

(b) In the Brown household electricity is used for lighting, oil for heating and gas for cooking. On the first day of the months February, May and October 1988, the number of units of each fuel used that day and the corresponding costs were recorded. This information is shown in Table 1.1.

Table 1.1

Date	Electricity (units)	Oil (units)	Gas (units)	Total co (pence)
1st Feb	6	4	3	97
1st May	3	5	2	86
1st Oct	9	3	5	138

The unit costs of electricity and oil remained unchanged throughout the year, but the unit cost of gas increased by 20% on 30th September 1988.

(i) Calculate the unit costs of electricity, oil and gas in February 1988.

(ii) On 1st December 1988, 5 units of electricity, 7 units of oil and 4 units of gas were used. Calculate the total cost of fuel for that day.

(N. Ireland Additional Mathematics, 1989)

17. An engineer carries out emergency repairs to factory machinery. For each repair job he charges a call-out fee, f pounds, which is doubled if the call is at night. In addition, he charges m pounds per mile for his travel to the factory and h pounds per hour for the time he spends at the job.

He made the following charges for 3 jobs.

Table 1.2

Time of call	Distance travelled (miles)	Time at job (hours)	Total charge (£)
day	20	2	39
night	12	6	95
day	40	5	80

(i) Use the data in Table 1.2 to write down three equations connecting f, m and h. Hence find the values of f, m and h.

(ii) Calculate how much he would charge for a call to a factory at night if he was required to travel 60 miles and work for 4 hours.

(N. Ireland Additional Mathematics (part question), 1987)

18. Solve the set of equations

$$x - 5y - z = 1$$
$$3x + 2y = 7$$
$$x + z = 10.$$

(N. Ireland Additional Mathematics, 1990)

19. A coal merchant sells three grades of coal, A, B and C, priced respectively at £x, £y and £z per bag. Table 1.3 gives the number of bags of each grade of coal purchased by each of three customers, and the amount paid by each.

Table 1.3

	Grade			Amount paid (£)
	A	B	C	
Customer 1	2	5	7	67
Customer 2	3	10	1	75
Customer 3	1	8	6	71

(i) Find the values of x, y and z.

(ii) A fourth customer spends £70 on the purchase of 4 bags of grade A coal, 2 bags of grade B coal and some bags of grade C. How many bags of grade C coal did he purchase?

(N. Ireland Additional Mathematics (part question), 1984)

20. A confectioner sells chocolate, toffee and fruit drops, priced respectively at x pence, y pence an z pence per 100 grams. Table 1.4 gives the quantity in grams of each type of confectio purchased, and the total amount paid, by each of three children.

Find the values of x, y and z.

A fourth child purchases 150 grams of chocolate, 200 grams of toffee and 50 grams of fruit dro from the same confectioner. How much does he pay?

Table 1.4

	Quantity purchased (in grams)			
	Chocolate	Toffee	Fruit drops	Total cost (pence)
1st child	100	200	300	80
2nd child	200	100	200	75
3rd child	100	100	400	75

(N. Ireland Additional Mathematics, 198

21. A sewing club makes patchwork quilts measuring 2 m by 1.5 m, the patterns of which consist squares of different coloured cloths. The length of the sides of the squares is 10 cm.

(i) Calculate the number of squares required to make one quilt.

Initially there are 1500 red squares, 1500 green squares and 1500 blue squares. Quilts are made three different designs which use the coloured squares in the proportions in Table 1.5.

(ii) Calculate the number of squares of each colour required for each one of the three designs.

On completion there remain 230 red squares, 70 green squares and no blue squares.

(iii) Calculate the number of quilts of each design which were made.

Table 1.5

	red	:	green	:	blue
design 1	3	:	2	:	5
design 2	1	:	2	:	3
design 3	5	:	5	:	2

(N. Ireland Additional Mathematics, 198

CHAPTER 2

Trigonometry

Leonard Euler's 1748 treatise *Introductio in analysin infinitorum* is the basis for trigonometry as we know it today. The concept of three basic trigonometrical functions as ratios of the lengths of sides of right-angled triangles dates from this text. The abbreviations *sin* and *cos* were used with the present day *tan* replaced by *tang*. The treatise established the triangle notation in which A, B and C are taken to be the angles of triangle ABC while a, b and c are interpreted as the lengths of the sides opposite vertices A, B and C respectively.

Euler was a Swiss mathematician who spent much of his working life in Russia. He was the son of a Calvinist pastor who instructed the young Euler in astronomy, medicine, mathematics, theology and oriental languages. He graduated from the University of Basel at the early age of sixteen but soon moved to Russia where he first became professor of mathematics and then director of the Academy of Science in St. Petersburg.

He lost the sight in one eye through overwork and, by his late sixties, was totally blind. An operation in 1771 restored his sight completely but only for a few days. Despite being blind, the ease with which he could mentally manipulate complex algebraic expressions prompted Francois Arago to note that Euler could perform intricate mathematical computations 'just as men breathe, as eagles sustain themselves in the air'.

Euler was the world's most prolific mathematician, having published 886 books and papers. He was twelve times the recipient of the prestigious biennial prize of the Paris Academy for essays on a wide range of subjects. He wrote in Latin, French, German and Russian. In 1739 he presented his theory of music, said to be too musical for mathematicians and too mathematical for musicians.

INTRODUCTION

n this chapter the trigonometrical principles used in the solution f right-angled triangles are extended to angles of any magnitude. t is instructive to begin by considering the graphs of $\sin x$, $\cos x$ nd $\tan x$ where x is no longer restricted to the range 0° to 90°. To this end, the reader is invited to complete Table 2.1 using a calculator and to confirm, by plotting the data in columns 2, 3 and 4, that the graphs of the three trigonometrical functions are as given in Figures 2.1–2.3.

Table 2.1 Values of $\sin x$, $\cos x$, and $\tan x$ for $-180° \leq x \leq 180°$

x	$\sin x$	$\cos x$	$\tan x$
− 180°	0.000		0.000
− 160°		− 0.940	
− 140°	− 0.643		0.839
− 120°		− 0.500	
− 100°	− 0.985		5.671
− 95°		− 0.087	
− 94°	− 0.998		14.301
− 93°		− 0.052	
− 92°	− 0.999		28.636
− 90°		0.000	
− 88°	− 0.999		− 28.636
− 87°		0.052	
− 86°	− 0.998		− 14.301
− 85°		0.087	
− 80°	− 0.985		− 5.671
− 60°		0.500	
− 40°	− 0.643		− 0.839
− 20°		0.940	
0°	0.000		0.000
20°		0.940	
40°	0.643		0.839
60°		0.500	
80°	0.985		5.671
85°		0.087	
86°	0.998		14.301
87°		0.052	
88°	0.999		28.636
90°		0.000	
92°	0.999		− 28.636
93°		− 0.052	
94°	0.998		− 14.301
95°		− 0.087	
100°	0.985		− 5.671
120°		− 0.500	
140°	0.643		− 0.839
160°		− 0.940	
180°	0.000		0.000

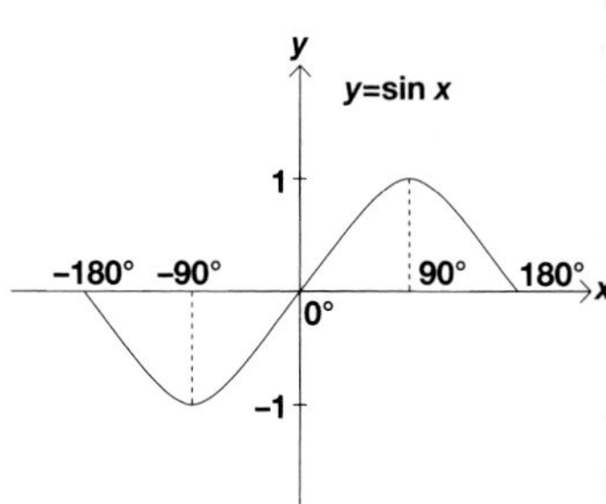

Figure 2.1

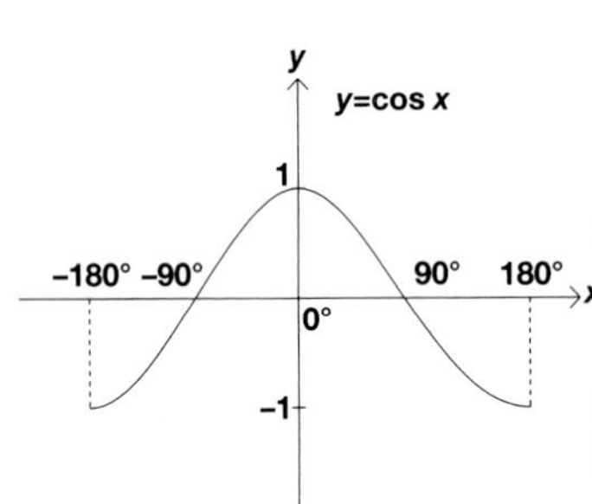

Figure 2.2

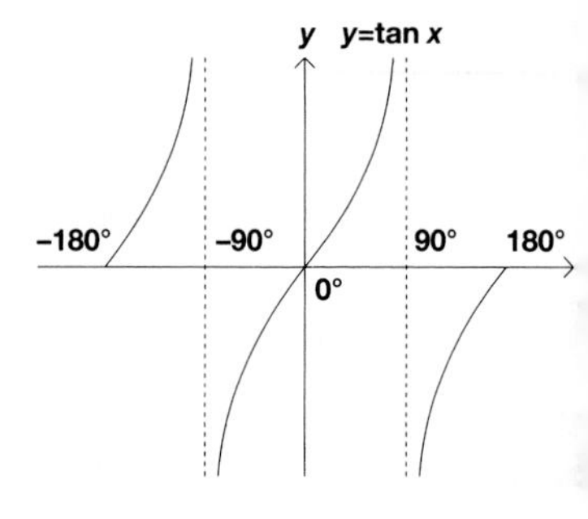

Figure 2.3

Figures 2.1–2.3 confirm the following.

(i) For angles between 0° and 90°, $\sin x$, $\cos x$ and $\tan x$ are all positive.
(ii) For angles between 90° and 180°, $\sin x$ is positive while $\cos x$ and $\tan x$ are negative.
(iii) For angles between $-90°$ and 0°, $\cos x$ is positive while $\sin x$ and $\tan x$ are negative.
(iv) For angles between $-180°$ and $-90°$, $\tan x$ is positive while $\sin x$ and $\cos x$ are negative.

Positive angles are measured as anti-clockwise rotations from OA as in Figure 2.4.

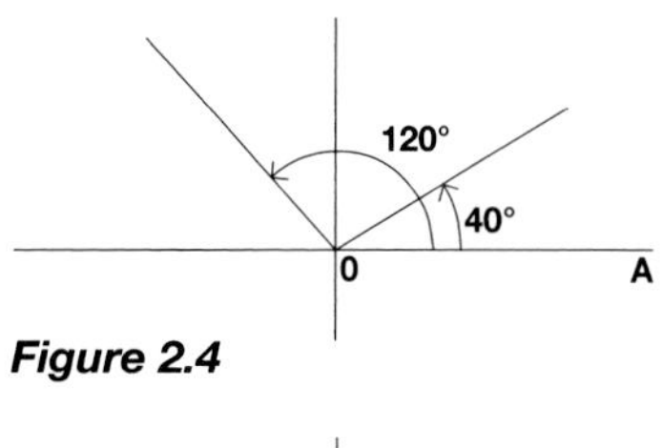

Figure 2.4

Negative angles are measured as clockwise rotations from OA as in Figure 2.5.

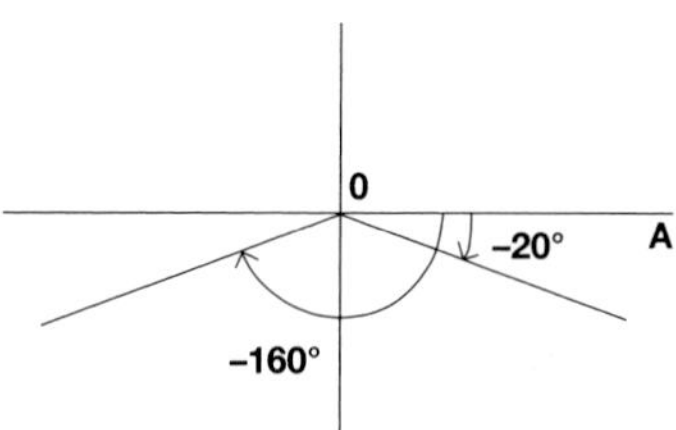

Figure 2.5

In trigonometry it is the convention to assign angles to four quadrants as in Figure 2.6.

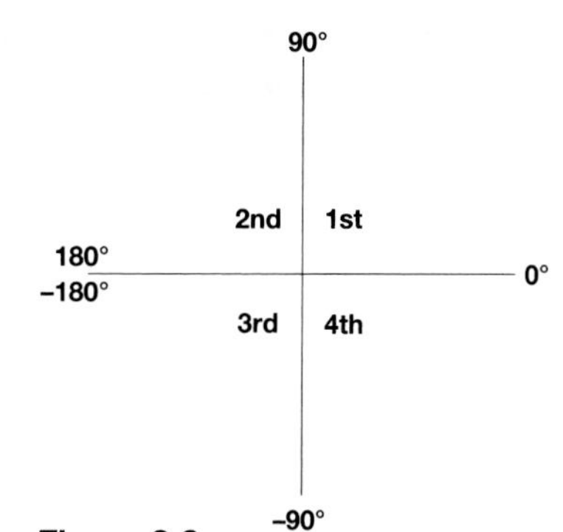

Figure 2.6

Angles such as 20°, 42°, 70° lie in the first quadrant;
angles such as 120°, 137°, 142°, 165° lie in the second quadrant;
angles such as $-20°$, $-52°$, $-78°$, $-84°$ lie in the fourth quadrant;
angles such as $-170°$, $-152°$, $-124°$, $-104°$ lie in the third quadrant.

The signs of the three trigonometric ratios ($\sin x$, $\cos x$ and $\tan x$) in these four quadrants are determined from a study of Figures 2.1–2.3 and the results summarised in Figure 2.7.

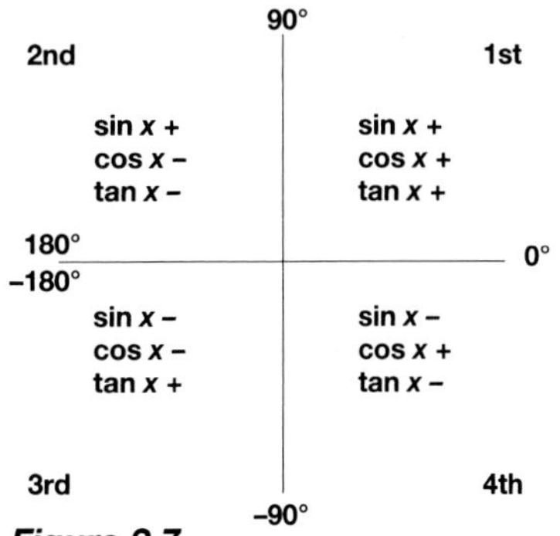

90°
2nd — sin x + cos x – tan x –
1st — sin x + cos x + tan x +
180° / –180° — 0°
3rd — sin x – cos x – tan x +
4th — sin x – cos x + tan x –
–90°

Figure 2.7

COSECANT, SECANT AND COTANGENT

The reciprocals of $\sin x$, $\cos x$ and $\tan x$ occur so frequently in mathematics that it is appropriate to name them cosecant x, secant x and cotangent x respectively. Their full names are usually abbreviated i.e.

$$\mathbf{cosec}\,x = \frac{1}{\sin x} \qquad \sec x = \frac{1}{\cos x} \qquad \cot x = \frac{1}{\tan x}$$

EXAMPLE 1

Evaluate (i) sec 42.7° (ii) cot 217.4° (iii) cosec 127°.

Solution 1

(i) $\sec 42.7° = \dfrac{1}{\cos 42.7°} = \dfrac{1}{0.7349} = 1.361$

(ii) $\cot 217.4° = \dfrac{1}{\tan 217.4°} = \dfrac{1}{0.7646} = 1.308$

(iii) $\operatorname{cosec} 127° = \dfrac{1}{\sin 127°} = \dfrac{1}{0.7986} = 1.252$

EXERCISE 2.1

Find the value of each of the following using a calculator.

1. $\operatorname{cosec} 34.2°$ **2.** $\sec 94.2°$ **3.** $\cot 34.1°$

4. $\operatorname{cosec} 192.3°$ **5.** $\sec 112.3°$ **6.** $\cot 112.4°$

7. $\operatorname{cosec} 712°$ **8.** $\sec 42.1°$ **9.** $\cot 213.72°$

10. $\sec(-47.4°)$

SOLVING SIMPLE TRIGONOMETRIC EQUATIONS

Many problems in mathematics reduce to the solution of an equation such as $\sin x = -0.7$, for example, so that mathematicians have had to develop methods for solving equations where the value of the angle is to be found given its sine, cosine or tangent. Unfortunately such equations have an infinite number of solutions and it has become the accepted practice that only solutions lying within a restricted range are listed. The range used throughout this book will be that between $-180°$ and $180°$. The solution of this type of equation is best illustrated through the examples below.

EXAMPLE 2

Solve $\sin\theta = -0.242$ for $-180° < \theta \leq 180°$.

Solution 2

Step 1 Ignore the negative sign for the moment – it will be used later (Step 3).
The equation $\sin\theta = 0.242$ yields the 'basic' angle θ.

Step 2 Solve $\sin\theta = 0.242$. Use of the $\sin^{-1}$ function on the calculator gives $\theta = 14°$.
The basic angle is therefore 14° and 14° is constructed with the horizontal in all four quadrants to produce Figure 2.8.

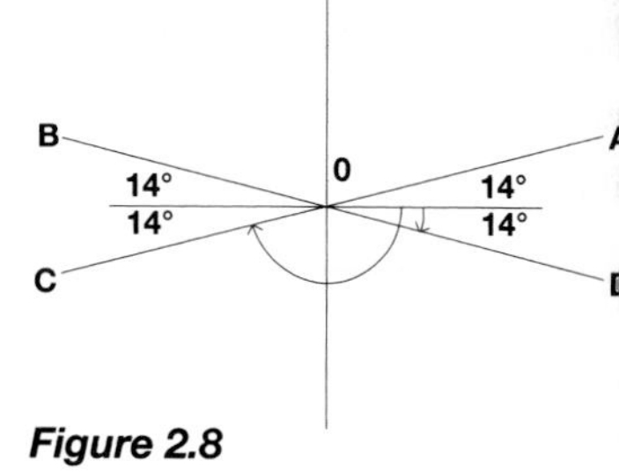

Figure 2.8

Step 3 Figure 2.7 assigns angles with negative sines to quadrants 3 and 4 and therefore interest focuses on OC and OD in Figure 2.8.
OD, being in the fourth quadrant, is specified by $-14°$ and OC by $-166°$.
Hence $\theta = -14°$ or $\theta = -166°$.

N.B. These solutions may be confirmed approximately by use of a rough sketch of the curve for $\sin x$ – see Figure 2.9.

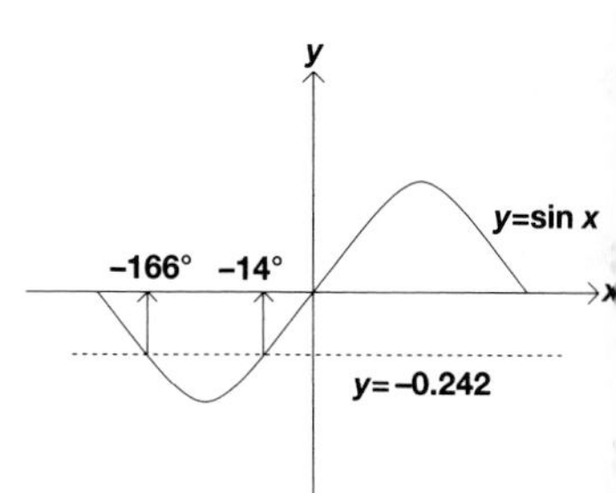

Figure 2.9

EXAMPLE 3

Solve $\cos\theta = 0.841$ for $-180° < \theta \leq 180°$.

Solution 3

Step 1 $\cos\theta = 0.841$ gives the basic angle.

Step 2 $\cos\theta = 0.841 \Rightarrow \theta = 32.75°$ (calculator).
As in Example 2, 32.75° is constructed with the horizontal in all four quadrants – see Figure 2.10.

Step 3 Since $\cos\theta$ is positive, and Figure 2.7 assigns angles with positive cosines to quadrants 1 and 4 then OA and OD are chosen in Figure 2.10. But OA corresponds to 32.75° and OD to $-32.75°$ and hence $\theta = -32.75°$ or $\theta = +32.75°$.

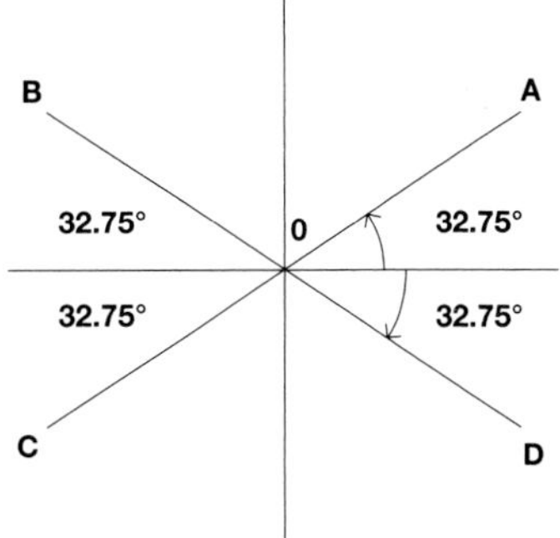

Figure 2.10

EXAMPLE 4

Solve $\tan\theta = -0.242$ for $-180° < \theta \leq 180°$.

Solution 4

Step 1 $\tan\theta = 0.242$ is the basic equation.

Step 2 $\tan\theta = 0.242 \Rightarrow \theta = 13.6°$ (calculator).
The angle 13.6° is constructed with the horizontal in all four quadrants as in Figure 2.11.

Step 3 Since $\tan\theta$ is negative, attention is focused on OB and OD.
Hence $\theta = 166.4°$ or $\theta = -13.6°$.

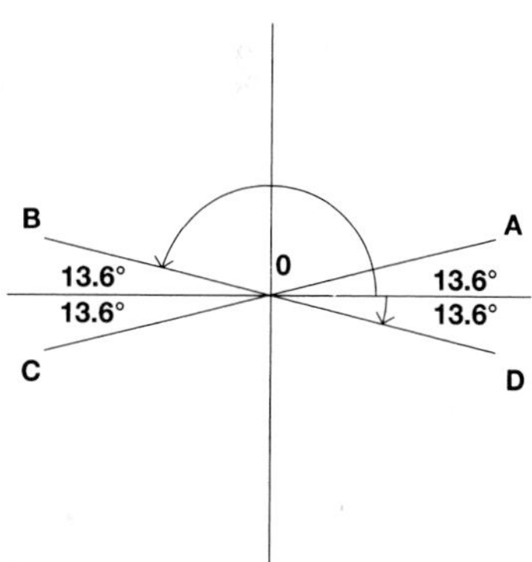

Figure 2.11

EXAMPLE 5

Solve $\text{cosec}\,\theta = -1.542$ for $-180° < \theta \leq 180°$.

Solution 5

Step 1 $\text{cosec}\,\theta = -1.542$

$$\Rightarrow \frac{1}{\sin\theta} = -1.542$$

$$\Rightarrow \sin\theta = \frac{1}{-1.542}$$

$$\Rightarrow \sin\theta = -0.6485$$

Now $\sin\theta = 0.6485$ will give the basic angle.

Step 2 $\sin\theta = 0.6485 \Rightarrow \theta = 40.43°$
Constructing this angle with the horizontal in all four quadrants yields Figure 2.12.

Step 3 Since $\sin\theta$ is negative, interest focuses on OC and OD.
Therefore $\theta = -40.43°$ or $\theta = -139.57°$.

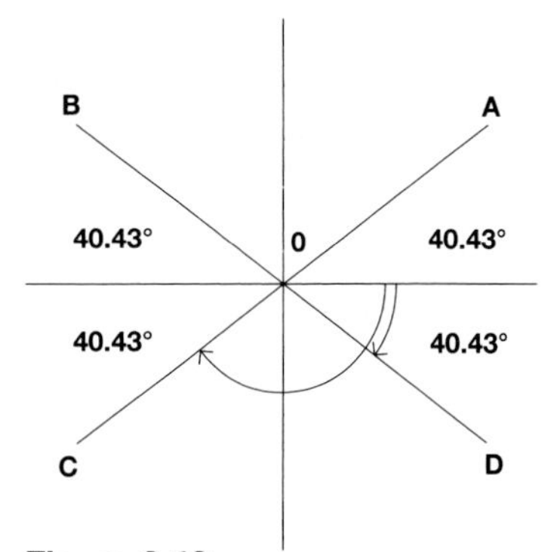

Figure 2.12

EXAMPLE 6

Solve $\sec\theta = -2.32$ for $-180° < \theta \leq 180°$.

Solution 6

Step 1 $\sec\theta = -2.32$

$$\Rightarrow \frac{1}{\cos\theta} = -2.32$$

$$\Rightarrow \cos\theta = \frac{1}{-2.32}$$

$$\Rightarrow \cos\theta = -0.4310$$

Now $\cos\theta = 0.4310$ will give the basic angle.

Step 2 $\cos\theta = 0.4310 \Rightarrow \theta = 64.47°$.
Constructing this angle with the horizontal in all four quadrants results in Figure 2.13.

Step 3 Since $\cos\theta$ is negative, attention is focused on OB and OC and so
$\theta = +115.53°$ or $\theta = -115.53°$.

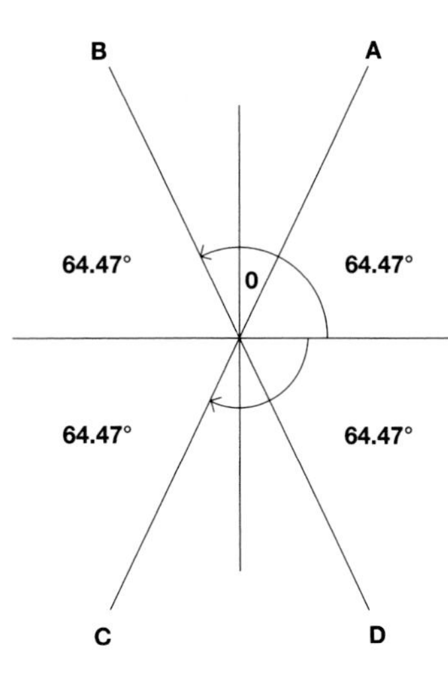

Figure 2.13

EXAMPLE 7

Solve $\cot\theta = 0.422$ for $-180° < \theta \leq 180°$.

Solution 7

Step 1 $\cot\theta = 0.422$

$$\Rightarrow \frac{1}{\tan\theta} = 0.422$$

$$\Rightarrow \tan\theta = \frac{1}{0.422}$$

$$\Rightarrow \tan\theta = 2.370$$

Step 2 $\tan\theta = 2.370 \Rightarrow \theta = 67.12°$ and therefore this angle is constructed with the horizontal in all four quadrants as in Figure 2.14.

Step 3 OA and OC are of interest since $\tan\theta$ is positive. Therefore $\theta = 67.12°$ or $\theta = -112.88°$.

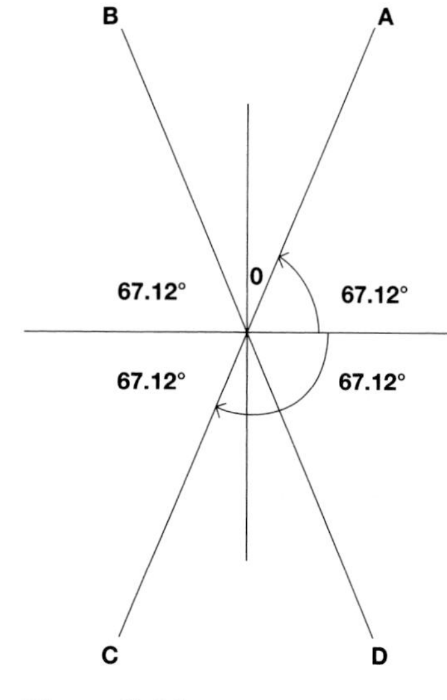

Figure 2.14

EXERCISE 2.2

Solve each of the following equations for θ in the range $-180° < \theta \leq 180°$.

1. (i) $\sin\theta = 0.788$ (ii) $\sin\theta = -0.5913$ (iii) $\sin\theta = 0.92$

2. (i) $\cos\theta = 0.809$ (ii) $\cos\theta = 0.309$ (iii) $\cos\theta = -0.839$

3. (i) $\tan\theta = -0.404$ (ii) $\tan\theta = 0.4663$ (iii) $\tan\theta = -2.605$

4. (i) $\operatorname{cosec}\theta = 1.4396$ (ii) $\operatorname{cosec}\theta = 1.269$ (iii) $\operatorname{cosec}\theta = -1.022$

5. (i) $\sec\theta = 8.206$ (ii) $\sec\theta = 1.071$ (iii) $\sec\theta = -1.031$

6. (i) $\cot\theta = 3.732$ (ii) $\cot\theta = -1.804$ (iii) $\cot\theta = -1.483$

THE TRIGONOMETRIC IDENTITIES

While the previous section equips the reader with the necessary skills for solving simple trigonometrical equations, the methods detailed in that section are not directly applicable to more complex equations such as $3\sec^2\theta - 5\tan\theta - 5 = 0$. As will become evident in the next section, the solution of such equations requires the use of four 'trigonometric identities'. A trigonometric **identity** is a relationship which is true for **any** angle. Pythagoras' theorem plays a significant role in the development of these four identities.

Consider the right-angled triangle in Figure 2.15.

Pythagoras' theorem for this triangle is

$$x^2 + y^2 = z^2$$

Dividing this equation in turn by z^2, y^2 and x^2 yields some interesting results.

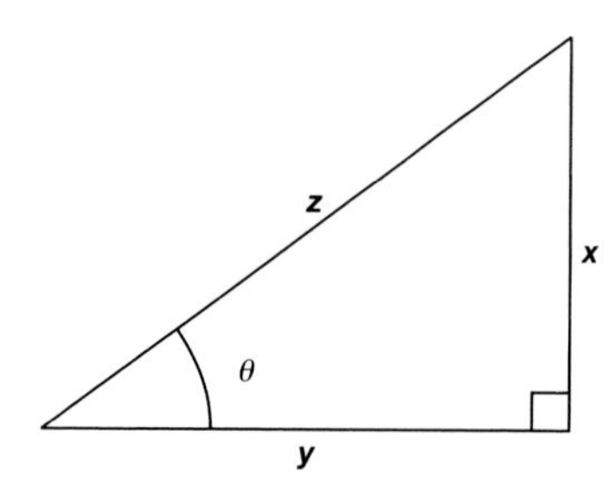

Figure 2.15

First, dividing by z^2

$$x^2 + y^2 = z^2 \Rightarrow \frac{x^2}{z^2} + \frac{y^2}{z^2} = 1$$

$$\Rightarrow \left(\frac{x}{z}\right)^2 + \left(\frac{y}{z}\right)^2 = 1$$

$$\Rightarrow (\sin\theta)^2 + (\cos\theta)^2 = 1 \text{ (see Figure 2.15).}$$

This first identity is commonly written as $\sin^2\theta + \cos^2\theta = 1$.

Then, dividing by y^2

$$x^2 + y^2 = z^2 \Rightarrow \frac{x^2}{y^2} + 1 = \frac{z^2}{y^2}$$

$$\Rightarrow \left(\frac{x}{y}\right)^2 + 1 = \left(\frac{z}{y}\right)^2$$

$$\Rightarrow \left(\frac{x}{y}\right)^2 + 1 = \left(\frac{1}{\left(\frac{y}{z}\right)}\right)^2$$

$$\Rightarrow (\tan\theta)^2 + 1 = \left(\frac{1}{\cos\theta}\right)^2$$

$$\text{or } \tan^2\theta + 1 = \sec^2\theta$$

s the second trigonometric identity.

Then, dividing by x^2

$$x^2 + y^2 = z^2 \Rightarrow 1 + \frac{y^2}{x^2} = \frac{z^2}{x^2}$$

$$\Rightarrow 1 + \left(\frac{y}{x}\right)^2 = \left(\frac{z}{x}\right)^2$$

$$\Rightarrow 1 + \left(\frac{1}{\left(\frac{x}{y}\right)}\right)^2 = \left(\frac{1}{\left(\frac{x}{z}\right)}\right)^2$$

$$\Rightarrow 1 + \left(\frac{1}{\tan\theta}\right)^2 = \left(\frac{1}{\sin\theta}\right)^2$$

$$\Rightarrow 1 + \cot^2\theta = \operatorname{cosec}^2\theta$$

is the third trigonometric identity.

Finally,

$$\frac{x}{z} \div \frac{y}{z} = \frac{x}{z} \times \frac{z}{y} = \frac{x}{y}$$

$$\text{or } \frac{\sin\theta}{\cos\theta} = \tan\theta.$$

These identities may be summarised:

$$\mathbf{\sin^2\theta + \cos^2\theta = 1}$$
$$\mathbf{\tan^2\theta + 1 = \sec^2\theta}$$
$$\mathbf{\cot^2\theta + 1 = \operatorname{cosec}^2\theta}$$
$$\mathbf{\tan\theta = \frac{\sin\theta}{\cos\theta}}$$

SOLVING QUADRATIC EQUATIONS INVOLVING SINE AND COSINE ONLY

EXAMPLE 8

Solve $6\cos^2\theta + 5\sin\theta - 7 = 0$ for $-180° < \theta \leq 180°$.

Solution 8

The first identity allows the equation to be written entirely in terms of $\sin\theta$.

Since $\sin^2\theta + \cos^2\theta = 1$

it follows that $\cos^2\theta = 1 - \sin^2\theta$

and hence $6\cos^2\theta + 5\sin\theta - 7 = 0$

$$\Rightarrow 6(1 - \sin^2\theta) + 5\sin\theta - 7 = 0$$
$$\Rightarrow 6 - 6\sin^2\theta + 5\sin\theta - 7 = 0$$
$$\Rightarrow 6\sin^2\theta - 5\sin\theta + 1 = 0$$

$$\Rightarrow \quad \sin\theta = \frac{5 \pm \sqrt{(5^2 - 4 \times 1 \times 6)}}{12}$$

$$\Rightarrow \quad \sin\theta = \frac{5 \pm 1}{12}$$

$$\Rightarrow \quad \sin\theta = \tfrac{1}{2} \text{ or } \sin\theta = \tfrac{1}{3}$$

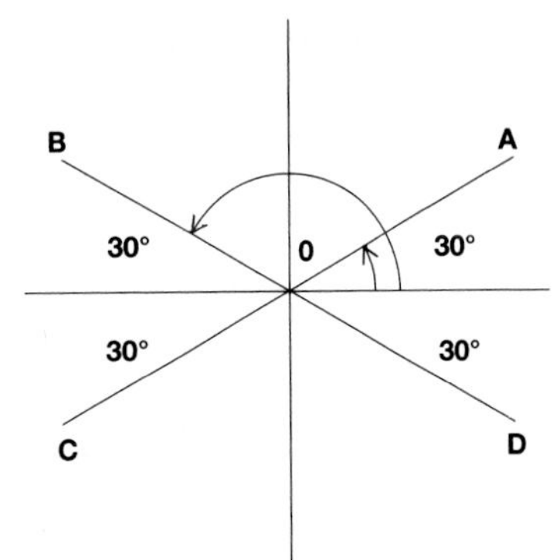

Figure 2.16

Consider first $\sin\theta = \frac{1}{2}$ which yields the basic angle 30° and the quadrant diagram shown in Figure 2.16.

As $\sin\theta$ is positive the solutions represented by OA and OB are of interest i.e. 30° and 150°.

Similarly $\sin\theta = \frac{1}{3}$ has solution $\theta = 19.47°$ or $\theta = 160.53°$.

In summary, the equation $6\cos^2\theta + 5\sin\theta - 7 = 0$ has four solutions: $\theta = 19.47°, 30°, 150°, 160.53°$.

EXAMPLE 9

Solve $\quad 20\sin^2\theta + 11\cos\theta - 16 = 0 \quad$ for $-180° < \theta \leq 180°$.

Solution 9

Using $\sin^2\theta + \cos^2\theta = 1$ in the form $\sin^2\theta = 1 - \cos^2\theta$ to eliminate the $\sin^2\theta$ term

$$20\sin^2\theta + 11\cos\theta - 16 = 0$$
$$\Rightarrow \quad 20(1 - \cos^2\theta) + 11\cos\theta - 16 = 0$$
$$\Rightarrow \quad 20\cos^2\theta - 11\cos\theta - 4 = 0.$$

$$\Rightarrow \quad \cos\theta = \frac{11 \pm \sqrt{(11^2 + 4 \times 4 \times 20)}}{40}$$

$$\Rightarrow \quad \cos\theta = \frac{11 \pm 21}{40}$$

$$\Rightarrow \quad \cos\theta = 0.8 \text{ or } -0.25$$

Using the methods shown earlier (p. 21)

$$\cos\theta = 0.8 \quad \Rightarrow \theta = \pm 36.87°$$
$$\cos\theta = -0.25 \Rightarrow \theta = \pm 104.48°$$

and therefore the complete set of solutions to the equation $20\sin^2\theta + 11\cos\theta - 16 = 0$ is $\theta = \pm 36.87°, \pm 104.48°$.

EXAMPLE 10

Solve $\quad 3 - 3\cos\theta = 2\sin^2\theta \quad$ for $-180° < \theta \leq 180°$.

Solution 10

Using $\sin^2\theta + \cos^2\theta = 1$ in the form $\sin^2\theta = 1 - \cos^2\theta$ to eliminate the $\sin^2\theta$ term

$$3 - 3\cos\theta = 2\sin^2\theta$$
$$\Rightarrow \quad 3 - 3\cos\theta = 2(1 - \cos^2\theta)$$

$$\Rightarrow \quad 2\cos^2\theta - 3\cos\theta + 1 = 0$$

$$\Rightarrow \quad \cos\theta = \frac{3 \pm \sqrt{(9-8)}}{4}$$

$$\Rightarrow \quad \cos\theta = \frac{3 \pm 1}{4}$$

$$\Rightarrow \quad \cos\theta = 1 \text{ or } 0.5.$$

Using the methods shown earlier (p. 21)

$$\cos\theta = 0.5 \Rightarrow \theta = 60^\circ \text{ or } \theta = -60^\circ$$

and, using Figure 2.2, $\cos\theta = 1$ at $\theta = 0^\circ$.

Equations such as $\sin\theta = 0$, $\sin\theta = \pm 1$, $\cos\theta = 0$, $\cos\theta = \pm 1$ are best solved using graphical methods.

The complete set of solutions to the equation $3 - 3\cos\theta = 2\sin^2\theta$ is therefore $\theta = 0^\circ, \pm 60^\circ$.

EXERCISE 2.3

Solve the following quadratic trigonometric equations for $-180^\circ < \theta \leq 180^\circ$.

1. $12\cos^2\theta - 17\sin\theta - 18 = 0$
2. $34 + \cos\theta = 40\sin^2\theta$
3. $18\cos^2\theta - 3\sin\theta - 8 = 0$
4. $18 = 17\sin\theta + 12\cos^2\theta$
5. $20\sin^2\theta = 23\cos\theta + 26$
6. $6\cos^2\theta + \sin\theta - 5 = 0$
7. $\cos^2\theta + \sin\theta + 1 = 0$

SOLVING QUADRATIC EQUATIONS INVOLVING TANGENT, COTANGENT, COSECANT AND SECANT

EXAMPLE 11

Solve $\quad 3\sec^2\theta - 5\tan\theta - 5 = 0 \quad$ for $-180^\circ < \theta \leq 180^\circ$.

Solution 11

This method involves using the identity $\sec^2\theta = 1 + \tan^2\theta$ to eliminate the $\sec^2\theta$ term and produce a quadratic equation in $\tan\theta$.

$$3\sec^2\theta - 5\tan\theta - 5 = 0$$
$$\Rightarrow \quad 3(1 + \tan^2\theta) - 5\tan\theta - 5 = 0$$
$$\Rightarrow \quad 3\tan^2\theta - 5\tan\theta - 2 = 0$$

$$\Rightarrow \quad \tan\theta = \frac{5 \pm \sqrt{(25+24)}}{6} = \frac{5 \pm 7}{6}$$

$$\Rightarrow \quad \tan\theta = 2 \text{ or } -\tfrac{1}{3}$$

$\tan\theta = 2 \quad \Rightarrow \theta = 63.43^\circ$ or $\theta = -116.57^\circ$ and
$\tan\theta = -\frac{1}{3} \Rightarrow \theta = 161.57^\circ$ or $\theta = -18.43^\circ$

In summary, the equation $3\sec^2\theta - 5\tan\theta - 5 = 0$ has solutions $\theta = -116.57^\circ, -18.43^\circ, 63.43^\circ, 161.57^\circ$.

EXAMPLE 12

Solve $\quad 8\,\mathrm{cosec}^2\theta - 2\cot\theta - 11 = 0 \quad$ for $\;-180° < \theta \leq 180°$.

Solution 12

Using the identity $\mathrm{cosec}^2\theta = 1 + \cot^2\theta$

$$\begin{aligned} & 8\,\mathrm{cosec}^2\theta - 2\cot\theta - 11 = 0 \\ \Rightarrow\quad & 8(1 + \cot^2\theta) - 2\cot\theta - 11 = 0 \\ \Rightarrow\quad & 8\cot^2\theta - 2\cot\theta - 3 = 0. \end{aligned}$$

$$\text{Now}\quad \cot\theta = \frac{2 \pm \sqrt{(4 + 96)}}{16} = \frac{2 \pm 10}{16} = 0.75 \text{ or } -0.5.$$

$$\Rightarrow\quad \cot\theta = 0.75 \text{ or } -0.5$$

$$\Rightarrow\quad \tan\theta = \frac{1}{0.75} \text{ or } \frac{1}{-0.5}$$

$$\Rightarrow\quad \tan\theta = \tfrac{4}{3} \text{ or } -2.$$

$\tan\theta = \frac{4}{3} \Rightarrow \theta = 53.13°$ or $\theta = -126.87°$ and
$\tan\theta = -2 \Rightarrow \theta = 116.57°$ or $\theta = -63.43°$.

In summary, the equation $8\,\mathrm{cosec}^2\theta - 2\cot\theta - 11 = 0$ has solutions $\theta = -126.9°,\ -63.43°,\ 53.13°,\ 116.6°$.

EXERCISE 2.4

Solve the following trigonometric equations for $-180° < \theta \leq 180°$.

1. $\tan^2\theta - 6\sec\theta + 9 = 0$

2. $2\cot^2\theta + 5\,\mathrm{cosec}\,\theta - 1 = 0$

3. $\sec^2\theta = 3\tan\theta - 1$

4. $\mathrm{cosec}^2\theta = 3 + \cot\theta$

5. $3\tan^2\theta + 5 = 7\sec\theta$

6. $2\cot^2\theta + 8 = 7\,\mathrm{cosec}\,\theta$

FURTHER PROBLEMS INVOLVING THE TRIGONOMETRIC IDENTITIES

EXAMPLE 13

Given $\cot\alpha = -\frac{24}{7}$ and that the angle α is obtuse find, *without the use of a calculator*, the value of $\sin\alpha$.

Solution 13

The identities $\mathrm{cosec}^2\alpha = 1 + \cot^2\alpha$ and $\mathrm{cosec}\,\alpha = \dfrac{1}{\sin\alpha}$ link $\cot\alpha$ to $\sin\alpha$.

$$\begin{aligned} \text{If}\quad & \cot\alpha = -\tfrac{24}{7} \\ \text{then}\quad & \mathrm{cosec}^2\alpha = 1 + \cot^2\alpha \\ \Rightarrow\quad & \mathrm{cosec}^2\alpha = 1 + \tfrac{576}{49} = \tfrac{625}{49} \\ \Rightarrow\quad & \mathrm{cosec}\,\alpha = \pm\tfrac{25}{7}. \end{aligned}$$

However, since α is obtuse, its sine (and consequently its cosecant) is positive.

Hence $\qquad \mathrm{cosec}\,\alpha = \frac{25}{7}$ or $\sin\alpha = \frac{7}{25}$.

EXAMPLE 14

Given $\cot\alpha = -\frac{4}{3}$ and that α is obtuse find, *without the use of a calculator*, the value of

$$\frac{\sin\alpha + \cos\alpha}{\operatorname{cosec}\alpha + \sec\alpha}$$

Solution 14

$$\begin{aligned}
\text{If}\quad & \cot\alpha = -\tfrac{4}{3} \\
& \operatorname{cosec}^2\alpha = 1 + \cot^2\alpha \\
\Rightarrow\quad & \operatorname{cosec}^2\alpha = 1 + \tfrac{16}{9} = \tfrac{25}{9} \\
\Rightarrow\quad & \operatorname{cosec}\alpha = \pm\tfrac{5}{3}.
\end{aligned}$$

But α is obtuse and $\sin\alpha$ is therefore positive.

$$\Rightarrow \operatorname{cosec}\alpha = \tfrac{5}{3} \text{ or } \sin\alpha = \tfrac{3}{5}$$

giving the first term in the numerator.

Using the identity $\sin^2\alpha + \cos^2\alpha = 1$ with $\sin\alpha = \frac{3}{5}$

$$\Rightarrow (\tfrac{3}{5})^2 + \cos^2\alpha = 1 \Rightarrow \cos\alpha = \pm\tfrac{4}{5}.$$

Now α is obtuse and therefore $\cos\alpha = -\frac{4}{5}$ since cosine is negative in the second quadrant. The second term in the numerator has now been found.

Finally, if $\cos\alpha = -\frac{4}{5}$ then $\sec\alpha = -\frac{5}{4}$.

In summary therefore

$$\frac{\sin\alpha + \cos\alpha}{\operatorname{cosec}\alpha + \sec\alpha} = \frac{\frac{3}{5} - \frac{4}{5}}{\frac{5}{3} - \frac{5}{4}} = -\frac{12}{25}$$

EXAMPLE 15

Find, *without the use of a calculator*, the value of

$$\cos\alpha + \cos\beta$$

given that $\tan\alpha = \frac{3}{4}$ and α is a reflex angle, and $\cot\beta = -\frac{12}{5}$ and β is an obtuse angle.

Solution 15

Using $\sec^2\alpha = 1 + \tan^2\alpha$ and $\tan\alpha = \frac{3}{4}$

$$\begin{aligned}
&\Rightarrow \sec^2\alpha = 1 + \tfrac{9}{16} = \tfrac{25}{16} \\
&\Rightarrow \sec\alpha = \pm\tfrac{5}{4}.
\end{aligned}$$

Since $\tan\alpha$ is positive and α is reflex (in quadrant 3 or 4) it follows that α lies in the third quadrant since only here is the tangent of α positive and α reflex.

It follows that $\sec\alpha = -\frac{5}{4}$ since cosine is negative in the 3rd quadrant.

It may therefore be concluded that $\cos\alpha = -\frac{4}{5}$.

Turning now to β:

$$\cot\beta = -\tfrac{12}{5} \Rightarrow \operatorname{cosec}^2\beta = 1 + (-\tfrac{12}{5})^2 = \tfrac{169}{25}$$
$$\Rightarrow \operatorname{cosec}\beta = \pm\tfrac{13}{5}.$$

But β is obtuse (in the 2nd quadrant) and so $\sin\beta$ (and, as a consequence, $\operatorname{cosec}\beta$) is positive.

Therefore

$$\operatorname{cosec}\beta = \tfrac{13}{5} \text{ or } \sin\beta = \tfrac{5}{13}.$$

Finally,

$$\cot\beta = -\tfrac{12}{5} \text{ so } \tan\beta = -\tfrac{5}{12}. \text{ Since } \tan\beta = \frac{\sin\beta}{\cos\beta}$$

it follows that

$$-\tfrac{5}{12} = \tfrac{5}{13} \div \cos\beta \Rightarrow -\tfrac{5}{12}\cos\beta = \tfrac{5}{13} \Rightarrow \cos\beta = -\tfrac{12}{13}.$$

In summary

$$\cos\alpha + \cos\beta = (-\tfrac{4}{5}) + (-\tfrac{12}{13}) = -\tfrac{112}{65}.$$

EXERCISE 2.5

Without using a calculator, find the trigonometric ratios listed in parts (i) and (ii) from the information supplied.

1. $\sin\theta = -\frac{7}{25}$; θ in 3rd quadrant, find (i) $\cos\theta$ and (ii) $\tan\theta$.

2. $\cos\theta = -\frac{3}{5}$; θ obtuse, find (i) $\sin\theta$ and (ii) $\cot\theta$.

3. $\tan\theta = -\frac{3}{4}$; θ reflex, find (i) $\sec\theta$ and (ii) $\sin\theta$.

4. $\cos\theta = \frac{12}{13}$; θ reflex, find (i) $\sin\theta$ and (ii) $\tan\theta$.

5. $\tan\theta = -\frac{24}{7}$; θ obtuse, find (i) $\sec\theta$ and (ii) $\cot\theta + \sin\theta$.

6. $\sin\theta = \frac{15}{17}$; θ obtuse, find (i) $\cos\theta$ and (ii) $\cot\theta$.

7. $\cot\theta = \frac{8}{15}$; θ reflex, find (i) $\sin\theta$ and (ii) $\sec\theta$.

8. $\sin\theta = \frac{24}{25}$; θ obtuse, find (i) $\cos\theta$ and (ii) $\tan\theta$.

9. $\cos\theta = -\frac{3}{5}$; θ obtuse, find (i) $\sin\theta$ and (ii) $\cot\theta$.

10. $\cos\theta = \frac{20}{29}$; θ acute, find (i) $\sin\theta$ and (ii) $\tan\theta$.

11. $\tan\theta = -\frac{24}{7}$; θ obtuse, find (i) $\sec\theta$ and (ii) $\sin\theta$.

12. $\sin\theta = \frac{8}{17}$; θ acute, find (i) $\cos\theta$ and (ii) $\cot\theta$.

13. $\cot\theta = -\frac{8}{15}$; θ reflex, find (i) $\sin\theta$ and (ii) $\sec\theta$.

14. $\cos\theta = -\frac{8}{17}$; θ obtuse, find (i) $\sin\theta$ and (ii) $\cot\theta$.

15. $\tan\theta = \frac{7}{24}$; θ reflex, find (i) $\sec\theta$ and (ii) $\sin\theta$.

16. $\cos\theta = \frac{4}{5}$; θ acute, find (i) $\sin\theta$ and (ii) $\tan\theta$.

17. $\tan\theta = -\frac{5}{12}$; θ obtuse, find (i) $\sec\theta$ and (ii) $\sin\theta$.

18. $\sin\theta = \frac{15}{17}$; θ acute, find (i) $\cos\theta$ and (ii) $\cot\theta$.

19. $\cot\theta = \frac{20}{21}$; θ reflex, find (i) $\sin\theta$ and (ii) $\sec\theta$.

THE SINE RULE

Notation: the length of the side opposite vertex A in Figure 2.17 is denoted a, the length of the side opposite vertex B is denoted b and the length of the side opposite vertex C is denoted c. Angles are denoted by capital letters.

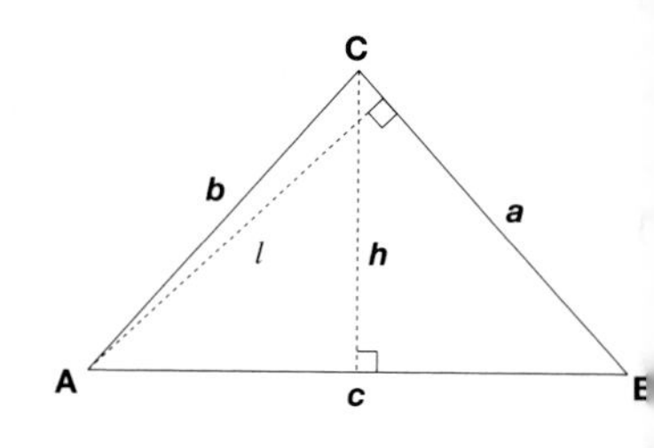

Figure 2.17

In Figure 2.17, h denotes the length of the perpendicular from C to AB and l the length of the perpendicular from A to CB.

Now from elementary trigonometry $h = b\sin A$ and $h = a\sin B$. Combining these equations gives $b\sin A = a\sin B$. Therefore

$$\frac{b}{\sin B} = \frac{a}{\sin A}$$

Also, $l = c\sin B$ and $l = b\sin C$, i.e. $c\sin B = b\sin C$, or

$$\frac{c}{\sin C} = \frac{b}{\sin B}$$

Combining the equations $\frac{b}{\sin B} = \frac{a}{\sin A}$ and $\frac{c}{\sin C} = \frac{b}{\sin B}$ gives

$$\boldsymbol{\frac{a}{\sin A} = \frac{b}{\sin B} = \frac{c}{\sin C}}$$

a result known as the **sine rule**.

EXAMPLE 16

Solve the triangle (i.e. find all the sides and angles) for which $a = 4.73\,\text{cm}$, $c = 3.58\,\text{cm}$, and angle ACB $= 42.2°$.

Solution 16

The triangle ABC is shown in Figure 2.18.

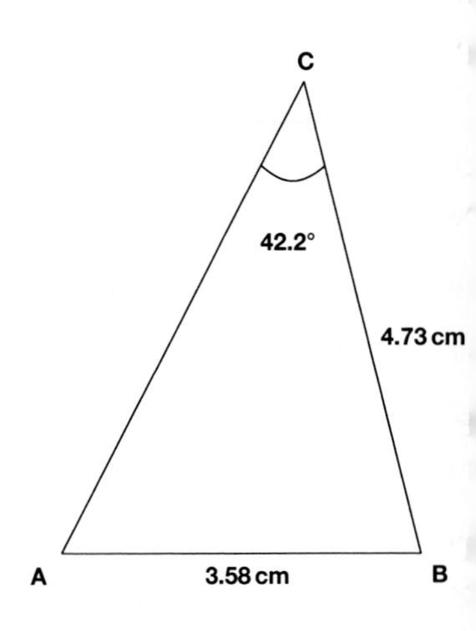

Figure 2.18

Using the sine rule

$$\frac{4.73}{\sin A} = \frac{b}{\sin B} = \frac{3.58}{\sin 42.2°}$$

Omitting the term involving the two unknowns, b and B, and solving for A,

$$\frac{4.73}{\sin A} = \frac{3.58}{\sin 42.2°}$$

$$\Rightarrow 3.58 \sin A = 4.73 \sin 42.2°$$
$$\Rightarrow A = 62.56°.$$

Angle B may now be computed since $A + B + C = 180°$.

$$B = 180° - (62.56° + 42.2°) = 75.24°.$$

The equation presented at the outset of this solution now becomes

$$\frac{4.73}{\sin 62.56°} = \frac{b}{\sin 75.24°} = \frac{3.58}{\sin 42.2°}$$

and

$$\frac{b}{\sin 75.24°} = \frac{3.58}{\sin 42.2°} \Rightarrow b = 5.154\,\text{cm}$$

The triangle is now solved as all its sides and angles are known.

EXERCISE 2.6

Solve the following triangles, i.e. find all the missing sides and angles, using the sine rule.

1. $A = 42°$ $B = 76°$ $c = 9$
2. $A = 39°$ $b = 24.3$ $C = 82°$
3. $a = 7.5$ $B = 67°$ $C = 69°$
4. $D = 29.3°$ $e = 112$ $F = 87.3°$
5. $d = 37$ $E = 46.3°$ $F = 57.4°$
6. $D = 56.3°$ $E = 56.3°$ $f = 7.5$
7. $P = 82.6°$ $q = 3.65$ $R = 34°$
8. $p = 73$ $Q = 62.82°$ $R = 47.3°$
9. $P = 34°$ $q = 29$ $R = 72.6°$
10. $P = 48°$ $Q = 54°$ $r = 11$

THE COSINE RULE

Consider Figure 2.19 where h is the perpendicular height and N is the foot of the perpendicular from B to AC.

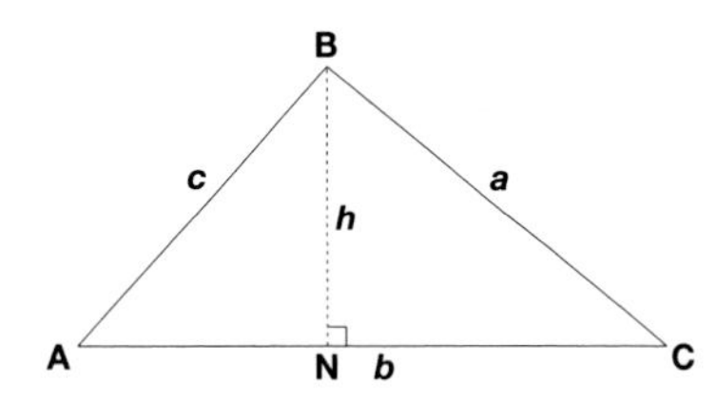

Figure 2.19

By Pythagoras' theorem

$$a^2 = h^2 + \text{NC}^2$$

but $h = c\sin A$ and $\text{AN} = c\cos A$ (by elementary trigonometry) so $\text{NC} = \text{AC} - \text{AN} = b - c\cos A$, and it follows that

$$\begin{aligned} a^2 &= (c\sin A)^2 + (b - c\cos A)^2 \\ &= c^2\sin^2 A + b^2 - 2bc\cos A + c^2\cos^2 A \\ &= c^2(\sin^2 A + \cos^2 A) + b^2 - 2bc\cos A. \end{aligned}$$

But, $\sin^2 A + \cos^2 A = 1$ and so

$$\boldsymbol{a^2 = b^2 + c^2 - 2bc\cos A}$$

This result is known as the **cosine rule**. Similarly the cosine rule may be derived as:

$$b^2 = a^2 + c^2 - 2ac\cos B \quad \text{or} \quad c^2 = a^2 + b^2 - 2ab\cos C$$

EXAMPLE 17

(i) In the triangle ABC, $a = 3.2$ cm, $b = 4.1$ cm and $C = 12.7°$ as shown in Figure 2.20. Calculate c.

(ii) In the triangle DEF (see Figure 2.21), $d = 4$ cm, $e = 3$ cm and $f = 2$ cm. Find the angle D.

C 12.7° 3.2 cm 4.1 cm B c A

Figure 2.20

Solution 17

(i) Using the cosine rule

$$\begin{aligned} c^2 &= (3.2)^2 + (4.1)^2 - 2(3.2)(4.1)\cos 12.7° \\ &= 10.24 + 16.81 - 25.60 \\ &= 1.45 \\ c &= 1.204\,\text{cm} \end{aligned}$$

(ii) Once again, applying the cosine rule

$$\begin{aligned} 4^2 &= 3^2 + 2^2 - 2(3)(2)\cos D \\ \Rightarrow 16 &= 9 + 4 - 12\cos D \\ \Rightarrow \cos D &= -0.25 \\ \Rightarrow D &= 104.5° \end{aligned}$$

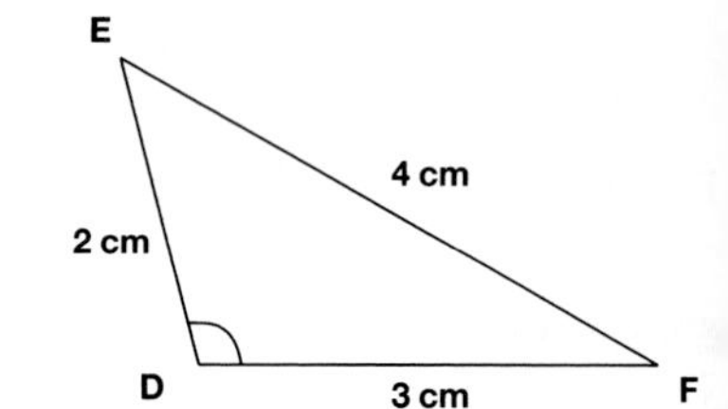

Figure 2.21

EXERCISE 2.7

Solve the following triangles using the cosine rule.

1.	$a = 6$	$b = 9$	$c = 12$	**2.**	$a = 4$	$b = 10$	$c = 7$
3.	$a = 24$	$b = 16$	$c = 29$	**4.**	$a = 23$	$b = 35$	$c = 21$
5.	$a = 27$	$b = 17$	$c = 15$	**6.**	$A = 55°$	$b = 9$	$c = 12$
7.	$a = 16$	$B = 70°$	$c = 22$	**8.**	$a = 32$	$b = 26$	$C = 82°$
9.	$a = 19$	$B = 110°$	$c = 24$	**10.**	$A = 134°$	$b = 14$	$c = 32$

EXERCISE 2.8

Solve the following triangles.

1.	$A = 40°$	$B = 60°$	$c = 9$	**2.**	$a = 7$	$b = 13$	$C = 62.3°$
3.	$a = 8$	$b = 10$	$c = 15$	**4.**	$A = 73.5°$	$a = 7$	$b = 7$
5.	$a = 9$	$B = 36°$	$C = 52°$	**6.**	$a = 6$	$b = 8$	$c = 13$
7.	$b = 6$	$c = 9$	$A = 60°$	**8.**	$a = 12$	$b = 18$	$c = 21$
9.	$c = 9.6$	$A = 113°$	$B = 31°$	**10.**	$a = 2.95$	$b = 4.86$	$c = 7.34$
11.	$a = 21.7$	$b = 29.2$	$B = 67°$	**12.**	$A = 52°$	$b = 6$	$C = 63°$
13.	$a = 3.6$	$B = 56°$	$c = 6.7$	**14.**	$A = 125°$	$b = 3.6$	$c = 6.5$
15.	$a = 2.5$	$b = 3.8$	$c = 5.2$	**16.**	$B = 63°$	$b = 11$	$c = 8$
17.	$b = 5$	$c = 4$	$C = 40°$	**18.**	$a = 4$	$b = 9$	$C = 53°$
19.	$a = 4.31$	$B = 62.5°$	$C = 74.5°$	**20.**	$b = 32.1$	$A = 27.2°$	$C = 69.3°$

SINE RULE – THE AMBIGUOUS CASE

Caution needs to be exercised in cases where two sides and a non-included angle of a triangle are given. In such cases there may be two possible solutions i.e. the question is ambiguous. The ambiguous case is best illustrated using an example.

EXAMPLE 18

Find the two solutions of the triangle with $c = 3.58\,\text{cm}$, $C = 42.2°$ and $a = 4.73\,\text{cm}$.

Solution 18

Figure 2.22 illustrates the two possible solutions, since both of the triangles (ABC and A′BC) meet the specification given.

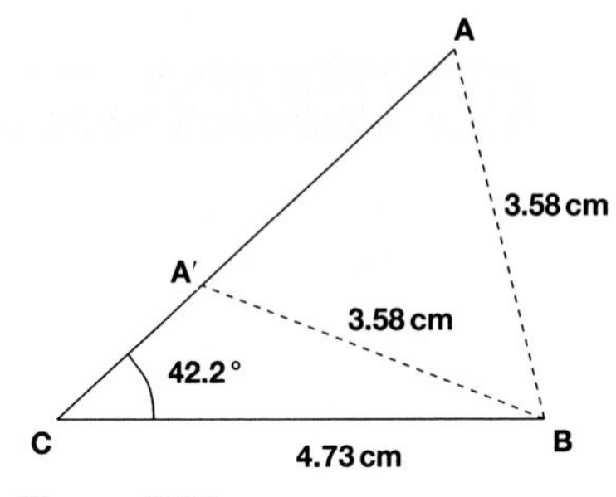

Figure 2.22

Note that C is a non-included angle – the included angle (included between the sides whose lengths are specified) is B.

Sine rule in its general form

$$\frac{a}{\sin A} = \frac{b}{\sin B} = \frac{c}{\sin C}$$

becomes

$$\frac{4.73}{\sin A} = \frac{b}{\sin B} = \frac{3.58}{\sin 42.2°}$$

$$\text{so} \quad \frac{4.73}{\sin A} = \frac{3.58}{\sin 42.2°}$$

$$\Rightarrow \sin A = 0.8875$$

Therefore $A = 62.56°$ or $117.4°$.

Hence, two separate solutions may be identified (not drawn to scale) in Figure 2.23.

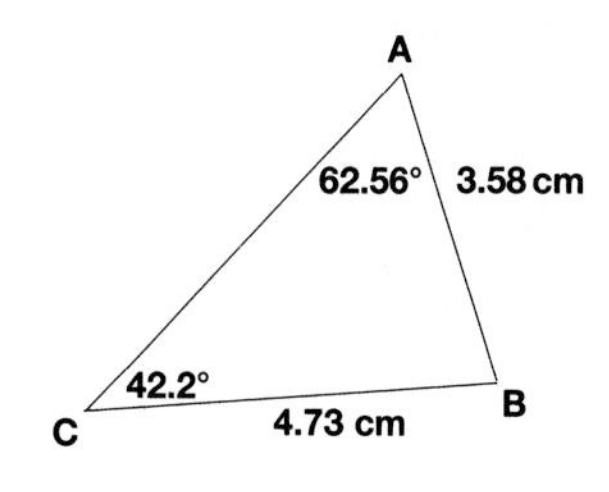

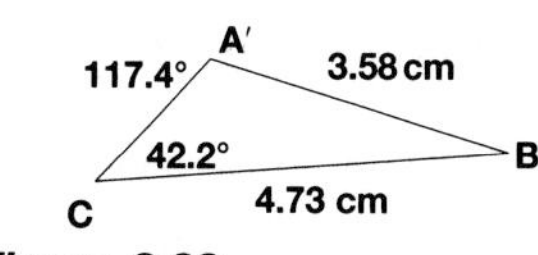

Figure 2.23

$B = 75.24°$ (top triangle) $\qquad B = 20.4°$ (lower triangle)

(since the sum of the angles of a triangle equals 180°).

And finally where calculation on left refers to top triangle and that on the right to the lower triangle:

$$\frac{3.58}{\sin 42.2°} = \frac{b}{\sin 75.24°} \qquad \frac{3.58}{\sin 42.2°} = \frac{b}{\sin 20.4°}$$

$$\Rightarrow b = 5.154\,\text{cm} \qquad \Rightarrow b = 1.858\,\text{cm}$$

In summary therefore the two solutions are:

$A = 62.56°$, $B = 75.24°$, $b = 5.154\,\text{cm}$

or

$A = 117.4°$, $B = 20.4°$, $b = 1.858\,\text{cm}$.

EXERCISE 2.9

Each case below is ambiguous. Find the two possible solutions.

1.	$a = 5.26$	$b = 4.49$	$B = 42.3°$	**2.**	$a = 9$	$c = 5.36$	$C = 32.2°$
3.	$a = 121$	$b = 139.5$	$A = 56.3°$	**4.**	$A = 11.3°$	$a = 24$	$c = 121.69$
5.	$b = 18.3$	$C = 49.3°$	$c = 16.7$	**6.**	$A = 29.5°$	$a = 10$	$b = 15$
7.	$A = 64°$	$a = 142$	$b = 151$	**8.**	$B = 37.1°$	$b = 6.5$	$c = 10.2$
9.	$B = 31°$	$b = 6$	$c = 9$	**10.**	$a = 7.29$	$b = 5$	$B = 35.8°$

THE AREA OF A TRIANGLE

The familiar 'half base times perpendicular height' formula for the area of a triangle may be translated to another form as follows.

Consider the triangle ABC in Figure 2.24.

The area of ΔABC is given by $\frac{1}{2}bh$ where h is the perpendicular height of the triangle.

But, by elementary trigonometry, $h = c\sin A$ therefore

Area ΔABC $= \frac{1}{2}b(c\sin A) = \frac{1}{2}\boldsymbol{bc}\ \mathbf{sin}\,A$.

The right-hand side of this formula may be committed to memory as: 'half the product of the lengths of two of the sides multiplied by the sine of the angle which they contain (or include)'.

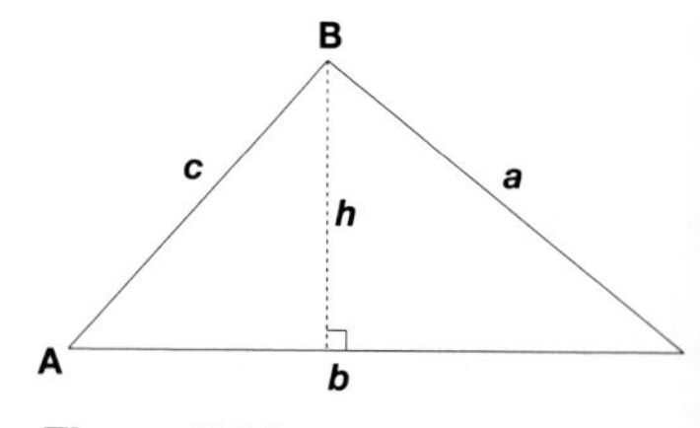

Figure 2.24

EXAMPLE 19

In the triangle ABC, the angle BAC is 37°, the side AB is of length 5 cm and the side BC is 3.5 cm long. Find the two possible areas of this triangle.

Solution 19

Figure 2.25 summarises the information given.

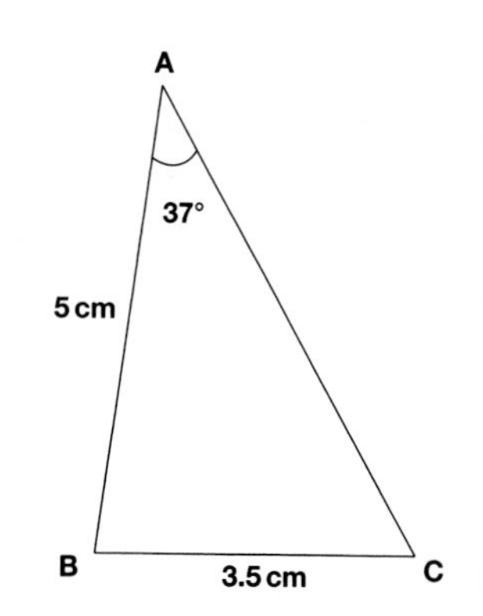

Figure 2.25

Using the sine rule

$$\frac{3.5}{\sin 37°} = \frac{5}{\sin C} \Rightarrow \sin C = 0.8597$$

$$\Rightarrow C = 59.29° \text{ or } 120.7°$$

If $C = 59.29°$, then $B = [180° - (37° + 59.29°)] = 83.71°$
$\Rightarrow$ area of the triangle is $\frac{1}{2}(5)(3.5)\sin 83.71°$ or
area ΔABC $= 8.697\,\text{cm}^2$.

If $C = 120.7°$, then $B = [180° - (37° + 120.7°)] = 22.30°$
$\Rightarrow$ area Δ ABC $= \frac{1}{2}(5)(3.5)\sin 22.30° = 3.320\,\text{cm}^2$.

The possible area is therefore $8.697\,\text{cm}^2$ or $3.320\,\text{cm}^2$.

EXERCISE 2.10

Find the area of each of the following triangles ABC.

1.	$b = 9$	$c = 6$	$A = 32°$	**2.**	$a = 8$	$c = 3$	$B = 23°$
3.	$a = 5$	$b = 8$	$C = 42°$	**4.**	$a = 7.1$	$c = 10.4$	$B = 56.2°$
5.	$b = 3.7$	$c = 1.4$	$A = 63.2°$	**6.**	$c = 4.49$	$b = 3.37$	$C = 67°$
7.	$b = 9$	$c = 14.3$	$C = 82.6°$	**8.**	$b = 47$	$c = 121$	$C = 54°$

EXERCISE 2.11

Each triangle ABC below may be drawn in two ways i.e. each is ambiguous. Calculate the two possible areas in each case.

1.	$c = 15$	$a = 17$	$C = 48°$	**2.**	$a = 8$	$c = 14$	$A = 28°$
3.	$c = 4.7$	$b = 8.1$	$C = 25.2°$	**4.**	$b = 8.1$	$c = 5.2$	$C = 32°$
5.	$b = 146$	$c = 153$	$B = 65.23°$	**6.**	$a = 121.69$	$c = 24.00$	$C = 11.3°$
7.	$a = 9$	$b = 6$	$B = 31°$	**8.**	$c = 5.00$	$b = 7.29$	$C = 35.8°$
9.	$a = 139.5$	$b = 121.0$	$B = 56.3°$	**10.**	$a = 4.729$	$b = 3.580$	$B = 42.2°$

CIRCULAR MEASURE

The radian is defined to be the angle subtended at the centre of a circle by an arc equal in length to the radius, as illustrated in Figure 2.26.

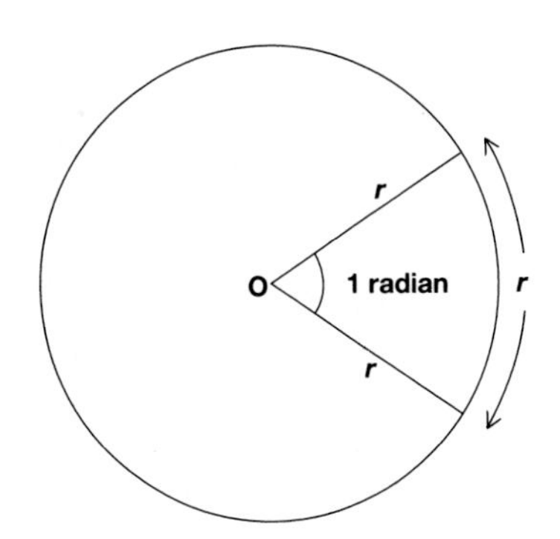

Figure 2.26

Now since there are 2π such arcs in one circumference, it follows that the complete angle at the centre of the circle is 2π radians, or

$$2\pi \text{ rads} = 360°$$

$$\Rightarrow \quad \boldsymbol{\pi} \textbf{ rads} = \mathbf{180°}$$

$$\Rightarrow \quad 1 \text{ rad} = \frac{360°}{2\pi} = \frac{180°}{\pi}$$

$\Rightarrow$ **1 rad = 57°** approximately.

Now consider Figure 2.27 which illustrates an arc of length l which subtends an angle of θ rads at the centre of a circle of radius r.

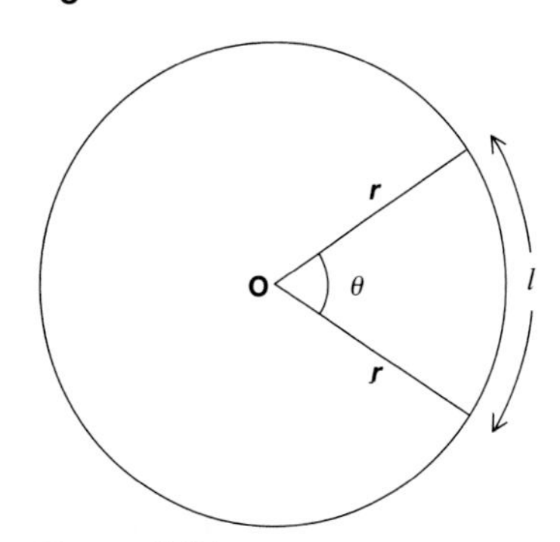

Figure 2.27

Using ratios: $l:2\pi r$ as $\theta:2\pi$, or

$$\frac{l}{2\pi r} = \frac{\theta}{2\pi}$$

$$\Rightarrow \quad l = r\theta$$

$\Rightarrow$ **length of arc =** $r\theta$

This result is known as the 'length of arc' formula.

Similarly, Figure 2.28 illustrates a sector of a circle of area A square units.

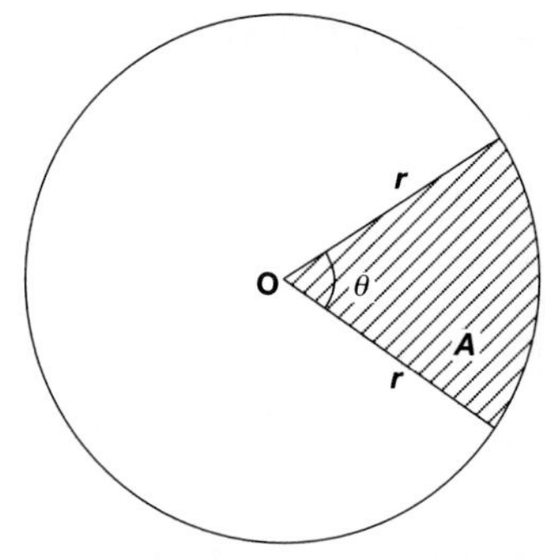

Figure 2.28

Once again, appealing to ratios, $A:\pi r^2$ as $\theta:2\pi$, i.e.

$$\frac{A}{\pi r^2} = \frac{\theta}{2\pi}$$

$$\Rightarrow \quad 2A = \theta r^2$$

$$\Rightarrow \quad \textbf{area of sector} = \tfrac{1}{2}r^2\theta$$

This result is known as the 'area of sector' formula.

Finally consider the shaded area (the segment) in Figure 2.29. This area can be found by subtracting the triangle area from the sector area.

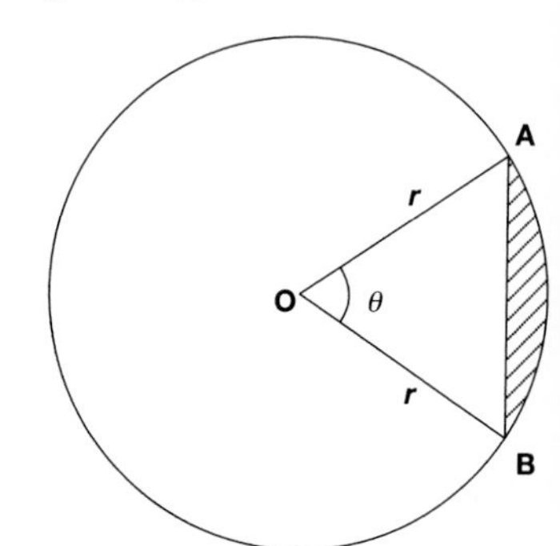

Figure 2.29

The area of sector OAB has been shown to be given by

area of sector OAB $= \frac{1}{2}r^2\theta$

and, using the formula derived in the previous section,

area of triangle OAB $= \frac{1}{2}r^2 \sin\theta$

Then the shaded area is given by

area of segment $= \frac{1}{2}r^2(\theta - \sin\theta)$

EXAMPLE 20

(i) Convert to radians (a) 34° (b) 314°.

(ii) Convert to degrees (a) 4 rads (b) 12.6 rads.

Solution 20

This solution illustrates how the important result, 180° = π rads, derived above, may be used to convert between degrees and radians. In what follows ' ÷ 180' translates as 'dividing by 180' and ' × 34' as 'multiplying by 34'.

(i) (a)

$$180° = \pi \text{ rads}$$

$$\div\ 180 \Rightarrow 1° = \frac{\pi}{180} \text{ rads}$$

$$\times\ 34 \Rightarrow 34° = \frac{34\pi}{180} \text{ rads}$$

$$34° = \frac{34 \times 3.142}{180} = 0.593 \text{ rads}$$

(b)

$$180° = \pi \text{ rads}$$

$$\div\ 180 \Rightarrow 1° = \frac{\pi}{180} \text{ rads}$$

$$\times 314 \Rightarrow 314° = \frac{314\pi}{180} \text{ rads}$$

$$314° = \frac{314 \times 3.142}{180} = 5.48 \text{ rads}$$

ii) (a)

$$\pi \text{ rads} = 180°$$

$$\div \pi \Rightarrow 1 \text{ rad} = \frac{180°}{\pi}$$

$$\times 4 \Rightarrow 4 \text{ rads} = \frac{4 \times 180°}{\pi}$$

$$4 \text{ rads} = \frac{4 \times 180°}{3.142} = 229.2°$$

(b)

$$\pi \text{ rads} = 180°$$

$$\div \pi \Rightarrow 1 \text{ rad} = \frac{180°}{\pi}$$

$$\times 12.6 \Rightarrow 12.6 \text{ rads} = \frac{12.6 \times 180°}{\pi}$$

$$12.6 \text{ rads} = \frac{12.6 \times 180°}{3.142} = 721.8°$$

EXAMPLE 21

Consider Figure 2.30 which illustrates a sector of a circle of radius 8 cm. The arc ACB subtends an angle of 24° at the circle's centre, O.

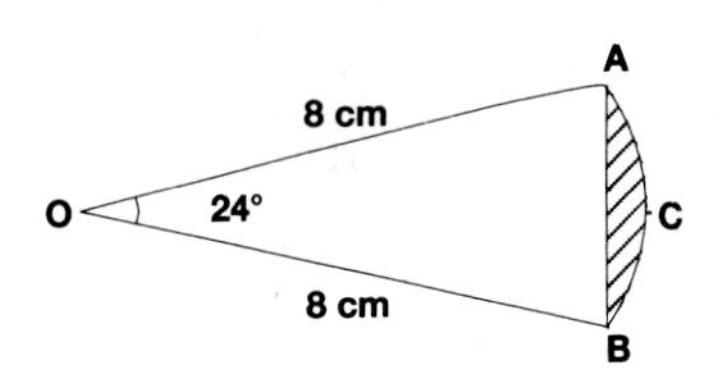

Figure 2.30

Calculate

i) the length of arc ACB,
ii) the area of sector OACB,
iii) the shaded area.

Solution 21

Firstly convert 24° to radians.

$$180° = \pi \text{ rads}$$

$$1° = \frac{\pi}{180} \text{ rads}$$

$$24° = \frac{24\pi}{180} \text{ rads}$$

$$24° = 0.4189 \text{ rads}$$

(i) Using the arc length formula:

$$\text{length of ACB} = r\theta = 8 \times 0.4189$$

The length of arc ACB is 3.351 cm.

(ii) Using the sector area formula:

$$\text{area of sector} = \tfrac{1}{2}r^2\theta = \tfrac{1}{2}(8)^2(0.4189)$$

The area of sector OACB = $13.40\,\text{cm}^2$.

(iii) The area of the triangle OAB is $\tfrac{1}{2}(8)^2 \sin 24°$

(or $\tfrac{1}{2}(8)^2 \sin(0.4189)$ with the calculator set to accept radians).

Both yield a triangle area of $13.02\,\text{cm}^2$.

Therefore the segment area is $(13.40 - 13.02)\,\text{cm}^2$

i.e. the shaded area is $0.38\,\text{cm}^2$.

EXAMPLE 22

A circle, centre O, has radius 3 m. A chord AB is such that its mid-point X is 1 m from O as illustrated in Figure 2.31.

Calculate

(i) the size of the angle AOB,
(ii) the length of the arc AB of the circle,
(iii) the area of the shaded portion shown in Figure 2.31.

(N. Ireland Additional Mathematics, 1991)

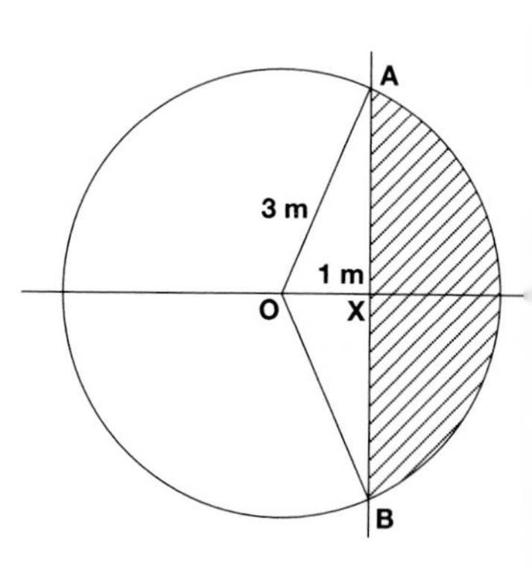

Figure 2.31

Solution 22

(i) $\cos \text{AOX} = \tfrac{1}{3} \quad \Rightarrow \quad \text{AOX} = 70.53°$.
Angle AOB is twice angle AOX, or AOB = 141.1°.

(ii)
$$180° = \pi \text{ rads}$$

$$1° = \frac{\pi}{180} \text{ rads}$$

$$141.1° = \frac{141.1 \times \pi}{180} \text{ rads}$$

$$141.1° = 2.463 \text{ rads}$$

Arc length = 3(2.463) = 7.389 m

(iii) Shaded area $= \tfrac{1}{2}r^2(\theta - \sin\theta)$
or $\tfrac{1}{2}(3)^2(2.463 - \sin 141.1°)$
$= 8.258\,\text{m}^2$.

EXERCISE 2.12

1. Convert to radians.

(i) 15°	(ii) 34°	(iii) 292°	(iv) 47°
(v) 362°	(vi) 193°	(vii) 56°	(viii) 314°
(ix) 4°	(x) 211°	(xi) 37.2°	(xii) 20°
(xiii) 306°	(xiv) 111°	(xv) 55.5°	

2. Convert to degrees.

(i)	5.48 rads	(ii)	6.318 rads	(iii)	3.211 rads
(iv)	0.733 rads	(v)	0.6318 rads	(vi)	0.937 rads
(vii)	2.59 rads	(viii)	5.43 rads	(ix)	3.968 rads
(x)	12.6 rads	(xi)	5 rads	(xii)	9.76 rads
(xiii)	0.85 rads	(xiv)	7.11 rads	(xv)	2.1 rads

3. In each case below the radius of a sector of a circle and the angle it subtends at the centre of the circle are given. Find the arc length and the area of the sector.

	Radius	Angle		Radius	Angle		Radius	Angle
(i)	2.75 m	1.8 rads	(ii)	8.2 m	0.937 rads	(iii)	4.13 m	1.24 rads
(iv)	2.5 m	4.2 rads	(v)	9.4 m	0.24 rads	(vi)	5.45 m	3.29 rads
(vii)	13.6 m	2.3 rads	(viii)	7.5 m	0.32 rads	(ix)	9.45 m	1.63 rads
(x)	2.36 m	1.67 rads	(xi)	9.2 m	54°	(xii)	3.2 m	4°
(xiii)	11.6 m	29°	(xiv)	4.42 m	32.4°	(xv)	3.92 m	56°
(xvi)	5.62 m	62.42°	(xvii)	7.5 m	74°	(xviii)	12 m	81.4°
(xix)	2.9 m	31°	(xx)	13.4 m	14.8°			

4. In each case below the radius of a sector of a circle and the angle it subtends at the centre of the circle are given. Find the segment area in each case.

	Radius	Angle		Radius	Angle		Radius	Angle
(i)	10 cm	37°	(ii)	15 cm	52°	(iii)	7 cm	88°
(iv)	13 cm	66°	(v)	11 cm	33°	(vi)	7 cm	46°
(vii)	21.6 cm	31°	(viii)	4.16 cm	74°	(ix)	8 cm	24°
(x)	22 cm	58°						

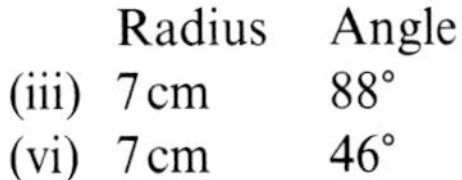

5. Figure 2.32 shows a semi-circle ABC, with centre O and radius 4 m, such that angle BOC = 90°. Given that CD is an arc of a circle, centre B, calculate

(i) the length of the arc CD,
(ii) the area of the shaded region.

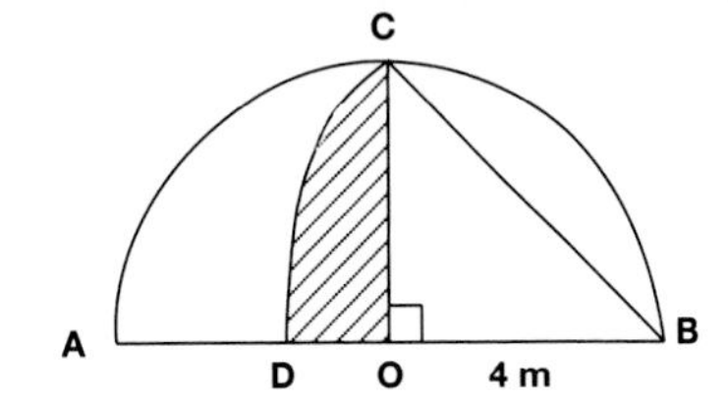

Figure 2.32

(U.C.L.E.S. Additional Mathematics, November 1991)

6. Figure 2.33 shows a sector OAB of a circle centre O, radius 5 cm. BN is perpendicular to OA. Given that BN = 3 cm, calculate

(i) angle BON in radians,
(ii) the perimeter of the shaded region,
(iii) the area of the shaded region.

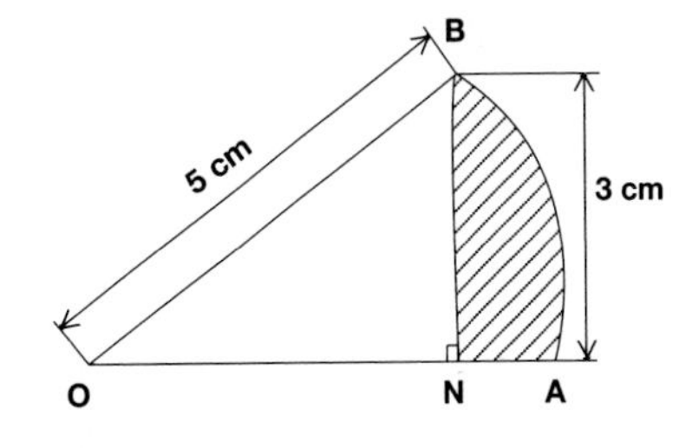

Figure 2.33

(U.C.L.E.S. Additional Mathematics, June 1992)

7. AB and DC are arcs of concentric circles, centre O, where OA = 2 m and AD = 4 m. Given that angle AOB = θ radians and that the perimeter of the figure ABCD, shown in Figure 2.34, is 12 m, calculate

(i) θ,
(ii) the shaded area ABCD.

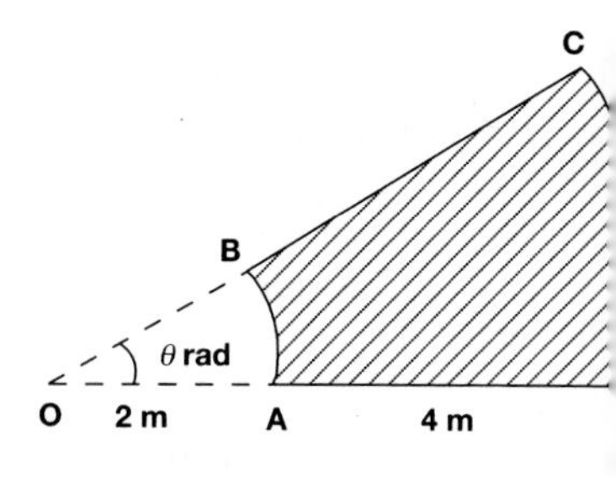

Figure 2.34

(U.C.L.E.S. Additional Mathematics, June 1991)

MORE DIFFICULT EXAMPLES

EXAMPLE 23

(a) O is the centre of a circle of radius 5 cm. A and B are points on the circumference of the circle such that AOB = 35°.

Calculate

(i) the area of the sector AOB,
(ii) the perimeter of this sector.

C and D are points on OA and OB respectively such that the area of the triangle COD is one half of the area of the sector AOB. Given that OC = OD, find the length of OC.

(b) Rewrite $7\sec\theta = 3(3\cos\theta - \tan\theta)$ as a quadratic equation in $\sin\theta$. Hence solve this equation, giving all the solutions which lie in the range $-180°$ to $180°$.

(N. Ireland Additional Mathematics, 1985)

Solution 23

(a) $180° = \pi$ rads

$$1° = \frac{\pi}{180} \text{ rads}$$

$$35° = \frac{35\pi}{180} \text{ rads} = 0.6109 \text{ rads}$$

(i) Area of sector $= \frac{1}{2}(5)^2(0.6109)$
$= 7.636\,\text{cm}^2$

(ii) Arc length $= (5)(0.6109) = 3.055$ cm
and hence the perimeter is given by:
perimeter $= 5 + 5 + 3.055 = 13.055$ cm

Figure 2.35 illustrates the positions of C and D.

The area of the triangle OCD is $\frac{1}{2}(\text{OC})^2 \sin 35°$ and so

$$\tfrac{1}{2}(\text{OC})^2 \sin 35° = \tfrac{1}{2}(7.64)$$

$$\Rightarrow \quad \text{OC} = 3.65\,\text{cm}$$

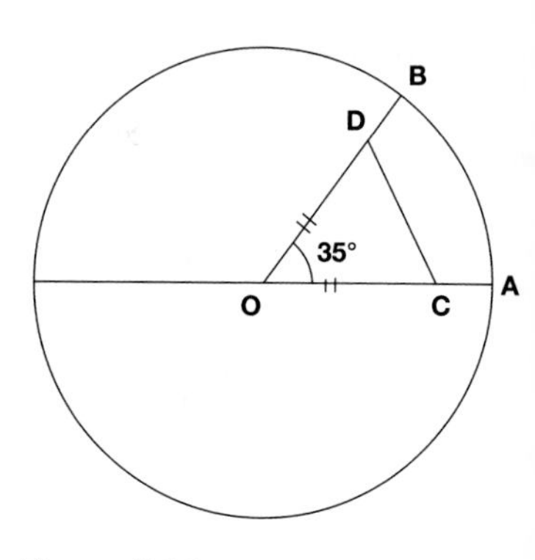

Figure 2.35

(b) To reach a quadratic in $\sin\theta$, multiply by $\cos\theta$ and then use $\sin^2\theta + \cos^2\theta = 1$.

$$7\sec\theta = 9\cos\theta - 3\tan\theta$$

may be multiplied by $\cos\theta$ to give

$$7 = 9\cos^2\theta - 3\sin\theta$$

or $7 = 9(1 - \sin^2\theta) - 3\sin\theta$

i.e. $9\sin^2\theta + 3\sin\theta - 2 = 0$

$$\Rightarrow \quad \sin\theta = \frac{-3 \pm \sqrt{(9 + 72)}}{18} = \frac{-3 \pm 9}{18}$$

Hence $\sin\theta = \frac{1}{3}$ or $\theta = -\frac{2}{3}$
Now $\sin\theta = \frac{1}{3} \Rightarrow \theta = 19.5°$ or $\theta = 160.5°$
and $\sin\theta = -\frac{2}{3} \Rightarrow \theta = -41.8°$ or $\theta = -138.2°$

Therefore $\theta = -138.2°, -41.8°, 19.5°, 160.5°$

EXAMPLE 24

(a) Rewrite $\sec^2\theta = 2(\tan\theta + 1)$ as a quadratic equation in $\tan\theta$. Hence solve this equation, giving all the solutions which lie in the range $-180°$ to $180°$.

(b) If $\sin\alpha < 0$ and $\cos\alpha = \frac{5}{13}$, calculate the value of
(i) $\sin\alpha$,
(ii) $\cot\alpha \sec\alpha$
without using tables or a calculator.

(c) Figure 2.36 shows the circular cross-section of a cylindrical oil tank of diameter 1 m. The length of the tank is 2 m.

The tank is fixed with the axis of the cylinder horizontal. The circular ends of the tank are plane surfaces.

If the depth of oil in the tank is 20 cm as shown in Figure 2.36, calculate the volume of oil in the tank.

(N. Ireland Additional Mathematics, 1984)

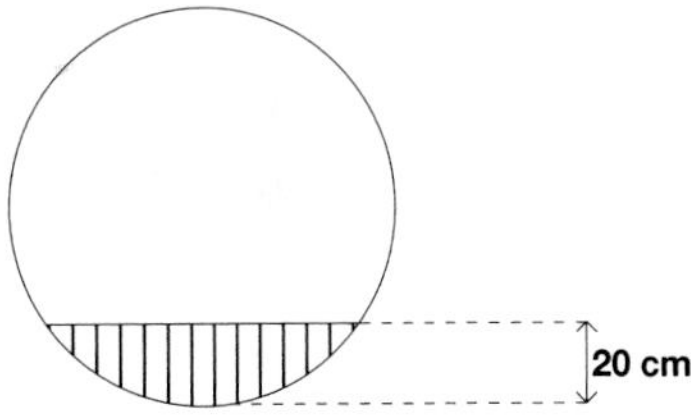

20 cm

Figure 2.36

Solution 24

(a) $\sec^2\theta = 2(\tan\theta + 1)$ must be written as a quadratic equation in $\tan\theta$.

Using the identity $\sec^2\theta = 1 + \tan^2\theta$ the above equation may be rewritten:

$$1 + \tan^2\theta = 2\tan\theta + 2$$

or $\tan^2\theta - 2\tan\theta - 1 = 0$

hence $\tan\theta = \dfrac{2 \pm \sqrt{8}}{2}$

i.e. $\tan\theta = 2.414$ or -0.414.
Now $\tan\theta = 2.414 \Rightarrow \theta = 67.5°$ or $\theta = -112.5°$
and $\tan\theta = -0.414 \Rightarrow \theta = -22.5°$ or $\theta = 157.5°$.

Summarising, the complete set of solutions is $-112.5°$, $-22.5°$, $67.5°$, $157.5°$.

(b) (i) Using the identity $\cos^2\alpha + \sin^2\alpha = 1$ with $\cos\alpha = \frac{5}{13}$ yields:

$$(\tfrac{5}{13})^2 + \sin^2\alpha = 1$$

or $$\sin^2\alpha = \frac{144}{169}$$

i.e. $$\sin\alpha = \pm\tfrac{12}{13}$$

But $\sin\alpha < 0$, therefore $\sin\alpha = -\frac{12}{13}$

(ii) $\cot\alpha\sec\alpha = \dfrac{1}{\sin\alpha} = -\frac{13}{12}$

(c) As the tank has radius 0.5 m, Figure 2.37 illustrates the geometry of the tank's cross-section.

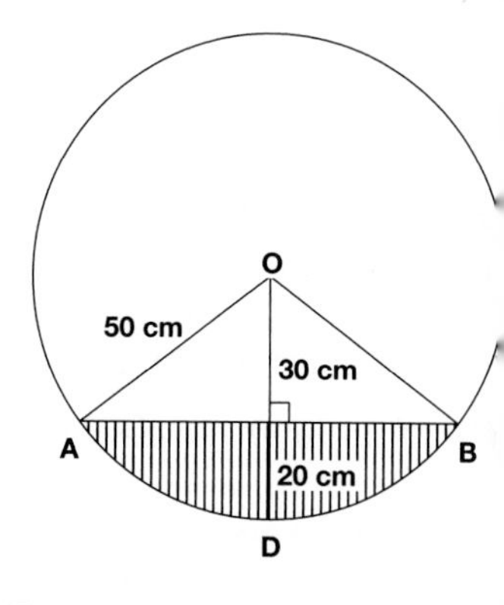

Figure 2.37

Now $\cos AOD = \frac{30}{50}$ and so $AOD = 53.13°$.
Hence angle $AOB = 106.26°$ (by symmetry).
Now convert $106.26°$ to radians.

$$180° = \pi \text{ rads}$$

$$1° = \frac{\pi}{180} \text{ rads}$$

$$106.26° = \frac{106.26\pi}{180} = 1.855 \text{ rads}$$

It follows that the shaded area is $\frac{1}{2}(50)^2(1.855 - \sin 106.26°)$
i.e. shaded area $= 1119\,\text{cm}^2$ ($= 0.1119\,\text{m}^2$).
Finally, the volume of oil $= 0.1119 \times 2 = 0.2238\,\text{m}^3$
or volume of oil $= 223.8$ litres

EXAMPLE 25

In the triangle ABC, the angle $BAC = 35°$, the side $AB = 6$ cm and the side $BC = 4$ cm. The area $S\,\text{cm}^2$ of the triangle lies in the range $4 < S < 6$.

Find (i) the angles ABC and ACB,
(ii) the length of AC.

The circle with centre B and radius 4 cm cuts AB produced at E.

Find (iii) the length of the minor arc CE,
(iv) the area of the minor sector CBE.

The line CB produced meets this circle again at F.

(v) Calculate the length of AF.

(N. Ireland Additional Mathematics, 1986)

Solution 25

The triangle ABC is illustrated in Figure 2.38.

(i) Applying sine rule

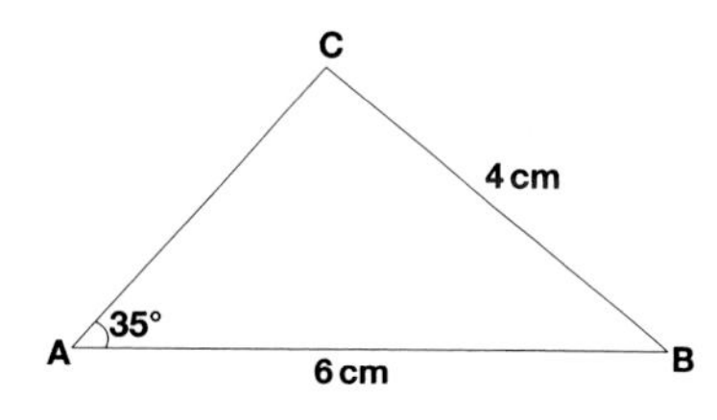

Figure 2.38

$$\frac{4}{\sin 35^\circ} = \frac{b}{\sin B} = \frac{6}{\sin C}$$

in particular

$$\frac{4}{\sin 35^\circ} = \frac{6}{\sin C}$$

i.e. $\sin C = 0.8604$
therefore $C = 59.36^\circ$ or 120.6°

If $C = 59.36^\circ$ then $B = 180^\circ - (59.36^\circ + 35^\circ) = 85.64^\circ$

Hence area $\Delta ABC = \frac{1}{2}(4)(6)\sin 85.64^\circ = 11.97\,\text{cm}^2$

Since this area does not lie in the range $4 < S < 6$, attention turns to the obtuse angle C.

$C = 120.6^\circ$ i.e. $B = 180^\circ - (120.6^\circ + 35^\circ) = 24.4^\circ$
Area of $\Delta ABC = \frac{1}{2}(4)(6)\sin 24.4^\circ = 4.957\,\text{cm}^2$.

Here $4 < 4.957 < 6$ and so the condition is satisfied.

(ii) Returning to the sine rule equation presented at the outset:

$$\frac{4}{\sin 35^\circ} = \frac{b}{\sin 24.4^\circ} = \frac{6}{\sin 120.6^\circ}$$

In particular

$$\frac{4}{\sin 35^\circ} = \frac{b}{\sin 24.4^\circ}$$

yields $b = 2.881$ cm.

In summary, angle ACB = 120.6°, angle ABC = 24.4° and AC = 2.881 cm.

(iii) Figure 2.39 illustrates the circle referred to in this section of the question.

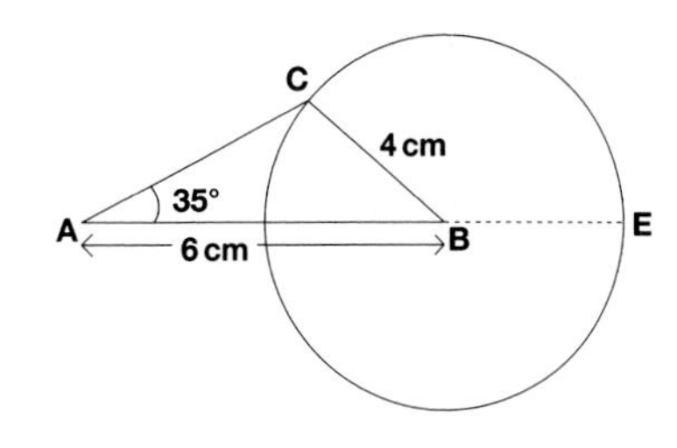

Figure 2.39

Now angle CBE = 180° − 24.4° = 155.6°

Converting this angle to radians

$180^\circ = \pi$ rads

$$1^\circ = \frac{\pi}{180} \text{ rads}$$

$$155.6^\circ = \frac{155.6\pi}{180} = 2.72 \text{ rads}$$

Hence the length of arc CE is given by

length CE = 4(2.72) = 10.88 cm

(iv) The area of sector CBE $= \frac{1}{2}(4)^2(2.72) = 21.76\,\text{cm}^2$

CB is produced to meet the circle at F as shown in Figure 2.40.

AF may be found using cosine rule with angle ABF = 155.6°

i.e. $AF^2 = 6^2 + 4^2 - 2(6)(4)\cos 155.6°$
or $AF = 9.783\,\text{cm}$.

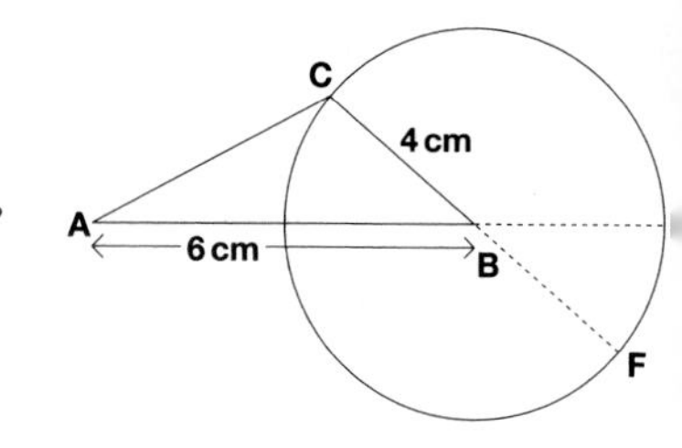

Figure 2.40

EXERCISE 2.13

1. (a) Solve the equation $3\sin^2\theta + 5\cos\theta = 1$ giving all the solutions which lie in the range $-180°$ to $180°$.

(b) *Note: A solution to this question by scale drawing will not be accepted.*

The lines XY and XZ are inclined to a line PQ at angles of 20° and 50° respectively, as in Figure 2.41. A circle, centre O and radius 5 cm touches the lines XY and XZ at A and B respectively.

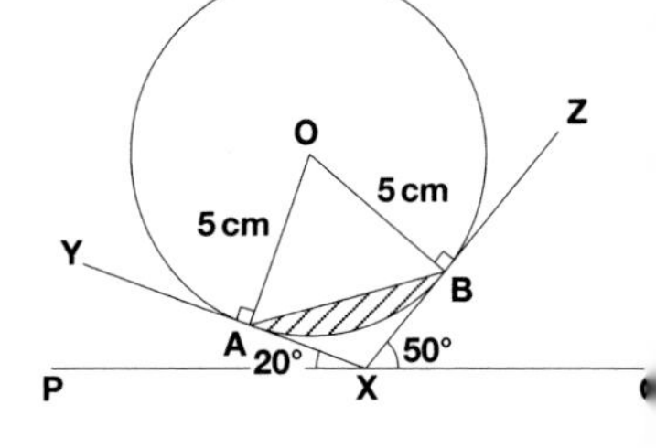

Figure 2.41

(i) Show that AOB = 70°.
(ii) Calculate the length of the chord AB.
(iii) Calculate the area of the triangle OAB.
(iv) Calculate the area of the shaded segment shown in Figure 2.41.

(N. Ireland Additional Mathematics, 1988)

2. In Egypt a tourist observed the Great Pyramid of Cheops, a cross-section of which is shown as ABC in Figure 2.42. A is the top of the pyramid, and B and C are the mid-points of opposite sides of the square base. From X, the door of his hotel, the angle of elevation of A was 17.16°. From Y, a point on the roof of his hotel 50 m vertically above X, the angles AYC and CYX were measured as 19.70° and 81.40° respectively. BCX is a straight line and A, B, C, X and Y all lie in the same plane. This information is shown in Figure 2.42.

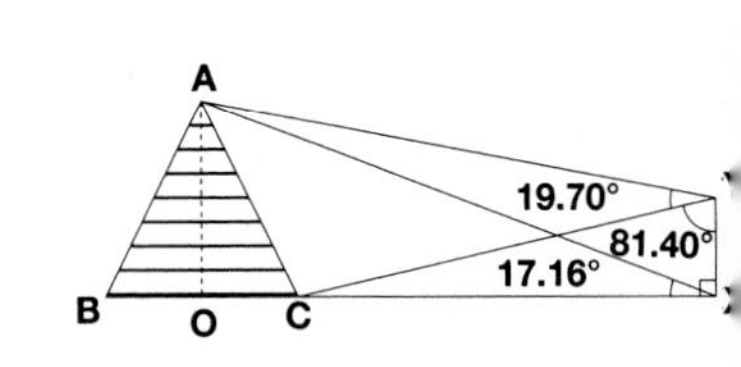

Figure 2.42

Calculate
(i) XC, the distance from the door of the hotel to the centre of the nearest side of the base of the pyramid,
(ii) the angles AXY and YAX,
(iii) the distance AX,
(iv) the length AC,
(v) the height AO of the pyramid and the length BC.

(N. Ireland Additional Mathematics, 1989)

3. A and B are the goalposts at one end of a football pitch and F is the corner flag as shown in Figure 2.43. A, B and F lie in a straight line with AB = 7.4 m and BF = 28 m.

   A player is at the position P, such that PBF = 85° and PFB = 25°.

   Calculate
   (i) the distance PF,
   (ii) the distance PA,
   (iii) the angle APB.

   The goalkeeper wishes to take up a position on the line AB at the point G which is x metres from the post A, such that APG = BPG, as shown in Figure 2.44.

   (iv) Calculate the distance x.

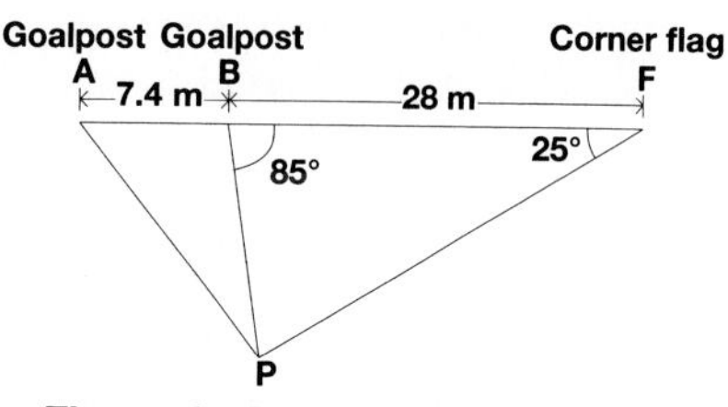

Figure 2.43

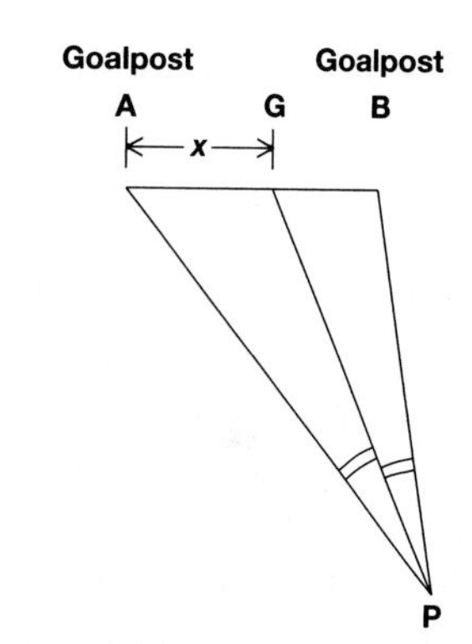

Figure 2.44

(N. Ireland Additional Mathematics, 1990)

4. A vertical lighthouse AL stands at the top edge of a vertical cliff, AB, as shown in Figure 2.45. A point S on a yacht is at sea level and 100 m from the base of the cliff. From S the angle of elevation of the top of the cliff is 42°. A point M on the mast of the yacht is 20 m vertically above S, and from M the angle subtended by the lighthouse is 9°, as shown in Figure 2.45.

   Calculate
   (i) the distance AS,
   (ii) the size of the angle ASM,
   (iii) the distance AM,
   (iv) the size of the angle MAS,
   (v) the size of the angle LAM,
   (vi) the length AL, the height of the lighthouse.

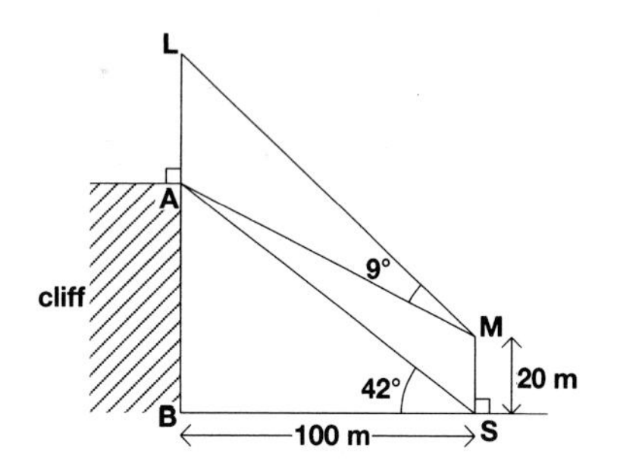

Figure 2.45

(N. Ireland Additional Mathematics, 1991)

TRIGONOMETRY IN THREE DIMENSIONS

While the trigonometry encountered to date will help the reader solve problems involving the angle between two lines, the investigation of the geometry of solid shapes often necessitates the calculation of two further angle types, namely, the angle between a line and a plane and the angle between two planes.

Calculating the angle between a line and a plane

Figure 2.46 illustrates the first such situation – the angle between the line PQ and the plane Π is to be found, where P lies in the plane.

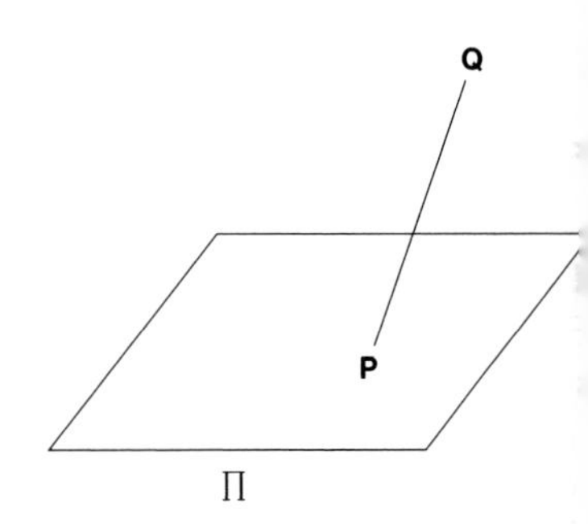

Figure 2.46

The stages in the calculation are as follows.

(i) Distinguish between the extremity of the line in the plane and that off the plane – in this case P is on the plane and Q is not.
(ii) Drop a perpendicular from Q, normal to the plane – QS in Figure 2.47.
(iii) Join that point on the original line in contact with the plane to the point where the normal touches the plane i.e. construct PS in Figure 2.48.
(iv) The angle required is that between the original line, PQ, and the line constructed, PS, i.e. α in Figure 2.49.

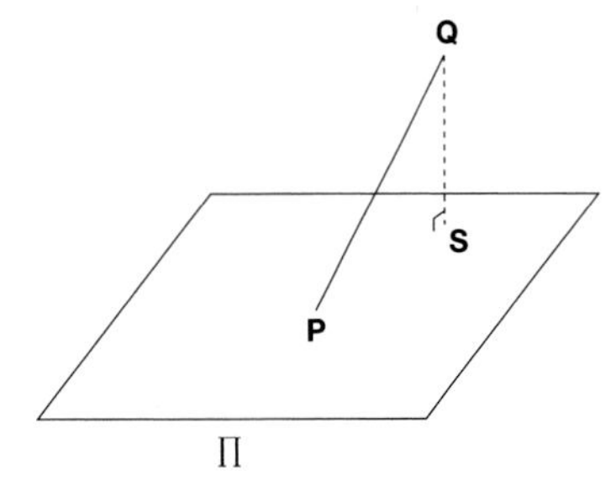

Figure 2.47

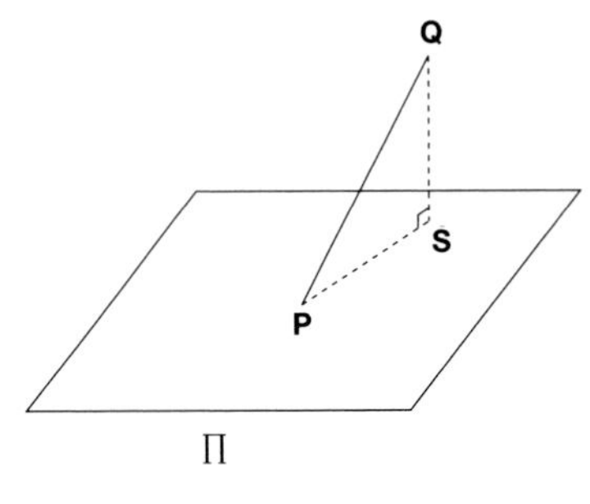

Figure 2.48

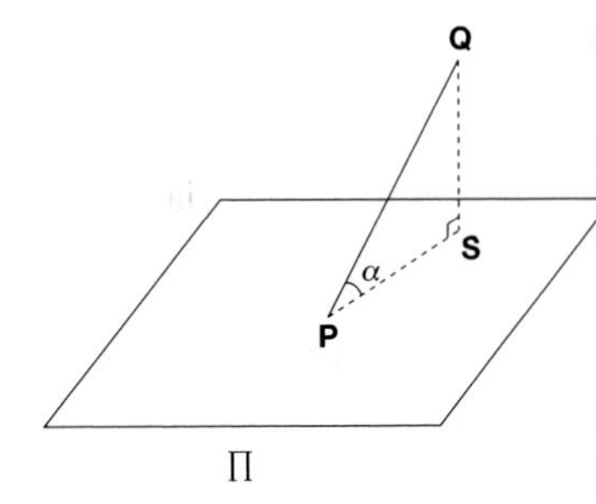

Figure 2.49

Calculating the angle between two planes

Now consider the angle between two planes. Figure 2.50 shows two planes Π_1 and Π_2 intersecting in a common line AB.

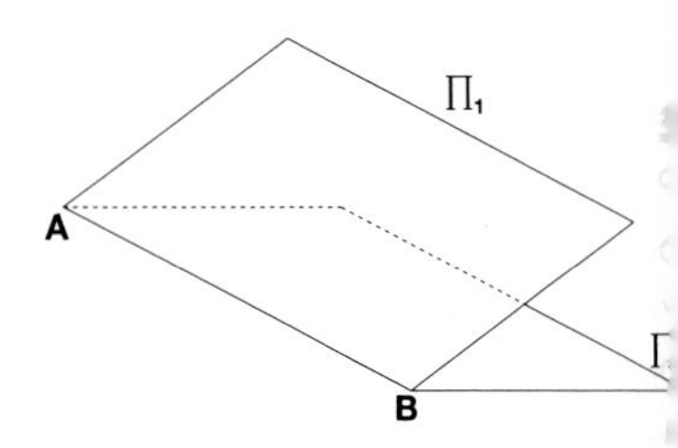

Figure 2.50

The following sequence of steps will assist the reader in calculating the angle between the two planes.

(i) Choose any convenient point on the common line – O (see Figure 2.51) is chosen in this case although the geometry of the figure will normally constrain the choice of point.
(ii) Construct two lines through O at right angles to the common line with one line in each plane – in this case OS is drawn in Π_1 and OT in Π_2 as illustrated in Figure 2.52.

ii) The required angle is that between the pair of lines constructed in (ii) – the angle α in Figure 2.52.

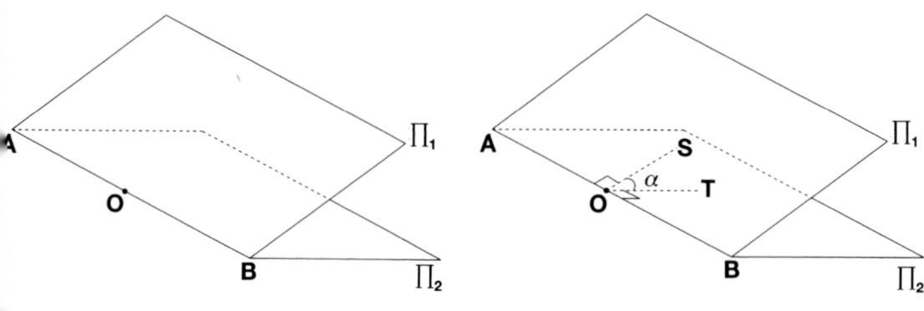

igure 2.51 *Figure 2.52*

XAMPLE 26

rom the base of a vertical tower of height 32 m, point A lies due outh and point B lies in a direction S 60°W with A and B both on he same horizontal level as the base of the tower. The top of the ower has angles of elevation of 14° from A and 16° from B. Find he distance from A to B.

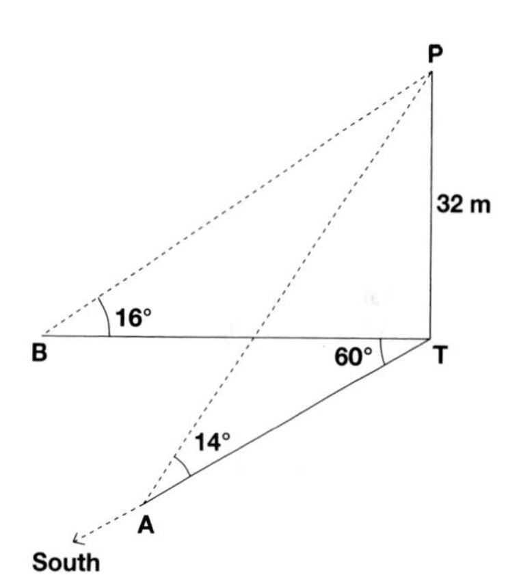

Figure 2.53

olution 26

igure 2.53 illustrates the information given in the example.

Now ATP is a right-angled triangle and so

$$\tan 14^\circ = \frac{32}{AT} \text{ or } AT = 128.3\,\text{m}$$

imilarly, turning to triangle BTP

$$\tan 16^\circ = \frac{32}{BT} \text{ or } BT = 111.6\,\text{m}$$

 more detailed specification of triangle ABT, illustrated in Figure .54, is therefore possible.

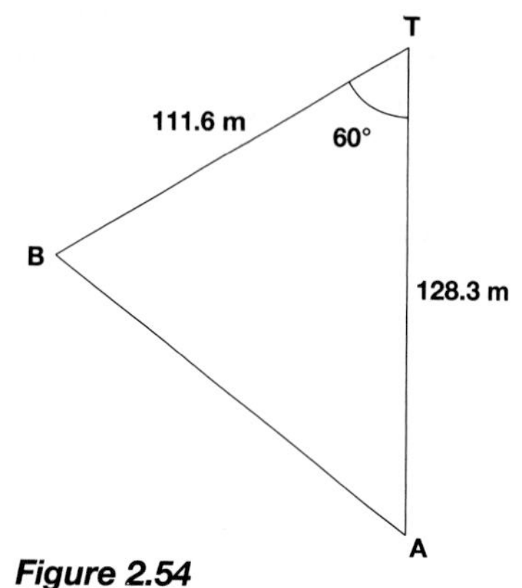

Figure 2.54

Jsing cosine rule:

$$AB^2 = 111.6^2 + 128.3^2 - 2(111.6)(128.3)\cos 60^\circ$$
$$\text{or} \quad AB = 120.8\,\text{m}$$

XAMPLE 27

oints A, B and C all lie in the same horizontal plane with B on a earing of 328° from A, and C on a bearing of 018° from A. A ertical tower of height 28 m stands at B and the angle of elevation f its top, from A, is 17°. A vertical tower of height 24 m stands at C nd the angle of elevation of the top of this tower from A is 22°. ind the distance from B to C and the bearing of C from B.

olution 27

n aerial view of the points A, B and C is shown in Figure 2.55.

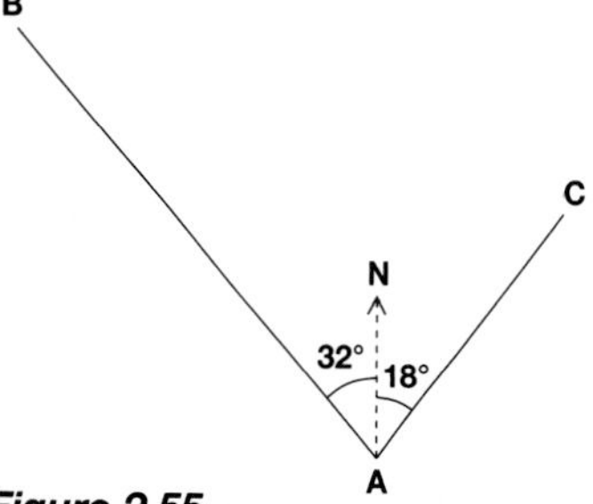

Figure 2.55

The information given on the angles of elevation of the two towers can be represented in the triangles of Figure 2.56.

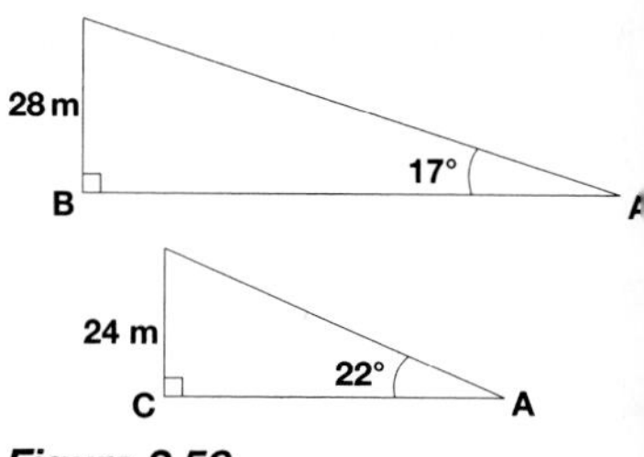

Figure 2.56

Simple trigonometry applied to these triangles in turn yields

$$\tan 17° = \frac{28}{AB} \quad \text{or} \quad AB = 91.58\,\text{m and}$$

$$\tan 22° = \frac{24}{AC} \quad \text{or} \quad AC = 59.40\,\text{m}$$

The plan given at the outset of this solution may be updated as in Figure 2.57.

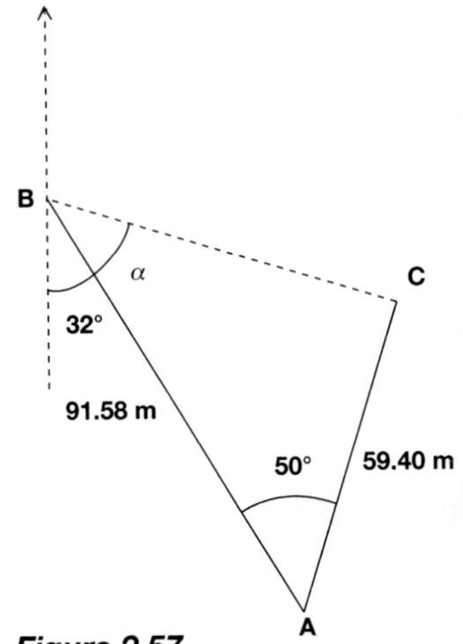

Figure 2.57

By cosine rule

$$BC^2 = 91.58^2 + 59.40^2 - 2 \times 91.58 \times 59.40 \cos 50°$$
$$\text{or} \quad BC = 70.16\,\text{m}$$

Sine rule yields

$$\frac{70.16}{\sin 50°} = \frac{59.40}{\sin \alpha}$$

$$\text{or} \quad \alpha = 40.43°$$

The bearing of C from B is hence 180° − 32° − 40.43° i.e. 107.6°.

EXAMPLE 28

From an observation point at sea level, an aircraft is observed at A on a bearing of 032°, elevation 24° and at height 580 m. The aircraft then flies due east for 900 m, maintaining this height. What will its bearing from the observation point be at the end of this flight?

Solution 28

The information of Example 28 is illustrated in Figure 2.58.

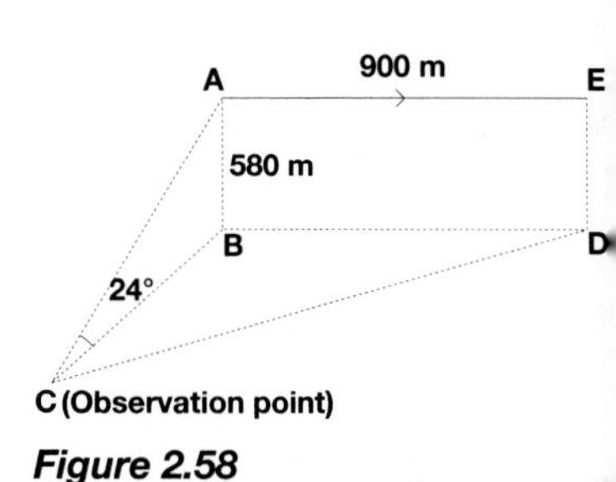

Figure 2.58

Figure 2.59 offers a 'bird's eye view'.

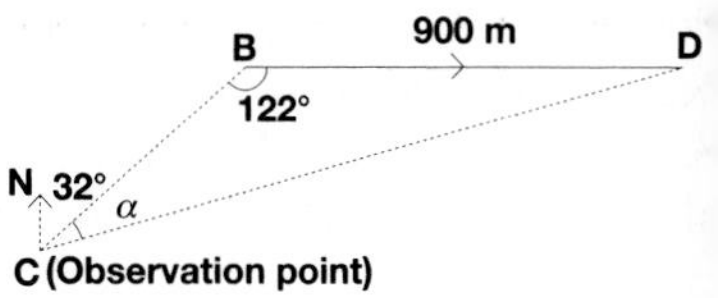

Using the fact that the angle of elevation is 24°

$$\tan 24° = \frac{580}{CB} \text{ or } CB = 1303\,\text{m}.$$

Examining Figure 2.59 therefore

$$CD^2 = 900^2 + 1303^2 - 2 \times 900 \times 1303 \cos 122°$$
$$\text{or} \quad CD = 1937\,\text{m}.$$

The application of sine rule to triangle BCD gives

$$\frac{1937}{\sin 122°} = \frac{900}{\sin \alpha}$$

$$\text{or} \quad \alpha = 23.21°$$

The bearing of D from C is therefore 32° + 23.21° = 055.21°.

B 900 m D
122°
N 32° α
C (Observation point)

Figure 2.59

EXAMPLE 29

Figure 2.60 shows a 'strange pyramid' which has a rectangular base 12 cm by 5 cm and vertex T vertically above corner Q of the rectangular base, with TR = 20 cm. Calculate the angle between faces TPS and PQRS.

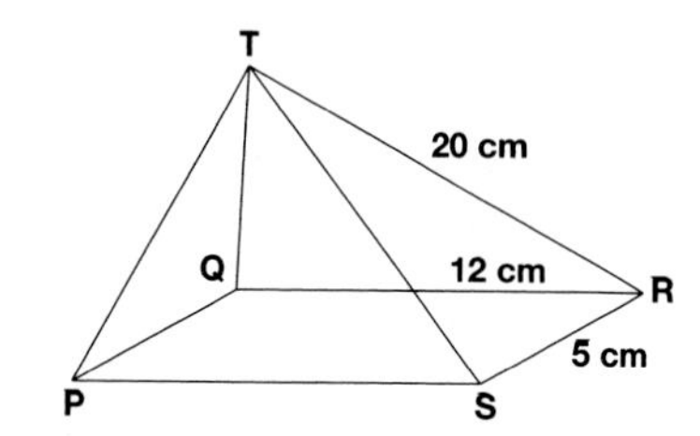

Figure 2.60

Solution 29

The common line between planes TPS and PQRS is obviously PS. Choose point P and (following the method given at the beginning of this section) construct two lines through P, at right angles to PS with a line in each plane. It becomes clear therefore that the angle required is that between PT and PQ – the reader is cautioned that this is no 'ordinary' pyramid; TPS = 90°.

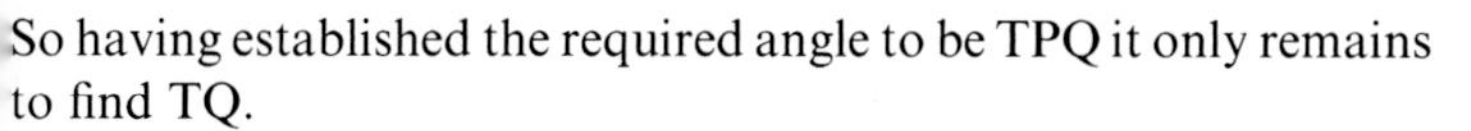

So having established the required angle to be TPQ it only remains to find TQ.

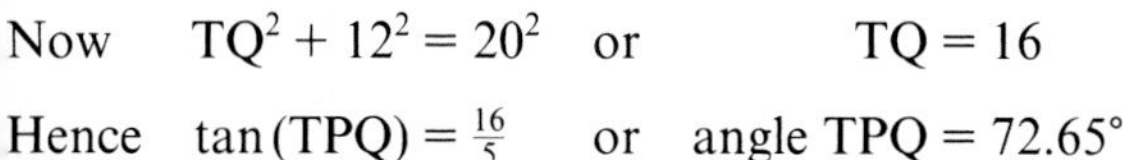

Now $\quad TQ^2 + 12^2 = 20^2 \quad$ or $\quad TQ = 16$

Hence $\quad \tan(TPQ) = \frac{16}{5} \quad$ or $\quad$ angle $TPQ = 72.65°$

EXERCISE 2.14

1. Figure 2.61 shows a wedge ABCDEF with CDEF a horizontal rectangle and ABCD a vertical rectangle. Given that AD = 5 cm and DE = 11 cm find

   (a) the length of EF given that the area of CDEF is $88\,\text{cm}^2$,
   (b) the length of the line joining A to F,
   (c) the angle between AF and the base CDEF,
   (d) the angle between the horizontal and plane ABFE.

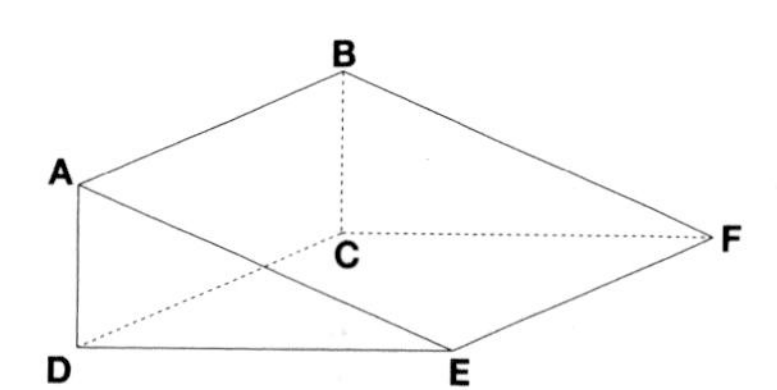

Figure 2.61

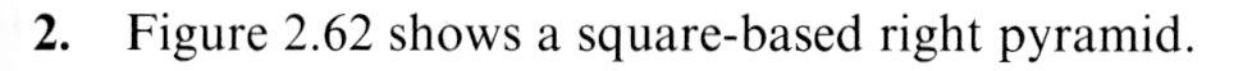

2. Figure 2.62 shows a square-based right pyramid.

   AB = 6 cm, BC = 5 cm and M is the mid-point of BC.

   Calculate

   (a) the length of AM,
   (b) the height of the pyramid,
   (c) the angle between the planes ABC and BCDE.

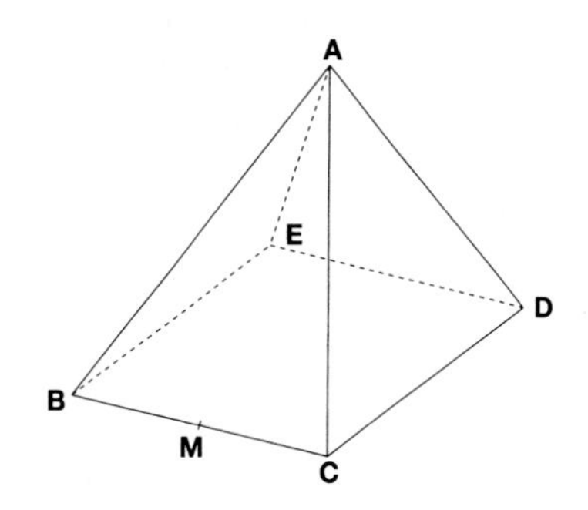

Figure 2.62

3. Figure 2.63 shows a right-triangular prism of length 12 cm. The rectangular base MNQP is horizontal. LMN and OPQ are vertical isosceles triangles with LM = LN = 6 cm and MN = 7 cm.

   Find

   (a) the height of LO above the base MNQP,
   (b) the angle between the plane OMN and the base,
   (c) the angle between the line ON and the base,
   (d) the angle between the plane LMPO and the base.

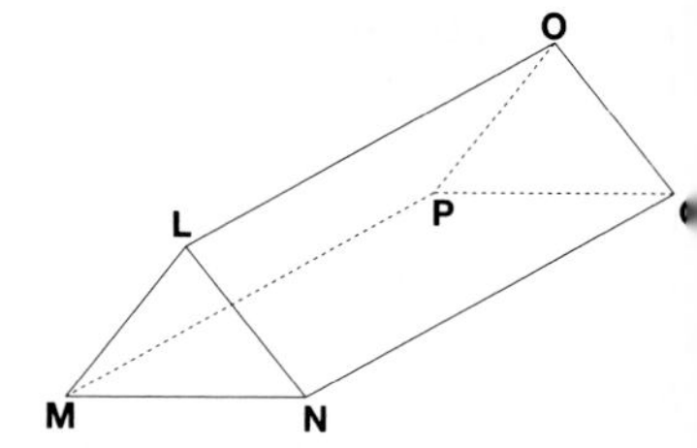

Figure 2.63

4. In Figure 2.64, ABCDEFGH is a cuboid with BC = 5 cm, AE = 3 cm and AB = 8 cm.

   Calculate

   (a) the length of BD,
   (b) the length of BH,
   (c) the angle BDC,
   (d) the angle DBH.

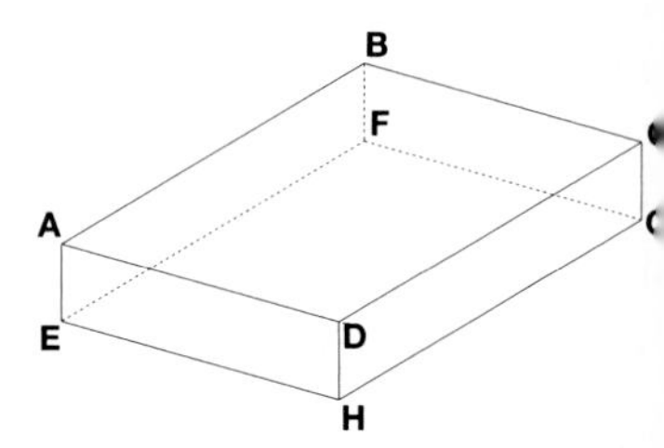

Figure 2.64

5. Figure 2.65 shows a solid where LMNO is a horizontal square base of side 12 cm and PQRS is a horizontal square top. The edges LP, MQ, NR and OS are all 5 cm long and make angles of 65° with the horizontal.

   Calculate

   (a) the length of OM,
   (b) the height of the solid,
   (c) the length of QS,
   (d) the length of the edge of the square top.

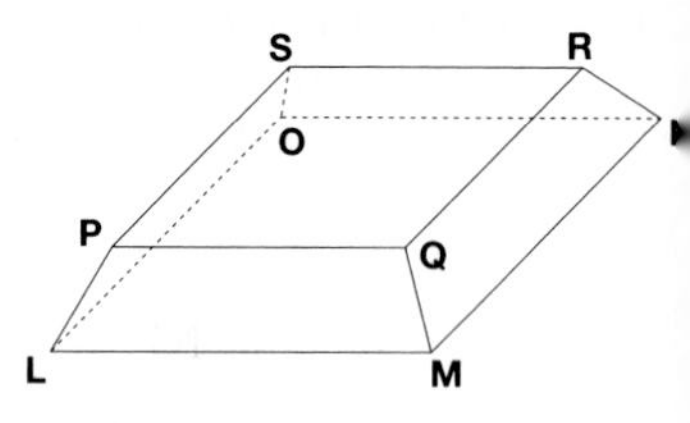

Figure 2.65

6. VWXYZ is a square based pyramid with WXYZ the square base of side 16 cm. V lies verticall above a point A, which is 8 cm from the line WZ and 5 cm from the line WX. I VW = VX = 18 cm, calculate

   (a) the angle between the plane VWX and the square base,
   (b) the height of the pyramid,
   (c) the angle VW makes with the base,
   (d) the angle between the plane VZY and the base.

7. A rectangular field LMNO lies in a horizontal plane. A pole of length 4 m is held vertically at N The top of the pole has an angle of elevation of 21° from M and 32° from O. Calculate the area o the field and find the angle of elevation of the top of the pole from L.

8. A vertical electricity pylon stands at some distance from a horizontal road XY which is 500 m long. The base of the pylon P is on the same level as the road. The angle of elevation of the top o the pylon from Y is 25°. The angle PXY is 32° and the angle XPY is a right angle. Calculate th height of the pylon and the distance of the base of the pylon from the road.

9. From a point X on the ground due west of a vertical aerial, the angle of elevation of the top of the aerial is 30° and from a point Y on the ground due south of the aerial, the angle of elevation of the top of the aerial is 24°. If the distance between X and Y is 840 m what is the height of the aerial?

10. F is the foot of a vertical tower which is 50 m high. X is a point due west of the tower from which the angle of elevation of the top of the tower is 21°. Y is a point due north of the tower from which the angle of elevation of the top of the tower is 31°. Given that X, Y and F are all in the same horizontal plane, calculate

(a) the distance of X from F,
(b) the distance of Y from F,
(c) if M is the point XY which is nearest to the foot of the tower, calculate the angle of elevation of the top of the tower from M.

11. L, M and N are three points in a horizontal plane with LM = 85 m. The angle MLN is 47° and the angle LMN is 68°. Calculate the distance of N from L. A vertical mast is positioned at L and from N the angle of elevation of the top of the mast is 32°. Calculate the height of the mast and the angle of elevation of the top of the mast from M.

12. X, Y and Z all lie in the same horizontal plane with Y on a bearing N42°W from X, and Z on a bearing N26°E from X. A vertical tower of height 25 m stands at Y and the angle of elevation of its top, from X, is 17°. A vertical tower of height 32 m stands at Z and the angle of elevation of its top from X is 22°. Find the distance between Y and Z.

13. Figure 2.66 shows a solid wedge with its face LMNO on horizontal ground. The faces LMNO and LMPQ are rectangles where LM = 5 cm, MN = 7 cm, MP = 7 cm and NP = 5 cm.

Calculate

(a) the angle PMN,
(b) the height of P above the plane LMNO,
(c) the angle which the line LP makes with the horizontal plane.

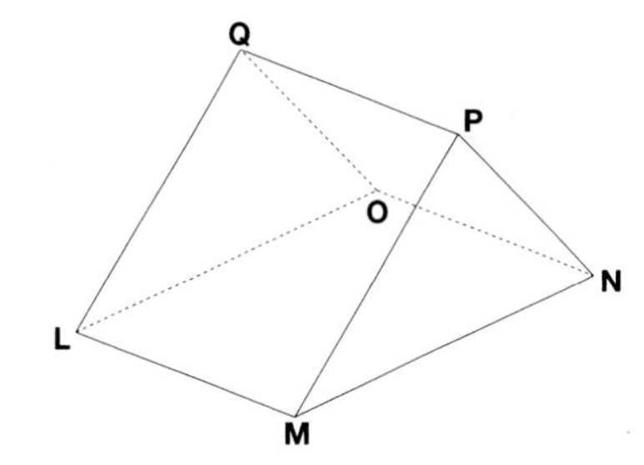

Figure 2.66

14. The door in the wall of a stable is a rectangle ABCD, where AB is the vertical hinge about which the door turns and A is the bottom corner. AB = 3 m and AD = 2 m. The door is opened to make an angle of 40° with the wall. If C_1 is the new position of C, calculate

(i) the distance CC_1,
(ii) the perpendicular distance of C_1 from the wall.

Deduce

(iii) the angle between AC and AC_1,
(iv) the angle AC_1 makes with the wall.

(Oxford and Cambridge Additional Mathematics, 1986)

15. A right pyramid OABCD of vertical height 1.2 m stands on a square base ABCD of side 1 m

Calculate

(i) the angle between the planes OAB and ABCD,
(ii) the length OC.

If the pyramid is then turned over so that the face OAB rests on a horizontal plane, calculate

(iii) the height of the edge CD above the plane,
(iv) the angle between the line OC and the plane.

(Oxford and Cambridge Additional Mathematics, 1985)

16. A, B, and C in Figure 2.67 are three points in a horizontal plane. From the point A a man observes a vertical flagpole CP whose height is known to be 12 m. The bearing of C from A is 066° and the angle of elevation of P from A is 22°. The man walks due south to B from where the bearing of C is 046°.

Calculate

(i) the distance AB,
(ii) the angle of elevation of P from B.

(Oxford and Cambridge Additional Mathematics, 1990)

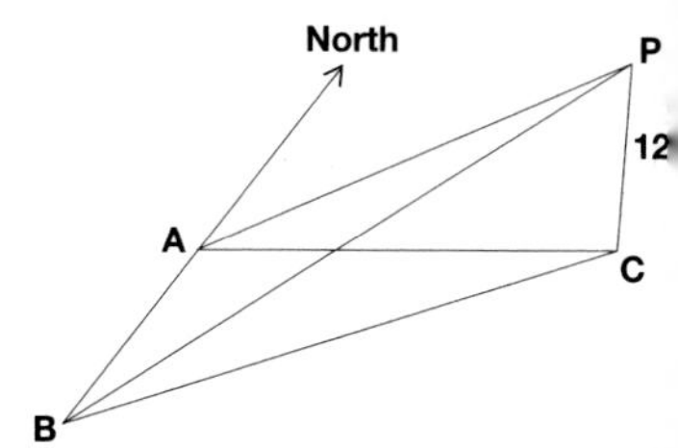

Figure 2.67

17. Figure 2.68 shows the plan of an open window A′B′, of width 90 cm, hinged at X, which has been opened through 60°, with A and B going to A′ and B′ respectively.

PQ is a metal rod which holds the window open. Both QP and XA are perpendicular to AB.

Calculate the lengths A′Q and PQ. Find also the distance of B′ from AB.

(Oxford and Cambridge Additional Mathematics, 1985)

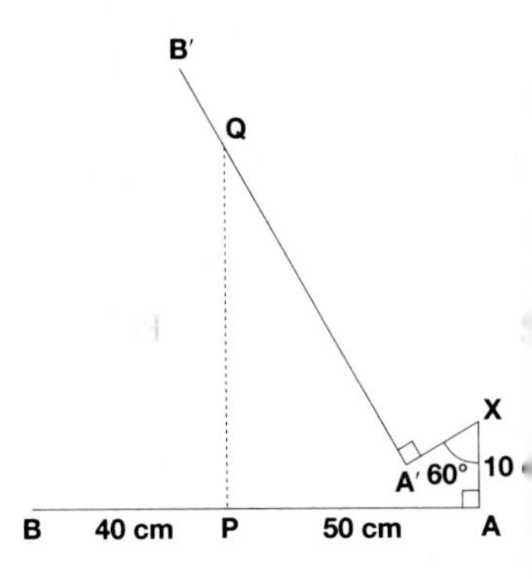

Figure 2.68

18. Figure 2.69 shows a right pyramid of height 9 cm. Its base is a regular pentagon ABCDE of side 2 cm and X is the centre of the base. P is the mid-point of AB.

Calculate

(i) the length PX,
(ii) the length AX,
(iii) the area of the base,
(iv) the angle between a sloping edge and the base,
(v) the angle between a sloping face and the base.

(Oxford and Cambridge Additional Mathematics, 1986)

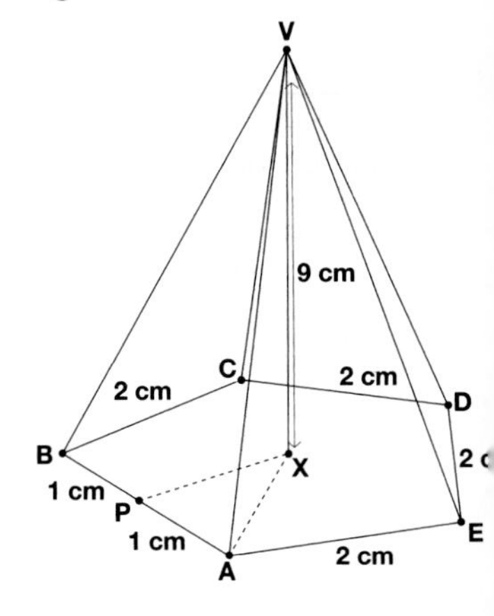

Figure 2.69

9. In Figure 2.70, O, A and B are three points on level horizontal ground. OP is a vertical flagpole of height 18 m and Q is its middle point. The flagpole is supported by two guy-ropes AQ and BP which are inclined at 30° and 60° to the ground respectively. The point B is on a bearing of 060° from O.

(i) OA and OB can be written in the form $a\sqrt{3}$ and $b\sqrt{3}$ respectively.
What are the values of a and b?
(ii) Prove that $AB^2 = 513$.
(iii) Calculate the bearing of B from A.

(Oxford and Cambridge Additional Mathematics, 1987)

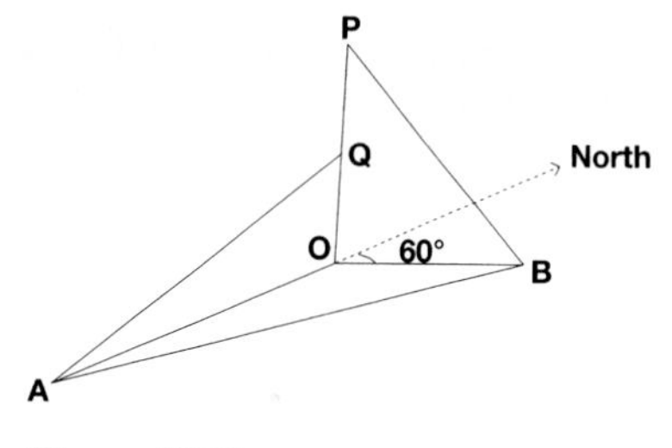

Figure 2.70

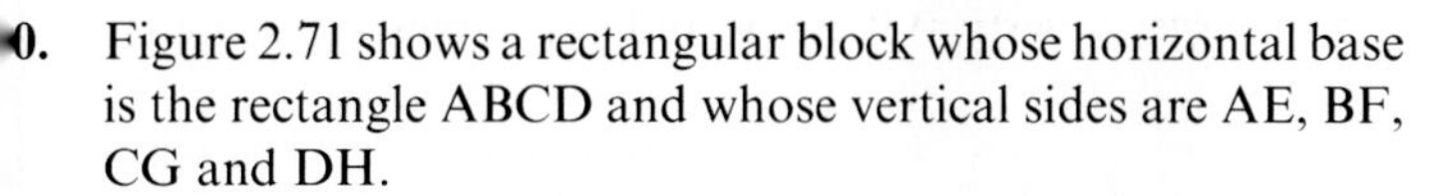

0. Figure 2.71 shows a rectangular block whose horizontal base is the rectangle ABCD and whose vertical sides are AE, BF, CG and DH.

The base measures 12 cm by 5 cm and the height is 6 cm as shown.

Calculate

(i) the angle FDB,
(ii) the perpendicular distance of B from AC,
(iii) the angle between the lines ED and DG,
(iv) the angle between the planes ABC and AFC.

(Oxford and Cambridge Additional Mathematics, 1987)

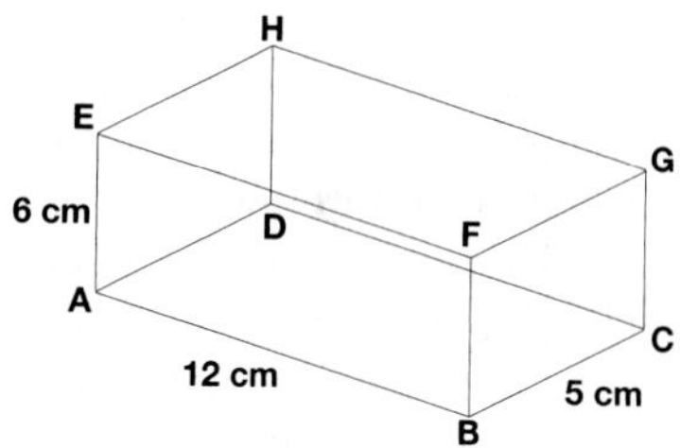

Figure 2.71

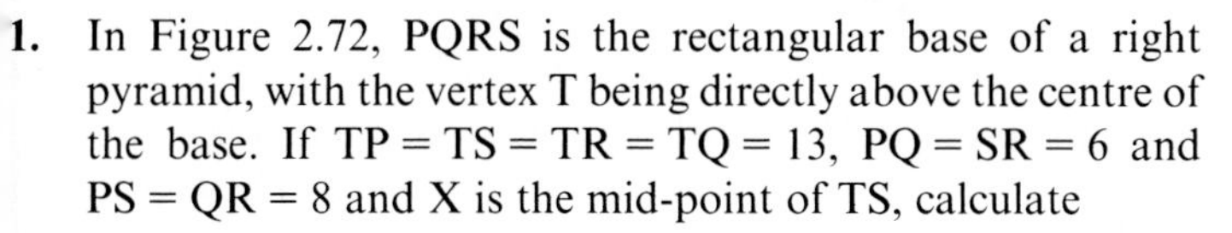

1. In Figure 2.72, PQRS is the rectangular base of a right pyramid, with the vertex T being directly above the centre of the base. If TP = TS = TR = TQ = 13, PQ = SR = 6 and PS = QR = 8 and X is the mid-point of TS, calculate

(i) the length of PR,
(ii) the angle TSP,
(iii) the length of PX,
(iv) the length of XR,
(v) the angle PXR.

(Oxford and Cambridge Additional Mathematics, 1990)

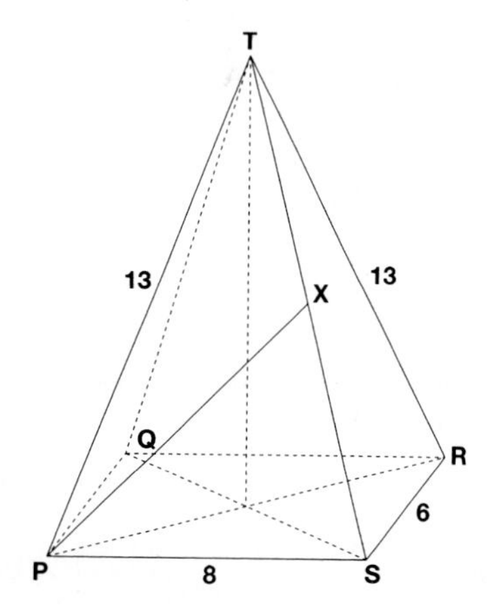

Figure 2.72

CHAPTER 3

Logarithms

Before the advent of the electronic calculator, calculations requiring much tedious multiplication and division were carried out using logarithmic tables which were the invention of the Scottish laird **John Napier** (1550–1617). Until the 1980s, the various examining boards had to take account of the time taken by candidates to 'look up' their logarithm booklets in order to compute cube roots, trigonometric functions and so on. In school examinations 'log tables', as they were known, were a godsend to the plagiarist who could surreptitiously fill the margins which enclosed the tables with vital formulae for use in the examination. The ideas which gave birth to the tables of logarithms were published in Napier's 1614 manuscript *Mirifici logarithmorum canonis descriptio*. His tables allowed the user to multiply two numbers by using a set of tables to convert each number in the product to a logarithm. These two logarithms were then added and the tables were used to convert this sum to the required product. The reader is reminded that before the calculator arrived in the classroom, pupils could add two numbers with much greater accuracy and speed than might be the case when multiplying the same two numbers, particularly where the numbers were written to a large number of decimal places. Henry Briggs completed the work of Napier following his death and, in 1624, published the first 'Briggian' logarithms in his *Arithmetica logarithmica*. These tables were accurate to 14 places of decimals but could not be used to compute the logarithm of any number between 20.000 and 90.000. This 'gap' was filled by Ezechiel De Decker who published the first complete set of logarithms in 1627.

INTRODUCTION

The logarithm of x to the base a is defined to be the power to which a must be raised to give x.

This may be expressed symbolically as

$$\log_a x = l \Leftrightarrow a^l = x. \qquad (1)$$

$\log_a x$ is read as 'the logarithm of x to the base a'.

Relationship (1) plays a central role in the development of the properties of logarithms. In each case the property is derived by determining the equation on the left-hand side when an equation of the form $a^l = x$ is inserted on the right-hand side.

The properties of logarithms

Consider $x = \log_a p$ and $y = \log_a q$.

From (1) above

$$\log_a p = x \Leftrightarrow p = a^x$$
$$\log_a q = y \Leftrightarrow q = a^y$$
$$\text{Now} \quad pq = a^x a^y = a^{x+y}$$

By reference to relation (1) the pattern

$$? \Leftrightarrow a^{x+y} = pq$$

may be completed as

$$\log_a pq = x + y \Leftrightarrow a^{x+y} = pq$$

But $x = \log_a p$ and $y = \log_a q$

hence

$$\mathbf{\log_a (pq) = \log_a p + \log_a q} \qquad \textbf{Property 1} \qquad (2)$$

imilarly $\dfrac{p}{q} = \dfrac{a^x}{a^y} = a^{x-y}$

By reference to relation (1) the pattern

$$? \Leftrightarrow a^{x-y} = \frac{p}{q}$$

may be completed as

$$\log_a \frac{p}{q} = x - y \Leftrightarrow a^{x-y} = \frac{p}{q}$$

or

$$\mathbf{\log_a \left(\frac{p}{q}\right) = \log_a p - \log_a q} \qquad \textbf{Property 2} \qquad (3)$$

Also $\quad p^n = (a^x)^n = a^{xn}$

Relation (1) can be used to complete the pattern

$$? \Leftrightarrow a^{xn} = p^n$$

s

$$\log_a p^n = xn \Leftrightarrow a^{xn} = p^n$$
$$\text{But} \qquad x = \log_a p$$
$$\text{or} \quad \mathbf{\log_a p^n = n \log_a p} \qquad \textbf{Property 3} \qquad (4)$$

A special case of (3) is where $p = q$ i.e.

$$\log_a\left(\frac{p}{p}\right) = \log_a p - \log_a p$$

or $\mathbf{\log_a 1 = 0}$ (5)

Also an interesting result may be derived from relation (1) when $l = 1$, $x = a$

$\log_a a = 1 \Leftrightarrow a^1 = a$

$\mathbf{\log_a a = 1}$ (6)

THE PROPERTIES OF COMMON LOGARITHMS

If the base of the logarithm is 10 it is often referred to as a 'common' logarithm and is written as '$\log_{10}$' or just simply 'log'.

The graph of log x

The graph for $y = 10^x$ is shown in Figure 3.1.

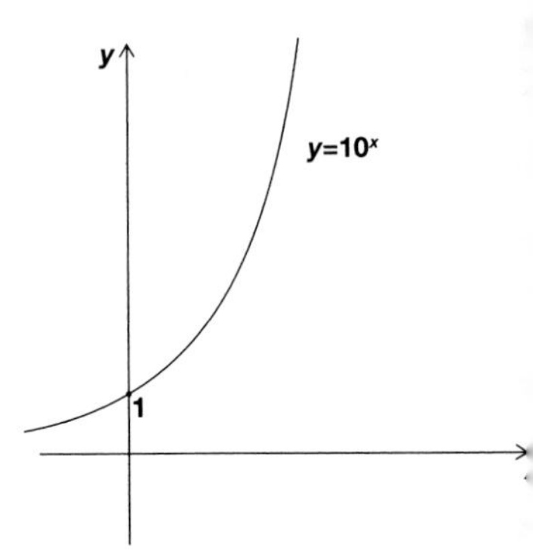

Figure 3.1

Now reflecting a curve in the line $y = x$ results in the interchange of x and y in the equation of the curve since reflection results in the interchange of the x and y coordinates – see Figure 3.2.

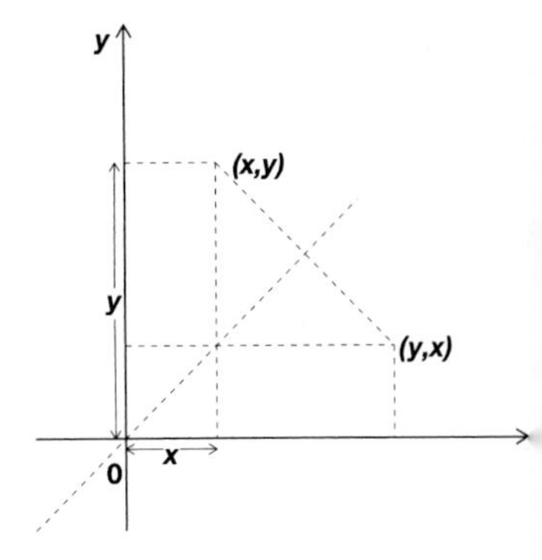

Figure 3.2

It follows that reflection of $y = 10^x$ in $y = x$ results in the graph of $x = 10^y$. But, according to (1) above, this may also be written $y = \log_{10}x$. The graph of $y = \log_{10}x$ is presented in Figure 3.3.

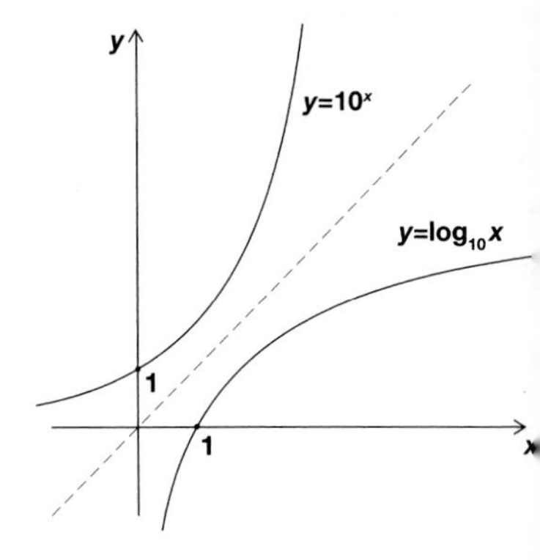

Figure 3.3

EXAMPLE 1

Find the logarithm to the base 10 of 2.42.

Solution 1

Log 2.42 is required. The key sequence for the calculator is

[2] [.] [4] [2] [log]

which reveals: log 2.42 = 0.3838

EXAMPLE 2

Solve for x: (i) $\log x = 3.1$ (ii) $\log x = -1.7$ (iii) $\log x = -12.71$.

Solution 2

(i) The 'inverse log' button or the '10^x' button on the calculator may be used.

The key sequence is

[3] [.] [1] [10^x]

It follows that $x = 1259$ to four significant figures.

ii) [1] [.] [7] [+/−] [10^x]

gives $x = 0.02$

iii) [1] [2] [.] [7] [1] [+/−] [10^x]

gives $x = 1.95 \times 10^{-13}$

EXAMPLE 3

Calculate, without using a calculator, the values of

(i) $\log\left(\frac{1}{1000}\right)$ (ii) $\log\frac{1}{\sqrt[3]{10}}$ (iii) $\log(1000\sqrt{10})$.

Solution 3

(i)
$$\frac{1}{1000} = 10^{-3}$$

hence $\log\left(\frac{1}{1000}\right) = \log_{10} 10^{-3}$

Making use of equation (4)

gives $\log\left(\frac{1}{1000}\right) = -3\log_{10} 10$

Finally, equation (6) with $a = 10$ gives

$$\log\left(\frac{1}{1000}\right) = -3$$

(ii) Similarly $\log\left(\frac{1}{\sqrt[3]{10}}\right) = \log(10^{-\frac{1}{3}}) = -\frac{1}{3}$

(iii)
$$\log(1000\sqrt{10}) = \log(10^3 \times 10^{\frac{1}{2}}) = \log 10^{3\frac{1}{2}}$$
$$= 3.5\log_{10} 10 = 3.5$$

EXAMPLE 4

Express the following in terms of $\log a$, $\log b$ and $\log c$.

(i) $\log\left(\frac{ac}{b}\right)$ (ii) $\log\sqrt{a}$ (iii) $\log(10c)$

Solution 4

(i) Using equation (3)

$$\log\left(\frac{ac}{b}\right) = \log(ac) - \log(b)$$

and by using equation (2) this may be further expanded as

$$\log\left(\frac{ac}{b}\right) = \log(a) + \log(c) - \log(b)$$

(ii) Since $\log\sqrt{a} = \log a^{\frac{1}{2}}$

then equation (4) yields

$$\log\sqrt{a} = \frac{\log a}{2}$$

(iii) By equation (2)

$$\log(10c) = \log 10 + \log c$$

Using equation (6) with $a = 10$ gives

$$\log(10c) = 1 + \log c$$

EXAMPLE 5

Express the following in terms of $\log a$, $\log b$ and $\log c$.

(i) $\log\left(\frac{\sqrt{a}}{bc^2}\right)$ (ii) $\log\left(\frac{b\sqrt{10}}{ac}\right)$ (iii) $\log\left(\frac{ab}{10\sqrt[3]{c}}\right)^2$

Solution 5

(i) Using equation (3)

$$\log\left(\frac{\sqrt{a}}{bc^2}\right) = \log\sqrt{a} - \log(bc^2)$$

Using equation (2) $\log bc^2$ may be written

$$\log(bc^2) = \log b + \log c^2$$

with further simplification using equation (4)

$$\log(bc^2) = \log b + 2\log c$$

In summary therefore

$$\log\left(\frac{\sqrt{a}}{bc^2}\right) = \log\sqrt{a} - \log b - 2\log c$$

so that, expressed in terms of $\log a$, $\log b$ and $\log c$

$$\log\left(\frac{\sqrt{a}}{bc^2}\right) = \tfrac{1}{2}\log a - \log b - 2\log c$$

(ii)
$$\log\left(\frac{b\sqrt{10}}{ac}\right) = \log(b\sqrt{10}) - \log(ac)$$

by equation (3)

or
$$\log\left(\frac{b\sqrt{10}}{ac}\right) = \log b + \log\sqrt{10} - \log a - \log c$$

Writing $\sqrt{10}$ as $10^{\frac{1}{2}}$ in conjunction with equation (4) yields

$$\log\left(\frac{b\sqrt{10}}{ac}\right) = \log b + \tfrac{1}{2}\log 10 - \log a - \log c$$

Now $\qquad \log 10 = \log_{10} 10 = 1$ (see equation (6))

so that, in terms of $\log a$, $\log b$ and $\log c$

$$\log\left(\frac{b\sqrt{10}}{ac}\right) = \tfrac{1}{2} + \log b - \log a - \log c$$

(iii) $$\log\left(\frac{ab}{10\sqrt[3]{c}}\right)^2 = 2\log\left(\frac{ab}{10\sqrt[3]{c}}\right)$$

Now

$$\log\left(\frac{ab}{10\sqrt[3]{c}}\right) = \log ab - \log 10c^{\frac{1}{3}}$$

or $$\log\left(\frac{ab}{10\sqrt[3]{c}}\right) = \log a + \log b - \log 10 - \log c^{\frac{1}{3}}$$

i.e. $$\log\left(\frac{ab}{10\sqrt[3]{c}}\right) = \log a + \log b - 1 - \tfrac{1}{3}\log c$$

From above

$$\log\left(\frac{ab}{10\sqrt[3]{c}}\right)^2 = 2\log\left(\frac{ab}{10\sqrt[3]{c}}\right)$$

and so

$$\log\left(\frac{ab}{10\sqrt[3]{c}}\right)^2 = 2(\log a + \log b - \tfrac{1}{3}\log c) - 2$$

EXAMPLE 6

Given $\log 2 = 0.3010$ and $\log 3 = 0.4771$, find, without the use of the log function on your calculator, the value of

(i) $\log\sqrt{8}$ (ii) $\log(30)^{\frac{1}{3}}$ (iii) $\log\sqrt{24}$.

Solution 6

(i) $\log\sqrt{8} = \log\sqrt{2^3} = \log 2^{\frac{3}{2}} = \tfrac{3}{2}\log 2$

hence $\log\sqrt{8} = (1.5)(0.3010) = 0.4515$

(ii) $\log(30)^{\frac{1}{3}} = \tfrac{1}{3}\log(30) = \tfrac{1}{3}(\log 10 + \log 3)$

i.e. $\log(30)^{\frac{1}{3}} = \tfrac{1}{3}(1 + 0.4771) = 0.4924$

(iii) $\log\sqrt{24} = \tfrac{1}{2}\log 24 = \tfrac{1}{2}(\log 8 + \log 3)$

or $\log\sqrt{24} = \tfrac{1}{2}(\log 2^3 + \log 3) = \tfrac{1}{2}(3\log 2 + \log 3)$

hence $\log\sqrt{24} = \tfrac{1}{2}[3(0.3010) + 0.4771] = 0.6901$

EXERCISE 3.1

1. Use your calculator to evaluate

(i) $\log 53$ (ii) $\log 99$ (iii) $\log 42$ (iv) $\log 84$ (v) $\log 0.1$

(vi) $\log 0.7$ (vii) $\log 2.7$ (viii) $\log 4.25$ (ix) $\log 101$ (x) $\log 476$

(xi) $\log 281$ (xii) $\log 912$ (xiii) $\log 1746$ (xiv) $\log 1024$ (xv) $\log 9999$

2. Evaluate *without* using your calculator

(i) $\log 10$ (ii) $\log 100$ (iii) $\log 1000$ (iv) $\log 10\,000$

(v) $\log\frac{1}{10}$ (vi) $\log\frac{1}{100}$ (vii) $\log\sqrt{10}$ (viii) $\log\sqrt[3]{10}$

(ix) $\log(\sqrt{10})^3$ (x) $\log(\sqrt[3]{10})^2$ (xi) $\log(10\sqrt{10})$ (xii) $\log(100\sqrt{10})$

(xiii) $\log\left(\frac{1}{\sqrt{10}}\right)$ (xiv) $\log\left(\frac{1}{10\sqrt{10}}\right)$ (xv) $\log\left(\frac{10}{\sqrt[3]{10}}\right)$ (xvi) $\log\left(\frac{100}{\sqrt[3]{10}}\right)$

3. Express the following in terms of $\log a$, $\log b$ and $\log c$.

(i) $\log(abc)$ (ii) $\log(ab^2c^3)$ (iii) $\log(a(bc)^2)$ (iv) $\log\left(\frac{ab}{c}\right)$

(v) $\log\left(\frac{a}{bc}\right)$ (vi) $\log\left(\frac{a^3}{b^2c}\right)$ (vii) $\log\left(\frac{(ab)^3}{c^2}\right)$ (viii) $\log(a\sqrt{b})$

(ix) $\log(a\sqrt{(bc)})$ (x) $\log(\sqrt{(abc)})$ (xi) $\log\left(\frac{ab}{\sqrt{c}}\right)$ (xii) $\log\left(\sqrt{\left(\frac{ab}{c}\right)}\right)$

(xiii) $\log\left(\frac{\sqrt{a}}{bc}\right)$ (xiv) $\log\left(a\sqrt{\left(\frac{b}{c}\right)}\right)$ (xv) $\log\left(b\sqrt{\left(\frac{a^3}{c}\right)}\right)$ (xvi) $\log(10abc)$

(xvii) $\log(10a^2bc^3)$ (xviii) $\log(10a)^2$ (xix) $\log\left(\frac{\sqrt{(10a)}}{bc}\right)$ (xx) $\log\left(\frac{a\sqrt{10}}{b\sqrt{c}}\right)$

4. Given that $\log 4 = 0.6$ and $\log 7 = 0.84$, find the values of the following without using the lo button on your calculator.

(i) $\log 1.75$ (ii) $\log 16$ (iii) $\log 49$ (iv) $\log 28$ (v) log

(vi) $\log\sqrt{7}$ (vii) $\log\sqrt{28}$ (viii) $\log 3.5$ (ix) $\log\frac{\sqrt{7}}{2}$

SOLVING EXPONENTIAL EQUATIONS USING LOGARITHMS

Consider the equation $3^x = 2$. How might x be found? By trial and error?

$x = 0$? $3^0 = 1$ Too small
$x = 2$? $3^2 = 9$ Too large
$x = 1$? $3^1 = 3$ Still too large

The solution clearly lies between $x = 0$ and $x = 1$.

The focus could be narrowed by trying $x = 0.1, 0.2, \ldots 0.9$ and so on. This iterative procedure will produce the answer eventually but fortunately a more direct approach is possible using logarithms.

Since $3^x = 2$

then $\log(3^x) = \log 2$

or $x \log 3 = \log 2$ using equation (4)

hence $x = \dfrac{\log 2}{\log 3} = \dfrac{0.3010}{0.4771} = 0.6309$

Logarithms are often appealed to in solving exponential equations where the variable, x, is a power.

EXAMPLE 7

Solve for x.

(i) $(5.2)^x = 7$ (ii) $7^{2x+1} = 4^{-x+3}$ (iii) $5^{2x+1} + 33(5^x) - 14 = 0$

Solution 7

(i) $(5.2)^x = 7$

hence $\log(5.2)^x = \log 7$

equation (4) gives

$$x \log(5.2) = \log 7$$

$$x = \frac{\log 7}{\log 5.2}$$

or $x = 1.18$

(ii) $7^{2x+1} = 4^{-x+3}$

It follows that

$$\log(7^{2x+1}) = \log(4^{-x+3})$$

and, via equation (4),

$$(2x+1)\log 7 = (-x+3)\log 4$$

simplifying

$$(2\log 7)x + \log 7 = -(\log 4)x + 3\log 4$$

or

$$x = \frac{3\log 4 - \log 7}{2\log 7 + \log 4}$$

$$x = \frac{3(0.6021) - (0.8451)}{2(0.8451) + (0.6021)}$$

$$x = 0.4193$$

(iii) $5^{2x+1} + 33(5^x) - 14 = 0$

letting $y = 5^x$ this equation may be written

$$5(5^{2x}) + 33(5^x) - 14 = 0$$

i.e. $5y^2 + 33y - 14 = 0$

using quadratic factorisation this equation may be written

$$(5y - 2)(y + 7) = 0$$

giving

$$y = \tfrac{2}{5} \text{ or } y = -7$$

But $y = 5^x$ and so

$$5^x = \tfrac{2}{5} \quad \text{or} \quad 5^x = -7$$

i.e. $\log(5^x) = \log(\tfrac{2}{5})$ or $\log(5^x) = \log(-7)$

$x\log 5 = \log(\tfrac{2}{5})$ or $x\log 5 = \log(-7)$

Clearly $5^x = -7$ has no solution and therefore only one value of x is possible.

$$x = \frac{\log(\tfrac{2}{5})}{\log 5} = \frac{-0.3979}{0.6990} = -0.5692$$

EXERCISE 3.2

1. Use your calculator to find x where

(i) $\log x = 3$ (ii) $\log x = -4.5$ (iii) $\log x = 7.2$ (iv) $\log x = 2.4$

(v) $\log x = 9$ (vi) $\log x = -8.3$ (vii) $\log x = 5.4$ (viii) $\log x = 0.7$

(ix) $\log x = 0.4$ (x) $\log x = -1.4$ (xi) $\log x = -5.25$ (xii) $\log x = 1.5$

(xiii) $\log x = 0.3$ (xiv) $\log x = 2.53$ (xv) $\log x = 2.32$ (xvi) $\log x = 0.58$

2. Solve for x

(i) $2^x = 9$ (ii) $4^x = 17$ (iii) $7^x = 60$ (iv) $3^x = 68$

(v) $5^{2x} = 32$ (vi) $8^{3x} = 256$ (vii) $6^{\frac{1}{2}x} = 59$ (viii) $2^{\frac{3}{2}x} = 56$

(ix) $5^{x+1} = 72$ (x) $4^{x-1} = 145$ (xi) $3^{2x+3} = 76$ (xii) $5^{7x+5} = 92$

(xiii) $3^{x+1} = 2^x$ (xiv) $4^{x-1} = 3^{x-2}$ (xv) $7^{x+1} = 8^{3-x}$ (xvi) $5^{x-1} = 2^{3x-1}$

(xvii) $2^{4x+5} = 5^{x+2}$ (xviii) $6^{3x+2} = 7^{2-x}$ (xix) $5^{5x+1} = 3^{1-2x}$ (xx) $9^{4x-1} = 2^{-3x}$

3. Solve for x

(i) $2^{2x} - 5(2^x) + 6 = 0$ (ii) $3^{2x} - 3^{x+2} + 8 = 0$ (iii) $2^{2x} - 7(2^x) + 12 = 0$

(iv) $5^{2x} + 6 = 5^{x+1}$ (v) $4^{2x} + 15 = 4^{x+2}$ (vi) $2^{2x} + 9 = 3(2^{x+1})$

(vii) $3^{2x} - 2(3^{x+2}) + 32 = 0$ (viii) $2^{2x} - 17(2^x) + 30 = 0$ (ix) $3^{2x} + 42 = 13(3^x)$

(x) $2^{2x} - 3(2^{x+1}) + 8 = 0$ (xi) $2^{2x+2} - 3(2^{x+3}) + 36 = 0$ (xii) $3^{2x+1} + 24 = 27(3^x)$

(xiii) $2^{2x+1} - 5(2^x) + 2 = 0$ (xiv) $5^{2x+1} - 7(5^x) + 2 = 0$ (xv) $3^{2x+1} - 8(3^x) + 4 = 0$

4. (a) Solve the equation

$$2\log x + \log 4 = \log(9x - 2).$$

(b) By means of the substitution $3^x = y$, solve the equation

$$3^{2x} - 3^{x+2} + 8 = 0,$$

giving answers to two decimal places where necessary.

(U.C.L.E.S. Additional Mathematics, June 1991, part question)

REDUCTION OF A LAW OF THE FORM P = kl^n TO LINEAR FORM USING LOGARITHMS

Suppose a table of values of quantity P is available at various values of quantity l. It is thought that P is related to l via the law $P = kl^n$, where k and n are to be determined. How can it be confirmed that the law holds and what are the values of k and n?

Suppose the logarithms of each side of the law are taken, thus

$$\log P = \log(kl^n)$$

or using equations (2) and (4)

$$\log P = \log k + \log l^n$$

$$\log P = \log k + n \log l \qquad (7)$$

If the table is reconstructed in terms of $\log P$ and $\log l$ and the tabulated pairs plotted, the straight line $Y = mX + c$ where $Y = \log P$, $X = \log l$, $c = \log k$, and $m = n$ results. This is easily confirmed by inserting these equations in equation (7) to give

$$Y = mX + c$$

It follows that the test which establishes a law of the form $P = kl^n$ consists of taking logarithms of the tabulated variables and plotting $\log P$ against $\log l$. If the result is approximately linear, the law is confirmed.

As noted above, the gradient of the line is n and the intercept yields k via $\log k = c$.

EXAMPLE 8

A fixed mass of gas undergoes an adiabatic expansion, its volume and pressure being related as in Table 3.1. Confirm that the expansion is governed by a law of the form $P = aV^m$ and confirm that $a = 24 \times 10^3$ and $m = -1.4$ approximately.

Table 3.1

P (cm of Hg)	29.47	22.61	18.10	14.93	12.61
V (cm^3)	120	145	170	195	220

Solution 8

Taking logarithm to the base 10 of P and V gives Table 3.2.

Table 3.2

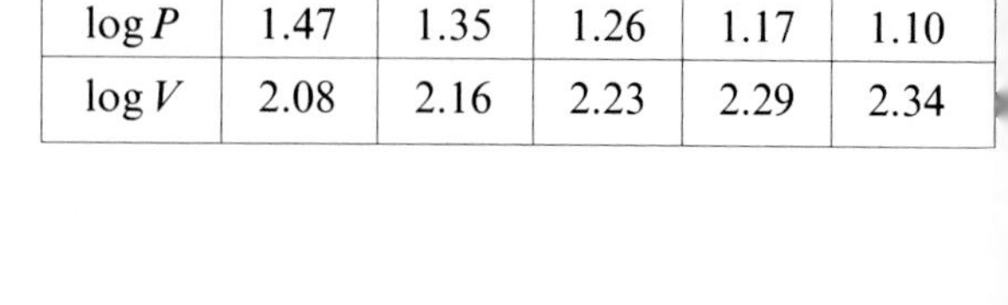

$\log P$	1.47	1.35	1.26	1.17	1.10
$\log V$	2.08	2.16	2.23	2.29	2.34

Consider the law $P = aV^m$ and take logs of each side.

$$\log P = \log aV^m$$
$$\Rightarrow \log P = \log a + m \log V$$

making use of the properties of logarithms.

Log P is plotted against log V as shown in Figure 3.4 and m can be estimated as

$$m = -\frac{1.47 - 1.17}{2.29 - 2.08} = -1.4$$

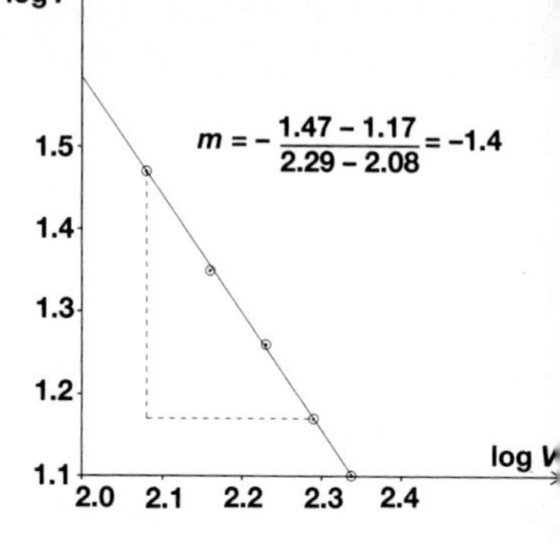

Figure 3.4

It follows that

$$\log P = \log a - 1.4 \log V$$

But, from the graph $\log P$ is 1.585 when $\log V$ is 2.

Therefore

$$1.585 = \log a - 1.4\,(2)$$

or $\log a = 4.385$
i.e. $a = 24266$
or $a = 24 \times 10^3$ to 2 significant figures
Hence $P = 24 \times 10^3 V^{-1.4}$ as required.

EXERCISE 3.3

1. The time of swing of a pendulum for various values of its length is given in Table 3.3.

Table 3.3

Time of swing T (seconds)	0.77	0.96	1.36	1.77	1.92
Length l (metres)	0.15	0.23	0.46	0.78	0.92

Find a law of the form $T = kl^n$.

2. The distance which a stone falls from rest for various values of the time of flight is given in Table 3.4.

Table 3.4

Distance D (metres)	1.25	7.2	26.45	72.2	105.8
Time T (seconds)	0.5	1.2	2.3	3.8	4.6

Find a law of the form $D = kT^n$.

3. The illumination of a bulb for various values of the distance from the bulb is given in Table 3.5.

Table 3.5

Illumination I (candela)	500	31.25	17.58	7.8	6.17
Distance d (metres)	0.3	1.2	1.6	2.4	2.7

Find a law of the form $I = kd^n$.

4. The power of the engine which drives a toy travelling on a level surface, for various speeds of the toy, is given in Table 3.6.

Table 3.6

Power P (watts)	0.76	3.37	7.5	9.98	30.72
Speed s (kilometres per hour)	14	23	30	33	48

Find a law of the form $P = ks^n$.

5. The resistance to the motion of a train for various values of the speed of the train is given in Table 3.7.

Table 3.7

Resistance R (newtons)	3671.3	4988.3	5387.9	6638	7887.2
Speed s (metres per second)	26	48	56	85	120

Find a law of the form $R = ks^n$.

6. The frequency of a note given by an organ pipe at various values of the length of the pipe is given in Table 3.8.

Find a law of the form $F = kl^n$.

Table 3.8

Frequency F (hertz)	1179	589	351	243	165
Length l (centimetres)	14	28	47	68	100

7. The safe speed at which a train can round a curve for various values of the radius of the curve is given in Table 3.9.

Table 3.9

Speed s (metres per second)	8.9	14.1	21.9	32.2	43.8
Radius r (metres)	20	50	120	260	480

Find a law of the form $s = kr^n$.

8. The frequency in kHz of a radio source at various wavelengths is given in Table 3.10.

Table 3.10

Frequency F (kHz)	1190	647	469	250	200
Wavelength λ (metres)	252	464	640	1200	1500

Find a law of the form $F = k\lambda^n$.

9. The extension produced in a stretched spring for various values of the tension in the spring is given in Table 3.11.

Table 3.11

Extension e (centimetres)	0.56	1.72	2.84	4.27	7.35
Tension T (newtons)	1.68	5.16	8.52	12.81	22.05

Find a law of the form $e = kT^n$.

10. The pressure of a certain mass of gas at various volumes is given in Table 3.12.

Table 3.12

Pressure P (pascals)	923	462	250	179	129
Volume V (m^3)	1.3	2.6	4.8	6.7	9.3

Find a law of the form $P = kV^n$.

11. T days is the time for a planet to complete one orbit around the Sun, and D million km is the distance of that planet from the Sun. Table 3.13 gives the values of T and D for 5 planets.

It is thought that a relationship of the form $T = kD^n$ exists between T and D, where k and n are constants. Verify this by drawing a graph of $\log T$ against $\log D$ and hence obtain values for k and n.

Table 3.13

Planet	T (days)	D (million km)
Mercury	88	57.9
Earth	365	149.6
Jupiter	4329	778.3
Uranus	30 600	2870.0
Pluto	90 670	5907.0

Using your values for k and n, find

(i) the time required by Mars to complete one orbit around the Sun, given that its distance from the Sun is 227.9 million km,

(ii) the distance of Venus from the Sun, given that it completes one orbit around the Sun in 225 days.

(N. Ireland Additional Mathematics, 1989)

12. An astronomer measured the temperature of the Sun's corona (the ring of atmosphere around the Sun) at various distances from the centre of the Sun. The results are given in Table 3.14.

Theory predicts that a relationship of the form $T = aR^k$ exists between T and R, where a and k are constants. Verify this by drawing the graph of $\log T$ against $\log R$ and hence obtain values for a and k.

Table 3.14

Distance from the centre of the Sun R (million km)	Temperature of the Sun's corona T (million °C)
2	0.75
3	0.67
5	0.58
10	0.47
20	0.39

(i) The Earth's surface is 150 million km from the centre of the Sun. Assuming that the formula is valid at this distance, use your values for a and k to find the temperature of the Sun's corona at the Earth's surface.

(ii) Calculate the distance from the centre of the Sun where the temperature of the corona would be 500 000 °C.

(N. Ireland Additional Mathematics, 1990)

13. The air resistance acting on a particular particle was measured while it moved through the atmosphere at various speeds. The results are given in Table 3.15.

It is believed that a relationship of the form $R = kv^n$ exists between R and v, where k and n are constants.

Verify this by drawing a suitable straight line graph, and hence obtain values for k and n.

Table 3.15

Speed v (m/s)	Air resistance R (newtons)
10	4.5
25	28.1
40	72.0
70	220.5
80	288.0

Using the formula $R = kv^n$ with your values for k and n, calculate

(i) the air resistance when $v = 35$ m/s,

(ii) the speed when the air resistance is 200.0 newtons.

(N. Ireland Additional Mathematics, 1991)

CHAPTER 4

Differential calculus

Sir Isaac Newton (1642–1727) was born the son of a Woolsthorpe squire and received a grammar school education in Grantham in England. He read mathematics at Trinity College Cambridge under Isaac Barrow who, in 1669, vacated the Lucasian Professorship to make way for his talented pupil, acknowledging that he had benefited infinitely more from their collaboration than Newton. By this time Newton had already been raised to fellowship of Trinity. Newton made frequent reference to the strain of academic life but remained forty years at Cambridge before becoming their Member of Parliament. In 1692, following a breakdown, he accepted the position of Warden and later Master of the Mint.

In 1687 Newton published what is arguably the world's most influential scientific work: *Philosophiae Naturalis Principia Mathematica*. It is of interest to ponder that civilisation might have been denied this mathematical masterpiece had it not been for the tireless efforts of the astronomer Edmond Halley who, as editor of the Royal Society, commissioned the work, and who later was forced to finance personally its dissemination when the Society failed to provide the necessary monies.

The years 1665 and 1666 saw Cambridge in the grip of the plague and Newton returned to Woolsthorpe to write and think. His output in these two years had a profound influence on the direction of science and his notebooks provide a unique insight into the mind of this great mathematician:

In the beginning of the year 1665 I found the method for approximating series and the rule for reducing any dignity of any binomial to such a series. The same year in May I found the method of tangents of Gregory and Sulzius, and in November had the direct method of Fluxions [differential calculus], and in the next year in January had the Theory of Colours, and in May following I had entrance into the inverse method of Fluxions [integral calculus], and in the same year I began to think of gravity extending to the orb of the 'Moon . . . and . . . compared the force requisite to keep the Moon in her orb with the force of gravity at the surface of the earth . . . All this was in the two years of 1665 and 1666, for in those years I was in the prime of my age for invention, and minded Mathematics and Philosophy more than at any time since.

THE GRADIENT OF A CURVE

The gradient, m, of a straight line may be calculated by choosing two points on the line, $P(x_1, y_1)$ and $Q(x_2, y_2)$, and using the standard formula from coordinate geometry

$$m = \frac{y_2 - y_1}{x_2 - x_1} \qquad (1)$$

The gradient of a straight line will be constant no matter which points P and Q are chosen.

The gradient of a curve at a specified point is considered to be the gradient of the tangent to the curve at that point. In Figure 4.1 the gradient of the curve at the point (x_1, y_1) is the gradient of the tangent shown.

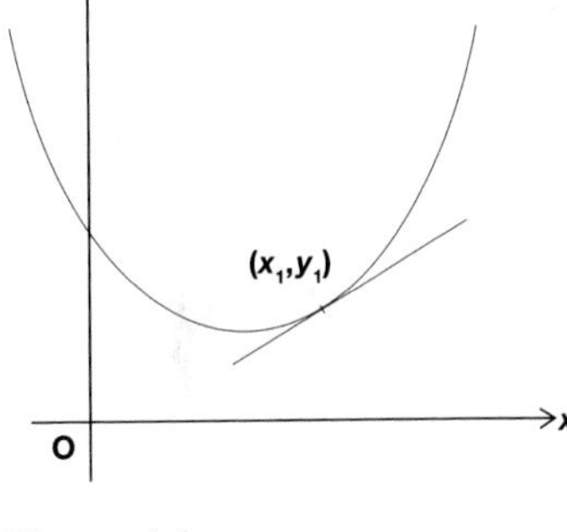

Figure 4.1

How is the gradient of a curve at given point to be calculated? Is the mathematician faced with the tedious task of constructing the tangent to the curve and then measuring its gradient? The following illustration demonstrates how the gradient of a curve may be found, without construction, provided the equation of the curve is known.

Consider the curve $y = x^2 - 2x$ shown in Figure 4.2. How might the gradient of this curve at P(3,3) be calculated?

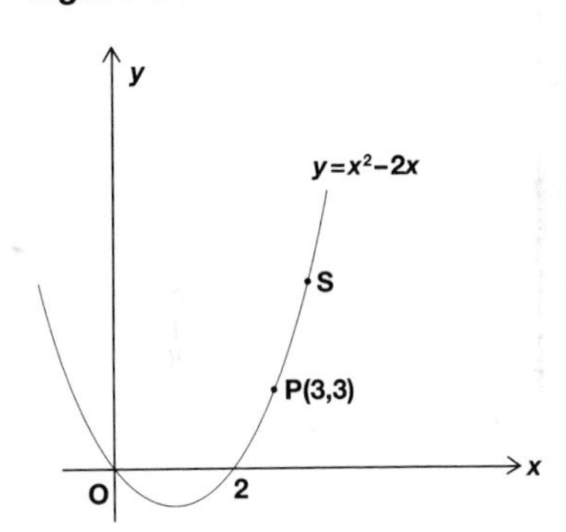

Figure 4.2

Consider the point S on the curve. The gradient of the line PS represents an approximation to the gradient of the tangent at P, with this approximation improving as S approaches P. Table 4.1 lists the gradient of PS for various positions of S. The gradient of PS in each case is computed using equation (1).

Table 4.1

S	Gradient PS
(4,8)	5
(3.9, 7.41)	4.9
(3.8, 6.84)	4.8
(3.6, 5.76)	4.6
(3.4, 4.76)	4.4
(3.2, 3.84)	4.2
(3.1, 3.41)	4.1
(3.05, 3.2025)	4.05
(3.005, 3.020025)	4.005
(3.0001, 3.00040001)	4.0001

It is clear that the gradient approaches 4 as S approaches P and hence the gradient of the tangent at P may be assigned the value 4.

Such a process is known as a 'limiting' process; S never actually becomes P. The gradient of PS is said to be 4 'in the limit' as S approaches P (written S → P)

or $\lim_{S \to P} (\text{grad PS}) = 4$

FIRST PRINCIPLES OF DIFFERENTIATION

Now consider the **general** case of a point $P(x, y)$ on the curve $y = x^2 - 2x$. Since $P(x, y)$ lies on $y = x^2 - 2x$, the point may be written $P(x, x^2 - 2x)$ as shown in Figure 4.3. Consider the neighbouring point S with x coordinate $x + h$. The y coordinate of S is given by the expression

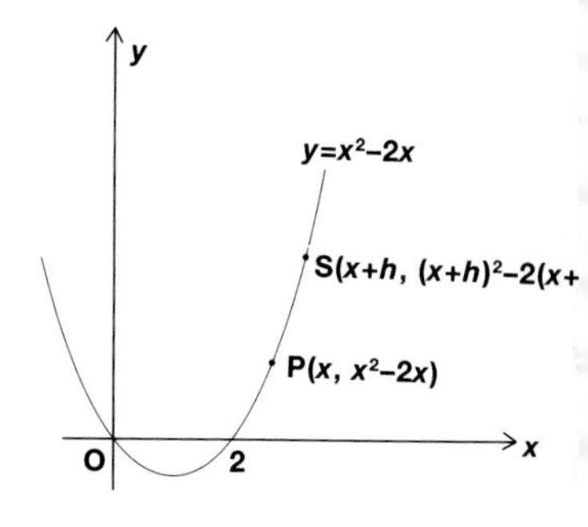

Figure 4.3

$$(x + h)^2 - 2(x + h) = x^2 + 2xh + h^2 - 2x - 2h$$

Using equation (1), the gradient of PS is given by

$$\text{gradient PS} = \frac{(x^2 + 2xh + h^2 - 2x - 2h) - (x^2 - 2x)}{x + h - x} \qquad (2)$$

$$= \frac{2xh + h^2 - 2h}{h} = 2x + h - 2$$

Using the reasoning of the previous section

$$\text{the gradient of tangent at P} = \lim_{S \to P} (2x + h - 2)$$

From an examination of the coordinates of S and P in Figure 4.3 it is clear that S → P is equivalent to $h \to 0$ since as $h \to 0$, S → P because $x + h$ (the x coordinate of S) $\to x$ (the x coordinate of P).

Accordingly, the tangent gradient may be re-expressed

$$\text{the gradient of tangent at P} = \lim_{h \to 0} (2x + h - 2)$$

But as $h \to 0$, $2x + h - 2 \to 2x - 2$ and hence the gradient of the tangent at P is $2x - 2$.

The gradient function is represented by the symbol $\frac{dy}{dx}$. Therefore $\frac{dy}{dx}$ is the gradient of the tangent to a curve at the general point (x, y) and equation (2) may be generalised

$$\frac{dy}{dx} = \lim_{h \to 0} \frac{y(x + h) - y(x)}{h} \qquad (3)$$

where $y(x + h)$ is the expression resulting from replacing each x on the right-hand side of the curve equation $y = \ldots$, by $x + h$. Note that $y(x + h)$ does not represent the product of y and $x + h$ but is a single entity read as 'y as a function of $x + h$'. $y(x)$, i.e. 'y as a function of x', is the right-hand side of the curve equation $y = \ldots$

In mathematics the general process of finding the gradient of a curve is referred to as 'differentiation' and the particular technique embodied in equation (3) is known as 'first principles differentiation'. The example which now follows revisits the problem posed in the first section of this chapter, using a first principles approach.

EXAMPLE 1

For $y = x^2 - 2x$ find

(i) $\dfrac{dy}{dx}$,

(ii) the gradient of the tangent to the curve (usually abbreviated to 'the gradient of the curve') at (3,3).

Solution 1

(i) $\dfrac{dy}{dx} = \lim_{h \to 0} \dfrac{y(x+h) - y(x)}{h}$

The first stage is to find the $y(x+h)$ and $y(x)$ of the numerator. Now $y(x)$ is available from the equation of the curve i.e.

$$y(x) = x^2 - 2x$$

$y(x+h)$ is found by replacing x, everywhere it appears on the right-hand side of the above equation, by $x + h$. Accordingly,

$$y(x+h) = (x+h)^2 - 2(x+h) = x^2 + 2xh + h^2 - 2x - 2h$$

Substituting in the first principles 'formula'

$$\frac{dy}{dx} = \lim_{h \to 0} \frac{x^2 + 2xh + h^2 - 2x - 2h - (x^2 - 2x)}{h}$$

$$= \lim_{h \to 0} \frac{2xh + h^2 - 2h}{h}$$

$$= \lim_{h \to 0} 2x + h - 2$$

$= 2x - 2$ (as h goes to zero in the expression $2x + h - 2$)

(ii) $\dfrac{dy}{dx} = 2x - 2$

represents the gradient of the tangent to the curve at any point (x, y) on the curve.

In particular, $x = 3$ at the point (3,3).

So $\dfrac{dy}{dx}$ at (3,3) is $2(3) - 2 = 4$ and the gradient of the curve at (3,3) is found to be 4, confirming the result found earlier.

EXAMPLE 2

Find $\frac{dy}{dx}$ from first principles given $y = 2x + 5$.

Solution 2

$$y(x) = 2x + 5$$
$$y(x+h) = 2(x+h) + 5 = 2x + 2h + 5$$
$$y(x+h) - y(x) = 2x + 2h + 5 - (2x+5) = 2h$$

Using the first principles formula

$$\frac{dy}{dx} = \lim_{h \to 0} \frac{2h}{h} = \lim_{h \to 0} 2$$
$$= 2 \text{ (since 2 has no } h \text{ dependence)}$$

Hence $\frac{dy}{dx} = 2$

EXAMPLE 3

Find $\frac{dy}{dx}$ from first principles given $y = x^3 + 4x^2 + 5$.

Solution 3

$$y(x) = x^3 + 4x^2 + 5$$
$$y(x+h) = (x+h)^3 + 4(x+h)^2 + 5$$
or $y(x+h) = x^3 + 3x^2h + 3xh^2 + h^3 + 4x^2 + 8xh + 4h^2 + 5$

Accordingly,

$$y(x+h) - y(x) = x^3 + 3x^2h + 3xh^2 + h^3 + 4x^2 + 8xh + 4h^2 + 5 - x^3 - 4x^2 - 5$$

or $y(x+h) - y(x) = 3x^2h + 3xh^2 + h^3 + 4h^2 + 8xh$

$$\frac{dy}{dx} = \lim_{h \to 0} \frac{3x^2h + 3xh^2 + h^3 + 4h^2 + 8xh}{h}$$
$$= \lim_{h \to 0} 3x^2 + 3xh + h^2 + 4h + 8x$$
$$= 3x^2 + 8x$$

(since $3xh$, h^2 and $4h$ all contain h as a factor). It follows that

$$\frac{dy}{dx} = 3x^2 + 8x$$

EXERCISE 4.1

Differentiate the following from first principles.

1. $y = 2x + 3$
2. $y = 5x$
3. $y = 12x - 7$

4. $y = x^2 + 3x - 2$

5. $y = 2x^2 + 4x + 7$

6. $y = -x^2 - 2x + 1$

7. $y = x^3 + 7$

8. $y = x^3 + 4x - 2$

9. $y = x^3 + 2x^2 + 7x + 4$

10. $y = 2x^3 + 5x^2 - 9x + 12$

DIFFERENTIATION

As has already been mentioned, the calculation of $\frac{dy}{dx}$ is known as differentiation. Alternatively, the process of finding $\frac{dy}{dx}$ may also be referred to as 'differentiating y with respect to x' or 'finding the first derivative'. To confuse matters further, $\frac{dy}{dx}$ is often assigned the title 'the first differential coefficient of y'.

Now the first principles procedure for finding $\frac{dy}{dx}$ is a tedious one and therefore a more direct method is called for.

Consider $y = 4x^2$. It is a straightforward exercise to confirm from first principles that $\frac{dy}{dx} = 8x$.

Similarly $\quad y = 7x^2 \Rightarrow \frac{dy}{dx} = 14x$

In addition $\quad y = 2x^3 \Rightarrow \frac{dy}{dx} = 6x^2$

and $\quad y = 7x^3 \Rightarrow \frac{dy}{dx} = 21x^2.$

Based upon these four results, a differentiation rule may be conjectured:

$$\text{if} \quad \mathbf{y = cx^n \Rightarrow \frac{dy}{dx} = cnx^{n-1}} \qquad (4)$$

The four results above clearly comply with the rule but what of cases where the power of x is 1 or 0? The rule will now be tested for two such cases.

A simple calculation will confirm that $y = 7x$ differentiates from first principles to give $\frac{dy}{dx} = 7$.

Using the symbolism of equation (4), $c = 7$ and $n = 1$.

Hence, using equation (4), $\dfrac{dy}{dx} = 7(1)x^{1-1} = 7x^0 = 7$

and the rule passes this test.

Now consider the line $y = 7$. Clearly the gradient of this line is zero (the line $y = 7$ is parallel to the x axis) and therefore $\dfrac{dy}{dx} = 0$.

Once again casting $y = 7$ in the symbols of equation (4)

$$y = 7x^0 \Rightarrow c = 7 \quad n = 0$$

If the rule is valid in this case then

$$\frac{dy}{dx} = (7)(0)x^{0-1} = 0$$

Based on the evidence analysed above, it would appear that equation (4) does indeed represent a general rule for differentiation, a rule which is easily extended to polynomials by using it to differentiate successive terms.

EXAMPLE 4

If $y = 4x^3 + 2x^2 - 7x - 2$, find $\dfrac{dy}{dx}$.

Solution 4

Using the rule

$$y = 4x^3 \Rightarrow \frac{dy}{dx} = 12x^2$$

$$y = 2x^2 \Rightarrow \frac{dy}{dx} = 4x$$

$$y = 7x \Rightarrow \frac{dy}{dx} = 7$$

$$y = 2 \Rightarrow \frac{dy}{dx} = 0$$

Hence $\dfrac{dy}{dx} = 12x^2 + 4x - 7$

EXAMPLE 5

If $y = \sqrt{x}$, find $\dfrac{dy}{dx}$.

Solution 5

$$y = x^{\frac{1}{2}}$$

$$\frac{dy}{dx} = \tfrac{1}{2}(x^{-\frac{1}{2}}) = \tfrac{1}{2}\left(\frac{1}{x^{\frac{1}{2}}}\right) = \frac{1}{2\sqrt{x}}.$$

Therefore $\dfrac{dy}{dx} = \dfrac{1}{2\sqrt{x}}$

EXAMPLE 6

If $y = \frac{1}{x}$, find $\frac{dy}{dx}$.

Solution 6

$y = x^{-1}$

$$\frac{dy}{dx} = -x^{-2} = -\frac{1}{x^2}$$

Therefore $\frac{dy}{dx} = -\frac{1}{x^2}$

EXAMPLE 7

If $y = \sqrt[3]{x}$, find $\frac{dy}{dx}$.

Solution 7

$y = x^{\frac{1}{3}}$

$$\frac{dy}{dx} = \tfrac{1}{3}(x^{-\frac{2}{3}}) = \tfrac{1}{3}\left(\frac{1}{x^{\frac{2}{3}}}\right) = \frac{1}{3\sqrt[3]{x^2}}$$

Therefore $\frac{dy}{dx} = \frac{1}{3\sqrt[3]{x^2}}$

EXAMPLE 8

If $y = \frac{1}{\sqrt{x}}$, find $\frac{dy}{dx}$.

Solution 8

$y = x^{-\frac{1}{2}}$

$$\frac{dy}{dx} = -\tfrac{1}{2}(x^{-1\frac{1}{2}}) = -\frac{1}{2}\left(\frac{1}{x^{\frac{3}{2}}}\right) = -\frac{1}{2x^{\frac{3}{2}}} = -\frac{1}{2\sqrt{x^3}}$$

Therefore $\frac{dy}{dx} = -\frac{1}{2\sqrt{x^3}}$

EXAMPLE 9

If $y = \frac{4}{3x^4}$, find $\frac{dy}{dx}$.

Solution 9

$$y = \frac{4}{3x^4} = \frac{4}{3}x^{-4}$$

$$\frac{dy}{dx} = -\frac{16}{3}(x^{-5}) = -\frac{16}{3}\left(\frac{1}{x^5}\right) = -\frac{16}{3x^5}$$

Therefore $\dfrac{dy}{dx} = -\dfrac{16}{3x^5}$

EXAMPLE 10

If $y = 8x^{\frac{3}{2}} - \dfrac{2}{\sqrt[3]{x}}$, find $\dfrac{dy}{dx}$.

Solution 10

$$y = 8x^{\frac{3}{2}} - 2\left(\frac{1}{x^{\frac{1}{3}}}\right) = 8x^{\frac{3}{2}} - 2x^{-\frac{1}{3}}$$

$$\frac{dy}{dx} = 12x^{\frac{1}{2}} + \frac{2}{3}(x^{-\frac{4}{3}}) = 12\sqrt{x} + \frac{2}{3}\left(\frac{1}{x^{\frac{4}{3}}}\right) = 12\sqrt{x} + \frac{2}{3\sqrt[3]{x^4}}$$

Therefore $\dfrac{dy}{dx} = 12\sqrt{x} + \dfrac{2}{3\sqrt[3]{x^4}}$

EXERCISE 4.2

Differentiate the following expressions using the rule

$$y = cx^n \Rightarrow \frac{dy}{dx} = cnx^{n-1}$$

1. x^4 **2.** x^7 **3.** x^9

4. $2x^3$ **5.** $4x^8$ **6.** $7x^4$

7. $\dfrac{x^3}{4}$ **8.** $\dfrac{x^6}{2}$ **9.** $\dfrac{3x^6}{4}$

10. $-3x^5$ **11.** $-9x^4$ **12.** $-5x^3$

13. mx^3 **14.** px^7 **15.** $-nx^4$

16. $2x$ **17.** $5x$ **18.** $9x$

19. 7 **20.** 1 **21.** 100

22. $x^2 + 7x + 4$ **23.** $x^4 - 2x^2 + 5$ **24.** $x^5 - 3x^2 + 7x$

25. $4x^3 + 7x^2 + 2x$ **26.** $2x^2 + 5x + 4$ **27.** $9x^3 + 2x^2 + 7x + 3$

28. $\dfrac{x^3}{2} + \dfrac{x^2}{4} + \dfrac{x}{2}$ **29.** $ax^2 + bx + c$ **30.** $(x + 7)^2$

31. $(x - 1)^2$ **32.** $(2x + 5)^2$ **33.** $(x - 3)(x + 4)$

34. $(x + 9)(x - 5)$ **35.** $(2x + 7)(x + 4)$ **36.** $(7x + 2)(2x + 9)$

37. $(3x + 8)(2x + 2)$ **38.** $(3x - 1)(5x + 7)$ **39.** $(4x - 3)(3x - 5)$

40. $(x^2 + 1)^2$ **41.** $(x^3 + 5)^2$ **42.** $(x^4 - 5)^2$

43. $(x + 7)^3$ **44.** $(x + 2)^4$ **45.** $(x + 3)(x - 6)(x + 2)$

46. $(x^2 - 3)^3$ **47.** $(x + 2)(x - 3)^2$ **48.** $(x + 4)(x + 3)(x + 2)(x + 1)$

EXERCISE 4.3

Differentiate the following expressions using the rule

$$y = cx^n \Rightarrow \frac{dy}{dx} = cnx^{n-1}$$

1. $x^{\frac{5}{2}}$ 2. $x^{\frac{2}{5}}$ 3. $x^{\frac{1}{3}}$ 4. $x^{\frac{4}{3}}$ 5. $x^{\frac{5}{6}}$ 6. x^{-3}

7. x^{-1} 8. x^{-2} 9. x^{-7} 10. x^{-5} 11. $2x^{-2}$ 12. $3x^{-3}$

13. $2x^{-4}$ 14. $5x^{-2}$ 15. $3x^{-2}$ 16. $\frac{1}{2x^3}$ 17. $\frac{1}{3x^3}$ 18. $\frac{1}{3x^2}$

19. $\frac{1}{4x^2}$ 20. $\frac{1}{5x^3}$

EXERCISE 4.4

Express each of the following in the form ax^n, and then find the first derivative.

1. $\sqrt[3]{x}$ 2. $\sqrt{x}$ 3. $\sqrt[3]{x^2}$ 4. $\sqrt[4]{x^3}$ 5. $\sqrt{x^3}$ 6. $\frac{1}{x^4}$

7. $\frac{1}{x^2}$ 8. $\frac{1}{x^3}$ 9. $\frac{1}{x^5}$ 10. $\frac{1}{x^6}$ 11. $\frac{2}{x}$ 12. $\frac{1}{x}$

13. $\frac{3}{x}$ 14. $\frac{2}{x^2}$ 15. $\frac{2}{x^3}$ 16. $\frac{1}{\sqrt{x}}$ 17. $\frac{1}{\sqrt[3]{x}}$ 18. $\frac{2}{3x^3}$

19. $\frac{4}{5x^2}$ 20. $\frac{1}{3x^4}$ 21. $\frac{1}{2x^{\frac{1}{2}}}$ 22. $\frac{1}{3x^{\frac{1}{3}}}$ 23. $\frac{1}{2\sqrt{x}}$ 24. $\frac{1}{2x}$

25. $\frac{2}{x^5}$

EXERCISE 4.5

Express each of the following as a sum of terms of the form ax^n and then find the first derivative.

1. $x + \frac{1}{x^2}$ 2. $\sqrt{x} + \frac{1}{x}$ 3. $3x - \frac{3}{x}$ 4. $x^2 + 5 - \frac{1}{x^2}$

5. $3x^2 - \frac{1}{4x^2}$ 6. $\frac{x}{3} + \frac{3}{x}$ 7. $\frac{x^2}{7} + \frac{7}{x^2}$ 8. $3x^4 + \frac{1}{2x^3}$

9. $7x^{\frac{1}{2}} + \frac{8}{x^{\frac{2}{3}}}$ 10. $x^2(1 + \sqrt{x})$ 11. $\sqrt[3]{x}(1 - \sqrt[3]{x})$ 12. $\left(x + \frac{1}{x}\right)^3$

13. $\left(x^3 - \frac{1}{x^3}\right)^2$ 14. $\frac{4 - x}{x}$ 15. $\frac{x - 3}{x}$ 16. $\frac{x^2 + 5x - 6}{x}$

17. $\frac{2x^4 - 3x^2 + 7}{x^2}$ 18. $\frac{x - 4}{\sqrt{x}}$ 19. $\frac{3 - x}{x^{\frac{1}{2}}}$ 20. $\frac{4x + 9}{x^3}$

21. $\dfrac{x+2}{4x^{\frac{1}{2}}}$ **22.** $\dfrac{(1+x)^3}{x^2}$ **23.** $\dfrac{(3-x)(2-x)}{x}$ **24.** $\dfrac{x^2-5x}{\sqrt{x}}$

25. $\dfrac{3x^3+x^2+2x+5}{3x^4}$ **26.** $\left(2x-\dfrac{4}{\sqrt{x}}\right)^2$ **27.** $\dfrac{(x+2)^3}{\sqrt{x}}$ **28.** $\left(x-\dfrac{1}{\sqrt{x}}\right)^2$

29. $\dfrac{2x^2+5x+7}{2\sqrt[3]{x}}$ **30.** $\dfrac{(x+1)^3}{2\sqrt{x}}$

THE EQUATION OF THE TANGENT TO A CURVE

This chapter began with an investigation of the gradient of the tangent to the curve $y = x^2 - 2x$ at (3,3).

The point (3,3) is denoted A in Figure 4.4 and further tangents at B, O, C and D are illustrated.

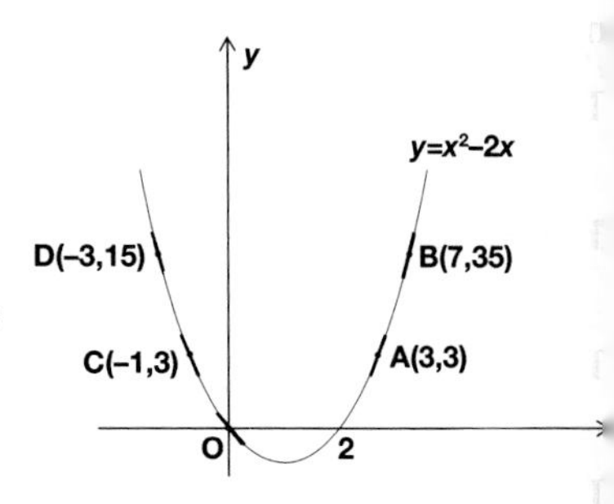

Figure 4.4

As has already been mentioned, the gradient of the curve at the general point (x, y) is found through considering $\dfrac{dy}{dx}$.

Here $\dfrac{dy}{dx} = 2x - 2$

The gradient of the tangent at A is therefore $2(3) - 2$ i.e. 4.

The tangent gradients at the remaining points are summarised in Table 4.2.

Table 4.2

Point	Coordinates	Tangent gradient
A	(3,3)	4
B	(7,35)	12
O	(0,0)	− 2
C	(− 1,3)	− 4
D	(− 3,15)	− 8

It is important to note that in the regions where y decreases as x increases (the neighbourhoods of D, C and O) $\dfrac{dy}{dx}$ is negative while in regions where y increases as x increases (the neighbourhoods of A and B) $\dfrac{dy}{dx}$ is positive.

This property is usually stated as

$\dfrac{dy}{dx}$ is positive for increasing y with x and

$\dfrac{dy}{dx}$ is negative for decreasing y with x.

The calculation of the gradient of the tangent at a specified point has been established but the **equation** of the tangent has not yet been considered.

The equation of the tangent may be written down by appealing to coordinate geometry (recalling that a tangent is a straight line).

The reader is reminded that the equation of the line through (a, b) of gradient m is given by

$$y - b = m(x - a)$$

Accordingly, the equations of the tangents illustrated in Figure 4.4 may be developed as follows.

Tangent at A

$y - 3 = 4(x - 3)$ or $y = 4x - 9$

Tangent at B

$y - 35 = 12(x - 7)$ or $y = 12x - 49$

Tangent at O

$y - 0 = -2(x - 0)$ or $y = -2x$

Tangent at C

$y - 3 = -4(x + 1)$ or $y = -4x - 1$

Tangent at D

$y - 15 = -8(x + 3)$ or $y = -8x - 9$

EXAMPLE 11

Find the equation of the tangent to the curve $y = 4x^2 - 2x$ at the point on the curve where $x = 1$.

Solution 11

At $x = 1$, $y = 4(1)^2 - 2(1) = 2$

Hence the point on the curve is (1,2).

The tangent is a straight line whose equation is

$y - b = m(x - a)$ where
$a = 1, b = 2, m = ?$

The gradient of the tangent is computed by differentiation.

$$\frac{dy}{dx} = 8x - 2$$

At (1,2) $m = 8(1) - 2 = 6$

Therefore the tangent equation is

$$y - 2 = 6(x - 1)$$
$$y = 6x - 4$$

EXAMPLE 12

(a) Find where the curve $y = x^2 - 3x + 2$ intersects

(i) the y axis,
(ii) the x axis.

(b) Find the equation of the tangent to this curve at the point where the curve intersects the y axis.

Solution 12

(a) (i) The y axis is the straight line $x = 0$ and therefore the point of intersection is found by reconciling the equations

$$x = 0 \text{ and}$$
$$y = x^2 - 3x + 2$$

Substituting $x = 0$ in $y = x^2 - 3x + 2$ yields $y = 2$. Clearly (0,2) lies on the curve and on the y axis and is therefore the point of intersection of curve and axis.

(ii) The x axis is the straight line $y = 0$ and so here the aim is to reconcile the equations

$$y = 0 \text{ and}$$
$$y = x^2 - 3x + 2$$

Once again, substituting the first equation in the second yields

$$x^2 - 3x + 2 = 0$$
$$\text{or} \quad (x - 2)(x - 1) = 0$$
$$\text{i.e.} \quad x = 1 \text{ or } x = 2$$

These points lie on the x axis and it follows that their coordinates are (1,0) and (2,0).

(b) $\dfrac{dy}{dx} = 2x - 3$ and so the tangent gradient at (0,2) is $2(0) - 3 = -3$.

The equation of the tangent is therefore

$$y - 2 = -3(x - 0) \quad \text{or} \quad y = -3x + 2$$

EXAMPLE 13

Find the point of intersection of the tangents to the curve $y = 2 - 3x + x^2$ at (1,0) and (3,2).

Solution 13

Tangent at (1,0)

$\dfrac{dy}{dx} = -3 + 2x$ and so the tangent gradient $= -3 + 2(1) = -1$

The tangent equation is therefore $y - 0 = -1(x - 1)$ or

$$y = -x + 1$$

Tangent at (3,2)

$\frac{dy}{dx} = -3 + 2x$ and so the tangent gradient $= -3 + 2(3) = 3$

The tangent equation is therefore $y - 2 = 3(x - 3)$ or $y = 3x - 7$

To discover the coordinates of the point of intersection of the tangents, the equations of the tangents are solved as two simultaneous linear equations i.e.

$y = -x + 1$ becomes $x + y = 1$
$y = 3x - 7$ becomes $-3x + y = -7$
i.e. $x = 2, y = -1$

The required point of intersection is $(2, -1)$.

EXAMPLE 14

Find the point on the curve $y = 2x^2 - 3x + 1$ at which the tangent gradient is 5.

Solution 14

The relationship between tangent gradient and $\frac{dy}{dx}$ has already been established.

Now $\frac{dy}{dx} = 4x - 3$ and so when the gradient is 5

$4x - 3 = 5$ i.e. $x = 2$

Where $x = 2$ on the curve $y = 2(2)^2 - 3(2) + 1 = 3$

The required point is therefore (2,3).

EXAMPLE 15

Find the points on the curve $y = 2x^3 - 9x^2 + 24x + 12$ at which the gradient of the tangent is 12.

Solution 15

$\frac{dy}{dx} = 6x^2 - 18x + 24$

Accordingly, x is required where

$6x^2 - 18x + 24 = 12$
or $6x^2 - 18x + 12 = 0$
or $x^2 - 3x + 2 = 0$
or $(x - 2)(x - 1) = 0$
i.e. $x = 1$ or $x = 2$
when $x = 1$ $y = 2(1)^3 - 9(1)^2 + 24(1) + 12 = 29$
when $x = 2$ $y = 2(2)^3 - 9(2)^2 + 24(2) + 12 = 40$

The required points are (1,29) and (2,40).

EXERCISE 4.6

Find the gradient, and hence the equation, of the tangent to each of the following curves at the give point.

1. $y = x^3$ at (2,8)
2. $y = 3x$ at $(-1, -3)$
3. $y = \sqrt{x}$ at (9,3)
4. $y = 4x^3$ at $(-2, -32)$
5. $y = x^{-1}$ at (1,1)
6. $y = \sqrt[3]{x}$ at (27,3)

EXERCISE 4.7

Find the equation of the tangent to the curve in each of the following.

1. $y = x^2$ at $x = -3$
2. $y = x^3$ at $x = 1$
3. $y = x^2 + 4$ at $x =$
4. $y = 1 - x^2$ at $x = 3$
5. $y = x^3 + 2x^2 + 6$ at $x = 2$
6. $y = 3x^4$ at $x = -$
7. $y = \sqrt[3]{x}$ at $x = 8$
8. $y = \dfrac{3}{x^2}$ at $x = 3$
9. $y = 5\sqrt{x}$ at $x = 16$
10. $y = (3x - 1)(2x - 2)$ at $x = 2$

EXERCISE 4.8

1. Show that there is one tangent to the curve $y = 3x^2 + 4$ which has gradient 12 and find i equation.

2. Find the coordinates of the point on the curve $y = 3x^2 + 2x + 6$ at which the tangent ha gradient 14.

3. Find the equation of the tangent to the curve $y = \dfrac{x^4}{9}$ at the point at which $x = 3$. Find where th tangent cuts the x and y axes.

4. Find the equation of the tangent to the curve $y = \dfrac{x^4}{2}$ at the point at which $x = 2$. Show that if th tangent cuts the axes at A and B, the mid-point of AB is $(\frac{3}{4}, -12)$.

5. Find the coordinates of the points on the curve $y = \dfrac{x-1}{x}$ at which the gradient of the curve is

6. Find the equation of the tangent to the curve $y = x^2 - 3x + 2$ at the point where it cuts the y axi

7. Find the points on the curve $y = \dfrac{2x^3}{3} - 2x^2$ at which the gradient of the tangent is 6. Find th equations of the tangents at these points.

8. Find the equation of the tangent to the parabola $x^2 = 4y$ at the point (6,9).

9. The tangent to the parabola $y = (x - 2)^2$ at the point (3,1) cuts the x axis and y axis at M and respectively. Find the coordinates of M and N and hence the length of MN.

10. Find the equations of the tangents to the curve $y = (2x - 1)(x + 1)$ at the points where the curve cuts the x axis. Find the point of intersection of these tangents.

11. Find the equations of the tangents to the parabola $y = x^2 + 2$ at the points P and Q whose x coordinates are 1 and -2 respectively. Show that the tangents intersect at a point on the x axis.

12. Find the equations of the tangents to the curve $y = 3x^4$ at the points $x = -1$ and $x = 1$ and find the point of intersection of these tangents.

13. (a) Differentiate $3x^{\frac{3}{2}} - x^{-\frac{2}{3}}$.

(b) Find the equation of the tangent to the curve

$$y = x^3 - 3x^2 + 7x + 1$$

at the point (1,6).

(N. Ireland Additional Mathematics, 1989)

14. (a) Find $\frac{dy}{dx}$ if $y = 5x^{-\frac{1}{4}}$.

(b) (i) Find the equation of the tangent to the curve

$$y = x^2 - 5x + 7$$

at the point (1,3).

(ii) Hence find the area of the triangle formed by this tangent, the x axis and the y axis.

(N. Ireland Additional Mathematics, 1990)

15. (i) Find $\frac{dy}{dx}$ if $y = 3x^2 - 2x^{-\frac{3}{2}}$.

(ii) Find the equation of the tangent to the curve

$$y = 3x^3 + 2x^2 - x - 8$$

at the point $(1, -4)$.

(N. Ireland Additional Mathematics, 1991)

16. A curve has an equation of the form $y = px + \frac{q}{x}$, where p and q are constants. Given that the curve passes through the points A(1,11) and B(4,$21\frac{1}{2}$),

(i) evaluate p and q,
(ii) obtain the equation of the tangent to the curve at the point where $x = 2$,
(iii) show that this tangent is parallel to AB.

(U.C.L.E.S. Additional Mathematics, November 1991, part question)

17. Calculate the coordinates of the point on the curve $y = 2x^2 - 3x + 2$ at which the gradient of the curve is 5.

Calculate the value of the constant k for which $y = 5x + k$ is a tangent to the curve.

(U.C.L.E.S. Additional Mathematics, June 1991)

MAXIMUM AND MINIMUM TURNING POINTS

Consider the curve $y = 4x^3 + 2x^2 + 7x + 2$.

$$\frac{dy}{dx} = 12x^2 + 4x + 7$$

Differentiation of $\frac{dy}{dx}$ leads to $\frac{d^2y}{dx^2}$ where

$$\frac{d^2y}{dx^2} = 24x + 4$$

(since differentiation of $12x^2 + 4x + 7$ yields $24x + 4$).

$\frac{d^2y}{dx^2}$ is known as the second differential coefficient of y with respect to x or the second derivative.

The second differential coefficient of y with respect to x is, of course, the first differential coefficient of $\frac{dy}{dx}$ with respect to x.

It has already been pointed out that $\frac{dy}{dx}$ is positive where y increases with increasing x and negative where y decreases with increasing x and it follows that $\frac{d^2y}{dx^2}$ is positive where $\frac{dy}{dx}$ (the gradient of the curve) increases with increasing x and negative where $\frac{dy}{dx}$ decreases with increasing x. Now consider the curve in Figure 4.5.

In the region LM, $\frac{dy}{dx}$ is positive since the gradient of any tangent in this region is positive. At the point M, $\frac{dy}{dx}$ is zero (the tangent is parallel to the x axis). In the region MP, $\frac{dy}{dx}$ is negative. At P, $\frac{dy}{dx} = 0$ and in the region PR, $\frac{dy}{dx}$ is again positive.

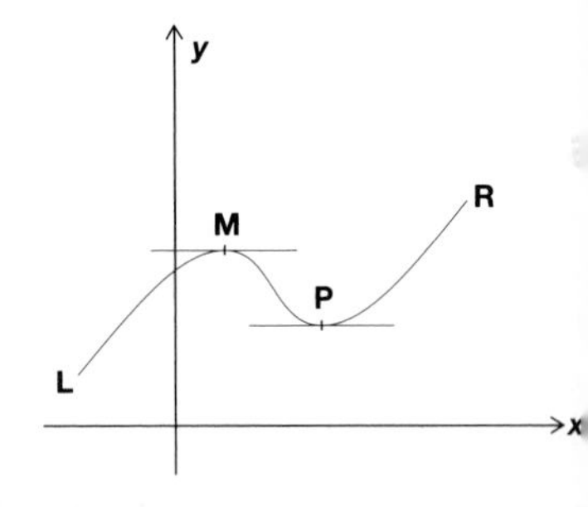

Figure 4.5

The y coordinate of M is greater than that of any of the points of the curve in the neighbourhood of M and M is called the **maximum point**. Similarly the y coordinate of P is smaller than that of any of the points of the curve in the neighbourhood of P and P is called a **minimum point**.

It is instructive to consider $\frac{dy}{dx}$ as x increases from left to right across M and across P. On the journey through M (from left to right) $\frac{dy}{dx}$ changes from positive, through zero (at M) to negative

.e. $\frac{dy}{dx}$ decreases with increasing x. It follows, from the definition of $\frac{d^2y}{dx^2}$ offered above, that $\frac{d^2y}{dx^2}$ will be negative at this maximum point. Similarly, approaching P from the left, passing through P and 'travelling' from P to R, the value of $\frac{dy}{dx}$ changes from negative, through zero (at P) to positive (in the region PR). Once again interpreting $\frac{d^2y}{dx^2}$ as above it follows that $\frac{d^2y}{dx^2}$ is positive at P since $\frac{dy}{dx}$ increases as x increases.

This reasoning leads to techniques for identifying the nature (maximum or minimum) of turning points on curves.

For a maximum turning point

$$\frac{dy}{dx} = 0 \text{ and } \frac{d^2y}{dx^2} < 0$$

For a minimum turning point

$$\frac{dy}{dx} = 0 \text{ and } \frac{d^2y}{dx^2} > 0$$

Turning points are sometimes referred to as the **stationary points** of a curve.

EXAMPLE 16

Find the turning point on the curve

$$y = 3 - 4x + x^2.$$

Solution 16

$$\frac{dy}{dx} = -4 + 2x$$

$$\frac{d^2y}{dx^2} = 2$$

The turning point is minimum since $\frac{d^2y}{dx^2}$ is positive.

$\frac{dy}{dx} = 0$ at the turning point and so

$$-4 + 2x = 0$$

or $\quad x = 2$

Since this point lies on the curve it follows that

$$y = 3 - 4(2) + (2)^2 = -1$$

Accordingly, there is a minimum turning point at $(2, -1)$.

EXAMPLE 17

Find the turning points on the curve

$$y = x^3 - 6x^2 + 9x + 3.$$

Solution 17

$$\frac{dy}{dx} = 3x^2 - 12x + 9$$

$$\frac{d^2y}{dx^2} = 6x - 12$$

It is not possible immediately to declare any point on the curve a maximum or minimum turning point since the sign of $\frac{d^2y}{dx^2}$ is dependent upon the value of x.

At turning points $\frac{dy}{dx} = 0$ and therefore $3x^2 - 12x + 9 = 0$

or $\quad x^2 - 4x + 3 = 0$
or $\quad (x - 3)(x - 1) = 0$
i.e. $\quad x = 1$ or $x = 3$

The points with x coordinates 1 and 3 lie on the curve and consequently their y coordinates may be determined by reference to the curve equation.

At $x = 1 \quad y = (1)^3 - 6(1)^2 + 9(1) + 3 = 7$
At $x = 3 \quad y = (3)^3 - 6(3)^2 + 9(3) + 3 = 3$

Is (1,7) a maximum or minimum turning point?

At $x = 1 \quad \frac{d^2y}{dx^2} = -6 \quad$ therefore (1,7) is a maximum point.

At $x = 3 \quad \frac{d^2y}{dx^2} = 6 \quad$ therefore (3,3) is a minimum turning point.

In summary, the curve has a maximum turning point at (1,7) and a minimum turning point at (3,3).

EXERCISE 4.9

Find the turning points on the curves defined below and determine the nature of each turning point.

1. $y = x^2$ **2.** $y = x^2 - 4x$ **3.** $y = 6 - 2x^2$ **4.** $y = 4x^2$

5. $y = x^3 - 3x$ **6.** $y = x^5 - 5x$ **7.** $y = (x - 3)(x + 2)$ **8.** $y = (x + 1)(2x + 3)$

9. $y = \frac{x^3}{3} - 2x^2 + 3x$ **10.** $y = 2x^3 - 8x^2 + 6x$

CURVE SKETCHING

There are three stages in sketching a curve

(i) **the location of the point of intersection of the curve and the y axis,**

(ii) **the location of the point(s) of intersection of the curve and the x axis,**

(iii) **the location of the turning point(s).**

This three stage process is illustrated in the following examples.

EXAMPLE 18

Sketch the curve $y = x^2 - 2x - 8$.

Solution 18

(i) The curve intersects the y axis where $x = 0$, i.e. at $(0, -8)$.

(ii) The curve intersects the x axis where $y = 0$, i.e. at values of x satisfying

$$x^2 - 2x - 8 = 0$$
or $$(x - 4)(x + 2) = 0$$
i.e. $$x = 4 \text{ or } x = -2$$

Hence the points of intersection of the curve and the x axis are $(4,0)$ and $(-2,0)$.

(iii) Now to the calculation of the turning points.

$$\frac{dy}{dx} = 2x - 2$$

$$\frac{d^2y}{dx^2} = 2$$ and so the turning point will be a minimum turning point.

$$\frac{dy}{dx} = 0 \Rightarrow x = 1$$

For $x = 1$ on the curve

$$y = (1)^2 - 2(1) - 8 = -9$$

The curve has a minimum turning point at $(1, -9)$ and is illustrated in Figure 4.6.

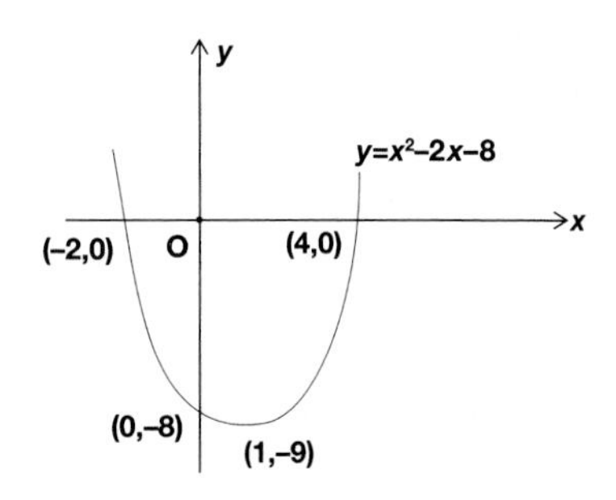

Figure 4.6

EXAMPLE 19

Sketch the curve $y = x^2 + 4x + 5$.

Solution 19

(i) The curve intersects the y axis where $x = 0$ i.e. at $(0,5)$.

(ii) The curve intersects the x axis where $y = 0$ i.e. at x values satisfying $x^2 + 4x + 5 = 0$.

$x^2 + 4x + 5 = 0$ is solved using the quadratic formula

$$x = \frac{-4 \pm \sqrt{(16-20)}}{2} = \frac{-4 \pm \sqrt{(-4)}}{2}$$

The coordinates of intersection are not real and therefore this curve does not intersect the x axis.

(iii) Now to the calculation of the turning points.

$$\frac{dy}{dx} = 2x + 4$$

$$\frac{d^2y}{dx^2} = 2$$ and so the turning point will be a minimum turning point.

$$\frac{dy}{dx} = 0 \Rightarrow x = -2$$

For $x = -2$ on this curve

$$y = (-2)^2 + 4(-2) + 5 = 1$$

In summary, the curve has a minimum turning point at $(-2,1)$.

The sketch of $y = x^2 + 4x + 5$ is given in Figure 4.7.

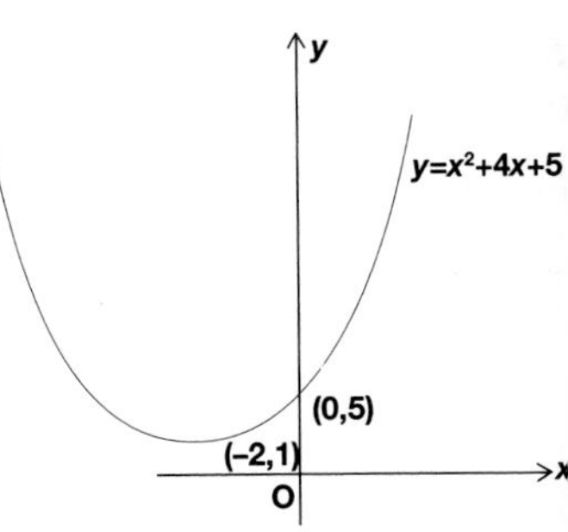

Figure 4.7

EXAMPLE 20

Sketch the curve $y = x^3 + 4x^2 - 3x$.

Solution 20

(i) The curve intersects the y axis where $x = 0$ i.e. at (0,0).

(ii) The curve intersects the x axis where $y = 0$ i.e. at x values which satisfy

$$x^3 + 4x^2 - 3x = 0$$

or $$x(x^2 + 4x - 3) = 0$$

This equation is satisfied if

$$x = 0 \text{ or } x^2 + 4x - 3 = 0$$

To solve the second of these the quadratic formula is applied.

$$x = \frac{-4 \pm \sqrt{(4^2 - 4(1)(-3))}}{2} = \frac{-4 \pm \sqrt{28}}{2} = \frac{-4 \pm 5.292}{2}$$

$$x = -4.65 \text{ or } x = 0.65$$

In summary, the curve intersects the x axis at points (0,0) $(-4.65,0)$ and $(0.65,0)$.

ii) Finally, the turning points are located.

$$\frac{dy}{dx} = 3x^2 + 8x - 3$$

$$\frac{d^2y}{dx^2} = 6x + 8$$

No immediate conclusions can be drawn as to the nature of the turning points as the sign of the right-hand side of the second derivative equation is dependent upon the value of x. Clearly, the x coordinate(s) of the turning point(s) must be found. To this end

$$\frac{dy}{dx} = 0 \Rightarrow 3x^2 + 8x - 3 = 0$$

or $(3x - 1)(x + 3) = 0$

Hence $x = \frac{1}{3}$ or $x = -3$

When $x = \frac{1}{3}$ $y = (\frac{1}{3})^3 + 4(\frac{1}{3})^2 - 3(\frac{1}{3}) = -\frac{14}{27}$

When $x = -3$ $y = (-3)^3 + 4(-3)^2 - 3(-3) = 18$

The turning points are $(\frac{1}{3}, -\frac{14}{27})$ and $(-3, 18)$.

What of the nature of these turning points?

At $x = \frac{1}{3}$, $\frac{d^2y}{dx^2} = 10$ indicating a minimum point.

At $x = -3$, $\frac{d^2y}{dx^2} = -10$ indicating a maximum point.

The curve is as in Figure 4.8.

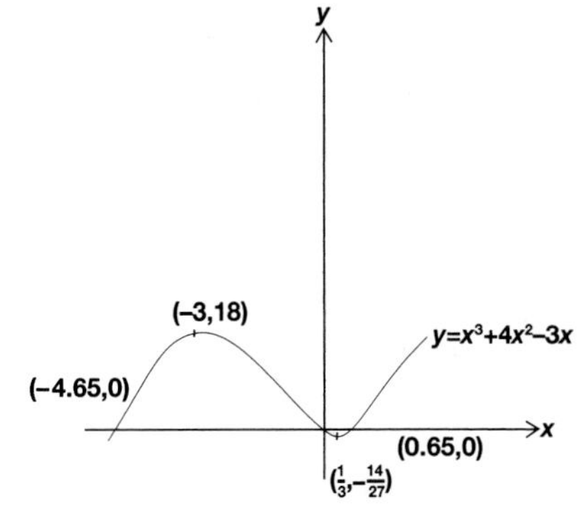

Figure 4.8

EXERCISE 4.10

Sketch these curves.

1. $y = x^2 - 6x$
2. $y = x^2 + x - 6$
3. $y = 9 - x^2$
4. $y = (2 - x)^2$
5. $y = x^2 + x + 4$
6. $y = -x^2 + 2x - 9$
7. $y = x(x - 2)^2$
8. $y = 4x - x^3$
9. $y = 4x^2 - x^3$
10. $y = x^4 - 5x^2 + 6$

THE APPLICATIONS OF DIFFERENTIATION

The examples which follow will illustrate the application of differentiation to 'real life' situations. The techniques used to find the maximum and minimum turning point on curves will be applied to situations where, for example, the maximum volume of a body is to be found subject to constraints on its surface area.

EXAMPLE 21

A 100 cm length of wire is bent to form two squares (see Figure 4.9), one of side x, the other of side y. Prove that the area enclosed by the squares is least when $x = y$.

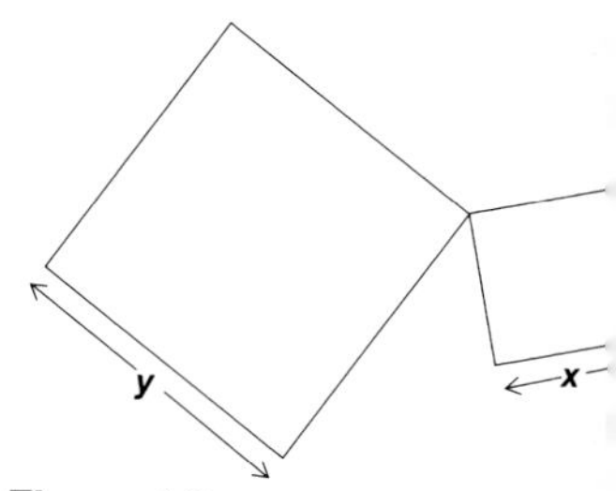

Figure 4.9

Solution 21

Since the total length of wire is 100 cm it follows that

$$4x + 4y = 100$$

or $$x + y = 25 \qquad (5)$$

The area enclosed is given by

$$A = x^2 + y^2$$

The right-hand side of this equation is the sum of squares of two variables. The aim is to find the minimum value of A and, to this end, A will be differentiated. In order to do so successfully the right-hand side of the above equation must be in terms of one variable only. If that variable be x, then y must be removed from the area expression using equation (5).

Accordingly,

$$A = x^2 + (25 - x)^2$$

or $$A = 2x^2 - 50x + 625$$

Since the least (minimum) value of A is required, it is instructive to differentiate A.

$$\frac{dA}{dx} = 4x - 50$$

$$\frac{d^2A}{dx^2} = 4 \quad \text{confirming a minimum area}$$

$$\frac{dA}{dx} = 0 \Rightarrow x = 12\tfrac{1}{2}$$

Now equation (5) yields $y = 12\frac{1}{2}$ thus confirming $x = y$.

EXAMPLE 22

A wastepaper bin is to be constructed of metal of negligible thickness in the shape of a cylinder closed at one end. (A cylinder of base radius r and height h has a volume $\pi r^2 h$ and a curved surface area $2\pi rh$.) It is required to have a volume of 3000 cm^3. If A denotes the surface area of metal required and r denotes the base radius, show that

$$A = \pi r^2 + \frac{6000}{r}$$

If the amount of metal required is to be minimum, find what radius r the base should be and hence what amount of metal (in cm^2) is required. Verify that this is a minimum value.

(N. Ireland Additional Mathematics, specimen paper, 1988)

Solution 22

Let h be the height and r the base radius.

Since the volume is to be 3000 cm^3

$$\pi r^2 h = 3000 \qquad (6)$$

Now $$A = \pi r^2 + 2\pi r h \qquad (7)$$

has two variables (r and h) on the right-hand side. However, h may be removed using equation (6) since this equation gives

$$h = \frac{3000}{\pi r^2}$$

Equation (7) now becomes

$$A = \pi r^2 + 2\pi r\left(\frac{3000}{\pi r^2}\right)$$

or $$A = \pi r^2 + \frac{6000}{r}$$ as required.

Now $$\frac{dA}{dr} = 2\pi r - \frac{6000}{r^2}$$

and $$\frac{d^2A}{dr^2} = 2\pi + \frac{12\,000}{r^3}$$

No matter what the value of r (r must be positive to be physically meaningful, of course) $\frac{d^2A}{dr^2}$ will be positive.

It follows that the r value resulting from the solution of $\frac{dA}{dr} = 0$ must be that appropriate to the minimum area, providing that it (the value of r) is positive.

$$\frac{dA}{dr} = 0$$

$$\Rightarrow \quad 2\pi r = \frac{6000}{r^2}$$

or $$r = \sqrt[3]{\left(\frac{3000}{\pi}\right)}$$

The required radius is 9.85 cm and the minimum area is given by

$$A = \pi(9.85)^2 + \frac{6000}{9.85} = 913.9$$

The required minimum surface area is 913.9 cm^2.

EXAMPLE 23

(a) Find the coordinates of the turning points of the curve $y = (2x - 3)(x^2 - 6)$ and determine whether each is a maximum or a minimum.

(b) A cylindrical container, open at one end, has height h cm and base radius r cm.

(i) Write down, in terms of h and r, expressions for

(A) the total surface area of the container, $S\,\text{cm}^2$,
(B) the volume of the container, $V\,\text{cm}^3$.

(ii) Given that S has the value 3π, show that

$$V = \tfrac{1}{2}\pi r(3 - r^2)$$

(iii) Hence find the value of r and the corresponding value of h which make V a maximum.

Solution 23

(a) $y = (2x - 3)(x^2 - 6)$ becomes

$$y = 2x^3 - 3x^2 - 12x + 18$$

$$\frac{dy}{dx} = 6x^2 - 6x - 12$$

$$\frac{d^2y}{dx^2} = 12x - 6$$

$$\frac{dy}{dx} = 0 \quad \Rightarrow \quad 6(x - 2)(x + 1) = 0 \quad \text{i.e. } x = 2 \text{ or } x = -1$$

$$x = 2 \quad \Rightarrow \quad y = -2$$
$$x = -1 \quad \Rightarrow \quad y = 25$$

Using $\frac{d^2y}{dx^2} = 12x - 6$ the following may be concluded.

$(2, -2)$ is a minimum turning point.
$(-1, 25)$ is a maximum turning point.

(b) (i) (A) $S = 2\pi rh + \pi r^2$
(B) $V = \pi r^2 h$

(ii) As $S = 3\pi$

$$2\pi rh + \pi r^2 = 3\pi$$
$$\text{or} \quad 2\pi rh = 3\pi - \pi r^2$$

$$\text{or} \quad h = \frac{3 - r^2}{2r} \qquad (8)$$

V may hence be written

$$V = \frac{\pi r^2(3 - r^2)}{2r} \quad \text{or}$$

$$V = \frac{\pi r(3 - r^2)}{2} \quad \text{as required.}$$

(iii) $\frac{dV}{dr} = \frac{3\pi}{2} - \frac{3\pi r^2}{2}$

$$\frac{d^2V}{dr^2} = -3\pi r$$

$$\frac{dV}{dr} = 0 \quad \Rightarrow \quad \frac{3\pi}{2} = \frac{3\pi r^2}{2} \quad \Rightarrow \quad r = 1$$

It follows that $\frac{d^2V}{dr^2} = -3\pi$, confirming that a maximum volume will result from $r = 1$ cm.

Substitution in equation (8) reveals $h = 1$ cm also.

EXAMPLE 24

The solid object illustrated in Figure 4.10 consists of a circular cylinder of height h and radius r surmounted by a hemisphere of the same radius.

Express the total surface area S of the object and the total volume V in terms of h and r.

If the total surface area S is 20π express h in terms of r and hence show that

$$V = 10\pi r - \frac{5\pi r^3}{6}$$

Find the value of r which makes V a maximum and calculate the maximum value of V, giving your answer in terms of π.

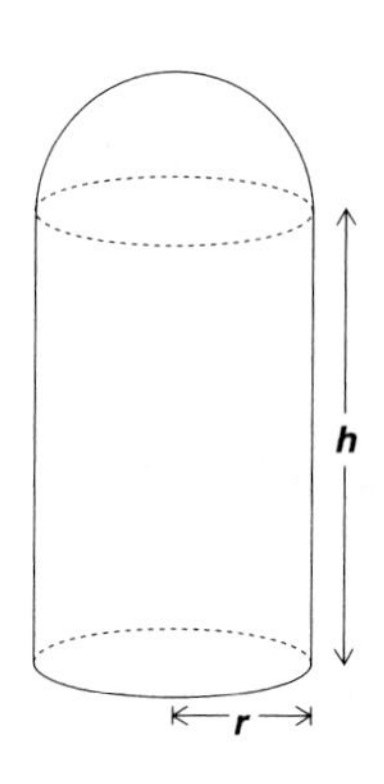

Figure 4.10

Solution 24

$S = 2\pi rh + 2\pi r^2 + \pi r^2$ (the surface area of a sphere is $4\pi r^2$)

$$V = \pi r^2 h + \frac{2}{3}\pi r^3 \qquad (9)$$

If $S = 20\pi$ then

$$20\pi = 2\pi rh + 3\pi r^2$$

or $h = \frac{10}{r} - \frac{3r}{2}$

Substituting this h expression in equation (9) yields

$$V = \pi r^2\left(\frac{10}{r} - \frac{3r}{2}\right) + \frac{2}{3}\pi r^3$$

or $V = 10\pi r - \frac{5\pi r^3}{6}$ as required.

The value of r which makes V maximum is required and so

$$\frac{dV}{dr} = 10\pi - \frac{5\pi r^2}{2}$$

$$\frac{d^2V}{dr^2} = -5\pi r$$

$$\frac{dV}{dr} = 0 \Rightarrow \frac{5\pi r^2}{2} = 10\pi \Rightarrow r^2 = 4 \Rightarrow r = 2$$

$$\frac{d^2V}{dr^2} = -10\pi \text{ for } r = 2, \text{ confirming the maximum.}$$

For $r = 2$

$$V = 10\pi(2) - \frac{5\pi(2)^3}{6}$$

i.e. $$V = \frac{40\pi}{3}$$

EXERCISE 4.11

1. A farmer erects a fence along three sides of a rectangle in order to make a sheepfold; the fourth side of the rectangle is provided by a hedge already in existence. Find the maximum area of the enclosure thus made if the total length of the fence is to be 80 m.

   (Oxford and Cambridge)

2. A square sheet of metal of side 12 cm has four equal square portions removed at the corners and the sides are then turned up to form an open rectangular box. Prove that, when the box has maximum volume, its depth is 2 cm.

   (Oxford and Cambridge)

3. The bottom of a rectangular tank is a square of side x m and the tank is open at the top. It is designed to hold 4 m^3 of liquid. Express in terms of x the total area of the bottom and four sides of the tank. Find the value of x for which this area is a minimum.

   (Oxford and Cambridge)

4. A closed rectangular box is made of very thin sheet metal, and its length is three times its width. If the volume of the box is 288 cm^3, show that its surface area is equal to

   $$\frac{768}{x} + 6x^2 \text{ cm}^2$$

   where x cm is the width of the box.

   Find by differentiation the dimensions of the box of least surface area.

   (Oxford and Cambridge)

5. A rectangular box without a lid is made of cardboard of negligible thickness. The sides of the base are $2x$ cm and $3x$ cm, and the height is y cm. If the total area of the cardboard is $200\,\text{cm}^2$, prove that

$$y = \frac{20}{x} - \frac{3x}{5}$$

Find the dimensions of the box when its volume is a maximum.

(Oxford and Cambridge)

6. A satellite is to be constructed in the shape of a cylinder of radius r m and length l m, with two hemispheres attached, one to each end of the cylinder, as shown in Figure 4.11.

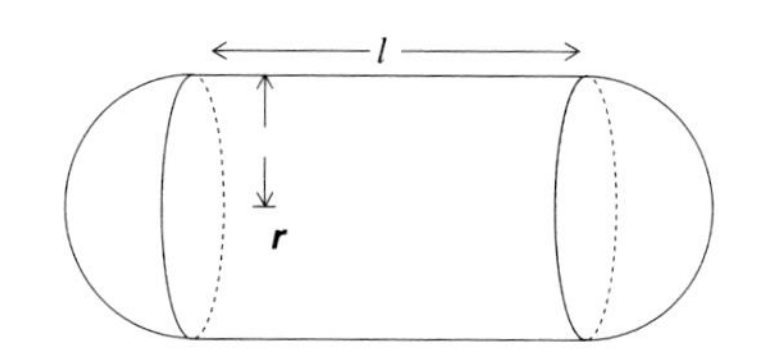

Figure 4.11

(i) If the total surface area of the satellite is $S\,\text{m}^2$, write down an expression for S in terms of r and l.

(ii) The satellite is designed to have a volume of $\frac{\pi}{6}\text{m}^3$. Show that the length of the cylinder is given by the expression $\frac{1-8r^3}{6r^2}$ m.

(iii) Given that the volume is to be $\frac{\pi}{6}\text{m}^3$, obtain an expression for the total surface area $S\,\text{m}^2$ in terms of r only. Simplify your expression as far as possible.

(iv) The surface of the satellite is to be coated with heat-resistant material which is expensive and so the amount of material used is to be kept to a minimum. Find the minimum surface area which will enclose a volume of $\frac{\pi}{6}\text{m}^3$.

(N. Ireland Additional Mathematics, 1989)

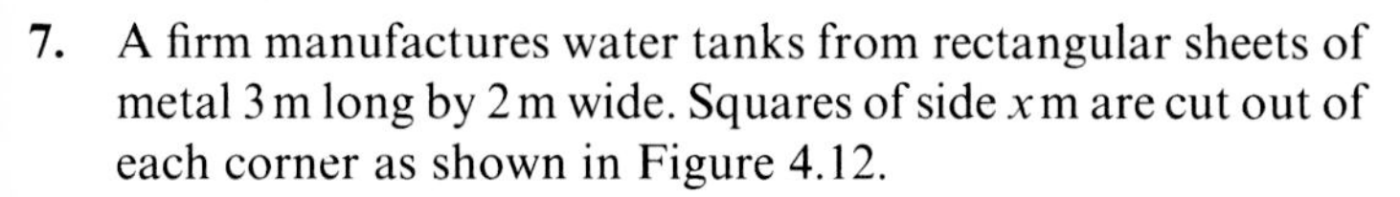

7. A firm manufactures water tanks from rectangular sheets of metal 3 m long by 2 m wide. Squares of side x m are cut out of each corner as shown in Figure 4.12.

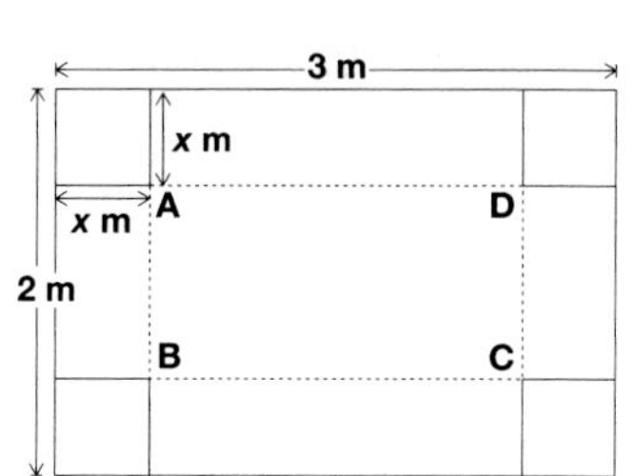

Figure 4.12

ABCD forms the base of the tank and the remaining parts are folded upwards and sealed to form the open tank.

Derive and simplify expressions in terms of x for

(i) the volume of water which the tank can hold,

(ii) the internal surface area of the tank.

Each sheet of metal costs the manufacturer £10. The internal surface of each tank is treated at a cost of £12.50 per m^2. The income obtained by selling a tank is £100 per m^3 of water it can hold.

(iii) Using this information and neglecting all other costs, derive an expression in terms of x for the profit obtained from the sale of each tank.

(iv) Calculate the maximum possible profit which can be obtained from the sale of each tank.

(N. Ireland Additional Mathematics, 1990)

8. A builder wishes to construct a window frame in the shape of a rectangle ABCD with a semicircular portion ASD at the top. The radius of the semicircular portion is r m and the height of the rectangular part is x m.

Write down in terms of x and r

(i) the total length, l, of the frame ABCDSA.

(ii) the area, a, enclosed by the frame.

The area enclosed by the frame is to be $4\,\text{m}^2$.

(iii) Show that

$$l = \left(\frac{\pi}{2} + 2\right)r + \frac{4}{r}$$

(iv) If the wood for the frame costs £2 per metre, calculate the minimum total cost of the wood such that the area enclosed by the frame is $4\,\text{m}^2$. Give the corresponding dimensions of the window correct to 2 decimal places.

(N. Ireland Additional Mathematics, 1991)

9. A cylindrical metal can, radius r cm and height h cm, without a lid is to be made. Write down a formula in terms of r and h for the area, $A\,\text{cm}^2$, of sheet metal required.

Its capacity is to be $64\pi\,\text{cm}^3$. Write down an equation in r and h which expresses this fact.

Hence show that $A = \pi r^2 + \dfrac{128\pi}{r}$.

Calculate the dimensions of the can if it is to require the minimum area of metal. Also state the area of metal needed.

(SMP Additional Mathematics)

10. The well-known brewery McNulty was informed by its market research department that the most popular size of can for export purposes would contain 3 litres of Best Bitter. For reasons of economy the brewery wishes each cylindrical can in which the beer will be exported to be made of the minimum amount of metal. If the radius of a 3-litre cylindrical can is r cm show that its surface area can be written as

$\left(2\pi r^2 + \dfrac{6000}{r}\right)\text{cm}^2$ and hence find the minimum area of metal required, showing that it is in fact a minimum.

[The volume of a cylinder is $\pi r^2 h$ and its curved surface area is $2\pi rh$. 1 litre = $1000\,\text{cm}^3$.]

(Oxford and Cambridge Additional Mathematics)

11. A rectangular sheet of paper ABCD, 18 cm by 35 cm, is folded over along the line PQ, where P is on CD and Q is on BC, so that the corner C lies on the edge AD as shown in Figure 4.13.

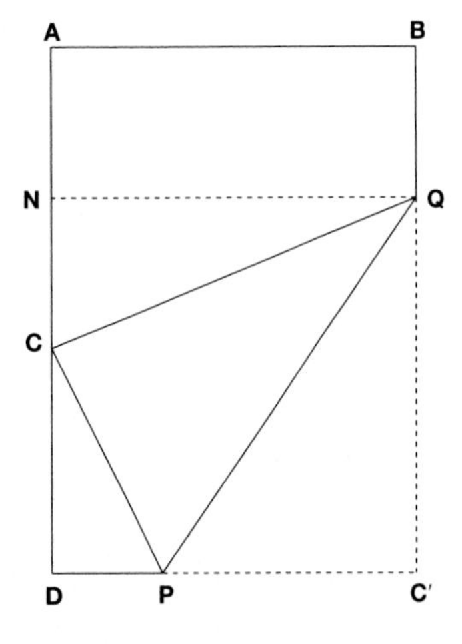

Figure 4.13

(i) (a) If P is 1 cm from the centre of C′D so that DP = 8 cm, state the length of CP and calculate the length CD.

(b) N is the point on AD such that QN is perpendicular to AD. Why must triangles QNC and CDP be the same shape? State the enlargement scale factor between them.

(c) Hence give the length QC and show that triangle QCP has area $150\,\text{cm}^2$.

(ii) Similarly, if P is x^2 cm from the centre, the area, T, of triangle QCP can be shown to be given by $T = \dfrac{3(9 + x^2)^2}{2x}$.

(a) Show that this may be written $T = 121.5x^{-1} + 27x + 1.5x^3$.

(b) Hence confirm that this area is a minimum when P is one-third of the way along the edge.

(SMP Additional Mathematics)

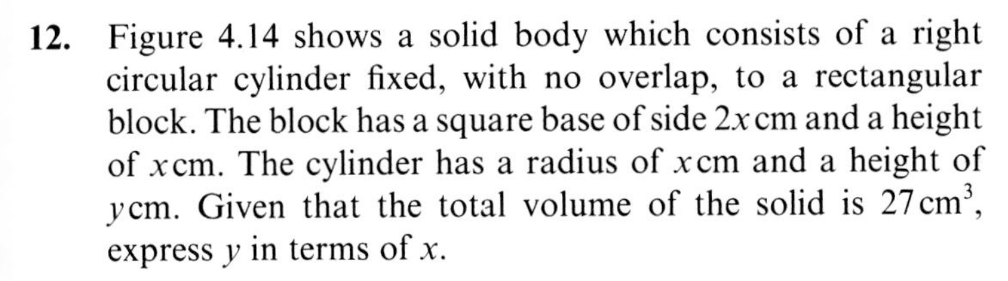

12. Figure 4.14 shows a solid body which consists of a right circular cylinder fixed, with no overlap, to a rectangular block. The block has a square base of side $2x$ cm and a height of x cm. The cylinder has a radius of x cm and a height of y cm. Given that the total volume of the solid is $27\,\text{cm}^3$, express y in terms of x.

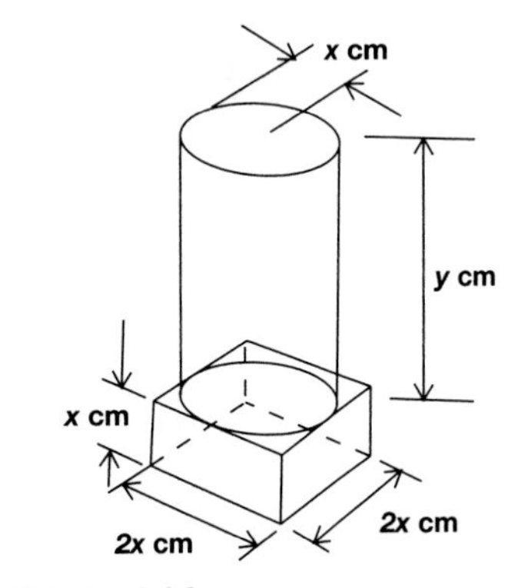

Figure 4.14

Hence show that the total surface area, $A\,\text{cm}^2$, of the solid is given by

$$A = \frac{54}{x} + 8x^2$$

Find

(i) the value of x for which A has a stationary value,
(ii) the values of A and of y corresponding to this value of x.

Determine whether the stationary value of A is a maximum or a minimum.

(U.C.L.E.S. Additional Mathematics, November 1991)

13. (a) Find the coordinates of the stationary point on the curve

$$y = \frac{16x^3 + 4x^2 + 1}{2x^2}.$$

(b) In the triangle ABC, angle ABC = 90°, AB = 5 cm and AC = 13 cm (see Figure 4.15). The rectangle BPQR is such that its vertices P, Q and R lie on BC, CA and AB respectively.

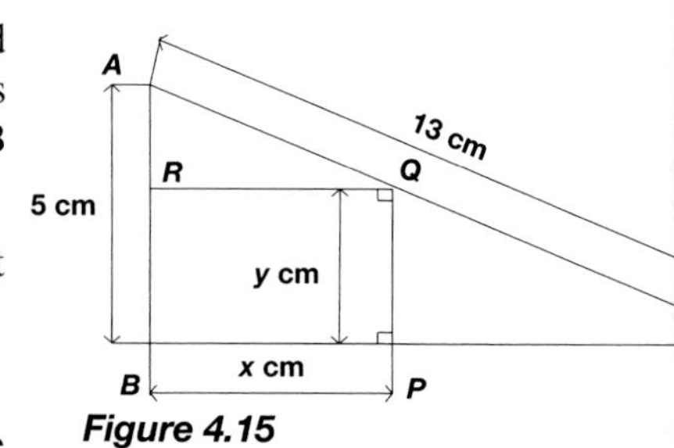

Figure 4.15

Given that BP = x cm and PQ = y cm, prove that $y = \frac{60 - 5x}{12}$.

Express the area of the rectangle in terms of x and hence calculate the maximum value of this area as x varies.

(U.C.L.E.S. Additional Mathematics, June 1992)

14. A parcel, in the form of a rectangular block, is held together by three pieces of tape as shown in Figure 4.16. The parcel has square ends of side x cm and is y cm in length. Express the total length of tape in terms of x and y. Given that the total length of tape is 450 cm, express the volume, V cm^3, of the parcel in terms of x. Find the value of x and of y for which V has a stationary value and determine whether this value is a maximum or a minimum.

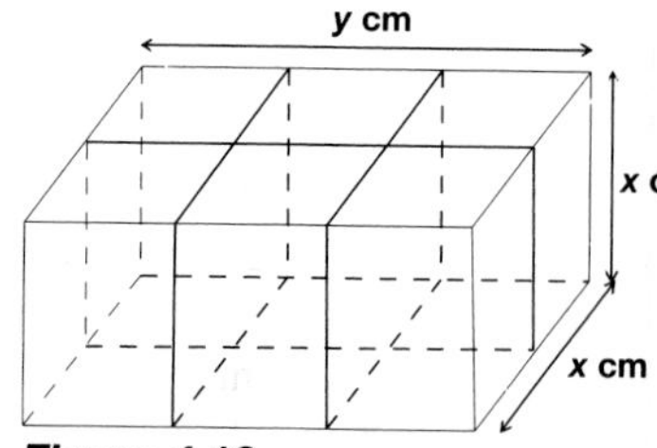

Figure 4.16

(U.C.L.E.S. Additional Mathematics, June 1991, part question)

CHAPTER 5

Matrices

The two mathematicians most closely associated with matrix algebra are **Arthur Cayley** (1821–1895) and **James Joseph Sylvester** (1814–1897) who were both born in England. Matrix notation has had a profound influence in all branches of mathematics, particularly quantum mechanics (the mathematical model governing the motion of very small bodies) and gravitation (the force which binds very large bodies together). Cayley showed early promise at school and was a Cambridge 'Wrangler' (a high-performing student in that university's mathematical *Tripos* papers). Unlike a number of contemporary Cambridge mathematics graduates who aspire to a financial career in the City of London, Cayley reluctantly abandoned mathematics to practise at the bar for fourteen years. However, his change of professional direction was not completely without benefit to mathematics for he published hundreds of mathematical papers while carrying out his legal responsibilities and encountered, through legal circles, another barrister of great mathematical talent, namely, Sylvester. They became lifelong friends despite a marked difference in temperament: Cayley was level-headed and cautious, Sylvester high-spirited and impetuous.

Sylvester had been sent down from the new University of London for threatening a fellow student with a table knife and later resigned his post as professor of mathematics at the University of Virginia because the university hierarchy refused to discipline a student who had insulted him. Upon return from America he became an actuary and eventually a barrister, supplementing his income by private teaching. Florence Nightingale was one of Sylvester's finest students.

INTRODUCTION

From time to time in mathematics, it is convenient to display numerical information in an array or **matrix**. The matrix **P** has two rows and three columns.

$$\mathbf{P} = \begin{pmatrix} 7 & 5 & 4 \\ 2 & 2 & 12 \end{pmatrix}$$

The number of rows and columns in a matrix determine its **order**, so that **P** is of order 2 × 3 (2 by 3). The matrix **Q** is an example of a **column** matrix, having only one column.

$$\mathbf{Q} = \begin{pmatrix} 2 \\ 7 \\ 3 \end{pmatrix}$$

Q is of order 3 × 1 since this matrix has three rows and one column. A matrix which has the same number of rows as columns is described as a **square** matrix.

The numbers contained in a matrix are referred to as the **elements** of the matrix. In matrix **P**, for example, the element in the first row and second column is 5, while the element in the second row and first column of **Q** is 7.

Two matrices are equal when corresponding elements are equal, thus

$$\begin{pmatrix} 2 & 1 & 7 \\ 1 & 0 & 4 \end{pmatrix} = \begin{pmatrix} 2 & 1 & 7 \\ 1 & 0 & 4 \end{pmatrix}$$

EXAMPLE 1

$$\mathbf{A} = \begin{pmatrix} 2 & 1 & 7 & 4 \\ 2 & 2 & 3 & -1 \\ 1 & 2 & 4 & -7 \end{pmatrix}$$

For matrix **A** state

(i) the number of rows,
(ii) the number of columns,
(iii) the order,
(iv) the element in row 2, column 4.

Solution 1

(i) There are 3 rows.
(ii) There are 4 columns.
(iii) 3 × 4
(iv) − 1

EXAMPLE 2

Given $\mathbf{A} = \begin{pmatrix} 2x + y & x - y \\ 5x + 4y & 7x - 2y \end{pmatrix}$

$\mathbf{B} = \begin{pmatrix} 1 & 5 \\ -2 & 20 \end{pmatrix}$ and $\mathbf{A} = \mathbf{B}$ find x and y.

Solution 2

Since these matrices are equal, corresponding elements are equal.

Therefore
$$\begin{aligned} 2x + y &= 1 \quad &(1)\\ x - y &= 5 \quad &(2)\\ 5x + 4y &= -2 \quad &(3)\\ 7x - 2y &= 20 \quad &(4) \end{aligned}$$

Consider equations (1) and (2).

$$\begin{aligned} 2x + y &= 1\\ x - y &= 5 \end{aligned}$$

Adding these equations yields $x = 2$ and substituting this x value in equation (1) gives $y = -3$.

The reader can confirm that $x = 2$, $y = -3$ satisfies all equations (1) to (4) inclusive.

EXERCISE 5.1

1. Consider the matrix $\begin{pmatrix} a & b & c \\ d & e & f \\ g & h & i \\ j & k & l \end{pmatrix}$

(a) State

(i) the number of rows,
(ii) the number of columns.

(b) List the elements in the third row.

(c) List the elements in the second column.

(d) Write down the entry in

(i) the first row and second column,
(ii) the second row and first column,
(iii) the third row and third column.

(e) State the rows and columns which describe the positions of these entries.

(i) a (ii) e (iii) h (iv) c (v) j (vi) l

2. For each of the following matrices, state the number of rows and columns, and the entry in the second row and first column.

(a) $\begin{pmatrix} 2 \\ -6 \end{pmatrix}$ (b) $\begin{pmatrix} 8 & 9 \\ 7 & -7 \end{pmatrix}$ (c) $\begin{pmatrix} p & q & r \\ s & t & u \end{pmatrix}$ (d) $\begin{pmatrix} 9 & 7 \\ 0 & 2 \\ 6 & 3 \end{pmatrix}$

(e) $\begin{pmatrix} 1 & -2 & 3 & -4 \\ 5 & 4 & 3 & 2 \end{pmatrix}$ (f) $\begin{pmatrix} 1 & 2 & 3 \\ 4 & 5 & 6 \\ 7 & 8 & 9 \end{pmatrix}$

3. Write down examples of matrices with numerical elements arranged in

(a) 2 rows and 3 columns, (b) 3 rows and 3 columns, (c) 4 rows and 1 column,

(d) 1 row and 5 columns.

4. State the order of each of the following matrices.

(a) $\begin{pmatrix} 7 & 5 & -1 & 0 \\ 2 & 8 & 9 & 4 \\ 6 & 3 & -2 & 1 \end{pmatrix}$ (b) $\begin{pmatrix} 6 & 9 & 2 \\ 8 & 7 & 4 \\ 3 & 6 & 1 \end{pmatrix}$ (c) $\begin{pmatrix} a & b \\ c & d \\ e & f \end{pmatrix}$ (d) $(1 \quad 2 \quad 3)$

(e) $\begin{pmatrix} -2 \\ 7 \\ 4 \\ 8 \end{pmatrix}$ (f) $\begin{pmatrix} p & q & r & s \\ t & u & v & w \end{pmatrix}$

5. Write down an example of a

(a) 2×2 matrix, (b) 3×2 matrix, (c) 4×3 matrix, (d) 1×1 matrix,

(e) 1×3 matrix, (f) 5×1 matrix.

6. List any equalities for pairs of the following matrices.

$\mathbf{A} = \begin{pmatrix} 1 & 2 & 3 \\ 4 & 5 & 6 \end{pmatrix}$ $\mathbf{B} = \begin{pmatrix} -7 & 0 \\ 2 & -6 \end{pmatrix}$ $\mathbf{C} = (-1 \quad 0 \quad 1)$ $\mathbf{D} = (1 \quad 0 \quad 1)$

$\mathbf{E} = \begin{pmatrix} -1 \\ 0 \\ 1 \end{pmatrix}$ $\mathbf{F} = \begin{pmatrix} 2 \\ 3 \end{pmatrix}$ $\mathbf{G} = \begin{pmatrix} 1 & 3 & 2 \\ 4 & 6 & 5 \end{pmatrix}$ $\mathbf{H} = \begin{pmatrix} -7 & 0 \\ -2 & 6 \end{pmatrix}$

$\mathbf{J} = (1 \quad 0 \quad -1)$ $\mathbf{K} = \begin{pmatrix} 1 & 2 & 3 \\ 4 & 5 & 6 \end{pmatrix}$ $\mathbf{L} = (-1 \quad 0 \quad 1)$ $\mathbf{M} = \begin{pmatrix} -7 & 0 \\ 2 & -6 \end{pmatrix}$

7. What is the order of each matrix in question 6?

8. Find x and y in each of the following.

(a) $(-4x \quad 3y) = (16 \quad 3)$ (b) $\begin{pmatrix} 2x+5 \\ 7-3y \end{pmatrix} = \begin{pmatrix} 9 \\ 1 \end{pmatrix}$ (c) $\begin{pmatrix} 2x+y \\ 5x-3y \end{pmatrix} = \begin{pmatrix} 7 \\ 12 \end{pmatrix}$

(d) $\begin{pmatrix} x^2 & y^3 \\ x^3 & y^2 \end{pmatrix} = \begin{pmatrix} 16 & -27 \\ 64 & 9 \end{pmatrix}$

THE ADDITION AND SUBTRACTION OF MATRICES

Matrix addition

Matrices are added by summing corresponding elements. If follows that incompatible matrices, i.e. matrices of differing order, may not be added.

Matrix addition is associative and commutative.

EXAMPLE 3

Given $\mathbf{A} = \begin{pmatrix} 2 & 1 & 7 \\ 1 & 2 & 4 \end{pmatrix}$ $\mathbf{B} = \begin{pmatrix} 2 & 1 & 2 \\ 1 & 0 & 7 \end{pmatrix}$ $\mathbf{C} = \begin{pmatrix} -1 & 2 \\ 4 & 3 \end{pmatrix}$

$\mathbf{D} = \begin{pmatrix} 0 & 4 \\ 1 & -2 \end{pmatrix}$ $\mathbf{E} = \begin{pmatrix} 1 & 0 & 2 \\ -1 & 7 & -4 \end{pmatrix}$

(i) Find, where possible, $\mathbf{A}+\mathbf{B}$, $\mathbf{A}+\mathbf{E}$, $\mathbf{C}+\mathbf{D}$, $\mathbf{B}+\mathbf{D}$, $\mathbf{D}+\mathbf{E}$, $\mathbf{E}+\mathbf{B}$.

(ii) Establish that $\mathbf{A} + \mathbf{B} = \mathbf{B} + \mathbf{A}$, i.e. for these matrices, addition is commutative.

(iii) Establish $\mathbf{A} + (\mathbf{B} + \mathbf{E}) = (\mathbf{A} + \mathbf{B}) + \mathbf{E}$, i.e. for these matrices, addition is associative.

Solution 3

(i) $\mathbf{A} + \mathbf{B} = \begin{pmatrix} 4 & 2 & 9 \\ 2 & 2 & 11 \end{pmatrix}$

$\mathbf{A} + \mathbf{E} = \begin{pmatrix} 3 & 1 & 9 \\ 0 & 9 & 0 \end{pmatrix}$

$\mathbf{C} + \mathbf{D} = \begin{pmatrix} -1 & 6 \\ 5 & 1 \end{pmatrix}$

$\mathbf{B} + \mathbf{D}$ is not possible as $\mathbf{B}$ and $\mathbf{D}$ differ in order.

$\mathbf{D} + \mathbf{E}$ is not possible as $\mathbf{D}$ and $\mathbf{E}$ differ in order.

$\mathbf{E} + \mathbf{B} = \begin{pmatrix} 3 & 1 & 4 \\ 0 & 7 & 3 \end{pmatrix}$

(ii) $\mathbf{B} + \mathbf{A} = \begin{pmatrix} 2 & 1 & 2 \\ 1 & 0 & 7 \end{pmatrix} + \begin{pmatrix} 2 & 1 & 7 \\ 1 & 2 & 4 \end{pmatrix} = \begin{pmatrix} 4 & 2 & 9 \\ 2 & 2 & 11 \end{pmatrix}$

$\mathbf{A} + \mathbf{B} = \begin{pmatrix} 2 & 1 & 7 \\ 1 & 2 & 4 \end{pmatrix} + \begin{pmatrix} 2 & 1 & 2 \\ 1 & 0 & 7 \end{pmatrix} = \begin{pmatrix} 4 & 2 & 9 \\ 2 & 2 & 11 \end{pmatrix}$

confirming that matrix addition is commutative in this case.

(iii)
$$\begin{aligned}
\mathbf{A} + (\mathbf{B} + \mathbf{E}) &= \begin{pmatrix} 2 & 1 & 7 \\ 1 & 2 & 4 \end{pmatrix} + \left(\begin{pmatrix} 2 & 1 & 2 \\ 1 & 0 & 7 \end{pmatrix} + \begin{pmatrix} 1 & 0 & 2 \\ -1 & 7 & -4 \end{pmatrix}\right) \\
&= \begin{pmatrix} 2 & 1 & 7 \\ 1 & 2 & 4 \end{pmatrix} + \begin{pmatrix} 3 & 1 & 4 \\ 0 & 7 & 3 \end{pmatrix} \\
&= \begin{pmatrix} 5 & 2 & 11 \\ 1 & 9 & 7 \end{pmatrix}
\end{aligned}$$

$$\begin{aligned}
(\mathbf{A} + \mathbf{B}) + \mathbf{E} &= \left(\begin{pmatrix} 2 & 1 & 7 \\ 1 & 2 & 4 \end{pmatrix} + \begin{pmatrix} 2 & 1 & 2 \\ 1 & 0 & 7 \end{pmatrix}\right) + \begin{pmatrix} 1 & 0 & 2 \\ -1 & 7 & -4 \end{pmatrix} \\
&= \begin{pmatrix} 4 & 2 & 9 \\ 2 & 2 & 11 \end{pmatrix} + \begin{pmatrix} 1 & 0 & 2 \\ -1 & 7 & -4 \end{pmatrix} \\
&= \begin{pmatrix} 5 & 2 & 11 \\ 1 & 9 & 7 \end{pmatrix}
\end{aligned}$$

confirming associativity in this case.

Matrix subtraction

The difference of two matrices is calculated by subtracting corresponding elements. As in the case of addition, incompatible matrices (those of differing order) may not be subtracted. Matrix subtraction is neither associative nor commutative.

EXAMPLE 4

Given

$$\mathbf{A} = \begin{pmatrix} 2 & 1 \\ 2 & 7 \end{pmatrix} \qquad \mathbf{B} = \begin{pmatrix} 2 & 1 & -4 \\ 11 & -2 & 9 \end{pmatrix}$$

$$\mathbf{C} = \begin{pmatrix} 2 & 1 \\ 7 & -4 \end{pmatrix} \qquad \mathbf{D} = \begin{pmatrix} 1 & 0 \\ -7 & -2 \end{pmatrix}$$

(i) Find, where possible, $\mathbf{A} - \mathbf{B}$, $\mathbf{A} - \mathbf{C}$, $\mathbf{D} - \mathbf{A}$.

(ii) Using matrices **A** and **C**, confirm that matrix subtraction is not commutative.

(iii) Using matrices **A**, **C** and **D**, confirm that matrix subtraction is not associative.

Solution 4

(i) $\mathbf{A} - \mathbf{B}$ is not possible as **A** and **B** are of differing order.

$$\mathbf{A} - \mathbf{C} = \begin{pmatrix} 2 & 1 \\ 2 & 7 \end{pmatrix} - \begin{pmatrix} 2 & 1 \\ 7 & -4 \end{pmatrix} = \begin{pmatrix} 0 & 0 \\ -5 & 11 \end{pmatrix}$$

$$\mathbf{D} - \mathbf{A} = \begin{pmatrix} 1 & 0 \\ -7 & -2 \end{pmatrix} - \begin{pmatrix} 2 & 1 \\ 2 & 7 \end{pmatrix} = \begin{pmatrix} -1 & -1 \\ -9 & -9 \end{pmatrix}$$

(ii) $$\mathbf{A} - \mathbf{C} = \begin{pmatrix} 2 & 1 \\ 2 & 7 \end{pmatrix} - \begin{pmatrix} 2 & 1 \\ 7 & -4 \end{pmatrix} = \begin{pmatrix} 0 & 0 \\ -5 & 11 \end{pmatrix}$$

$$\mathbf{C} - \mathbf{A} = \begin{pmatrix} 2 & 1 \\ 7 & -4 \end{pmatrix} - \begin{pmatrix} 2 & 1 \\ 2 & 7 \end{pmatrix} = \begin{pmatrix} 0 & 0 \\ 5 & -11 \end{pmatrix}$$

clearly $\mathbf{A} - \mathbf{C} \neq \mathbf{C} - \mathbf{A}$

(iii) $$(\mathbf{A} - \mathbf{C}) - \mathbf{D} = \left(\begin{pmatrix} 2 & 1 \\ 2 & 7 \end{pmatrix} - \begin{pmatrix} 2 & 1 \\ 7 & -4 \end{pmatrix}\right) - \begin{pmatrix} 1 & 0 \\ -7 & -2 \end{pmatrix}$$

$$= \begin{pmatrix} 0 & 0 \\ -5 & 11 \end{pmatrix} - \begin{pmatrix} 1 & 0 \\ -7 & -2 \end{pmatrix}$$

$$= \begin{pmatrix} -1 & 0 \\ 2 & 13 \end{pmatrix}$$

$$\mathbf{A}-(\mathbf{C}-\mathbf{D})=\begin{pmatrix}2&1\\2&7\end{pmatrix}-\left(\begin{pmatrix}2&1\\7&-4\end{pmatrix}-\begin{pmatrix}1&0\\-7&-2\end{pmatrix}\right)$$
$$=\begin{pmatrix}2&1\\2&7\end{pmatrix}-\begin{pmatrix}1&1\\14&-2\end{pmatrix}$$
$$=\begin{pmatrix}1&0\\-12&9\end{pmatrix}$$

clearly $(\mathbf{A}-\mathbf{C})-\mathbf{D}\neq\mathbf{A}-(\mathbf{C}-\mathbf{D})$

EXERCISE 5.2

Find the sums of the matrices in questions 1 to 15.

1. $\begin{pmatrix}7\\2\end{pmatrix}+\begin{pmatrix}3\\8\end{pmatrix}$ **2.** $\begin{pmatrix}1\\3\end{pmatrix}+\begin{pmatrix}4\\-2\end{pmatrix}$ **3.** $\begin{pmatrix}3x\\2y\end{pmatrix}+\begin{pmatrix}-5x\\y\end{pmatrix}$

4. $\begin{pmatrix}2a\\b\end{pmatrix}+\begin{pmatrix}3\\4\end{pmatrix}$ **5.** $\begin{pmatrix}a\\b\end{pmatrix}+\begin{pmatrix}2c\\d\end{pmatrix}$ **6.** $\begin{pmatrix}-3m\\n\end{pmatrix}+\begin{pmatrix}3m\\-n\end{pmatrix}$

7. $(1\quad 2)+(3\quad 4)$ **8.** $(7\quad -5)+(-2\quad 0)$ **9.** $\begin{pmatrix}1&0\\0&1\end{pmatrix}+\begin{pmatrix}5&2\\3&8\end{pmatrix}$

10. $\begin{pmatrix}3&2\\1&4\end{pmatrix}+\begin{pmatrix}2&0\\1&5\end{pmatrix}$ **11.** $\begin{pmatrix}3&2\\1&4\end{pmatrix}+\begin{pmatrix}-2&0\\1&-5\end{pmatrix}$

12. $\begin{pmatrix}3&-4\\5&1\end{pmatrix}+\begin{pmatrix}1&4\\-5&0\end{pmatrix}$ **13.** $\begin{pmatrix}7&9&6\\4&3&2\end{pmatrix}+\begin{pmatrix}8&-6&7\\-1&0&-2\end{pmatrix}$

14. $\begin{pmatrix}3x&y\\5x&-y\end{pmatrix}+\begin{pmatrix}x&2y\\-4x&y\end{pmatrix}$ **15.** $\begin{pmatrix}a&2b\\-a&-b\end{pmatrix}+\begin{pmatrix}3a&2b\\-a&3b\end{pmatrix}$

16. (a) $\mathbf{A}=\begin{pmatrix}4&7\\2&3\end{pmatrix}$ $\mathbf{B}=\begin{pmatrix}2&1\\-3&0\end{pmatrix}$ $\mathbf{C}=\begin{pmatrix}3&1\\1&-5\end{pmatrix}$

Find these matrices.

(i) $\mathbf{A}+\mathbf{B}$
(ii) $\mathbf{B}+\mathbf{C}$
(iii) $(\mathbf{A}+\mathbf{B})+\mathbf{C}$
(iv) $\mathbf{A}+(\mathbf{B}+\mathbf{C})$

(b) Is it true generally that $(\mathbf{A}+\mathbf{B})+\mathbf{C}=\mathbf{A}+(\mathbf{B}+\mathbf{C})$?
What law for addition of matrices does this result suggest?

17. Write down the negative of each of the following matrices.

(a) $\begin{pmatrix}-4\\6\end{pmatrix}$ (b) $\begin{pmatrix}3\\2\\1\end{pmatrix}$ (c) $\begin{pmatrix}8&-5\\-3&2\end{pmatrix}$ (d) $(2\quad -3)$ (e) $\begin{pmatrix}4&2&-1\\-3&8&2\end{pmatrix}$

18. Given that **A** is a 3×2 matrix, solve the equation

$$\mathbf{A} + \begin{pmatrix} 1 & 3 \\ 5 & -7 \\ 0 & 9 \end{pmatrix} = \mathbf{0}$$

where **0** is the 3×2 matrix whose every element is equal to zero, the so-called 'null' matrix.

19. Simplify each of the following.

(a) $\begin{pmatrix} 7 \\ 6 \end{pmatrix} - \begin{pmatrix} 5 \\ 3 \end{pmatrix}$ (b) $\begin{pmatrix} 5 \\ 2 \end{pmatrix} - \begin{pmatrix} -2 \\ 2 \end{pmatrix}$ (c) $\begin{pmatrix} -3 \\ 4 \end{pmatrix} - \begin{pmatrix} -3 \\ -5 \end{pmatrix}$ (d) $\begin{pmatrix} -2p \\ q \end{pmatrix} - \begin{pmatrix} p \\ -3q \end{pmatrix}$

(e) $\begin{pmatrix} a+3 \\ 5-y \end{pmatrix} - \begin{pmatrix} 5 \\ 3+2y \end{pmatrix}$ (f) $\begin{pmatrix} r \\ s \end{pmatrix} - \begin{pmatrix} t \\ 2u \end{pmatrix}$

20. Simplify

(a) $\begin{pmatrix} 7 & 11 \\ 5 & 2 \end{pmatrix} - \begin{pmatrix} 4 & 6 \\ 3 & 2 \end{pmatrix}$ (b) $\begin{pmatrix} 3 & -4 \\ 7 & 5 \end{pmatrix} - \begin{pmatrix} 0 & -4 \\ -3 & 0 \end{pmatrix}$

(c) $\begin{pmatrix} 1 & 4 \\ -3 & 1 \end{pmatrix} - \begin{pmatrix} 4 & -3 \\ -1 & 3 \end{pmatrix}$ (d) $\begin{pmatrix} 2a & 3 \\ 5 & 4b \end{pmatrix} - \begin{pmatrix} a & -1 \\ 2 & -b \end{pmatrix}$

21. Given

$$\mathbf{A} = \begin{pmatrix} 2 & -3 \\ -1 & 4 \end{pmatrix} \quad \mathbf{B} = \begin{pmatrix} 3 & 5 \\ -2 & 1 \end{pmatrix} \quad \text{and } \mathbf{C} = \begin{pmatrix} 2 & 4 \\ 1 & 0 \end{pmatrix}$$

find in simplest form

(a) $\mathbf{A} + \mathbf{B}$ (b) $\mathbf{B} + \mathbf{C}$ (c) $\mathbf{A} + \mathbf{B} + \mathbf{C}$ (d) $\mathbf{A} - \mathbf{C}$

(e) $\mathbf{B} - \mathbf{C}$ (f) $\mathbf{C} - \mathbf{A}$ (g) $(\mathbf{A} + \mathbf{B}) + (\mathbf{B} + \mathbf{C})$ (h) $(\mathbf{A} + \mathbf{B}) - (\mathbf{A} + \mathbf{C})$

22. Solve the equation

$$\mathbf{X} + \begin{pmatrix} 6 \\ -1 \end{pmatrix} = \begin{pmatrix} -2 \\ 4 \end{pmatrix}$$

for the 2×1 matrix **X**.

23. If

$$(a \quad b \quad c) - (2 \quad -6 \quad 0) = (3 \quad 8 \quad 2)$$

find a, b, c.

24. Solve each of the following equations for the 2×2 matrix **X**.

(a) $\mathbf{X} + \begin{pmatrix} 1 & 0 \\ 0 & 1 \end{pmatrix} = \begin{pmatrix} 0 & 3 \\ 9 & 6 \end{pmatrix}$ (b) $\begin{pmatrix} 6 & 5 \\ 1 & 3 \end{pmatrix} + \mathbf{X} = \begin{pmatrix} 3 & -3 \\ 4 & 2 \end{pmatrix}$

(c) $\mathbf{X} - \begin{pmatrix} 2 & -3 \\ -1 & 4 \end{pmatrix} = \begin{pmatrix} -1 & 5 \\ 3 & 9 \end{pmatrix}$ (d) $\begin{pmatrix} 2 & 5 \\ 7 & 4 \end{pmatrix} - \mathbf{X} = \begin{pmatrix} -3 & 8 \\ 7 & 2 \end{pmatrix}$

25. Find a, b, c, d in each of the following.

(a) $\begin{pmatrix} a & b \\ c & d \end{pmatrix} - \begin{pmatrix} 2 & 1 \\ -1 & 0 \end{pmatrix} = \begin{pmatrix} 2 & -1 \\ 3 & 4 \end{pmatrix}$ (b) $\begin{pmatrix} 2 & 5 \\ -9 & 8 \end{pmatrix} - \begin{pmatrix} a & b \\ c & d \end{pmatrix} = \begin{pmatrix} -1 & 0 \\ 4 & 5 \end{pmatrix}$

26. If

$$\mathbf{X} = \begin{pmatrix} 2 & -1 \\ 1 & 0 \\ 3 & 4 \end{pmatrix} \quad \mathbf{Y} = \begin{pmatrix} 2 & 3 \\ -1 & 4 \\ 0 & 5 \end{pmatrix} \quad \text{and } \mathbf{Z} = \begin{pmatrix} 3 & 1 \\ -2 & 3 \\ 1 & 4 \end{pmatrix}$$

simplify these.

(a) $\mathbf{X} + \mathbf{Y}$ (b) $\mathbf{Y} - \mathbf{Z}$ (c) $(\mathbf{X} + \mathbf{Y}) - \mathbf{Z}$ (d) $\mathbf{X} + (\mathbf{Y} - \mathbf{Z})$

MATRIX MULTIPLICATION

Multiplication by a scalar

Multiplication of a matrix by a scalar is achieved by multiplying each element of the matrix by the scalar, i.e.

$$\mathbf{A} = \begin{pmatrix} 2 & 1 & 7 \\ 1 & 0 & -4 \end{pmatrix} \Rightarrow 3\mathbf{A} = \begin{pmatrix} 6 & 3 & 21 \\ 3 & 0 & -12 \end{pmatrix}$$

Matrix multiplication

Before multiplying matrices a 'compatibility check' must be carried out. For the product **AB** to exist the number of columns of **A** must equal the number of rows of **B**. Hence **A** could be of order 2×7 and **B** of order 7×4. Multiplication is possible here since **A** has 7 columns and **B** has 7 rows.

$(2 \times 7).(7 \times 4)$
↓ ↓
compatible

Further examples of compatible pairs are

$(2 \times 5).(5 \times 4)$
$(3 \times 2).(2 \times 4)$
$(1 \times 3).(3 \times 1)$

The outer numbers in each case above give the order of the product i.e.

the product $(2 \times 7).(7 \times 4)$ results in a 2×4 matrix
the product $(2 \times 5).(5 \times 4)$ results in a 2×4 matrix
the product $(3 \times 2).(2 \times 4)$ results in a 3×4 matrix
the product $(1 \times 3).(3 \times 1)$ results in a 1×1 matrix.

The next step is to demonstrate how such products are carried out.

Consider

$$\mathbf{A} = \begin{pmatrix} 2 & 1 & 7 & 4 \\ 3 & 2 & 1 & 2 \end{pmatrix} \quad \mathbf{B} = \begin{pmatrix} 1 & 2 & 1 \\ 2 & 1 & 0 \\ 4 & 7 & 2 \\ 1 & 2 & 1 \end{pmatrix}$$

Since **A** is of order 2×4 and **B** of order 4×3, these matrices are compatible for multiplication and yield a product of order 2×3 i.e. $(2 \times 4).(4 \times 3)$ produces a matrix of order 2×3.

↓ ↓
compatible

This may be represented

$$\begin{pmatrix} 2 & 1 & 7 & 4 \\ 3 & 2 & 1 & 2 \end{pmatrix} \begin{pmatrix} 1 & 2 & 1 \\ 2 & 1 & 0 \\ 4 & 7 & 2 \\ 1 & 2 & 1 \end{pmatrix} = \begin{pmatrix} ? & ? & ? \\ ? & ? & ? \end{pmatrix}$$

To find the number in the lth row and mth column of the product matrix the lth row of **A** is multiplied by the mth column of **B**.

Therefore to evaluate the element in the first row and second column of the 2×3 matrix, the first row of **A** i.e. (2 1 7 4) is multiplied by the second column of **B** i.e. (2 1 7 2), resulting in

element at row 1 and column 2 of product
$= 2 \times 2 + 1 \times 1 + 7 \times 7 + 4 \times 2 = 62$

Similarly, the element in row 2 and column 3 of the (2×3) product is found by multiplying row 2 of **A** i.e. (3 2 1 2) by column 3 of **B** i.e. (1 0 2 1) giving

element at row 2 and column 3 of product
$= 3 \times 1 + 2 \times 0 + 1 \times 2 + 2 \times 1 = 7$

Summarising the product thus far

$$\begin{pmatrix} 2 & 1 & 7 & 4 \\ 3 & 2 & 1 & 2 \end{pmatrix} \begin{pmatrix} 1 & 2 & 1 \\ 2 & 1 & 0 \\ 4 & 7 & 2 \\ 1 & 2 & 1 \end{pmatrix} = \begin{pmatrix} ? & 62 & ? \\ ? & ? & 7 \end{pmatrix}$$

It is left as an exercise for the reader to complete the product and confirm that

$$\begin{pmatrix} 2 & 1 & 7 & 4 \\ 3 & 2 & 1 & 2 \end{pmatrix} \begin{pmatrix} 1 & 2 & 1 \\ 2 & 1 & 0 \\ 4 & 7 & 2 \\ 1 & 2 & 1 \end{pmatrix} = \begin{pmatrix} 36 & 62 & 20 \\ 13 & 19 & 7 \end{pmatrix}$$

latrix multiplication is associative but not commutative.

i.e. $\mathbf{A}(\mathbf{B}.\mathbf{C}) = (\mathbf{A}.\mathbf{B})\mathbf{C}$

ut $\mathbf{A}.\mathbf{B} = \mathbf{B}.\mathbf{A}$ is not always true.

latrix division is not defined.

XAMPLE 5

iven

$$\mathbf{X} = \begin{pmatrix} 2 & -1 & 7 \\ 1 & 2 & -4 \end{pmatrix} \text{ and } \mathbf{Y} = \begin{pmatrix} 1 & -2 & 0 \\ -4 & 2 & 5 \end{pmatrix}$$

nd $3\mathbf{X} + 2\mathbf{Y}$.

olution 5

$$3\mathbf{X} = \begin{pmatrix} 6 & -3 & 21 \\ 3 & 6 & -12 \end{pmatrix}$$

$$2\mathbf{Y} = \begin{pmatrix} 2 & -4 & 0 \\ -8 & 4 & 10 \end{pmatrix}$$

herefore $3\mathbf{X} + 2\mathbf{Y} = \begin{pmatrix} 8 & -7 & 21 \\ -5 & 10 & -2 \end{pmatrix}$

XAMPLE 6

$$\mathbf{A} = \begin{pmatrix} 2 & -1 \\ 4 & 3 \end{pmatrix} \quad \mathbf{B} = \begin{pmatrix} 2 \\ -5 \end{pmatrix} \quad \mathbf{C} = (1 \quad 7)$$

ind (where possible) the matrix products **AB**, **AC** and **BC**.

olution 6

xamine the orders of the matrices forming the product **AB**.

$(2 \times 2).\ (2 \times 1)$ resulting in a product of order 2×1.

↓ ↓

compatible

he calculation reduces to

$$\begin{pmatrix} 2 & -1 \\ 4 & 3 \end{pmatrix}\begin{pmatrix} 2 \\ -5 \end{pmatrix} = \begin{pmatrix} ? \\ ? \end{pmatrix}$$

he element in the first row and first column of the product is ound by multiplying the first row of **A** i.e. $(2 \quad -1)$ by the first olumn of **B** i.e. $(2 \quad -5)$. This element is computed as

$$(2 \times 2) + (-1 \times -5) = 9$$

imilarly the second row and first column element of the product is omputed as

$$(4 \times 2) + (3 \times -5) = -7$$

In summary

$$\begin{pmatrix} 2 & -1 \\ 4 & 3 \end{pmatrix}\begin{pmatrix} 2 \\ -5 \end{pmatrix} = \begin{pmatrix} 9 \\ -7 \end{pmatrix}$$

Now consider the product **AC**. The compatibility check reveals

$(2 \times 2).(1 \times 2)$
↓ ↓
incompatible

Therefore this product is impossible.

Finally, the product **BC**. Consideration of order yields

$(2 \times 1).(1 \times 2)$
↓ ↓
compatible

and that the order of the product is 2×2.

The framework for the calculation is therefore

$$\begin{pmatrix} 2 \\ -5 \end{pmatrix}(1 \quad 7) = \begin{pmatrix} ? & ? \\ ? & ? \end{pmatrix}$$

The element at row 1 and column 1 is $2 \times 1 = 2$
The element at row 1 and column 2 is $2 \times 7 = 14$
The element at row 2 and column 1 is $-5 \times 1 = -5$
The element at row 2 and column 2 is $-5 \times 7 = -35$

Therefore

$$\begin{pmatrix} 2 \\ -5 \end{pmatrix}(1 \quad 7) = \begin{pmatrix} 2 & 14 \\ -5 & -35 \end{pmatrix}$$

Powers of matrices and the unit matrix

The matrix $\mathbf{X}^2$ merely indicates the product **X.X**. Similarly $\mathbf{X}^3 = \mathbf{X.X.X}$.

In order to comply with the compatibility requirement, it is only possible to calculate powers of square matrices. In order to illustrate this, suppose **A** was of order (2×3). The matrix $\mathbf{A}^2$ requires **A** to be multiplied by **A**. In terms of matrix order this may be represented

$(2 \times 3).(2 \times 3)$
↓ ↓
incompatible

An $n \times n$ unit matrix (or identity matrix) has 1s along its leading diagonal and zeros elsewhere. For example the order (3×3) unit matrix is

$$\begin{pmatrix} 1 & 0 & 0 \\ 0 & 1 & 0 \\ 0 & 0 & 1 \end{pmatrix}$$

This matrix is normally denoted **I**, its name (identity matrix or unit matrix) deriving from the property

$\mathbf{A}.\mathbf{I} = \mathbf{I}.\mathbf{A} = \mathbf{A}$ for any square matrix $\mathbf{A}$

Consider

$$\mathbf{A} = \begin{pmatrix} 2 & 1 \\ 2 & 7 \end{pmatrix}$$

$$\mathbf{A}.\mathbf{I} = \begin{pmatrix} 2 & 1 \\ 2 & 7 \end{pmatrix}\begin{pmatrix} 1 & 0 \\ 0 & 1 \end{pmatrix} = \begin{pmatrix} 2 & 1 \\ 2 & 7 \end{pmatrix}$$

$$\mathbf{I}.\mathbf{A} = \begin{pmatrix} 1 & 0 \\ 0 & 1 \end{pmatrix}\begin{pmatrix} 2 & 1 \\ 2 & 7 \end{pmatrix} = \begin{pmatrix} 2 & 1 \\ 2 & 7 \end{pmatrix}$$

Clearly $\mathbf{A}.\mathbf{I} = \mathbf{I}.\mathbf{A} = \mathbf{A}$.

EXAMPLE 7

Given $\mathbf{X} = \begin{pmatrix} 2 & 1 \\ 4 & -1 \end{pmatrix}$ and $\mathbf{I} = \begin{pmatrix} 1 & 0 \\ 0 & 1 \end{pmatrix}$

prove $\mathbf{X}^3 + 7\mathbf{I} = \begin{pmatrix} 27 & 7 \\ 28 & 6 \end{pmatrix}$.

Solution 7

$$\mathbf{X}^2 = \begin{pmatrix} 2 & 1 \\ 4 & -1 \end{pmatrix}\begin{pmatrix} 2 & 1 \\ 4 & -1 \end{pmatrix} = \begin{pmatrix} 8 & 1 \\ 4 & 5 \end{pmatrix}$$

$$\mathbf{X}^3 = \mathbf{X}^2.\mathbf{X} = \begin{pmatrix} 8 & 1 \\ 4 & 5 \end{pmatrix}\begin{pmatrix} 2 & 1 \\ 4 & -1 \end{pmatrix} = \begin{pmatrix} 20 & 7 \\ 28 & -1 \end{pmatrix}$$

Also $7\mathbf{I} = 7\begin{pmatrix} 1 & 0 \\ 0 & 1 \end{pmatrix} = \begin{pmatrix} 7 & 0 \\ 0 & 7 \end{pmatrix}$

So $\mathbf{X}^3 + 7\mathbf{I} = \begin{pmatrix} 20 & 7 \\ 28 & -1 \end{pmatrix} + \begin{pmatrix} 7 & 0 \\ 0 & 7 \end{pmatrix} = \begin{pmatrix} 27 & 7 \\ 28 & 6 \end{pmatrix}$

EXAMPLE 8

Solve

$$\begin{pmatrix} 2 & -1 \\ 1 & 5 \end{pmatrix}\begin{pmatrix} x \\ y \end{pmatrix} = \begin{pmatrix} 7 \\ -2 \end{pmatrix}.$$

Solution 8

Consider the product

$$\begin{pmatrix} 2 & -1 \\ 1 & 5 \end{pmatrix}\begin{pmatrix} x \\ y \end{pmatrix}$$

Examining the orders of the two matrices

$$\underbrace{(2 \times 2).(2 \times 1)}_{\text{compatible}}$$

(2 × 2). (2 × 1)
↓ ↓
compatible

Also

$$\begin{pmatrix} 2 & -1 \\ 1 & 5 \end{pmatrix}\begin{pmatrix} x \\ y \end{pmatrix} = \begin{pmatrix} ? \\ ? \end{pmatrix}$$

The element in row 1 and column 1 of the product is $(2)(x) + (-1)(y) = 2x - y$.

The element in row 2 and column 1 is $(1)(x) + (5)(y) = x + 5y$.

Hence

$$\begin{pmatrix} 2 & -1 \\ 1 & 5 \end{pmatrix}\begin{pmatrix} x \\ y \end{pmatrix} = \begin{pmatrix} 2x - y \\ x + 5y \end{pmatrix}$$

It follows that

$$\begin{pmatrix} 2x - y \\ x + 5y \end{pmatrix} = \begin{pmatrix} 7 \\ -2 \end{pmatrix}$$

leading to the simultaneous equations

$$2x - y = 7$$
$$x + 5y = -2$$

which solve as $x = 3$, $y = -1$.

EXERCISE 5.3

1. Simplify the following.

(a) $2\begin{pmatrix} 5 & 3 \\ 0 & 13 \end{pmatrix}$ (b) $3\begin{pmatrix} 2 \\ 5 \end{pmatrix}$ (c) $2\begin{pmatrix} 1 & -2 & 0 \\ 3 & -5 & 2 \end{pmatrix}$

(d) $3\begin{pmatrix} 9 & 5 \\ 2 & 8 \end{pmatrix}$ (e) $-\frac{1}{2}\begin{pmatrix} 6 \\ -4 \\ 0 \end{pmatrix}$ (f) $-4(1 \quad 2 \quad 3)$

(g) $3\begin{pmatrix} 2 & 1 & -3 \\ 5 & 4 & 0 \end{pmatrix}$ (h) $-2\begin{pmatrix} 1 & -3 & 2 \\ -2 & 3 & 4 \end{pmatrix}$ (i) $4\begin{pmatrix} a & -b & 3c \\ 2a & -7b & 5c \end{pmatrix}$

2. $\mathbf{A} = \begin{pmatrix} 2 & -1 & 3 \\ 5 & 0 & -3 \end{pmatrix}$ and $\mathbf{B} = \begin{pmatrix} 3 & 1 & -2 \\ 1 & 4 & 0 \end{pmatrix}$

Find each of these in its simplest form.

(a) $\mathbf{A} + \mathbf{B}$ (b) $3(\mathbf{A} + \mathbf{B})$ (c) $3\mathbf{A}$ (d) $3\mathbf{B}$ (e) $3\mathbf{A} + 3\mathbf{B}$ (f) $6\mathbf{A}$ (g) $2(3\mathbf{A})$

(h) $9\mathbf{B}$ (i) $3(3\mathbf{B})$

3. Copy and complete the entries in each of the following.

(a) $\begin{pmatrix} 6 & 9 \\ 12 & 3 \end{pmatrix} = 3\begin{pmatrix} ? & ? \\ ? & ? \end{pmatrix}$ (b) $\begin{pmatrix} 6 & -2 \\ -8 & 0 \end{pmatrix} = 2\begin{pmatrix} ? & ? \\ ? & ? \end{pmatrix}$

(c) $\begin{pmatrix} -4 & 8 & 0 \\ 12 & 0 & -8 \end{pmatrix} = -4\begin{pmatrix} ? & ? & ? \\ ? & ? & ? \end{pmatrix}$ (d) $\begin{pmatrix} 2 & -2 & 0 \\ 6 & 1 & 4 \end{pmatrix} = \frac{1}{2}\begin{pmatrix} ? & ? & ? \\ ? & ? & ? \end{pmatrix}$

4. $\mathbf{X} = \begin{pmatrix} 3 & 1 \\ -4 & 2 \end{pmatrix}$ and $\mathbf{Y} = \begin{pmatrix} -5 & 10 \\ 8 & 9 \end{pmatrix}$

Simplify

(a) $2\mathbf{X}$ (b) $3\mathbf{Y}$ (c) $4\mathbf{X}$ (d) $5\mathbf{Y}$ (e) $\mathbf{X}+\mathbf{Y}$ (f) $3(\mathbf{X}+\mathbf{Y})$

(g) $\mathbf{X}-\mathbf{Y}$ (h) $2(\mathbf{X}-\mathbf{Y})$ (i) $3\mathbf{X}+3\mathbf{Y}$ (j) $2\mathbf{X}-2\mathbf{Y}$ (k) $5\mathbf{X}+2\mathbf{Y}$ (l) $6\mathbf{X}-\mathbf{Y}$

5. Simplify

(a) $4\begin{pmatrix} 3 & 4 & -1 \\ -2 & 1 & 0 \end{pmatrix} + 3\begin{pmatrix} 1 & -1 & 3 \\ 2 & -7 & 5 \end{pmatrix}$ (b) $4\begin{pmatrix} 2 & -2 & 0 \\ 6 & 1 & 4 \end{pmatrix} - 2\begin{pmatrix} 1 & -2 & 0 \\ 5 & -2 & 3 \end{pmatrix}$

6. Solve each of the following equations for the 2×2 matrix $\mathbf{A}$.

(a) $2\mathbf{A} = \begin{pmatrix} 4 & 6 \\ 8 & 2 \end{pmatrix}$ (b) $3\mathbf{A} + \begin{pmatrix} 3 & 15 \\ 2 & 5 \end{pmatrix} = \begin{pmatrix} 12 & 3 \\ 8 & 2 \end{pmatrix}$

(c) $\begin{pmatrix} 2 & -4 \\ -4 & 2 \end{pmatrix} + 5\mathbf{A} = \begin{pmatrix} 17 & 16 \\ 6 & 12 \end{pmatrix}$ (d) $\begin{pmatrix} 3 & 6 \\ 15 & 1 \end{pmatrix} - 2\mathbf{A} = \begin{pmatrix} 9 & 2 \\ 3 & -3 \end{pmatrix}$

7. Find the matrix $\mathbf{A}$ in each of the following.

(a) $2\begin{pmatrix} 6 \\ 2 \\ 0 \end{pmatrix} + \mathbf{A} = 5\begin{pmatrix} 2 \\ 5 \\ -1 \end{pmatrix}$ (b) $3\begin{pmatrix} 2 & -2 & 2 \\ 4 & 6 & 0 \end{pmatrix} - 2\mathbf{A} = 4\begin{pmatrix} 1 & 2 & 1 \\ 2 & -1 & 3 \end{pmatrix}$

8. Given that

$$3\begin{pmatrix} a & b \\ c & d \end{pmatrix} + \begin{pmatrix} -7 & 1 \\ 3 & 15 \end{pmatrix} = \begin{pmatrix} 5 & 22 \\ 18 & 3 \end{pmatrix}$$

find the entries a, b, c, d.

9. Find the following matrix products.

(a) $(2 \quad 5)\begin{pmatrix} 4 \\ 1 \end{pmatrix}$ (b) $(4 \quad 6)\begin{pmatrix} 5 \\ 7 \end{pmatrix}$ (c) $(7 \quad -1)\begin{pmatrix} 3 \\ -2 \end{pmatrix}$

(d) $(1 \quad 2 \quad 3)\begin{pmatrix} 4 \\ 5 \\ 6 \end{pmatrix}$ (e) $(2 \quad -4 \quad 3)\begin{pmatrix} 2 \\ 3 \\ 1 \end{pmatrix}$ (f) $(a \quad b \quad c)\begin{pmatrix} d \\ e \\ f \end{pmatrix}$

10. Find x in each of the following.

(a) $\begin{pmatrix} x & 2 \end{pmatrix}\begin{pmatrix} 3 \\ 4 \end{pmatrix} = (11)$ (b) $\begin{pmatrix} 3 & x \end{pmatrix}\begin{pmatrix} 2 \\ 5 \end{pmatrix} = (26)$ (c) $\begin{pmatrix} 5 & x \end{pmatrix}\begin{pmatrix} 1 \\ x \end{pmatrix} = (9$

(d) $\begin{pmatrix} x & 3 \end{pmatrix}\begin{pmatrix} x \\ -4 \end{pmatrix} = (13)$

11. Find each of the following products in its simplest form.

(a) $\begin{pmatrix} 1 & 2 \\ 3 & 4 \end{pmatrix}\begin{pmatrix} 5 \\ 6 \end{pmatrix}$ (b) $\begin{pmatrix} 7 & 0 \\ 1 & 8 \end{pmatrix}\begin{pmatrix} 7 \\ 4 \end{pmatrix}$ (c) $\begin{pmatrix} 0 & 8 \\ 4 & -6 \end{pmatrix}\begin{pmatrix} 6 \\ -5 \end{pmatrix}$

(d) $\begin{pmatrix} -7 & 9 \\ 2 & -4 \end{pmatrix}\begin{pmatrix} 4 \\ 0 \end{pmatrix}$ (e) $\begin{pmatrix} 6 & -6 \\ 2 & 2 \end{pmatrix}\begin{pmatrix} 4 \\ -4 \end{pmatrix}$ (f) $\begin{pmatrix} 6 & 1 \\ 5 & 5 \end{pmatrix}\begin{pmatrix} 2 \\ 9 \end{pmatrix}$

(g) $\begin{pmatrix} 1 & 0 \\ 1 & 0 \end{pmatrix}\begin{pmatrix} 7 \\ -2 \end{pmatrix}$ (h) $\begin{pmatrix} 0 & 1 \\ -1 & 0 \end{pmatrix}\begin{pmatrix} 2a \\ -2a \end{pmatrix}$ (i) $\begin{pmatrix} 2 & 3 \\ -3 & 1 \end{pmatrix}\begin{pmatrix} a \\ -3a \end{pmatrix}$

12. $\mathbf{X} = \begin{pmatrix} -1 & 2 \\ 3 & -4 \end{pmatrix}$ $\mathbf{Y} = \begin{pmatrix} -2 \\ 1 \end{pmatrix}$ $\mathbf{Z} = \begin{pmatrix} 5 & 7 \end{pmatrix}$

Which of the products **XY**, **YX**, **YZ**, **ZY**, **XZ**, **ZX** are possible?

Simplify those products which exist.

13. Carry out matrix multiplication, where possible, for each of the following.

(a) $\begin{pmatrix} 2 & 1 \\ 1 & 2 \end{pmatrix}\begin{pmatrix} 3 \\ 1 \end{pmatrix}$ (b) $\begin{pmatrix} 3 \\ 1 \end{pmatrix}\begin{pmatrix} 2 & 1 \\ 1 & 2 \end{pmatrix}$ (c) $\begin{pmatrix} 7 & 5 \end{pmatrix}\begin{pmatrix} 2 \\ 3 \end{pmatrix}$

(d) $\begin{pmatrix} 1 & 2 & 3 \\ 4 & 5 & 6 \\ 7 & 8 & 9 \end{pmatrix}\begin{pmatrix} 1 \\ 2 \\ 3 \end{pmatrix}$ (e) $\begin{pmatrix} 2 & -1 \\ 0 & 3 \\ 1 & 2 \end{pmatrix}\begin{pmatrix} 4 \\ 5 \\ 2 \end{pmatrix}$ (f) $\begin{pmatrix} 5 & 3 & 2 \\ 1 & -2 & 0 \\ 2 & 4 & 1 \end{pmatrix}\begin{pmatrix} 1 \\ 0 \\ 4 \end{pmatrix}$

(g) $\begin{pmatrix} 7 & 1 \\ 3 & 2 \end{pmatrix}\begin{pmatrix} 3 & 2 \end{pmatrix}$ (h) $\begin{pmatrix} 3 & 2 \end{pmatrix}\begin{pmatrix} 7 & 1 \\ 3 & 2 \end{pmatrix}$

14. In each of the following, find a system of equations in x and y. Hence find x and y.

(a) $\begin{pmatrix} 1 & -1 \\ 1 & 1 \end{pmatrix}\begin{pmatrix} x \\ y \end{pmatrix} = \begin{pmatrix} 5 \\ 11 \end{pmatrix}$ (b) $\begin{pmatrix} 1 & -1 \\ 1 & 1 \end{pmatrix}\begin{pmatrix} x \\ y \end{pmatrix} = \begin{pmatrix} 0 \\ 8 \end{pmatrix}$ (c) $\begin{pmatrix} 3 & -4 \\ 5 & 1 \end{pmatrix}\begin{pmatrix} x \\ y \end{pmatrix} = \begin{pmatrix} 18 \\ 7 \end{pmatrix}$

(d) $\begin{pmatrix} 5 & 3 \\ 7 & 6 \end{pmatrix}\begin{pmatrix} x \\ -y \end{pmatrix} = \begin{pmatrix} 9 \\ 9 \end{pmatrix}$ (e) $\begin{pmatrix} 2 & -1 \\ 5 & -6 \end{pmatrix}\begin{pmatrix} x \\ y \end{pmatrix} = \begin{pmatrix} 10 \\ -45 \end{pmatrix}$ (f) $\begin{pmatrix} x & y \\ y & -x \end{pmatrix}\begin{pmatrix} 2 \\ 1 \end{pmatrix} = \begin{pmatrix} 11 \\ 2 \end{pmatrix}$

15. Find the following products in their simplest form.

(a) $\begin{pmatrix} 2 & 9 \\ 3 & 5 \end{pmatrix}\begin{pmatrix} 1 & 2 \\ 3 & 2 \end{pmatrix}$ (b) $\begin{pmatrix} 7 & 6 \\ 3 & 4 \end{pmatrix}\begin{pmatrix} 1 & 3 \\ 2 & 2 \end{pmatrix}$ (c) $\begin{pmatrix} 9 & 4 \\ 5 & 2 \end{pmatrix}\begin{pmatrix} 3 & 1 \\ 1 & 4 \end{pmatrix}$

(d) $\begin{pmatrix} 1 & 2 \\ 3 & 4 \end{pmatrix}\begin{pmatrix} 2 & 1 \\ 4 & 3 \end{pmatrix}$ (e) $\begin{pmatrix} 2 & 0 \\ -1 & 4 \end{pmatrix}\begin{pmatrix} 2 & 5 & 3 \\ 4 & 2 & -1 \end{pmatrix}$

(f) $\begin{pmatrix} 1 & 0 \\ -1 & 0 \end{pmatrix}\begin{pmatrix} 7 & 3 & -5 & 2 \\ 2 & -1 & 0 & 3 \end{pmatrix}$

16. Find the following products.

(a) $\begin{pmatrix} 2 & -3 \\ 5 & 6 \end{pmatrix}\begin{pmatrix} 1 & 0 \\ 0 & 1 \end{pmatrix}$ (b) $\begin{pmatrix} 1 & 0 \\ 0 & 1 \end{pmatrix}\begin{pmatrix} 7 & -2 \\ 1 & 5 \end{pmatrix}$ (c) $\begin{pmatrix} p & q \\ r & s \end{pmatrix}\begin{pmatrix} 1 & 0 \\ 0 & 1 \end{pmatrix}$

(d) $\begin{pmatrix} 1 & 0 \\ 0 & 1 \end{pmatrix}\begin{pmatrix} p & q \\ r & s \end{pmatrix}$ (e) $\begin{pmatrix} 1 & 0 \\ 0 & 1 \end{pmatrix}\begin{pmatrix} 1 & 0 \\ 0 & 1 \end{pmatrix}$ (f) $\begin{pmatrix} 1 & 0 \\ 0 & 1 \end{pmatrix}\begin{pmatrix} 1 & 1 \\ 1 & 1 \end{pmatrix}$

17. Given that $\mathbf{X} = \begin{pmatrix} 2 & -1 \\ 0 & 3 \end{pmatrix}$, find the matrices $\mathbf{X}^2$ and $\mathbf{X}^3$.

18. If $\begin{pmatrix} 2 & 5 \\ 1 & 3 \end{pmatrix}\begin{pmatrix} p & q \\ r & s \end{pmatrix} = \begin{pmatrix} 1 & 0 \\ 0 & 1 \end{pmatrix}$

find p, q, r, s.

19. If $\begin{pmatrix} w & x \\ y & z \end{pmatrix}\begin{pmatrix} 2 & 3 \\ 3 & 5 \end{pmatrix} = \begin{pmatrix} 1 & 0 \\ 0 & 1 \end{pmatrix}$

find w, x, y, z.

20. If $\mathbf{X} = \begin{pmatrix} 3 & 2 \\ 1 & 0 \end{pmatrix}$ $\mathbf{Y} = \begin{pmatrix} -1 & 2 \\ -2 & 1 \end{pmatrix}$ $\mathbf{Z} = \begin{pmatrix} 3 & 4 \\ 5 & -2 \end{pmatrix}$

find each of these in its simplest form.

(a) **XY** (b) **YX** (c) **YZ** (d) **ZY** (e) **XZ** (f) **ZX**

21. Given that $\mathbf{A} = \begin{pmatrix} 1 & 3 \\ 5 & 2 \\ 4 & -1 \end{pmatrix}$ $\mathbf{B} = \begin{pmatrix} 2 & 2 \\ 1 & 3 \\ 5 & 6 \end{pmatrix}$ $\mathbf{C} = \begin{pmatrix} 3 & 4 \\ 1 & 2 \end{pmatrix}$

find each of these.

(a) **AC** (b) **BC** (c) $\mathbf{A} + \mathbf{B}$

22. $\mathbf{P} = \begin{pmatrix} 1 & 2 \\ -2 & 3 \end{pmatrix}$ $\mathbf{Q} = \begin{pmatrix} 3 & -4 \\ 2 & -1 \end{pmatrix}$

Find each of these.

(a) $\mathbf{P} + \mathbf{Q}$ (b) $\mathbf{P} - \mathbf{Q}$ (c) $(\mathbf{P} + \mathbf{Q})(\mathbf{P} - \mathbf{Q})$ (d) $\mathbf{P}^2$ (e) $\mathbf{Q}^2$

Is it true that $(\mathbf{P} + \mathbf{Q})(\mathbf{P} - \mathbf{Q}) = \mathbf{P}^2 - \mathbf{Q}^2$?

23. $\mathbf{P} = \begin{pmatrix} 3 & -1 \\ 2 & -5 \end{pmatrix}$ $\mathbf{Q} = \begin{pmatrix} 3 & 2 \\ 7 & 5 \end{pmatrix}$

Find each of these.

(a) $\mathbf{P} + \mathbf{Q}$ (b) $(\mathbf{P} + \mathbf{Q})^2$ (c) $\mathbf{P}^2$ (d) $2\mathbf{PQ}$ (e) $\mathbf{Q}^2$

Is it true that $(\mathbf{P} + \mathbf{Q})^2 = \mathbf{P}^2 + 2\mathbf{PQ} + \mathbf{Q}^2$?

24. $\mathbf{X} = \begin{pmatrix} 2 & -3 \\ -2 & 1 \end{pmatrix}$

Verify that $\mathbf{X}^2 - 3\mathbf{X} - 4\mathbf{I} = \mathbf{0}$, where $\mathbf{I}$ is the unit matrix of order 2.

25. Show that the matrix

$$\mathbf{X} = \begin{pmatrix} -1 & 3 \\ 2 & 1 \end{pmatrix}$$

satisfies the equation $\mathbf{X}^2 - 7\mathbf{I} = \mathbf{0}$

26. If $\mathbf{X} = \begin{pmatrix} 3 & -1 \\ -3 & 2 \end{pmatrix}$,

find a, b such that $\mathbf{X}^2 = a\mathbf{X} + b\mathbf{I}$

27. $\mathbf{A} = \begin{pmatrix} 2 & 1 \\ 3 & 2 \\ 4 & 3 \end{pmatrix}$ $\mathbf{B} = \begin{pmatrix} 1 & -2 & 3 \\ -3 & 2 & 1 \end{pmatrix}$

Find **AB** and **BA**.

28. $\mathbf{A} = \begin{pmatrix} 1 & 2 & 3 \\ -1 & 0 & 1 \\ 2 & 3 & 1 \end{pmatrix}$ $\mathbf{B} = \begin{pmatrix} 0 & 2 & -1 \\ 3 & 2 & 1 \\ 2 & 1 & 2 \end{pmatrix}$

Find **AB** and **BA**.

THE INVERSE OF A (2×2) MATRIX

The **determinant,** Δ, of the matrix

$$\begin{pmatrix} a & b \\ c & d \end{pmatrix}$$

is defined to be $ad - bc$. It follows therefore that Δ for

$$\begin{pmatrix} 2 & 1 \\ 7 & 4 \end{pmatrix}$$

is given by $\Delta = 2 \times 4 - 1 \times 7 = 1$.

When $\mathbf{AB} = \mathbf{BA} = \mathbf{I}$, $\mathbf{B}$ is defined to be the 'inverse' of $\mathbf{A}$ and written $\mathbf{B} = \mathbf{A}^{-1}$.

Consider the matrices

$$\mathbf{A} = \begin{pmatrix} a & b \\ c & d \end{pmatrix} \quad \mathbf{B} = \begin{pmatrix} \dfrac{d}{\Delta} & \dfrac{-b}{\Delta} \\ \dfrac{-c}{\Delta} & \dfrac{a}{\Delta} \end{pmatrix}$$

Now $\mathbf{AB} = \begin{pmatrix} a & b \\ c & d \end{pmatrix} \begin{pmatrix} \frac{d}{\Delta} & \frac{-b}{\Delta} \\ \frac{-c}{\Delta} & \frac{a}{\Delta} \end{pmatrix}$

$$= \begin{pmatrix} \frac{ad - bc}{\Delta} & \frac{-ab + ab}{\Delta} \\ \frac{cd - cd}{\Delta} & \frac{-bc + da}{\Delta} \end{pmatrix}$$

or $\mathbf{AB} = \begin{pmatrix} \frac{\Delta}{\Delta} & \frac{0}{\Delta} \\ \frac{0}{\Delta} & \frac{\Delta}{\Delta} \end{pmatrix} = \begin{pmatrix} 1 & 0 \\ 0 & 1 \end{pmatrix} = \mathbf{I}$

Similarly

$$\mathbf{BA} = \begin{pmatrix} \frac{d}{\Delta} & \frac{-b}{\Delta} \\ \frac{-c}{\Delta} & \frac{a}{\Delta} \end{pmatrix} \begin{pmatrix} a & b \\ c & d \end{pmatrix}$$

$$= \begin{pmatrix} \frac{da - bc}{\Delta} & \frac{db - bd}{\Delta} \\ \frac{-ac + ac}{\Delta} & \frac{-cb + ad}{\Delta} \end{pmatrix}$$

or $\mathbf{BA} = \begin{pmatrix} \frac{\Delta}{\Delta} & \frac{0}{\Delta} \\ \frac{0}{\Delta} & \frac{\Delta}{\Delta} \end{pmatrix} = \begin{pmatrix} 1 & 0 \\ 0 & 1 \end{pmatrix} = \mathbf{I}$

Clearly $\mathbf{A}^{-1} = \begin{pmatrix} \frac{d}{\Delta} & \frac{-b}{\Delta} \\ \frac{-c}{\Delta} & \frac{a}{\Delta} \end{pmatrix}$

Applying the rule for multiplying a matrix by a scalar (in reverse)

$$\mathbf{A^{-1} = \frac{1}{\Delta}\begin{pmatrix} \boldsymbol{d} & \boldsymbol{-b} \\ \boldsymbol{-c} & \boldsymbol{a} \end{pmatrix}}$$

Consider the matrix $\mathbf{L} = \begin{pmatrix} 2 & 1 \\ 4 & -2 \end{pmatrix}$

For this matrix $\Delta = 2 \times -2 - 1 \times 4 = -4 - 4 = -8$

Hence $\mathbf{L}^{-1} = -\frac{1}{8}\begin{pmatrix} -2 & -1 \\ -4 & 2 \end{pmatrix} = \frac{1}{8}\begin{pmatrix} 2 & 1 \\ 4 & -2 \end{pmatrix}$

Checking $\mathbf{LL}^{-1} = \begin{pmatrix} 2 & 1 \\ 4 & -2 \end{pmatrix} \times \frac{1}{8}\begin{pmatrix} 2 & 1 \\ 4 & -2 \end{pmatrix}$

$$= \begin{pmatrix} 2 & 1 \\ 4 & -2 \end{pmatrix} \begin{pmatrix} \frac{1}{4} & \frac{1}{8} \\ \frac{1}{2} & -\frac{1}{4} \end{pmatrix} = \begin{pmatrix} 1 & 0 \\ 0 & 1 \end{pmatrix}$$

and

$$\mathbf{L}^{-1}\mathbf{L} = \frac{1}{8}\begin{pmatrix} 2 & 1 \\ 4 & -2 \end{pmatrix} \begin{pmatrix} 2 & 1 \\ 4 & -2 \end{pmatrix} = \frac{1}{8}\begin{pmatrix} 8 & 0 \\ 0 & 8 \end{pmatrix} = \begin{pmatrix} 1 & 0 \\ 0 & 1 \end{pmatrix}$$

Singular matrices

A matrix is said to be singular if its determinant is zero i.e. if $\Delta = 0$.

It follows that it will not be possible to compute the inverse of a singular matrix, as this would involve division by zero.

For example, consider

$$\mathbf{A} = \begin{pmatrix} 4 & -2 \\ 2 & -1 \end{pmatrix}$$

Here $\Delta = 4 \times (-1) - (-2) \times 2 = -4 + 4 = 0$

The inverse of **A** does not exist.

In summary therefore singular matrices have zero determinant and the inverse of a singular matrix cannot be found.

EXAMPLE 9

For the matrices

$$\mathbf{A} = \begin{pmatrix} 2 & 1 \\ 7 & 2 \end{pmatrix} \quad \mathbf{B} = \begin{pmatrix} 0 & -1 \\ 1 & 4 \end{pmatrix}$$

prove that $(\mathbf{AB})^{-1} = \mathbf{B}^{-1}\mathbf{A}^{-1}$

Solution 9

$$\mathbf{AB} = \begin{pmatrix} 2 & 1 \\ 7 & 2 \end{pmatrix} \begin{pmatrix} 0 & -1 \\ 1 & 4 \end{pmatrix} = \begin{pmatrix} 1 & 2 \\ 2 & 1 \end{pmatrix}$$

$(\mathbf{AB})^{-1} = ?$

For **AB** $\quad \Delta = 1 - 4 = -3$

Therefore $\quad (\mathbf{AB})^{-1} = -\frac{1}{3}\begin{pmatrix} 1 & -2 \\ -2 & 1 \end{pmatrix} = \begin{pmatrix} -\frac{1}{3} & \frac{2}{3} \\ \frac{2}{3} & -\frac{1}{3} \end{pmatrix}$

$$\mathbf{B} = \begin{pmatrix} 0 & -1 \\ 1 & 4 \end{pmatrix}$$

hence, for **B** $\quad \Delta = 0 \times 4 - (-1)(1) = 1$

and $\quad \mathbf{B}^{-1} = \frac{1}{1}\begin{pmatrix} 4 & 1 \\ -1 & 0 \end{pmatrix}$

$$\mathbf{A} = \begin{pmatrix} 2 & 1 \\ 7 & 2 \end{pmatrix}$$

and so $\Delta = 4 - 7 = -3$

therefore $\mathbf{A}^{-1} = -\frac{1}{3}\begin{pmatrix} 2 & -1 \\ -7 & 2 \end{pmatrix} = \begin{pmatrix} -\frac{2}{3} & \frac{1}{3} \\ \frac{7}{3} & -\frac{2}{3} \end{pmatrix}$

Finally $\mathbf{B}^{-1}\mathbf{A}^{-1} = \begin{pmatrix} 4 & 1 \\ -1 & 0 \end{pmatrix}\begin{pmatrix} -\frac{2}{3} & \frac{1}{3} \\ \frac{7}{3} & -\frac{2}{3} \end{pmatrix} = \begin{pmatrix} -\frac{1}{3} & \frac{2}{3} \\ \frac{2}{3} & -\frac{1}{3} \end{pmatrix}$

Clearly $(\mathbf{AB})^{-1} = \mathbf{B}^{-1}\mathbf{A}^{-1}$ as required.

EXAMPLE 10

Solve for the (2 × 2) matrix **X**.

$$\begin{pmatrix} 2 & 1 \\ 4 & 7 \end{pmatrix}\mathbf{X} = \begin{pmatrix} 1 & 2 \\ -3 & 14 \end{pmatrix}$$

Solution 10

It is instructive to find the inverse of

$$\begin{pmatrix} 2 & 1 \\ 4 & 7 \end{pmatrix}$$

for reasons which will be apparent later.

$$\Delta = 2 \times 7 - 1 \times 4 = 10$$

So the inverse of $\begin{pmatrix} 2 & 1 \\ 4 & 7 \end{pmatrix}$ is $\frac{1}{10}\begin{pmatrix} 7 & -1 \\ -4 & 2 \end{pmatrix}$

If the equation $\begin{pmatrix} 2 & 1 \\ 4 & 7 \end{pmatrix}\mathbf{X} = \begin{pmatrix} 1 & 2 \\ -3 & 14 \end{pmatrix}$

is multiplied from the left by the inverse of

$$\begin{pmatrix} 2 & 1 \\ 4 & 7 \end{pmatrix}$$

then

$$\frac{1}{10}\begin{pmatrix} 7 & -1 \\ -4 & 2 \end{pmatrix}\begin{pmatrix} 2 & 1 \\ 4 & 7 \end{pmatrix}\mathbf{X} = \frac{1}{10}\begin{pmatrix} 7 & -1 \\ -4 & 2 \end{pmatrix}\begin{pmatrix} 1 & 2 \\ -3 & 14 \end{pmatrix}$$

hence $\frac{1}{10}\begin{pmatrix} 10 & 0 \\ 0 & 10 \end{pmatrix}\mathbf{X} = \frac{1}{10}\begin{pmatrix} 10 & 0 \\ -10 & 20 \end{pmatrix}$

or $\begin{pmatrix} 1 & 0 \\ 0 & 1 \end{pmatrix}\mathbf{X} = \begin{pmatrix} 1 & 0 \\ -1 & 2 \end{pmatrix}$

or $\mathbf{IX} = \begin{pmatrix} 1 & 0 \\ -1 & 2 \end{pmatrix}$

therefore $\mathbf{X} = \begin{pmatrix} 1 & 0 \\ -1 & 2 \end{pmatrix}$

EXERCISE 5.4

In questions 1–6, show that each matrix is the inverse of the other.

1. $\begin{pmatrix} 1 & 5 \\ 1 & 6 \end{pmatrix}$ and $\begin{pmatrix} 6 & -5 \\ -1 & 1 \end{pmatrix}$

2. $\begin{pmatrix} 8 & 3 \\ 5 & 2 \end{pmatrix}$ and $\begin{pmatrix} 2 & -3 \\ -5 & 8 \end{pmatrix}$

3. $\begin{pmatrix} 1 & 2 \\ 4 & 9 \end{pmatrix}$ and $\begin{pmatrix} 9 & -2 \\ -4 & 1 \end{pmatrix}$

4. $\begin{pmatrix} 5 & 2 \\ 7 & 3 \end{pmatrix}$ and $\begin{pmatrix} 3 & -2 \\ -7 & 5 \end{pmatrix}$

5. $\begin{pmatrix} 8 & -3 \\ -5 & 2 \end{pmatrix}$ and $\begin{pmatrix} 2 & 3 \\ 5 & 8 \end{pmatrix}$

6. $\begin{pmatrix} 4 & -3 \\ -1 & 1 \end{pmatrix}$ and $\begin{pmatrix} 1 & 3 \\ 1 & 4 \end{pmatrix}$

Study the pattern in the entries of the pairs of matrices in questions 1–6. Use this pattern to write down the inverse of each of the matrices in questions 7–14 and confirm your answers by multiplication.

7. $\begin{pmatrix} 3 & 5 \\ 1 & 2 \end{pmatrix}$

8. $\begin{pmatrix} 5 & 4 \\ 6 & 5 \end{pmatrix}$

9. $\begin{pmatrix} 7 & 4 \\ 5 & 3 \end{pmatrix}$

10. $\begin{pmatrix} 5 & -2 \\ -2 & 1 \end{pmatrix}$

11. $\begin{pmatrix} 5 & -9 \\ -1 & 2 \end{pmatrix}$

12. $\begin{pmatrix} 10 & 3 \\ 3 & 1 \end{pmatrix}$

13. $\begin{pmatrix} -1 & -1 \\ 1 & 0 \end{pmatrix}$

14. $\begin{pmatrix} -10 & -11 \\ 11 & 12 \end{pmatrix}$

15. $\mathbf{P} = \begin{pmatrix} 6 & 4 \\ 1 & 2 \end{pmatrix}$ $\mathbf{Q} = \begin{pmatrix} 1 & -2 \\ -\frac{1}{2} & 3 \end{pmatrix}$

    Show that $\mathbf{PQ} = 4\mathbf{I} = \mathbf{QP}$, and hence that $\mathbf{Q} = 4\mathbf{P}^{-1}$.

State whether each of the matrices in questions 16–30 has an inverse. If the inverse exists, find it.

16. $\mathbf{A} = \begin{pmatrix} 7 & 3 \\ 1 & 1 \end{pmatrix}$

17. $\mathbf{A} = \begin{pmatrix} 2 & 2 \\ 2 & 3 \end{pmatrix}$

18. $\mathbf{A} = \begin{pmatrix} 8 & -3 \\ -7 & 2 \end{pmatrix}$

19. $\mathbf{A} = \begin{pmatrix} 4 & 8 \\ 2 & 4 \end{pmatrix}$

20. $\mathbf{A} = \begin{pmatrix} 4 & 3 \\ 9 & 7 \end{pmatrix}$

21. $\mathbf{A} = \begin{pmatrix} 1 & 1 \\ 1 & 2 \end{pmatrix}$

22. $\mathbf{A} = \begin{pmatrix} 6 & 9 \\ 2 & 3 \end{pmatrix}$

23. $\mathbf{A} = \begin{pmatrix} 1 & 5 \\ 4 & 9 \end{pmatrix}$

24. $\mathbf{A} = \begin{pmatrix} 0 & 1 \\ 1 & 1 \end{pmatrix}$

25. $\mathbf{A} = \begin{pmatrix} 3 & -4 \\ 2 & 8 \end{pmatrix}$

26. $\mathbf{A} = \begin{pmatrix} 1 & 9 \\ 2 & 7 \end{pmatrix}$

27. $\mathbf{A} = \begin{pmatrix} 7 & 3 \\ -2 & 2 \end{pmatrix}$

28. $\mathbf{A} = \begin{pmatrix} 4 & -3 \\ -3 & 2 \end{pmatrix}$

29. $\mathbf{A} = \begin{pmatrix} -4 & 5 \\ 2 & -1 \end{pmatrix}$

30. $\mathbf{A} = \begin{pmatrix} -2 & 0 \\ 3 & 1 \end{pmatrix}$

31. If $\mathbf{A} = \begin{pmatrix} 2 & 3 \\ 2 & 4 \end{pmatrix}$ $\mathbf{B} = \begin{pmatrix} 3 & 7 \\ 2 & 5 \end{pmatrix}$

    Calculate

    (a) $\mathbf{AB}$ (b) $\mathbf{BA}$ (c) $\mathbf{A}^{-1}$ (d) $\mathbf{B}^{-1}$ (e) $(\mathbf{AB})^{-1}$ (f) $\mathbf{A}^{-1}\mathbf{B}^{-1}$ (g) $\mathbf{B}^{-1}\mathbf{A}^{-1}$ (h) $(\mathbf{BA})^{-1}$

    List any equalities between pairs of these matrices.

32. $\mathbf{A} = \begin{pmatrix} 4 & 3 \\ 3 & 2 \end{pmatrix}$ $\mathbf{B} = \begin{pmatrix} 2 & -3 \\ -1 & 2 \end{pmatrix}$

Verify that $(\mathbf{AB})^{-1} = \mathbf{B}^{-1}\mathbf{A}^{-1}$.

33. Solve the matrix equation

$$\begin{pmatrix} 2 & 3 \\ 1 & 4 \end{pmatrix} \mathbf{X} = \begin{pmatrix} 3 & 5 \\ 4 & 0 \end{pmatrix}$$

for the 2×2 matrix $\mathbf{X}$ by pre-multiplying each side by the inverse of matrix

$$\begin{pmatrix} 2 & 3 \\ 1 & 4 \end{pmatrix}$$

SOLVING SIMULTANEOUS EQUATIONS BY MATRIX METHODS

Consider the simultaneous equations

$$x - 3y = -24$$
$$5x - 2y = -29$$

Writing each side as a 2×1 matrix

$$\begin{pmatrix} x - 3y \\ 5x - 2y \end{pmatrix} = \begin{pmatrix} -24 \\ -29 \end{pmatrix}$$

It may be seen that the matrix on the left-hand side is the result of the product

$$\begin{pmatrix} 1 & -3 \\ 5 & -2 \end{pmatrix}\begin{pmatrix} x \\ y \end{pmatrix}$$

and so the equations may be recast in a matrix format as

$$\begin{pmatrix} 1 & -3 \\ 5 & -2 \end{pmatrix}\begin{pmatrix} x \\ y \end{pmatrix} = \begin{pmatrix} -24 \\ -29 \end{pmatrix}$$

Now the inverse of $\begin{pmatrix} 1 & -3 \\ 5 & -2 \end{pmatrix}$ is $\frac{1}{13}\begin{pmatrix} -2 & 3 \\ -5 & 1 \end{pmatrix}$

Multiplying from the left (pre-multiplying) by this inverse

$$\frac{1}{13}\begin{pmatrix} -2 & 3 \\ -5 & 1 \end{pmatrix}\begin{pmatrix} 1 & -3 \\ 5 & -2 \end{pmatrix}\begin{pmatrix} x \\ y \end{pmatrix} = \frac{1}{13}\begin{pmatrix} -2 & 3 \\ -5 & 1 \end{pmatrix}\begin{pmatrix} -24 \\ -29 \end{pmatrix}$$

$$\frac{1}{13}\begin{pmatrix} 13 & 0 \\ 0 & 13 \end{pmatrix}\begin{pmatrix} x \\ y \end{pmatrix} = \frac{1}{13}\begin{pmatrix} -39 \\ 91 \end{pmatrix} = \begin{pmatrix} -3 \\ 7 \end{pmatrix}$$

$$\begin{pmatrix} 1 & 0 \\ 0 & 1 \end{pmatrix}\begin{pmatrix} x \\ y \end{pmatrix} = \begin{pmatrix} -3 \\ 7 \end{pmatrix}$$

i.e. $\quad \mathbf{I}\begin{pmatrix} x \\ y \end{pmatrix} = \begin{pmatrix} -3 \\ 7 \end{pmatrix}$ or $\begin{pmatrix} x \\ y \end{pmatrix} = \begin{pmatrix} -3 \\ 7 \end{pmatrix}$

Hence $\quad x = -3$
$\qquad\quad y = 7$

since equal matrices have corresponding elements equal.

EXERCISE 5.5

Find the solution sets of the systems of equations in questions 1–12 by matrix methods.

1. $x + y = 10$, $x - y = 6$ **2.** $x - y = 0$, $x + y = 6$ **3.** $4x + 3y = 34$, $x - 3y = 1$ **4.** $7x + 2y = 19$, $4x + 3y = 22$

5. $3x + 4y = 10$, $2x + 3y = 8$ **6.** $x + 5y = 7$, $3x - y = -11$ **7.** $x + 5y = 9$, $2x + 3y = 11$ **8.** $2x + y = 7$, $5x - 3y = 12$

9. $6x - 5y = 8$, $4x - 3y = 6$ **10.** $2x + y = 5$, $3x - y = 5$ **11.** $x - y = 7$, $3x - y = 31$ **12.** $x + 3y + 7 = 0$, $x + y + 1 = 0$

13. Find the inverse of the matrix $\begin{pmatrix} 2 & 1 \\ 3 & -2 \end{pmatrix}$ and use it to solve the following systems of linear simultaneous equations.

(a) $2x + y = 12$, $3x - 2y = 25$ (b) $4x + 2y - 8 = 0$, $6x - 4y + 2 = 0$ (c) $2x + y = 0$, $3x - 2y = 7$

Let **A**, **B** and **X** be square matrices of the same order such that $\mathbf{AX} = \mathbf{B}$. Then, if the inverse, $\mathbf{A}^{-1}$, of **A** exists, **X** can be found as follows.

$$\mathbf{AX} = \mathbf{B} \Leftrightarrow \mathbf{A}^{-1}\mathbf{AX} = \mathbf{A}^{-1}\mathbf{B} \Leftrightarrow \mathbf{IX} = \mathbf{A}^{-1}\mathbf{B} \Leftrightarrow \mathbf{X} = \mathbf{A}^{-1}\mathbf{B}$$

Use the fact that if $\mathbf{AX} = \mathbf{B}$ and $\mathbf{A}^{-1}$ exists, then $\mathbf{X} = \mathbf{A}^{-1}\mathbf{B}$ to solve the matrix equations in questions 14–17.

14. $\begin{pmatrix} 3 & 2 \\ 1 & 1 \end{pmatrix}\mathbf{X} = \begin{pmatrix} 0 & 2 \\ 2 & 7 \end{pmatrix}$ **15.** $\begin{pmatrix} 5 & -7 \\ -2 & 3 \end{pmatrix}\mathbf{X} = \begin{pmatrix} 4 & 3 \\ 5 & 2 \end{pmatrix}$ **16.** $\begin{pmatrix} 1 & 4 \\ 3 & 6 \end{pmatrix}\mathbf{X} = \begin{pmatrix} 2 & -3 \\ 4 & -5 \end{pmatrix}$

17. $\begin{pmatrix} -5 & 3 \\ -7 & 6 \end{pmatrix}\mathbf{X} = \begin{pmatrix} 1 & 2 \\ -4 & -3 \end{pmatrix}$

TRANSFORMATION OF THE PLANE USING MATRICES

Area considerations

The square OABC where O is (0,0), A is (1,0), B is (1,1) and C is (0,1) is transformed by matrix

$$\mathbf{L} = \begin{pmatrix} a & b \\ c & d \end{pmatrix}$$

to the quadrilateral OA′B′C′, as shown in the Figure 5.1.

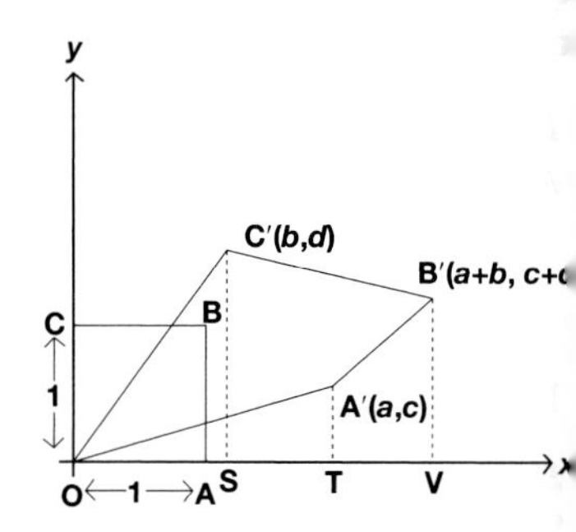

Figure 5.1

It is instructive to ask: how is the area of the image (OA′B′C′) related to the area of the object figure (OABC)? The coordinates of each vertex are multiplied by **L** to give the vertices of the transformed figure.

$$\begin{pmatrix} a & b \\ c & d \end{pmatrix}\begin{pmatrix} 0 \\ 0 \end{pmatrix} = \begin{pmatrix} 0 \\ 0 \end{pmatrix} \qquad \text{i.e. O is mapped to O}$$

$$\begin{pmatrix} a & b \\ c & d \end{pmatrix}\begin{pmatrix} 1 \\ 0 \end{pmatrix} = \begin{pmatrix} a \\ c \end{pmatrix} \qquad \text{i.e. A is mapped to A}'$$

$$\begin{pmatrix} a & b \\ c & d \end{pmatrix}\begin{pmatrix} 1 \\ 1 \end{pmatrix} = \begin{pmatrix} a+b \\ c+d \end{pmatrix} \qquad \text{i.e. B is mapped to B}'$$

$$\begin{pmatrix} a & b \\ c & d \end{pmatrix}\begin{pmatrix} 0 \\ 1 \end{pmatrix} = \begin{pmatrix} b \\ d \end{pmatrix} \qquad \text{i.e. C is mapped to C}'$$

The area of square OABC is, of course, 1 square unit.

In order to find the area of OA′B′C′, perpendiculars from A′, B′ and C′ are dropped to the x axis, meeting that axis at T, V and S respectively.

Now

$$\text{area OA}'\text{B}'\text{C}' = \text{area OC}'\text{S} + \text{area SC}'\text{B}'\text{V} - \text{area OA}'\text{T} - \text{area TA}'\text{B}'\text{V}$$

or

$$\begin{aligned} \text{area OA}'\text{B}'\text{C}' &= \tfrac{1}{2}bd + \tfrac{1}{2}(c+2d)a - \tfrac{1}{2}ac - \tfrac{1}{2}(2c+d)b \\ &= \tfrac{1}{2}bd + \tfrac{1}{2}ac + ad - \tfrac{1}{2}ac - cb - \tfrac{1}{2}bd \\ &= ad - bc \\ &= \Delta, \text{ the determinant of } \mathbf{L} \end{aligned}$$

It follows that

the area of the transformed (image) figure
$= \Delta \times$ the area of the object figure

since the area of that figure is unity. This result is often stated in the form: **the area scale factor associated with a given transformation is equal to the determinant of the matrix representing the transformation**.

EXAMPLE 11

A plane figure has area $4\,\text{cm}^2$. It is transformed by the matrix

$$\begin{pmatrix} 12 & 1 \\ 7 & 4 \end{pmatrix}$$

Find the area of the image figure.

Solution 11

Since $\Delta = 12 \times 4 - 7 \times 1 = 41$
then the area of image figure is $41 \times 4 = 164\,\text{cm}^2$.

EXERCISE 5.6

1. Find the determinant, Δ, of each of the following matrices.

(a) $\begin{pmatrix} 4 & 0 \\ 3 & 2 \end{pmatrix}$ (b) $\begin{pmatrix} 2 & 3 \\ 1 & 5 \end{pmatrix}$ (c) $\begin{pmatrix} 3 & 5 \\ -2 & 0 \end{pmatrix}$ (d) $\begin{pmatrix} 3 & 1 \\ 5 & 4 \end{pmatrix}$

(e) $\begin{pmatrix} 3 & 5 \\ 5 & 9 \end{pmatrix}$ (f) $\begin{pmatrix} 2 & -2 \\ 3 & 3 \end{pmatrix}$ (g) $\begin{pmatrix} 7 & 2 \\ 0 & 1 \end{pmatrix}$ (h) $\begin{pmatrix} 2 & 0 \\ -4 & 4 \end{pmatrix}$

(i) $\begin{pmatrix} 5 & 5 \\ -3 & 1 \end{pmatrix}$ (j) $\begin{pmatrix} 4 & 0 \\ 0 & 4 \end{pmatrix}$ (k) $\begin{pmatrix} 0 & 1 \\ -2 & 4 \end{pmatrix}$ (l) $\begin{pmatrix} -3 & 5 \\ -2 & 1 \end{pmatrix}$

(m) $\begin{pmatrix} 2 & -1 \\ 5 & -2 \end{pmatrix}$ (n) $\begin{pmatrix} -2 & 4 \\ -4 & 1 \end{pmatrix}$ (o) $\begin{pmatrix} 3 & 8 \\ -2 & -3 \end{pmatrix}$ (p) $\begin{pmatrix} -3 & 1 \\ 0 & -2 \end{pmatrix}$

(q) $\begin{pmatrix} -7 & 5 \\ 5 & -3 \end{pmatrix}$ (r) $\begin{pmatrix} -2 & 3 \\ -1 & -7 \end{pmatrix}$ (s) $\begin{pmatrix} 2 & -5 \\ -3 & 7 \end{pmatrix}$ (t) $\begin{pmatrix} 4 & -3 \\ -2 & 5 \end{pmatrix}$

(u) $\begin{pmatrix} -3 & 6 \\ -4 & 8 \end{pmatrix}$

2. An object plane figure has an area of $2\,\text{cm}^2$ and it is transformed using the matrix

$$\begin{pmatrix} 7 & 1 \\ 8 & 2 \end{pmatrix}$$

Find
(a) the area scale factor of the transformation,
(b) the area of the image.

3. An object plane figure with an area of $4\,\text{cm}^2$ is transformed using the matrix

$$\begin{pmatrix} 8 & 3 \\ 7 & 3 \end{pmatrix}$$

Find
(a) the area scale factor of the transformation,
(b) the area of the image.

4. The matrix

$$\begin{pmatrix} 4 & 2 \\ 2 & 3 \end{pmatrix}$$

transforms an object plane figure of area $3\,\text{cm}^2$.

Find the area of its image.

5. An object plane figure of area $10\,\text{cm}^2$ is transformed by the matrix

$$\begin{pmatrix} 1 & 0 \\ 1 & \frac{1}{2} \end{pmatrix}$$

Find the area of its image.

6. An object plane figure of area $16\,cm^2$ is transformed by the matrix

$$\begin{pmatrix} 0 & -\frac{1}{2} \\ \frac{1}{2} & 0 \end{pmatrix}$$

What is the area of its image?

7. The matrix

$$\begin{pmatrix} 6 & 9 \\ 2 & 4 \end{pmatrix}$$

transforms an object plane figure so that its image has an area of $12\,cm^2$. What is the area of the object plane figure?

8. The matrix

$$\begin{pmatrix} 5 & 5 \\ 2 & 3 \end{pmatrix}$$

transforms an object plane figure so that its image has an area of $20\,cm^2$. What is the area of the object plane figure?

9. If an object plane figure maps on to its image, under a transformation given by the matrix

$$\begin{pmatrix} 5 & 1 \\ 3 & 2 \end{pmatrix}$$

so that the area of the image is $35\,cm^2$, what is the area of the object plane figure?

10. The matrix

$$\begin{pmatrix} \frac{1}{4} & 0 \\ 1 & 1 \end{pmatrix}$$

maps an object plane figure on to an image of area $3\,cm^2$. Find the area of the object plane figure.

11. The matrix

$$\begin{pmatrix} \frac{1}{3} & -\frac{1}{3} \\ 1 & 1 \end{pmatrix}$$

produces an image of area $3\,cm^2$.

What is the area of the object plane figure?

12. A parallelogram has vertices (1,0), (– 1,6), (2,3) and (4, – 3). It is transformed by the matrix

$$\begin{pmatrix} 1 & \frac{1}{3} \\ 1 & 1 \end{pmatrix}$$

(a) Find the positions of the vertices of its image and draw the image on squared paper.
(b) Find the area scale factor of the transformation and hence find the area of the original parallelogram.

13. A parallelogram with vertices (0,3), (− 3,6), (− 6,12), (− 3,9) is transformed by the matrix

$$\begin{pmatrix} 6 & 3 \\ 3 & 3 \end{pmatrix}$$

(a) Calculate the vertices of its image and draw both the parallelogram and its image on the same diagram.

(b) Find the area scale factor of the transformation, and hence find the area of the parallelogram.

Transformations of the plane using a singular matrix

A plane figure, transformed by a singular matrix (i.e. a matrix with $\Delta = 0$) will be 'collapsed' on to a straight line. That is to say, the image of each vertex of the object figure lies on the same straight line.

EXAMPLE 12

The plane figure ABCD with A (2,7), B (1,6), C (5,2) and D (0, − 2) is transformed using the matrix

$$\begin{pmatrix} 2 & 8 \\ 1 & 4 \end{pmatrix}$$

Prove that the image lies on the line $y = \frac{1}{2}x$.

Solution 12

$$\begin{pmatrix} 2 & 8 \\ 1 & 4 \end{pmatrix}\begin{pmatrix} 2 \\ 7 \end{pmatrix} = \begin{pmatrix} 60 \\ 30 \end{pmatrix}$$

$$\begin{pmatrix} 2 & 8 \\ 1 & 4 \end{pmatrix}\begin{pmatrix} 1 \\ 6 \end{pmatrix} = \begin{pmatrix} 50 \\ 25 \end{pmatrix}$$

$$\begin{pmatrix} 2 & 8 \\ 1 & 4 \end{pmatrix}\begin{pmatrix} 5 \\ 2 \end{pmatrix} = \begin{pmatrix} 26 \\ 13 \end{pmatrix}$$

$$\begin{pmatrix} 2 & 8 \\ 1 & 4 \end{pmatrix}\begin{pmatrix} 0 \\ -2 \end{pmatrix} = \begin{pmatrix} -16 \\ -8 \end{pmatrix}$$

All the points (60,30), (50,25), (26,13) and (− 16, − 8) lie on $y = \frac{1}{2}x$.

Since $\Delta = 0$ it should come as no surprise that the image encloses no area.

EXERCISE 5.7

1. A matrix which has a zero determinant is said to be singular. Which of these matrices are singular?

(a) $\begin{pmatrix} 6 & 3 \\ 8 & 4 \end{pmatrix}$ (b) $\begin{pmatrix} 7 & 2 \\ 4 & 3 \end{pmatrix}$ (c) $\begin{pmatrix} 1 & 0 \\ 0 & -1 \end{pmatrix}$ (d) $\begin{pmatrix} 4 & 2 \\ 10 & 5 \end{pmatrix}$

(e) $\begin{pmatrix} 3 & -6 \\ -2 & 4 \end{pmatrix}$ (f) $\begin{pmatrix} -4 & 8 \\ 2 & -4 \end{pmatrix}$ (g) $\begin{pmatrix} 2 & 3 \\ -4 & 6 \end{pmatrix}$ (h) $\begin{pmatrix} 3 & 2 \\ -1 & 0 \end{pmatrix}$

(i) $\begin{pmatrix} -1 & 2 \\ -4 & 8 \end{pmatrix}$ (j) $\begin{pmatrix} 4 & 10 \\ 2 & -5 \end{pmatrix}$ (k) $\begin{pmatrix} -10 & -5 \\ 2 & 1 \end{pmatrix}$ (l) $\begin{pmatrix} -3 & 6 \\ 2 & -4 \end{pmatrix}$

2. The parallelogram with vertices (0,1), (1,1), (1,2) (2,2) is transformed by the matrix

$$\begin{pmatrix} 3 & 1 \\ 9 & 3 \end{pmatrix}$$

(a) Find the vertices of its image. Labelling the x axis from 0 to 10 and the y axis from 0 to 25, draw both the parallelogram and its image on the same diagram.

(b) What is the area of the parallelogram, the area of its image and the area scale factor of the transformation?

(c) Find the determinant of the matrix.

(d) This transformation could be called a collapse. What is the equation of the line on which the image lies?

3. The triangle with vertices (1,0), (1,1) and (2,1) is transformed by the matrix

$$\begin{pmatrix} 4 & 2 \\ 8 & 4 \end{pmatrix}$$

(a) Find the vertices of its image. Draw the triangle and its image on axes as in question 2 above.

(b) What is the area of the triangle, the area of its image and the area scale factor of the transformation?

(c) Find the determinant of the matrix.

(d) What is the equation of the line on which the image lies?

4. The triangle with vertices (1,1), (2,1), (2,2) is transformed by the matrix

$$\begin{pmatrix} 4 & 8 \\ 2 & 4 \end{pmatrix}$$

(a) Labelling the x axis from 0 to 25 and the y axis from 0 to 15, draw both the triangle and its image on the same diagram.

(b) What is the area scale factor of the transformation? What is the determinant of the matrix?

(c) Find the equation of the line on which the image lies.

5. The rectangle with vertices (1,1), (4,1), (4,2), (1,2) is transformed by the matrix

$$\begin{pmatrix} 1 & 1 \\ -3 & -3 \end{pmatrix}$$

(a) Draw both the rectangle and its image onto axes labelled from -5 to 10.

(b) What is the area scale factor of the transformation, and the determinant of the matrix?

(c) Find the equation of the line on which the image lies.

A summary of the standard transformations in matrix format

Transformation	Matrix
Reflection in $y = 0$	$\begin{pmatrix} 1 & 0 \\ 0 & -1 \end{pmatrix}$
Reflection in $x = 0$	$\begin{pmatrix} -1 & 0 \\ 0 & 1 \end{pmatrix}$
Reflection in $y = x$	$\begin{pmatrix} 0 & 1 \\ 1 & 0 \end{pmatrix}$
Reflection in $y = -x$	$\begin{pmatrix} 0 & -1 \\ -1 & 0 \end{pmatrix}$
Rotation of 90° anticlockwise about the origin	$\begin{pmatrix} 0 & -1 \\ 1 & 0 \end{pmatrix}$
Rotation of 180° anticlockwise about the origin	$\begin{pmatrix} -1 & 0 \\ 0 & -1 \end{pmatrix}$
Rotation of 270° anticlockwise about the origin	$\begin{pmatrix} 0 & 1 \\ -1 & 0 \end{pmatrix}$
Rotation of $\alpha°$ anticlockwise about the origin	$\begin{pmatrix} \cos\alpha° & -\sin\alpha° \\ \sin\alpha° & \cos\alpha° \end{pmatrix}$
One-way stretch, scale factor k, with $x = 0$ as the invariant line	$\begin{pmatrix} k & 0 \\ 0 & 1 \end{pmatrix}$
One-way stretch, scale factor k, with $y = 0$ as the invariant line	$\begin{pmatrix} 1 & 0 \\ 0 & k \end{pmatrix}$
Enlargement, scale factor k, with centre the origin	$\begin{pmatrix} k & 0 \\ 0 & k \end{pmatrix}$
Shear, shear factor k, with $x = 0$ as the invariant line	$\begin{pmatrix} 1 & 0 \\ k & 1 \end{pmatrix}$
Shear, shear factor k, with $y = 0$ as the invariant line	$\begin{pmatrix} 1 & k \\ 0 & 1 \end{pmatrix}$

Combining reflections

The single equivalent transformation corresponding to two reflections, one following the other, is

(i) a translation (where the axes of reflection are parallel),
(ii) a rotation (where the axes of reflection are not parallel).

EXAMPLE 13

(i) Determine the transformation matrix **M** which transforms the point (1,0) to the point $(\frac{3}{5}, \frac{4}{5})$, and the point (0,1) to the point $(\frac{4}{5}, -\frac{3}{5})$.

(ii) A is the point $(-1,2)$, B is the point $(0,10)$ and C is the point $(5,5)$. Calculate the coordinates of A′, B′, C′, the images of A, B and C respectively under the transformation **M**. Draw accurately on graph paper a diagram showing the triangles ABC and A′B′C′.

(iii) Describe geometrically the transformation represented by **M**.

(iv) Write down the matrix **N** which represents a reflection in the line $y = x$.

(v) Calculate the matrix product **MN** and describe geometrically the transformation represented by **MN**.

(N. Ireland Additional Mathematics, 1989)

Solution 13

(i) Let the required matrix $\mathbf{M} = \begin{pmatrix} a & b \\ c & d \end{pmatrix}$

So $\begin{pmatrix} a & b \\ c & d \end{pmatrix}\begin{pmatrix} 1 \\ 0 \end{pmatrix} = \begin{pmatrix} \frac{3}{5} \\ \frac{4}{5} \end{pmatrix}$

or $\begin{pmatrix} a \\ c \end{pmatrix} = \begin{pmatrix} \frac{3}{5} \\ \frac{4}{5} \end{pmatrix}$

hence $a = \frac{3}{5}$ $c = \frac{4}{5}$

Also $\begin{pmatrix} a & b \\ c & d \end{pmatrix}\begin{pmatrix} 0 \\ 1 \end{pmatrix} = \begin{pmatrix} \frac{4}{5} \\ -\frac{3}{5} \end{pmatrix}$

$\begin{pmatrix} b \\ d \end{pmatrix} = \begin{pmatrix} \frac{4}{5} \\ -\frac{3}{5} \end{pmatrix}$

or $b = \frac{4}{5}$ $d = -\frac{3}{5}$

It follows that $\mathbf{M} = \begin{pmatrix} \frac{3}{5} & \frac{4}{5} \\ \frac{4}{5} & -\frac{3}{5} \end{pmatrix}$

(ii) $\begin{pmatrix} \frac{3}{5} & \frac{4}{5} \\ \frac{4}{5} & -\frac{3}{5} \end{pmatrix}\begin{pmatrix} -1 \\ 2 \end{pmatrix} = \begin{pmatrix} 1 \\ -2 \end{pmatrix}$

$\begin{pmatrix} \frac{3}{5} & \frac{4}{5} \\ \frac{4}{5} & -\frac{3}{5} \end{pmatrix}\begin{pmatrix} 0 \\ 10 \end{pmatrix} = \begin{pmatrix} 8 \\ -6 \end{pmatrix}$

$\begin{pmatrix} \frac{3}{5} & \frac{4}{5} \\ \frac{4}{5} & -\frac{3}{5} \end{pmatrix}\begin{pmatrix} 5 \\ 5 \end{pmatrix} = \begin{pmatrix} 7 \\ 1 \end{pmatrix}$

It follows that the coordinates of A′, B′ and C′ are respectively $(1, -2)$, $(8, -6)$ and $(7,1)$.

The triangle ABC and its image A′B′C′ are shown in Figure 5.2.

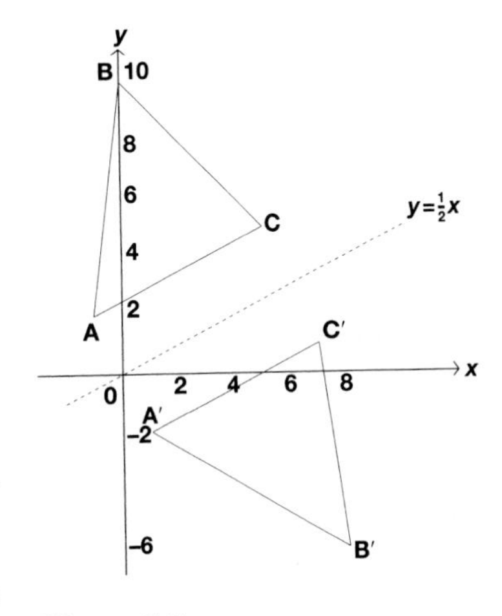

Figure 5.2

(iii) From Figure 5.2 it can be seen that the transformation matrix **M** reflects in the line $y = \frac{1}{2}x$.

(iv) $\mathbf{N} = \begin{pmatrix} 0 & 1 \\ 1 & 0 \end{pmatrix}$

(v) $\mathbf{MN} = \begin{pmatrix} \frac{3}{5} & \frac{4}{5} \\ \frac{4}{5} & -\frac{3}{5} \end{pmatrix}\begin{pmatrix} 0 & 1 \\ 1 & 0 \end{pmatrix} = \begin{pmatrix} \frac{4}{5} & \frac{3}{5} \\ -\frac{3}{5} & \frac{4}{5} \end{pmatrix}$

Since a reflection followed by a reflection is equivalent to a rotation, it is instructive to calculate the angle of rotation. Using the matrix (see the list of standard transformation matrices above) associated with a general rotation, it follows that

$$\begin{pmatrix} \cos\alpha & -\sin\alpha \\ \sin\alpha & \cos\alpha \end{pmatrix} = \begin{pmatrix} \frac{4}{5} & +\frac{3}{5} \\ -\frac{3}{5} & \frac{4}{5} \end{pmatrix}$$

Hence $\cos\alpha = \frac{4}{5}$
$\sin\alpha = -\frac{3}{5}$

Therefore $\alpha = 323.1°$ and **MN** represents a rotation of 323.1° anticlockwise or 36.9° clockwise.

EXERCISE 5.8

1. M_1 and M_2 are transformations defined by the matrices $\mathbf{M}_1 = \begin{pmatrix} 0 & 1 \\ -1 & 0 \end{pmatrix}$ and $\mathbf{M}_2 = \begin{pmatrix} 1 & 0 \\ 3 & 1 \end{pmatrix}$

   Determine the images of the vertices of the triangle O (0,0), A (1,3), B (5,3) when M_1M_2 is applied to them, illustrating your results on a diagram. Repeat the process for M_2M_1. Is the effect of M_1M_2 the same as that of M_2M_1?

   Determine the area of triangle OAB. Is the area preserved under either (or both) of the above transformations?

2. The matrix $\mathbf{A} = \begin{pmatrix} 5 & -3 \\ 2 & -2 \end{pmatrix}$ represents the transformation

   $$T: \begin{pmatrix} x \\ y \end{pmatrix} \to \begin{pmatrix} 5 & -3 \\ 2 & -2 \end{pmatrix}\begin{pmatrix} x \\ y \end{pmatrix}$$

   Find

   (a) the coordinates of the points onto which (4,2) and (−4,2) are mapped by the transformation T,

   (b) the inverse of **A**,

   (c) the coordinates of the point which is mapped to (12,8) under the transformation T,

   (d) the matrix $\mathbf{A}^2$.

3. A transformation X is described by the matrix $\begin{pmatrix} 0 & 1 \\ 1 & 0 \end{pmatrix}$ and a translation Y is described by the vector $\begin{pmatrix} 2 \\ -2 \end{pmatrix}$

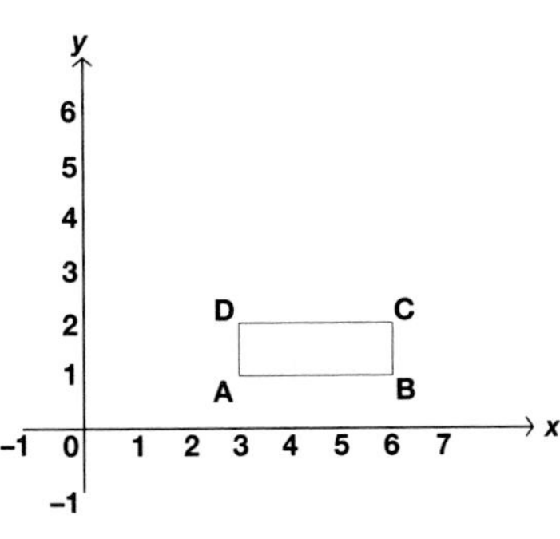

Figure 5.3

(a) Copy Figure 5.3 and show the image of the rectangle ABCD under the transformation X. Label the image A′B′C′D′. Describe this transformation in geometrical terms.

(b) Show the image A′B′C′D′ under the translation Y. Label this image A″B″C″D″. Describe in geometrical terms the single transformation that maps ABCD onto A″B″C″D″.

4. A transformation X is described by the matrix $\begin{pmatrix} 0 & 1 \\ 1 & 0 \end{pmatrix}$ and a translation Y is described by the vector $\begin{pmatrix} 3 \\ -3 \end{pmatrix}$.

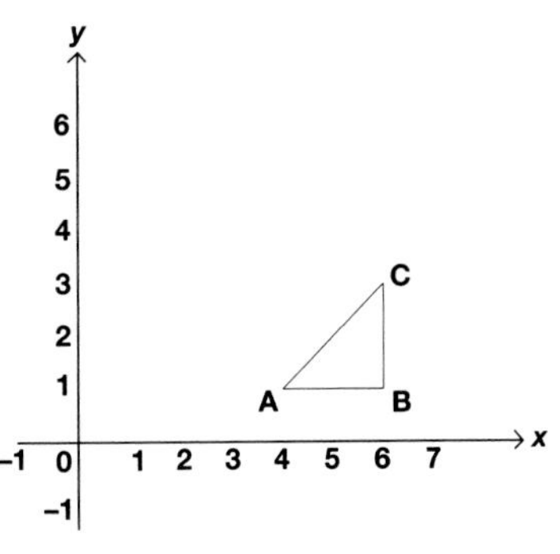

Figure 5.4

(a) Copy Figure 5.4 and show the image of the triangle ABC under the transformation X. Label the image A′B′C′. Describe this transformation in geometrical terms.

(b) Show the image A′B′C′ under the translation Y. Label this image A″B″C″. Describe in geometrical terms the single transformation that maps ABC onto A″B″C″.

5. The vertices of a triangle ABC have coordinates A (2,1), B (3,3) and C (5,2). Triangle ABC is mapped onto triangle A′B′C′ by the transformation represented by the matrix **T** where

$$\mathbf{T} = \begin{pmatrix} 0 & 1 \\ -1 & 0 \end{pmatrix}$$

(a) Find the coordinates of A′, B′ and C′.

(b) Draw and label the triangles ABC and A′B′C′ on graph paper.

(c) Describe the transformation represented by **T**.

6. Copy Figure 5.5.

(a) Write down the matrix representing a 180° rotation about the origin.

(b) Draw the image of the triangle P after it has been rotated through 180° about the origin. Label it Q.

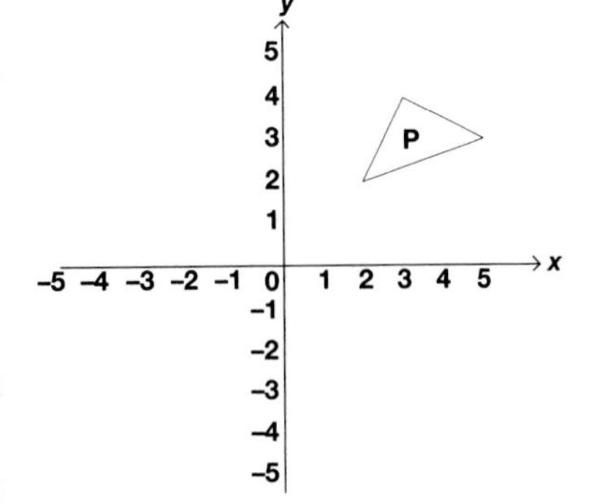

Figure 5.5

(c) Draw the image of the triangle P after the translation described by the vector $\begin{pmatrix} 0 \\ -6 \end{pmatrix}$ and label it R.

(d) Draw the image of the triangle Q after it has been rotated through 180° about the point (0, − 3). Label it S.

(e) What do you notice about the triangles labelled R and S?

7. Copy Figure 5.6.

(a) Write down the matrix representing reflection in the x axis.

(b) Draw the image of the triangle ABC after it is reflected in the x axis. Label it A′B′C′.

(c) Write down the matrix representing a quarter turn anticlockwise rotation about the origin and draw the image of A′B′C′ after a rotation through one quarter of a turn anticlockwise about the origin. Label it A″B″C″.

(d) Describe the single transformation that will map ABC onto A″B″C″ and write down the matrix which represents this transformation.

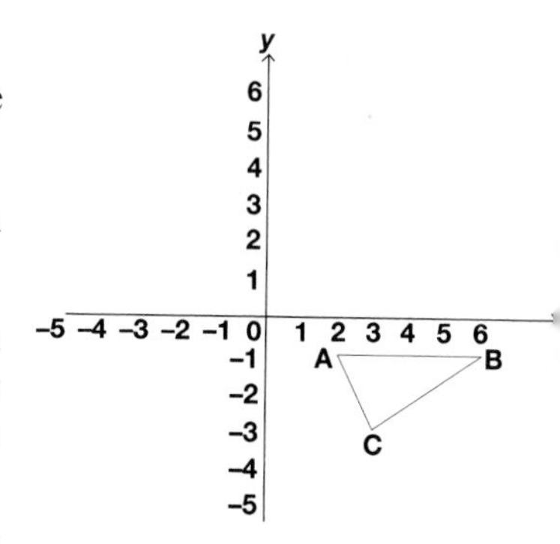

Figure 5.6

8. Copy Figure 5.7.

(a) Write down the matrix which represents reflection in the line $y = x$.

(b) Draw the reflection of the triangle ABC in the line $y = x$, and label it $A_1B_1C_1$.

(c) Draw the reflection of $A_1B_1C_1$ in the x axis and label it $A_2B_2C_2$.

(d) Describe the single transformation that will map ABC on to $A_2B_2C_2$ and confirm your choice of transformation using matrix methods.

(e) Draw the reflection of $A_1B_1C_1$ in the y axis and label it $A_3B_3C_3$.

(f) Describe the single transformation that will map ABC on to $A_3B_3C_3$ and confirm your choice of transformation by matrix methods.

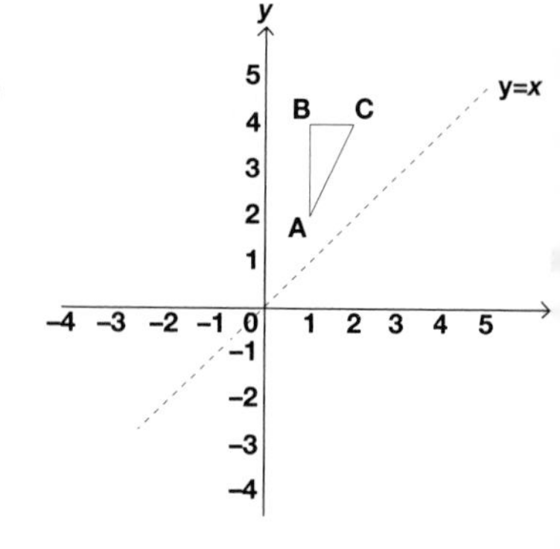

Figure 5.7

9. Copy Figure 5.8.

(a) Write down the matrix representing reflection in the line $y = -x$.

(b) Draw the reflection of the triangle ABC in the line $y = -x$ and label it $A_1B_1C_1$.

(c) Draw the reflection of the triangle $A_1B_1C_1$ in the y axis, and label it $A_2B_2C_2$.

(d) Describe the single transformation which will transform ABC on to $A_2B_2C_2$ and confirm your choice of transformation by matrix methods.

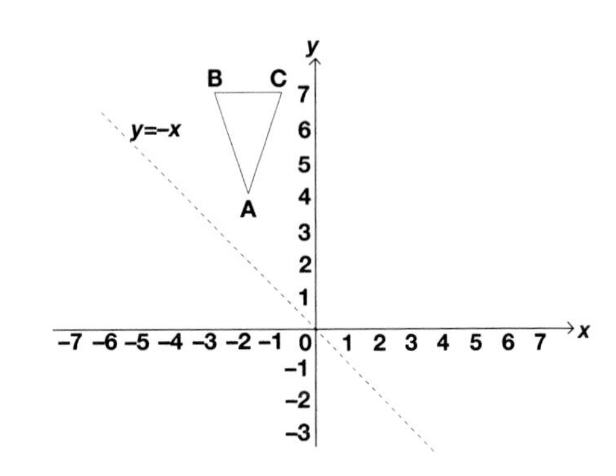

Figure 5.8

10. (a) Find the inverse of the matrix

$$\begin{pmatrix} 2 & 1 \\ 4 & 3 \end{pmatrix}$$

Hence solve the equation

$$\begin{pmatrix} 2 & 1 \\ 4 & 3 \end{pmatrix}\begin{pmatrix} x \\ y \end{pmatrix} = \begin{pmatrix} 4 \\ 6 \end{pmatrix}$$

(b) Find the values of a, b, c and d in the following matrix equation.

$$\begin{pmatrix} 2 & -1 \\ c & 1 \end{pmatrix}\begin{pmatrix} 4 & b \\ a & 0 \end{pmatrix} = \begin{pmatrix} 13 & 4 \\ d & 6 \end{pmatrix}$$

(N. Ireland Additional Mathematics, 1989)

11. The matrices, **A**, **B** and **C** are defined by

$$\mathbf{A} = \begin{pmatrix} 3 & -1 \\ 0 & 2 \end{pmatrix} \quad \mathbf{B} = \begin{pmatrix} 7 & 3 \\ -2 & 0 \end{pmatrix} \quad \mathbf{C} = \begin{pmatrix} -1 & -1 \\ 1 & -1 \end{pmatrix}$$

(i) Find $\mathbf{A}^{-1}$, the inverse of **A**.

(ii) Find the matrix **D** such that $\mathbf{AD} = \mathbf{B}$.

(iii) Calculate the matrix $\mathbf{M} = \mathbf{C} + \mathbf{D}$ and describe fully the geometrical transformation represented by **M**.

(N. Ireland Additional Mathematics, 1990)

12. A triangle OAB has vertices O (0,0), A (3,3) and B (1,3).

The rotation matrix

$$\mathbf{R} = \begin{pmatrix} \frac{1}{\sqrt{2}} & \frac{1}{\sqrt{2}} \\ -\frac{1}{\sqrt{2}} & \frac{1}{\sqrt{2}} \end{pmatrix}$$

maps the points A and B to S and T respectively.

(i) Find the coordinates of S and T.

(ii) Find the area of the triangle OST.

A triangle OPQ is mapped to the triangle OST by the transformation matrix

$$\mathbf{L} = \begin{pmatrix} -2 & 0 \\ 0 & -2 \end{pmatrix}$$

(iii) Find the transformation matrix **M** which maps the triangle OST to the triangle OPQ.

(iv) Hence find the transformation matrix which maps the triangle OAB to the triangle OPQ.

(v) Hence find the area of the triangle OPQ.

(N. Ireland Additional Mathematics, 1990)

13. The matrices **A** and **B** are defined by

$$\mathbf{A} = \begin{pmatrix} 2 & 0 \\ 1 & -3 \end{pmatrix} \quad \mathbf{B} = \begin{pmatrix} \frac{3}{2} & 0 \\ \frac{1}{6} & \frac{2}{3} \end{pmatrix}$$

(i) Find $\mathbf{A}^{-1}$, the inverse of **A**.

(ii) Find the matrix **C** such that $(\mathbf{B} + \mathbf{C})^{-1} = \mathbf{A}$.

(iii) Describe geometrically the transformation represented by **C**.

(N. Ireland Additional Mathematics, 1991)

14. The matrix

$$\mathbf{M} = \begin{pmatrix} a & b \\ c & d \end{pmatrix}$$

maps the square ABCD with vertices A (0,0), B (2,0), C (2,2) and D (0,2) on to the parallelogram A′B′C′D′ with vertices A′ (0,0), B′ (6,2), C′ (8,6) and D′ (2,4).

(i) Find the matrix **M**.

(ii) Using the matrix **M** determine the area of the parallelogram A′B′C′D′.

(iii) Show that the matrix

$$\mathbf{R} = \begin{pmatrix} 0 & 1 \\ -1 & 0 \end{pmatrix}$$

represents a clockwise rotation of 90° about the origin.

(iv) A point Q with coordinates (x, y) is transformed to a point Q′ by the matrix **M**. The point Q′ is then rotated clockwise through 90° about the origin to the point P (5,5). Determine the matrix **Z** which will map P back on to Q and hence find the coordinates of Q.

(N. Ireland Additional Mathematics, 1991)

CHAPTER 6

Integral calculus

The German mathematician **Gottfried Wilhelm Leibniz** (1646–1716) was born in Leipzig, the son of a professor of moral philosophy and graduated from university at seventeen. Leibniz was one of the great polymaths with extensive knowledge of history, linguistics, theology, biology, mathematics, inventing, diplomacy and geology. He declined the chair of law at the University of Altdorf to spend the greater part of his professional life as a highly accomplished lawyer and diplomat. He was an incurable optimist and made attempts to unite Germany and to merge the Catholic and Protestant faiths.

On a diplomatic visit to Paris he encountered Huygens, who interested him in mathematics, an interest that was to become an obsessional 'hobby'. He invented the first 'computer' capable of handling the four arithmetic operations and calculating square roots; Pascal's earlier 'gears' computer could only add and subtract.

The scientific community now acknowledge that Newton's 'Method of Fluxions' and the *Nova Methodus pro Maximis et Minimis, Itemque Tangentibus, qua nec Irrationales Quantitates Moratur* of Leibniz, are reports of entirely independent discoveries of the calculus. Newton discovered calculus in the period 1665–1666 while Leibniz came independently to the same conclusions in the years 1673–1676. However, Leibniz published his findings in 1684–1686 while Newton still hadn't completed the documentation of the fluxion theory by the beginning of the 1700s.

Where the fluxion notation is extremely unwieldy, Leibniz, a master of mathematical symbolism, cast calculus in a form which has survived unchanged down the centuries.

Neither man knew of the other's work until Wallis observed to Newton that Dutch mathematicians considered Leibniz the inventor of the calculus. A wedge was driven between British mathematicians and their Continental colleagues when Nicolas Fatio de Duillier accused Leibniz of plagiarism in 1699. The rather vague report of a specially convened committee of the Royal Society was interpreted as corroborating de Duillier's claim. Mathematical historians believe that Newton, who was a ruthless scheming man, brought pressure to bear on the committee in order that it might find in his favour. This accusation, of having plagiarised the most powerful concept in all mathematics, stayed with Leibniz to his death. Newton added insult to injury by deleting all references to Leibniz from the third edition of *Principia*.

INDEFINITE INTEGRATION

Throughout Chapter 4, questions were encountered in which y was given and $\frac{dy}{dx}$ was to be found. This process was identified as 'differentiation'. The reverse process, **integration**, involves finding y, given $\frac{dy}{dx}$. As an illustration, suppose it is given that $\frac{dy}{dx} = 3x^2$. Familiarity with the differentiation process would indicate that y could be, for instance

$$x^3 + 2 \quad \text{or} \quad x^3 - 7 \quad \text{or} \quad x^3 + 4 \quad \text{or} \quad x^3 + 14 \quad \text{or} \quad x^3 - 72$$

since all of these expressions for y would yield $\frac{dy}{dx} = 3x^2$. There is some measure of uncertainty in finding y; y is said to be known 'up to an arbitrary constant' i.e. the answer is written in the form $y = x^3 +$ 'something'. Mathematicians use a standard notation

$$\frac{dy}{dx} = 3x^2 \Rightarrow y = x^3 + c$$

where c is unknown and cannot be identified without further information. The constant, c, plays an important role in integration and is known as the **constant of integration.**

As has already been mentioned, this 'reverse differentiation' process is known as integration. A symbolism has been developed to deal with integration and attention now turns to that symbolism.

Given

$$\frac{dy}{dx} = 3x^2$$

y is said to be equal to **the integral** of $3x^2$ with the following symbolism

$$y = \int 3x^2 dx$$

The solution of

$$\frac{dy}{dx} = 3x^2$$

has been demonstrated to be $y = x^3 + c$ and therefore

$$y = \int 3x^2 dx = x^3 + c$$

Reversing the differentiation process in the equations $\frac{dy}{dx} = x^n$ $(n = 0,1,2,3,4)$ yields

$$\frac{dy}{dx} = 1 \quad \Rightarrow y = \int 1\,dx = x + c$$

$$\frac{dy}{dx} = x \quad \Rightarrow \quad y = \int x\,dx = \frac{x^2}{2} + c$$

$$\frac{dy}{dx} = x^2 \quad \Rightarrow \quad y = \int x^2\,dx = \frac{x^3}{3} + c$$

$$\frac{dy}{dx} = x^3 \quad \Rightarrow \quad y = \int x^3\,dx = \frac{x^4}{4} + c$$

$$\frac{dy}{dx} = x^4 \quad \Rightarrow \quad y = \int x^4\,dx = \frac{x^5}{5} + c$$

Clearly, the following rule may be conjectured from these results.

$$\int x^n dx = \frac{x^{n+1}}{n+1} + c \qquad (n \neq -1)$$

The general result just proposed can be tested for the five integrals which precede it; all five certainly attest to its validity. The requirement that $n \neq -1$ ensures that the denominator on the right-hand side can never equal zero. Furthermore, reversing the differentiation process in three further cases

$$\int 3x^4 dx = \frac{3x^5}{5} + c$$

$$\int 3x^7 dx = \frac{3x^8}{8} + c$$

$$\int 2x^6 dx = \frac{3x^7}{7} + c$$

The search for a pattern in these results prompts the conjecture

$$\int \boldsymbol{ax^n dx} = \frac{\boldsymbol{ax^{n+1}}}{\boldsymbol{n+1}} + \boldsymbol{c} \qquad (n \neq -1) \tag{1}$$

Equation (1) may be confirmed to be valid for all integrals considered hitherto and is the widely used 'rule' for performing integration. It is instructive to consider examples of its use.

EXAMPLE 1

Find $\int \sqrt{x}\,dx$.

Solution 1

$$\int \sqrt{x}\,dx = \int x^{\frac{1}{2}} dx = \frac{x^{\frac{3}{2}}}{\frac{3}{2}} + c \quad \text{using equation (1) with } a = 1,\ n = \tfrac{1}{2}$$

So $\int \sqrt{x}\,dx = \dfrac{2x^{\frac{3}{2}}}{3} + c$

EXAMPLE 2

Find $\displaystyle\int \frac{2}{x^2} dx$.

Solution 2

$$\int \frac{2}{x^2}\mathrm{d}x = \int 2x^{-2}\mathrm{d}x$$

$$= 2\left(\frac{x^{-1}}{-1}\right) + c \quad \text{using equation (1) with } a = 2 \text{ and } n = -2$$

$$= -2x^{-1} + c = -\frac{2}{x} + c$$

EXAMPLE 3

Find $\int \frac{4}{\sqrt[3]{x}}\mathrm{d}x$.

Solution 3

$$\int \frac{4}{\sqrt[3]{x}}\mathrm{d}x = \int 4x^{-\frac{1}{3}}\mathrm{d}x$$

$$= 4\left(\frac{x^{\frac{2}{3}}}{\frac{2}{3}}\right) + c$$

$$= 4\left(\frac{3x^{\frac{2}{3}}}{2}\right) + c$$

$$= 6x^{\frac{2}{3}} + c$$

EXAMPLE 4

Find $\int (4x^4 + 2x^2 + 3x + 2)\mathrm{d}x$.

Solution 4

$$\int (4x^4 + 2x^2 + 3x + 2)\mathrm{d}x = \frac{4x^5}{5} + \frac{2x^3}{3} + \frac{3x^2}{2} + 2x + c$$

EXAMPLE 5

Find $\int \frac{x+3}{x^4}\mathrm{d}x$.

Solution 5

$$\int \frac{x+3}{x^4}\mathrm{d}x = \int \left(\frac{1}{x^3} + \frac{3}{x^4}\right)\mathrm{d}x$$

$$= \int (x^{-3} + 3x^{-4})\mathrm{d}x$$

$$= \frac{x^{-2}}{-2} + 3\left(\frac{x^{-3}}{-3}\right) + c$$

$$= -\frac{1}{2x^2} - \frac{1}{x^3} + c$$

EXAMPLE 6

Find $\int \frac{x+1}{\sqrt{x}}\mathrm{d}x$.

Solution 6

$$\int \frac{x+1}{\sqrt{x}}\,dx = \int \left(\frac{x}{\sqrt{x}} + \frac{1}{\sqrt{x}}\right)dx$$
$$= \int (x^{\frac{1}{2}} + x^{-\frac{1}{2}})\,dx$$
$$= \frac{2x^{\frac{3}{2}}}{3} + 2x^{\frac{1}{2}} + c$$

EXAMPLE 7

Find $\displaystyle\int \frac{\sqrt{x}+5}{x^2}\,dx$.

Solution 7

$$\int \frac{\sqrt{x}+5}{x^2}\,dx = \int \frac{x^{\frac{1}{2}}+5}{x^2}\,dx$$
$$= \int \left(\frac{x^{\frac{1}{2}}}{x^2} + \frac{5}{x^2}\right)dx$$
$$= \int (x^{-\frac{3}{2}} + 5x^{-2})\,dx$$
$$= \frac{x^{-\frac{1}{2}}}{-\frac{1}{2}} + 5\left(\frac{x^{-1}}{-1}\right) + c$$
$$= -2x^{-\frac{1}{2}} - 5x^{-1} + c$$
$$= -\frac{2}{\sqrt{x}} - \frac{5}{x} + c$$

EXERCISE 6.1

Integrate the following expressions and check your answers using differentiation.

1. (a) $1(=x^{\circ})$ (b) x (c) x^2 (d) x^3 (e) x^6

2. (a) 2 (b) $4x$ (c) $6x^2$ (d) $-8x^3$ (e) $8x^7$

3. (a) x^{-2} (b) x^{-3} (c) $\dfrac{1}{x^4}$ (d) $\dfrac{1}{x^5}$ (e) $-\dfrac{5}{x^6}$

4. (a) $x^{\frac{1}{2}}$ (b) $x^{\frac{1}{3}}$ (c) $x^{\frac{2}{3}}$ (d) $x^{-\frac{1}{2}}$ (e) $\dfrac{1}{2\sqrt{x}}$

Find

5. $\int 3\,dx$

6. $\int x\,dx$

7. $\int (x+3)\,dx$

8. $\int (x^2-2)\,dx$

9. $\int (2-8x^3)\,dx$

10. $\int (9x^2+6x+7)\,dx$

11. $\int (12x^2-5)\,dx$

12. $\int (15x^4+12x^3)\,dx$

13. $\int (x-3)(x+3)\,dx$

14. $\int (x-2)^2\,dx$

15. $\int (3-6x)^2\,dx$

16. $\int x(x-1)(2x+3)\,dx$

17. $\int\left(1+\frac{1}{x^2}\right)\mathrm{d}x$

18. $\int\frac{9x+2x^2}{x}\mathrm{d}x$

19. $\int x(1+x^2)^2\mathrm{d}x$

20. $\int\left(x+\frac{1}{x}\right)^2\mathrm{d}x$

21. $\int(x^{\frac{3}{2}}-x^{\frac{2}{3}})\mathrm{d}x$

22. $\int x^{\frac{1}{2}}(5x+9)\mathrm{d}x$

23. $\int\frac{1+\sqrt{x}}{x^2}\mathrm{d}x$

24. $\int\frac{6x^4-2}{x^2}\mathrm{d}x$

25. $\int x^2\left(1+\frac{2}{x^4}\right)\mathrm{d}x$

26. $\int\left(x^2-\frac{1}{x^2}\right)\mathrm{d}x$

27. $\int\frac{x^4+x^3+\sqrt{x}}{x^2}\mathrm{d}x$

28. $\int\frac{1-9x}{\sqrt{x}}\mathrm{d}x$

29. $\int\left(x-\frac{1}{x}\right)^2\mathrm{d}x$

30. $\int\frac{(x-2)^2}{x^4}\mathrm{d}x$

31. $\int\frac{x^2+1}{\sqrt{x}}\mathrm{d}x$

32. $\int\left(\sqrt{x}+\frac{2}{x}\right)^2\mathrm{d}x$

FINDING y AS A FUNCTION OF x GIVEN $\frac{dy}{dx}$ AND y AT A PARTICULAR VALUE OF x

This section will illustrate, through a series of examples, how the arbitrary constant may be determined provided the value of y is specified at some particular value of x.

EXAMPLE 8

Given

$$\frac{\mathrm{d}y}{\mathrm{d}x}=6x+2 \text{ and } y=7 \text{ when } x=0$$

find y as a function of x.

Solution 8

Since

$$\frac{\mathrm{d}y}{\mathrm{d}x}=6x+2$$

then $y=\int(6x+2)\mathrm{d}x$

or $y=3x^2+2x+c$

But $y=7$ when $x=0$ or

$$7=3(0)^2+2(0)+c\Rightarrow c=7$$

Therefore

$$y=3x^2+2x+7$$

EXAMPLE 9

Given

$$\frac{dy}{dx} = 3x^2 + 8x + 3 \text{ and } y = 5 \text{ when } x = -1$$

Find y as a function of x.

Solution 9

Since

$$\frac{dy}{dx} = 3x^2 + 8x + 3$$

Then $y = \int(3x^2 + 8x + 3)dx$

Or $y = x^3 + 4x^2 + 3x + c$

But $y = 5$ when $x = -1$ or

$$5 = (-1)^3 + 4(-1)^2 + 3(-1) + c \Rightarrow c = 5$$

Therefore

$$y = x^3 + 4x^2 + 3x + 5$$

EXAMPLE 10

Given

$$\frac{dy}{dx} = 2x - \frac{1}{x^2} \text{ and } y = 4 \text{ when } x = 1$$

Find y as a function of x.

Solution 10

Since

$$\frac{dy}{dx} = 2x - \frac{1}{x^2}$$

Then $y = \int\left(2x - \frac{1}{x^2}\right)dx$

Or $y = x^2 + \frac{1}{x} + c$

But $y = 4$ when $x = 1$ or

$$4 = 1^2 + \frac{1}{1} + c \Rightarrow c = 2$$

Therefore

$$y = x^2 + \frac{1}{x} + 2$$

EXAMPLE 11

The gradient function for a curve is $5x^2 - 2$ and the curve passes through (0,2). Find the equation of the curve.

Solution 11

Since the gradient function is identified with $\dfrac{dy}{dx}$

$$\frac{dy}{dx} = 5x^2 - 2$$

$$\Rightarrow \quad y = \int (5x^2 - 2)\,dx$$

$$\text{or} \quad y = \frac{5x^3}{3} - 2x + c.$$

Furthermore, since the curve passes through (0,2)

$$2 = \tfrac{5}{3}(0)^3 - 2(0) + c$$

$$\text{or} \quad 2 = c$$

It follows that

$$y = \frac{5x^3}{3} - 2x + 2$$

EXAMPLE 12

Find y such that y satisfies the following conditions.

(a) $\dfrac{d^2y}{dx^2} = 24x - 4$

(b) $y = 10$ when $x = 1$

(c) $\dfrac{dy}{dx} = 1$ when $x = 0$

Solution 12

Since $\dfrac{d^2y}{dx^2}$ is found by differentiating $\dfrac{dy}{dx}$, then setting $\dfrac{dy}{dx} = y'$ and writing $\dfrac{d^2y}{dx^2} = \dfrac{dy'}{dx}$ results in $\dfrac{dy'}{dx} = 24x - 4$

It follows that

$$y' = \int (24x - 4)\,dx$$

$$\text{i.e.} \quad y' = 12x^2 - 4x + c$$

But when $x = 0$, $y'\left(\text{i.e. } \dfrac{dy}{dx}\right) = 1$ and so

$$1 = 12(0)^2 - 4(0) + c$$

$$\text{or} \quad 1 = c$$

Now $\quad y' = 12x^2 - 4x + 1$

$\Rightarrow \quad \dfrac{dy}{dx} = 12x^2 - 4x + 1$

ntegrating again

$$y = \int(12x^2 - 4x + 1)dx$$

or $\quad y = 4x^3 - 2x^2 + x + c$

But $\quad y = 10$ when $x = 1$

i.e. $\quad 10 = 4(1)^3 - 2(1)^2 + 1 + c$

or $\quad c = 7$

Accordingly

$$y = 4x^3 - 2x^2 + x + 7$$

EXERCISE 6.2

1. Given $\dfrac{dy}{dx} = 6x$ and that $y = 5$ when $x = 2$, find y as a function of x.

2. Given $\dfrac{dy}{dx} = 2x - 3$ and that $y = 2$ when $x = 3$, find y as a function of x.

3. Given $\dfrac{dy}{dx} = 3x^2$ and that $y = 0$ when $x = 1$, find y as a function of x.

4. Given $\dfrac{dy}{dx} = 4x - \dfrac{6}{x^4}$ and that $y = 4$ when $x = 1$, find y as a function of x.

5. Given $\dfrac{dy}{dx} = 2 + \dfrac{2}{\sqrt{x}}$ and that $y = 17$ when $x = 4$, find y as a function of x.

6. Given $\dfrac{dy}{dx} = x(3 - x)$ and that $y = 3$ when $x = 2$, find y as a function of x.

7. In parts (a) to (e) the gradient function for a curve is given together with a point on the curve. The equation of the curve is required in each case.

(a) $9x^2 + 6x + 7 \quad (1,2)$

(b) $\left(x + \dfrac{1}{x}\right)^2 \quad (-3,4)$

(c) $x^{\frac{3}{2}} - x^{\frac{2}{3}} \quad (1,0)$

(d) $x^{-2} + x^{-\frac{3}{2}} \quad (\frac{1}{4},0)$

(e) $x + \dfrac{4}{x^2} + \dfrac{4}{\sqrt{x}} \quad (4,24)$

8. Given $\dfrac{d^2y}{dx^2} = 10$ and also that $\dfrac{dy}{dx} = 17$ when $x = 1$ and $y = 36$ when $x = 2$, find y as a function of x.

9. Given $\frac{d^2y}{dx^2} = 24x + 4$ and also that $\frac{dy}{dx} = 7$ when $x = 0$ and $y = 15$ when $x = 1$, find y as a function of x.

10. Given $\frac{d^2y}{dx^2} = 12x - 8$ and also that $\frac{dy}{dx} = 16$ when $x = -1$ and $y = 1$ when $x = 2$, find y as a function of x.

11. Given $\frac{d^2y}{dx^2} = 48x^2 + 12x - 6$ and also that $\frac{dy}{dx} = 17$ when $x = 0$ and $y = -2$ when $x = 0$, find y as a function of x.

12. Given $\frac{d^2y}{dx^2} = -\frac{3}{4x^{\frac{5}{2}}}$ and also that $\frac{dy}{dx} = \frac{1}{2}$ when $x = 1$ and $y = 1$ when $x = 1$, find y as a function of x.

13. The curve for which $\frac{dy}{dx} = 3x^2 + ax + b$, where a and b are constants, has stationary points a (1,0) and (− 3,32). Find
   (i) the value of a and of b,
   (ii) the equation of the curve.

(U.C.L.E.S. Additional Mathematics, June 1992

DEFINITE INTEGRATION AND THE AREA UNDER A CURVE

In order to introduce some of the notation necessary to establish the relationship between the area under a curve and the process of integration, it is instructive to briefly revisit the first principles of differentiation.

Let δx (pronounced 'delta x') represent a small increase in x, δx being a single entity not to be confused with the product of δ and x. Similarly δy represents a small increase in y. The point S with coordinates $(x + \delta x, y + \delta y)$ is therefore located very close to the point P(x, y) in Figure 6.1.

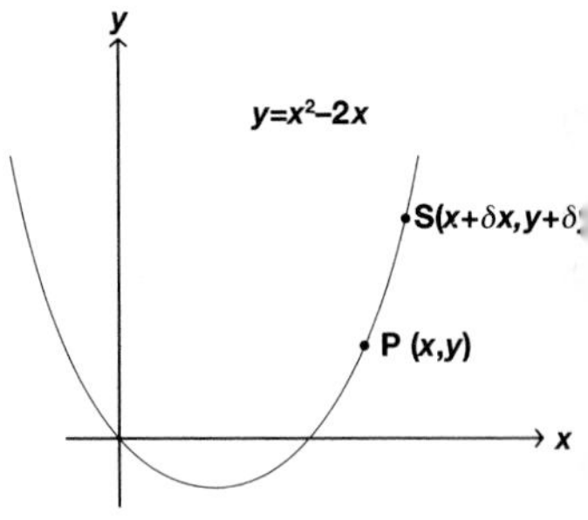

Figure 6.1

In addressing the task of finding $\frac{dy}{dx}$ by first principles, $\frac{dy}{dx}$ was interpreted as the gradient of the tangent to the curve at the point (x, y). The gradient of the tangent was understood as being the limiting value of the gradient of chord PS as S tends towards P i.e.

$$\frac{dy}{dx} = \lim_{\delta x \to 0} \text{grad PS}$$

or
$$\frac{dy}{dx} = \lim_{\delta x \to 0} \frac{y + \delta y - y}{x + \delta x - x}$$

hence $\dfrac{dy}{dx} = \lim_{\delta x \to 0} \dfrac{\delta y}{\delta x}$ (2)

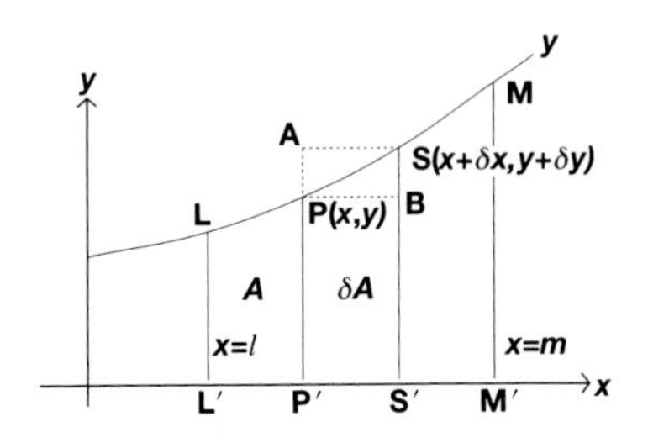

Figure 6.2

Now consider Figure 6.2 in which y is some function of x. The points L, P, S and M lie on the curve and lines through L, P, S and M parallel to the y axis meet the x axis at L′, P′, S′ and M′ respectively. P has coordinates (x, y) and $S(x + \delta x, y + \delta y)$.

The points L′, P′, S′ and M′ have coordinates as follows

L′$(l,0)$, P′$(x,0)$, S′$(x + \delta x,0)$ and M′$(m,0)$

The area enclosed between the curve, LL′, L′P′ and P′P is denoted A and clearly as x increases from $x = l$ to $x = m$, A increases from zero to a value equal to the area enclosed by the curve and the three lines LL′, L′M′ and M′M. It is this latter area, i.e. the area of the figure LMM′L′, which is of primary interest. Since P′S′$(= \delta x)$ is small then the area enclosed by the curve, PP′, P′S′ and S′S may be assumed small and, in keeping with convention, denoted δA. The dotted lines AS and PB are parallel to the x axis and PA is parallel to the y axis.

Now clearly

area PBS′P′ $< \delta A <$ area AP′S′S

or $y\delta x < \delta A < (y + \delta y)\delta x$

Dividing by $\delta x(> 0)$ gives

$$y < \frac{\delta A}{\delta x} < (y + \delta y)$$

As $\delta x \to 0$ then $\dfrac{\delta A}{\delta x}$ becomes $\dfrac{dA}{dx}$ (from equation (2))

and $y + \delta y$ tends to y (as S approaches P)

The final inequality above shows $\dfrac{\delta A}{\delta x}$ to be sandwiched between y and $y + \delta y$ with δy becoming ever smaller. As a result $\dfrac{\delta A}{\delta x}$ becomes sandwiched between y and y in the limit.

i.e. $y < \dfrac{dA}{dx} < y$

or $\dfrac{dA}{dx} = y$

It follows therefore from page 136 that

$$A = \int y\,dx$$

Since y is a function of x, the right-hand side of this equation will become a function of x, say $f(x) + c$.

Hence $\quad A = \mathrm{f}(x) + c$

A is a function of x and so the previous equation may be written more fully as

$$A(x) = \mathrm{f}(x) + c \tag{3}$$

Now $\quad A = 0$ when $x = l$

or $\quad A(l) = 0 = \mathrm{f}(l) + c$

or $\quad c = -\mathrm{f}(l)$

Hence equation (3) may be simplified to

$$A(x) = \mathrm{f}(x) - \mathrm{f}(l) \tag{4}$$

But the area LL′M′M = $A(m)$.

Therefore

$$\text{area LL′M′M} = \mathrm{f}(m) - \mathrm{f}(l) \quad \text{(from equation (4))}$$

But from equation (3)

$$\mathrm{f}(x) = A(x) - c$$

or $\quad$ area LL′M′M $= (A(m) - c) - (A(l) - c)$
$\qquad\qquad = A(m) - A(l)$

However, it has already been established that

$$A = \int y\,\mathrm{d}x$$

therefore

$$A(m) = \int y\,\mathrm{d}x \text{ evaluated at } x = m$$

and $\quad A(l) = \int y\,\mathrm{d}x$ evaluated at $x = l$

Accordingly

area LL′M′M
$\quad = (\int y\,\mathrm{d}x$ evaluated at $x = m) - (\int y\,\mathrm{d}x$ evaluated at $x = l)$

This equation may be rewritten with the right-hand side replaced by the symbolism

$$\int_l^m y\,\mathrm{d}x$$

so that the area of LL′M′M becomes

$$\text{area LL′M′M} = \int_l^m y\,\mathrm{d}x$$

where $\int_l^m y\,\mathrm{d}x$ is known as a **definite limit integral** with

$$\int_l^m y\,\mathrm{d}x = (\int y\,\mathrm{d}x \text{ evaluated at } x = m) - (\int y\,\mathrm{d}x \text{ evaluated at } x = l)$$

It is clear that the area under a curve may be found by appealing to definite limit integration and therefore the remainder of this section is given over to illustrating how such integrals are evaluated.

EXAMPLE 13

Evaluate $\int_{-1}^{3} (x-3)(x+5)\mathrm{d}x$.

Solution 13

$$\int_{-1}^{3} (x-3)(x+5)\mathrm{d}x = \int_{-1}^{3} (x^2+2x-15)\mathrm{d}x$$

$$= \text{the value of } \left(\frac{x^3}{3} + x^2 - 15x + c\right) \text{ at } x = 3$$

$$- \text{the value of } \left(\frac{x^3}{3} + x^2 - 15x + c\right) \text{ at } x = -1$$

$$= (9 + 9 - 45 + c) - (-\tfrac{1}{3} + 1 + 15 + c)$$

$$= -42\tfrac{2}{3}$$

In integral calculus the shorthand notation $\Big[\, y \,\Big]_a^b$ indicates 'the value of the expression y at $x = b$ minus the value of y at $x = a$'.

Therefore

$$\int_{-1}^{3} (x-3)(x+5)\mathrm{d}x$$

$$= \left[\frac{x^3}{3} + x^2 - 15x\right]_{-1}^{3}$$

$$= -42\tfrac{2}{3}$$

EXAMPLE 14

Evaluate $\int_{2}^{6} \frac{12}{x^2}\mathrm{d}x$.

Solution 14

$$\int \frac{12}{x^2}\mathrm{d}x = \int 12x^{-2}\mathrm{d}x = -12x^{-1} + c$$

Therefore

$$\int_{2}^{6} \frac{12}{x^2}\mathrm{d}x = \Big[-12x^{-1}\Big]_{2}^{6} = -2 + 6 = 4$$

Hence

$$\int_{2}^{6} \frac{12}{x^2}\mathrm{d}x = 4$$

EXAMPLE 15

Evaluate $\int_4^9 (\sqrt{x}+2)^2 \mathrm{d}x$.

Solution 15

$$\int(\sqrt{x}+2)^2 \mathrm{d}x = \int(x+4\sqrt{x}+4)\mathrm{d}x$$
$$= \frac{x^2}{2}+\frac{8x^{\frac{3}{2}}}{3}+4x+c$$

Therefore

$$\int_4^9 (\sqrt{x}+2)^2 \mathrm{d}x = \left[\frac{x^2}{2}+\frac{8x^{\frac{3}{2}}}{3}+4x\right]_4^9$$
$$= \left(\frac{81}{2}+72+36\right)-\left(8+\frac{64}{3}+16\right) = 103\tfrac{1}{6}$$

Hence

$$\int_4^9 (\sqrt{x}+2)^2 \mathrm{d}x = 103\tfrac{1}{6}$$

EXAMPLE 16

Evaluate $\int_4^9 \frac{1}{x\sqrt{x}} \mathrm{d}x$.

Solution 16

$$\int \frac{1}{x\sqrt{x}} \mathrm{d}x = \int x^{-1\frac{1}{2}} \mathrm{d}x$$
$$= \frac{x^{-\frac{1}{2}}}{-\frac{1}{2}}+c$$
$$= -2x^{-\frac{1}{2}}+c$$

Therefore

$$\int_4^9 \frac{1}{x\sqrt{x}} \mathrm{d}x = \left[-\frac{2}{\sqrt{x}}\right]_4^9$$
$$= (-\tfrac{2}{3})-(-1) = \tfrac{1}{3}$$

Hence

$$\int_4^9 \frac{1}{x\sqrt{x}} \mathrm{d}x = \tfrac{1}{3}$$

EXAMPLE 17

Evaluate $\int_1^4 \left(\frac{\sqrt{x}}{2}-x\right)(x^2+2)\mathrm{d}x$.

Solution 17

$$\int\left(\frac{\sqrt{x}}{2}-x\right)(x^2+2)\,dx=\int\left(\frac{x^{\frac{5}{2}}}{2}+x^{\frac{1}{2}}-x^3-2x\right)dx=\left(\frac{x^{\frac{7}{2}}}{7}+\frac{2x^{\frac{3}{2}}}{3}-\frac{x^4}{4}-x^2\right)+c$$

Therefore

$$\int_1^4\left(\frac{\sqrt{x}}{2}-x\right)(x^2+2)\,dx=\left[\frac{x^{\frac{7}{2}}}{7}+\frac{2x^{\frac{3}{2}}}{3}-\frac{x^4}{4}-x^2\right]_1^4=\left(\tfrac{128}{7}+\tfrac{16}{3}-64-16\right)-\left(\tfrac{1}{7}+\tfrac{2}{3}-\tfrac{1}{4}-1\right)=-55\tfrac{79}{84}$$

Hence

$$\int_1^4\left(\frac{\sqrt{x}}{2}-x\right)(x^2+2)\,dx=-55\tfrac{79}{84}$$

EXERCISE 6.3

Evaluate the following.

1. $\int_2^4 x\,dx$
2. $\int_{-1}^1 9x^2\,dx$
3. $\int_4^9 \frac{1}{\sqrt{x}}\,dx$
4. $\int_0^1 x^{\frac{3}{2}}\,dx$
5. $\int_{-1}^1 -\frac{6}{x^4}\,dx$
6. $\int_{-2}^2 (x+5)\,dx$
7. $\int_1^2 3(x^2-3)\,dx$
8. $\int_0^1 (x+1)(x+2)\,dx$
9. $\int_{-3}^3 (x^2+2x+4)\,dx$
10. $\int_{-1}^1 (10x^4+9x^2)\,dx$
11. $\int_0^1 (x+7)^2\,dx$
12. $\int_{-1}^1 (1+3x)^2\,dx$
13. $\int_0^1 (7-4x-3x^2)\,dx$
14. $\int_1^2 (1-x)(5-3x)\,dx$
15. $\int_{-1}^1 \left(2x-\frac{1}{x^2}\right)dx$
16. $\int_{-1}^1 \left(x+\frac{2}{x^2}\right)dx$
17. $\int_1^2 2x(2x^2-1)\,dx$
18. $\int_0^1 (3+x^{\frac{1}{2}})^2\,dx$
19. $\int_0^1 \frac{1-9x}{\sqrt{x}}\,dx$
20. $\int_1^2 4x(x+3)(x-3)\,dx$
21. (a) Find

   $$\int(x^{\frac{1}{2}}-x^{-\frac{2}{3}})\,dx$$

   (b) Simplify

   $$\frac{x^3+x^2+2x+2}{x+1}$$

   and hence evaluate

   $$\int_0^1 \frac{(x^3+x^2+2x+2)}{(x+1)}\,dx$$

(N. Ireland Additional Mathematics, 1990)

The area under a curve

Before addressing examples in which the areas under curves are computed, it is instructive to revisit the principal finding of this section, namely, that the area shown in Figure 6.3 may be evaluated through calculating the integral

$$A = \int_l^m y\,\mathrm{d}x$$

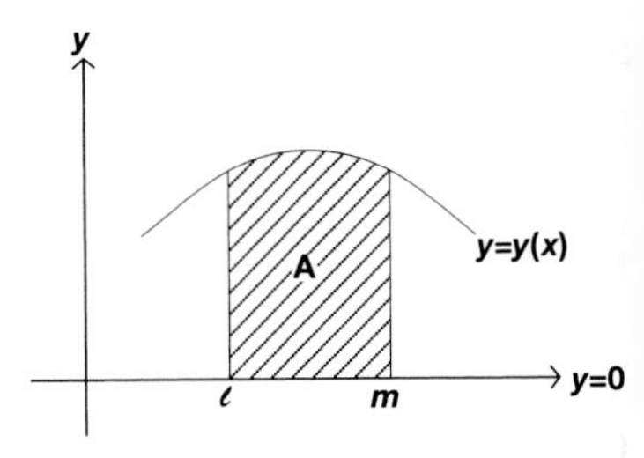

Figure 6.3

Much can be gained by viewing this area as enclosed by lines $x = l$ and $x = m$, both parallel to the y axis, together with an upper boundary (the curve) and a lower boundary (the line $y = 0$). Furthermore, since

$$\int_l^m y\,\mathrm{d}x = \int_l^m (y - 0)\,\mathrm{d}x$$

the situation illustrated in Figure 6.3 may be viewed as a special case of a more general formula for the area bounded by any two curves and lines $x = m$ and $x = l$.

Loosely stated, that formula is

$$\text{area} = \int_l^m (\text{upper boundary curve} - \text{lower boundary curve})\,\mathrm{d}x$$

This general formula, when applied to the area illustrated in Figure 6.4, would yield

$$A = \int_s^t [f(x) - g(x)]\,\mathrm{d}x \tag{5}$$

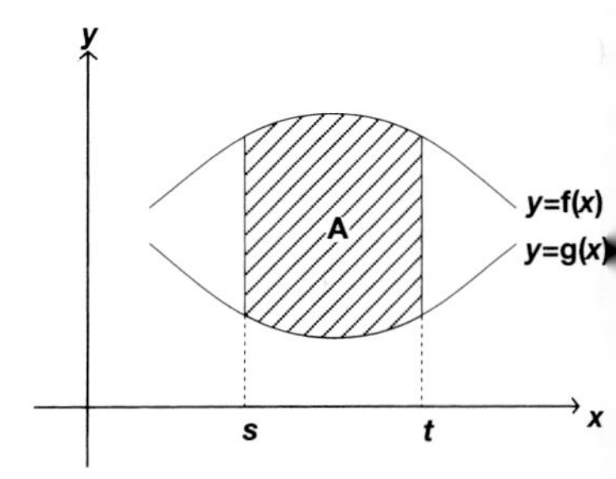

Figure 6.4

The application of this general formula to the calculation of area is made clear in the following examples.

EXAMPLE 18

Figure 6.5 shows the curve

$$y = x^2 - 5x + 4$$

A_1 is the area bounded by the coordinate axes and the curve, A_2 is the area bounded by the curve and the x axis and A_3 is the area bounded by the curve, the x axis and the line $x = 5$.

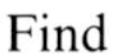

Find

(i) A_1
(ii) A_2
(iii) A_3.

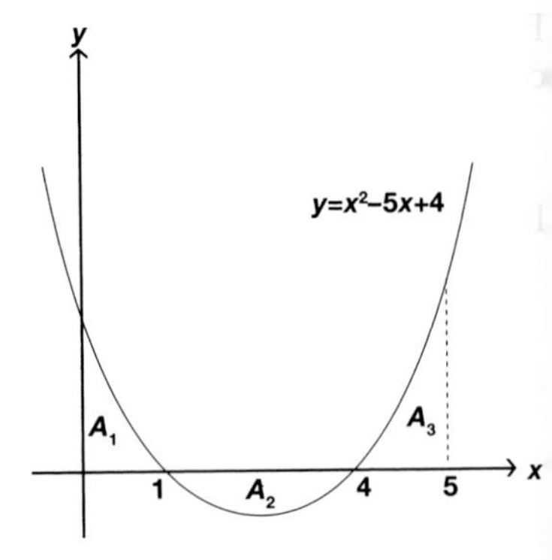

Figure 6.5

Solution 18

(i) The use of equation (5) to determine A_1 necessitates a comparison between A_1, as illustrated in Figure 6.5, and A, as illustrated in Figure 6.4. This comparison yields

$s = 0$, $t = 1$, $f(x) = x^2 - 5x + 4$ and $g(x) = 0$

Hence, using equation (5)

$$A_1 = \int_0^1 [(x^2 - 5x + 4) - 0]dx$$

or $A_1 = \frac{11}{6}$

(ii) The area A_2 is bounded above by $y = 0$, below by the curve $y = x^2 - 5x + 4$ and is sandwiched between $x = 1$ and $x = 4$. Here

$s = 1$, $t = 4$, $f(x) = 0$ and $g(x) = x^2 - 5x + 4$

Accordingly

$$A_2 = \int_1^4 [0 - (x^2 - 5x + 4)]dx$$

$$= \int_1^4 (-x^2 + 5x - 4)dx$$

or $A_2 = \frac{9}{2}$

(iii) Finally, A_3 is determined using equation (5) with

$s = 4$, $t = 5$, $f(x) = x^2 - 5x + 4$ and $g(x) = 0$

Thus $$A_3 = \int_4^5 [(x^2 - 5x + 4) - 0]dx$$

$$= \int_4^5 (x^2 - 5x + 4)dx$$

or $A_3 = \frac{11}{6}$

EXAMPLE 19

Find the areas A_1, A_2 and A_3 in Figure 6.6.

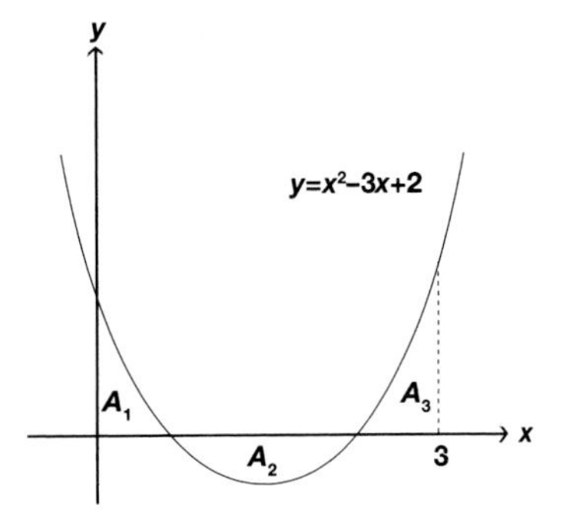

Figure 6.6

Solution 19

First, the points of intersection of the curve and the x axis are computed.

$x^2 - 3x + 2 = 0 \Rightarrow x = 1$ or $x = 2$

It follows that

$$A_1 = \int_0^1 (x^2 - 3x + 2 - 0)dx = \frac{5}{6}$$

$$A_2 = \int_1^2 (0 - (x^2 - 3x + 2))dx = \int_1^2 (-x^2 + 3x - 2)dx = \frac{1}{6}$$

$$A_3 = \int_2^3 (x^2 - 3x + 2 - 0)dx = \frac{5}{6}$$

EXAMPLE 20

Find the areas A_1, A_2 and A_3 in Figure 6.7.

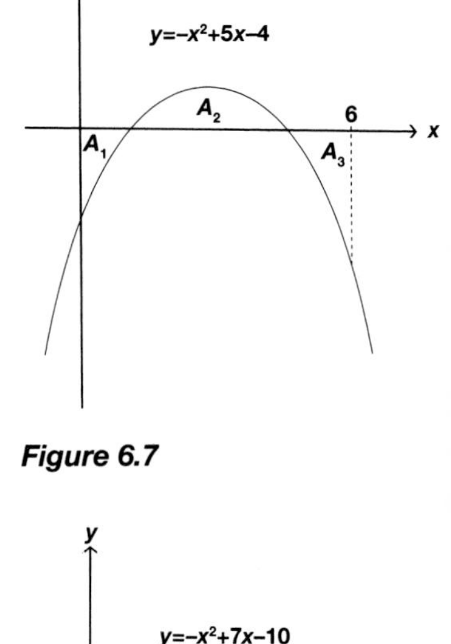

Figure 6.7

Solution 20

The curve $y = -x^2 + 5x - 4$ intersects the x axis at $x = 1$ and $x = 4$.

$$A_1 = \int_0^1 (0 - (-x^2 + 5x - 4))\mathrm{d}x = \int_0^1 (x^2 - 5x + 4)\mathrm{d}x = 1\tfrac{5}{6}$$

$$A_2 = \int_1^4 (-x^2 + 5x - 4 - 0)\mathrm{d}x = 4\tfrac{1}{2}$$

$$A_3 = \int_4^6 (0 - (-x^2 + 5x - 4))\mathrm{d}x = \int_4^6 (x^2 - 5x + 4))\mathrm{d}x = 8\tfrac{2}{3}$$

EXAMPLE 21

Find A_1, A_2 and A_3 in Figure 6.8.

Solution 21

The curve $y = -x^2 + 7x - 10$ intersects the x axis at $x = 2$ and $x = 5$.

$$A_1 = \int_0^2 0 - (-x^2 + 7x - 10)\mathrm{d}x = \int_0^2 (x^2 - 7x + 10)\mathrm{d}x = 8\tfrac{2}{3}$$

$$A_2 = \int_2^5 (-x^2 + 7x - 10 - 0)\mathrm{d}x = 4\tfrac{1}{2}$$

$$A_3 = \int_5^6 (0 - (-x^2 + 7x - 10))\mathrm{d}x = \int_5^6 (x^2 - 7x + 10)\mathrm{d}x = \tfrac{11}{6}$$

Figure 6.8

EXERCISE 6.4

1. Calculate the shaded areas in Figure 6.9 by integration.

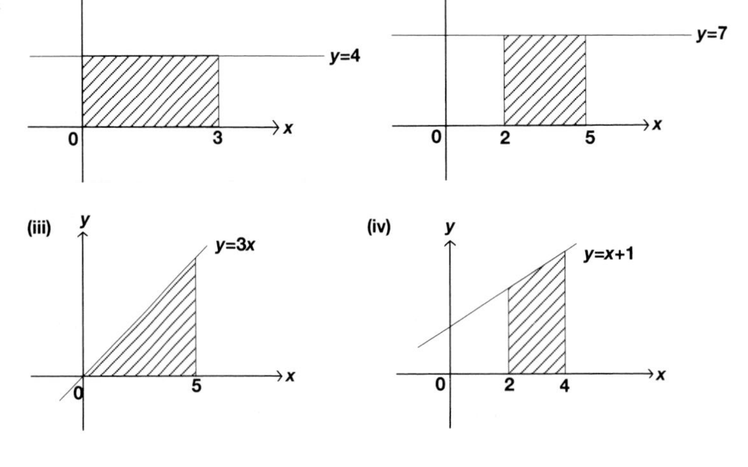

Figure 6.9

2. Calculate the shaded areas in Figure 6.10 by integration.

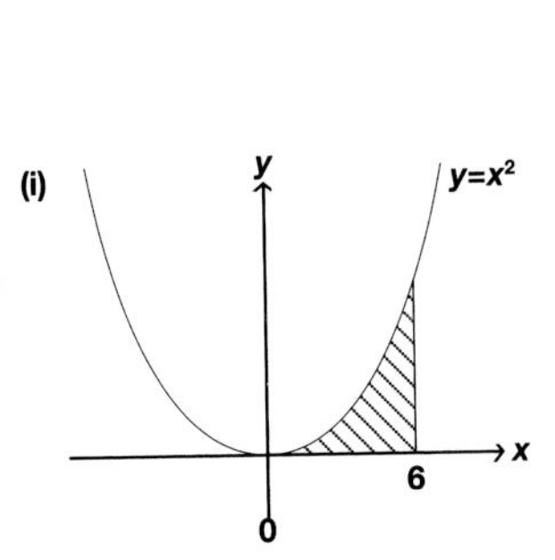

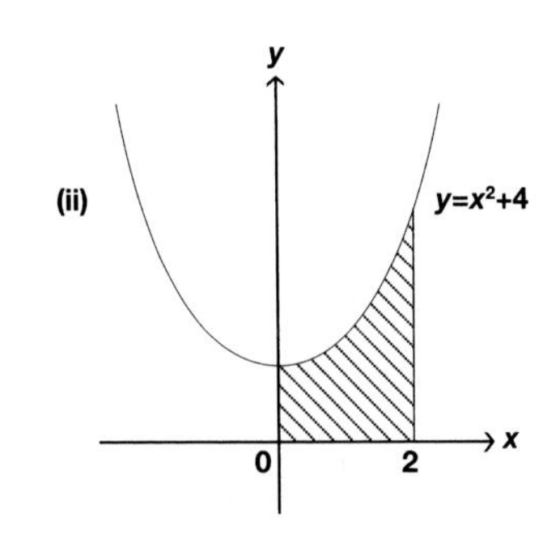

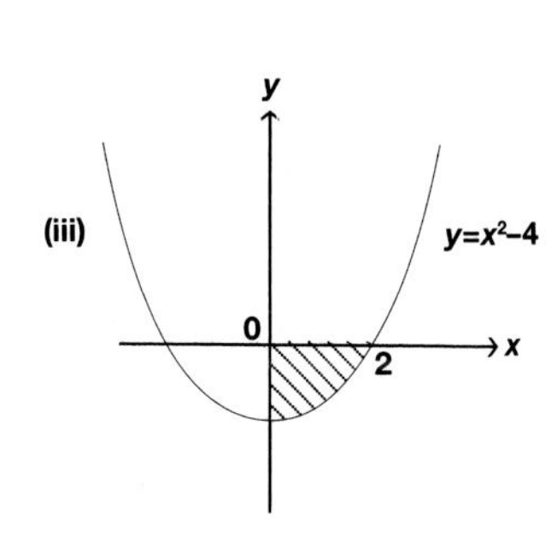

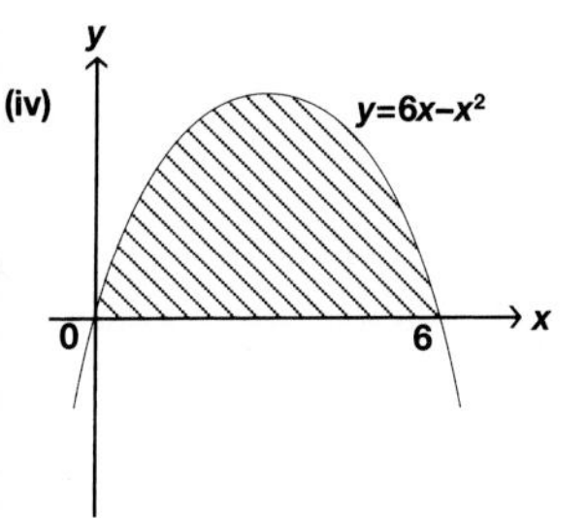

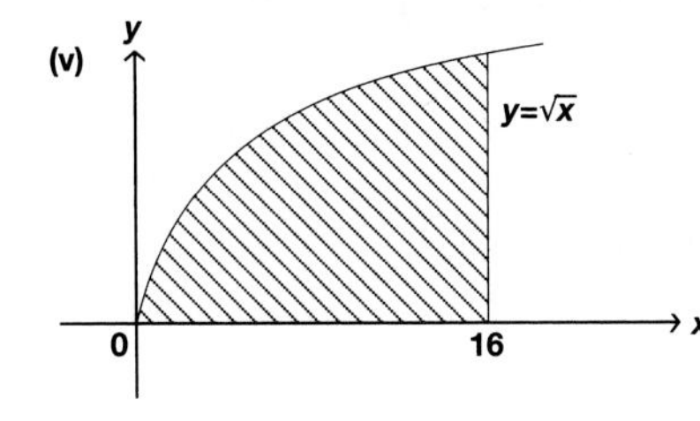

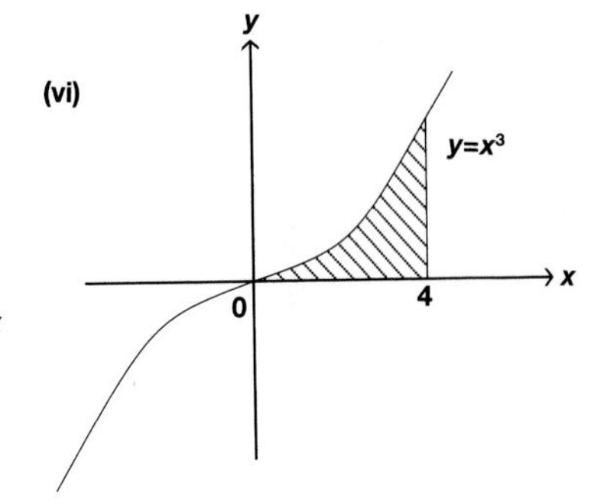

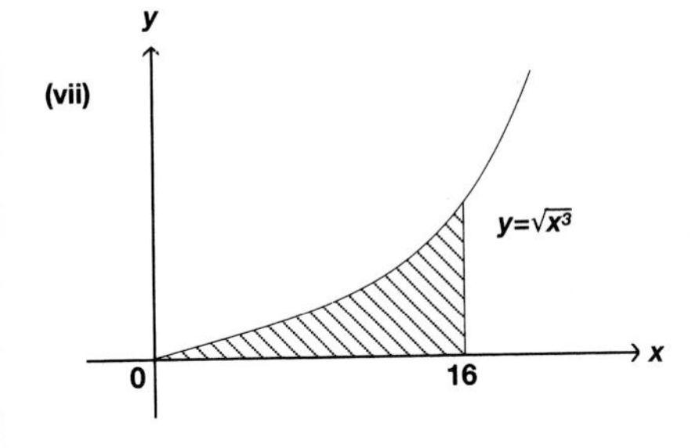

Figure 6.10

3. Calculate the area bounded by each of these curves and the x axis.

   (a) $y = 4 - x^2$ (b) $y = x(x - 8)$ (c) $y = x^3 - 6x^2$

4. Sketch the curve $y = (x - 4)^2$ and calculate the area bounded by the curve and both axes.

5. The triangle formed by the coordinate axes and the lines $y = 6x$ and $x = 6$ is divided into two parts by the curve $y = x^2$. Draw a diagram and calculate the area of each part.

6. Calculate

$$\int_0^4 (x^3 - 6x^2 + 8x)\,dx$$

Illustrate your answer with a diagram.

7. Find the points of intersection of the curve $y = x^3 - 9x$ with the x axis and calculate the area bounded by the curve and this axis.

8. Repeat question 7 for the curve $y = x(4 - x^2)$.

9. Repeat question 7 for the curve $y = x^3 - 3x^2 - 4x$.

10. The curve $y = x^2 - 6x + 5$ cuts the y axis at A and the x axis at B and C as shown in Figure 6.11. Find the coordinates of A, B and C, and calculate the total area of the shaded parts.

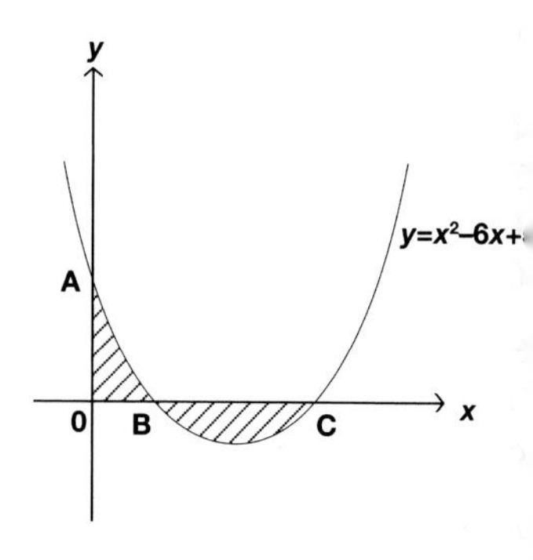

Figure 6.11

11. (a) Find $\int(3x^3 - \sqrt{x})\,dx$.
(b) Figure 6.12 is a sketch of the curve $y = x^2 - 4x + 3$.

Calculate the area of the shaded region.

(N. Ireland Additional Mathematics, 1989)

12. Determine the area of the region bounded by the curve $y = (1 + x)(1 - x)$ and the x axis.

(U.C.L.E.S. Additional Mathematics, November 1991)

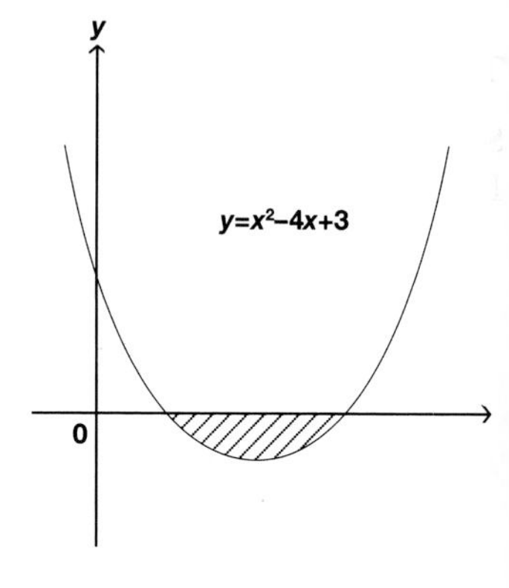

Figure 6.12

THE AREA BETWEEN TWO CURVES

The calculation of the area between two curves is a three stage process.

Step 1 **Sketch both curves.**

Step 2 **Calculate s and t, the x coordinates of the points of intersection of the curves.**

Step 3 **Use equation (5) to calculate the area.**

EXAMPLE 22

Calculate the area enclosed by the curves $y = 4 - x^2$ and $y = x^2 - 2x$.

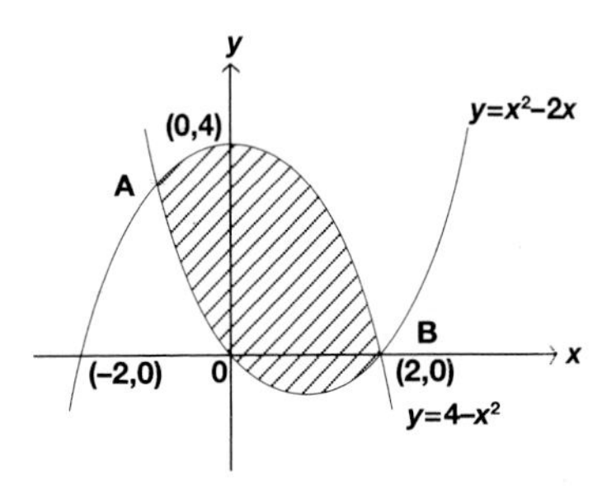

Figure 6.13

Solution 22

Step 1
Using the methods detailed in the curve sketching section of Chapter 4, the enclosed area is as shown in Figure 6.13.

Step 2
The x coordinates of the points A and B (the points at which the curves intersect) in the figure are found as follows.

At intersection

$$4 - x^2 = x^2 - 2x$$
$$\text{or} \quad 2x^2 - 2x - 4 = 0$$
$$\text{or} \quad x = 2 \text{ or } -1$$

The x coordinate of A is therefore -1 and that of B is 2.

Step 3
Finally, using the symbolism of equation (5)

$s = -1$, $t = 2$, $f(x) = 4 - x^2$ and $g(x) = x^2 - 2x$, and so

$$\text{area} = \int_{-1}^{2} [(4 - x^2) - (x^2 - 2x)]\,dx$$
$$= \int_{-1}^{2} (-2x^2 + 2x + 4)\,dx = 9$$

EXAMPLE 23

Calculate the area enclosed by the curve $y = x^2 - 6x + 2$ and the line $y = -x + 2$.

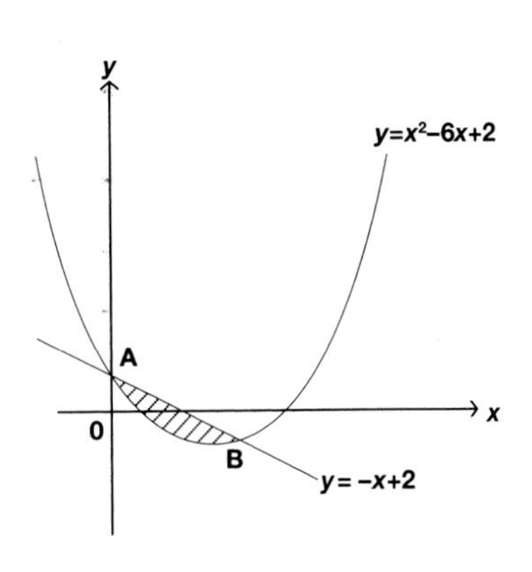

Figure 6.14

Solution 23

Step 1
Using standard curve-sketching techniques the required area may be illustrated as in Figure 6.14.

Step 2

In order to compute the x coordinates of points A and B the right-hand sides of the equations of curve and line are equated i.e.

$$x^2 - 6x + 2 = -x + 2$$

or $\quad x^2 - 5x = 0$

i.e. $\quad x = 5 \text{ or } 0$

It follows that A has x coordinate 0 and B has x coordinate 5.

Step 3

The required area is therefore given by

$$\int_0^5 [(-x+2) - (x^2 - 6x + 2)]\,dx = \int_0^5 (-x^2 + 5x)\,dx = 20\tfrac{5}{6}$$

since the diagram clearly shows $y = -x + 2$ to be the upper boundary curve and $y = x^2 - 6x + 2$ the lower.

EXERCISE 6.5

Find the area enclosed by the following.

1. $y = x^2 - 2x + 2$
$y = 5$

2. $y = x^2 - 6x + 9$
$y = 1$

3. $y = -x^2 + 3x -$
$y = -4$

4. $y = x(x - 2)$
$y = x$

5. $y = 4 - 3x - x^2$
$y = -2x - 2$

6. $y = x(x - 1)$
$y = x(2 - x)$

7. $y = x(x + 3)$
$y = x(5 - x)$

8. $y = x^2 - 5x$
$y = 3x^2 - 6x$

9. $y = x^2 - 3x - 7$
$y = 5 - x - x^2$

10. $y = 2x^2 + 7x + 3$
$y = 9 + 4x - x^2$

11. (i) Find $\int (2x^{-3} - 7x^{\frac{3}{4}})\,dx$.

(ii) The curves $y = \dfrac{4}{x^2}$ and $y = -x^2 + 5$ intersect at the points A (1,4) and B (2,1) as shown in Figure 6.15. Calculate the area enclosed by the curves between these points.

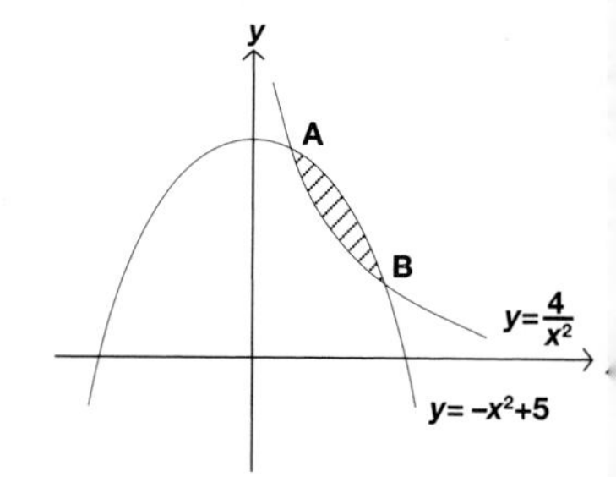

Figure 6.15

(N. Ireland Additional Mathematics, 1991

CHAPTER 7

Motion with constant acceleration

The Italian mathematician **Galileo Galilei** (1564–1642) was born in Pisa. He demonstrated that all bodies fall with the same acceleration. Galileo is recognised as the first scientist to understand that acceleration is the consequence of changing velocity. Although he embarked on a medical career he was fascinated by astronomy and, by 1589, was professor of mathematics at Pisa. He is acknowledged as the father of mathematical physics, believing that theories only acquire standing when confirmed by experiment. His 1610 publication, *The Sidereal Messenger*, is a tribute to his reputation as a great astronomer. He was the first scientist to observe the mountains of the Moon and he identified the four satellites of Jupiter.

An example of his adherence to what has since become known as the 'scientific method' is his experimental determination of the laws governing falling bodies. He timed small bodies falling down inclined planes, using the planes to 'dilute' gravity and increase the trajectory time in order that the crude clocks of the period could reliably record the time.

Galileo is best remembered for his great struggle with the Inquisition in support of Copernicus. Until Copernicus published his theory that planets orbit the Sun, the Ptolomaic model was accepted. Ptolomy's model had the Earth at the centre of the universe, orbited by the Sun, the stars, the Moon, Mercury, Jupiter, Venus, Saturn and Mars, in eight great spheres. Beyond the fixed stars in the outermost sphere was the unknown. Since this void seemed a good place to locate heaven and hell, the Catholic Church enthusiastically declared the model to accord with scripture. The Copernican model, on the other hand, had the Sun at the centre of things and was in keeping with the astronomical measures of Galileo and his contemporaries.

His support for Copernicus, in both his teachings and writings, brought Galileo a mild rebuke from Monsignor Dini: 'write freely but keep outside the sacristy'. Nevertheless, in 1616 the Catholic Church formally declared Copernicanism 'false and erroneous'.

Despite Galileo's best efforts, his 1632 masterpiece *Dialogue on the Two Chief World Systems, Ptolomaic and Copernican* was construed as pro-Copernican and he was summoned before the Inquisition. He was sentenced to house arrest in Arcetri where he became blind. However, in the margin of his copy of the *Dialogue on the Two Chief World Systems*, the following is scribbled.

... And who can doubt that it will lead to the worst disorders when minds created free by God are compelled to submit slavishly to an outside will? When we are told to deny our senses and subject them to the whim of others? When people devoid of whatsoever competence are made judges over experts and are granted authority to treat them as they please? These are the novelties which are apt to bring about the ruin of commonwealths and the subversion of the state.

INTRODUCTION

A robot is programmed to move along a line AB from south to north. The robot leaves A at 2.00 p.m. and moves at a steady rate northwards for 8 minutes, covering a distance of 50 m in that time. It stops for 4 minutes and then moves at a steady pace southwards for 70 m, arriving at its destination at 2.18 p.m. Finally it moves back to A at a steady rate, arriving at 2.24 p.m. A graphical representation of this robot's journey is given in Figure 7.1.

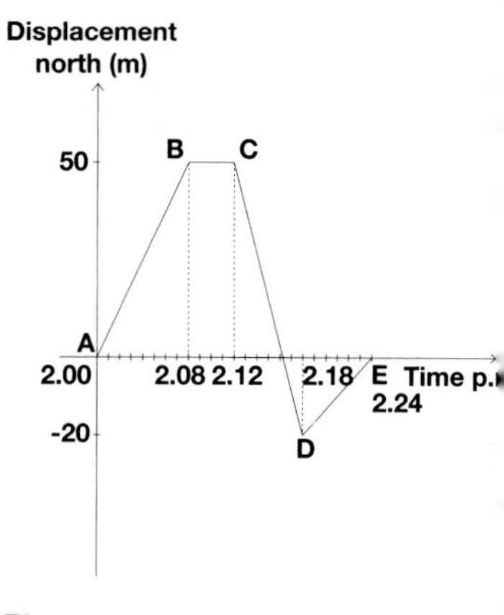

Figure 7.1

Figure 7.1 is known as a **displacement–time graph**. Displacement is an important concept. It is similar to distance but has the additional property of direction – in this case, points north of A are considered to have positive displacements, points south of A to have negative displacements and A itself to have zero displacement. The distinction between distance and displacement may be illuminated by pointing to the fact that at E the displacement is zero while the distance travelled to get to E is $50 + 0 + 70 + 20 = 140$ m.

Consider each portion of Figure 7.1 in turn. Since the robot moves steadily for 8 minutes and travels 50 m in that time, it is reasonable to conclude that it has travelled 12.5 m at 2.02 p.m., 25 m at 2.04 p.m., 37.5 m at 2.06 p.m. and so on. The graph representing its path for the first 8 minutes should therefore be a straight line, AB in Figure 7.1. In mechanics this type of motion is termed motion with 'uniform velocity'; bodies moving with uniform velocity or uniform speed in a straight line cover equal distances in equal units of time. Indeed the uniform velocity with which the robot travels from A to B may be easily computed. Since it travels 50 m in 8 minutes then

$$V_{AB} = \frac{50}{8} = 6.25\,\text{m/min}$$

and so the path AB is travelled at a steady velocity of 6.25 metres per minute northwards.

The gradient of AB in Figure 7.1 may be used as an alternative representation of the velocity.

$$\text{gradient of AB} = \frac{50}{8} = 6.25\,\text{m/min}$$

The relationship between gradient and velocity is confirmed for the BC section of the path; the gradient of BC is zero, in keeping with the information given above i.e. that the robot is at rest for these four minutes.

Turning to the CD section

$$\text{the gradient of CD} = -\frac{70}{6} = -11\tfrac{2}{3}$$

or $$V_{CD} = -11\tfrac{2}{3}\,\text{m/min}$$

The velocity during the journey from C to D, V_{CD}, indicates that the robot travels at $11\frac{2}{3}$ m/min southwards.

Finally

$$\text{the gradient of DE} = \frac{20}{6} = 3\tfrac{1}{3}$$

or $$V_{DE} = 3\tfrac{1}{3}\,\text{m/min}$$

To complete this section, average velocity and average speed are considered. Average velocity is defined as

$$\textbf{average velocity} = \frac{\textbf{total displacement}}{\textbf{total time}}$$

It follows that the average velocity for the robot's complete journey from A to E on the graph is

$$\text{average velocity} = \frac{0}{24} = 0\,\text{m/min}$$

On the other hand, the average speed is defined as

$$\textbf{average speed} = \frac{\textbf{total distance}}{\textbf{total time}}$$

and so the robot's average speed is

$$\frac{50 + 70 + 20}{24} = 5.83\,\text{m/min}$$

VELOCITY–TIME GRAPHS

A body starts from rest (velocity = 0) and, with steadily increasing speed, reaches a speed of 12 m/s in 10 seconds. It moves at this steady speed for 25 seconds and then comes to rest in a further 6 seconds. A graphical representation of the body's motion is shown in Figure 7.2 with velocity now plotted vertically.

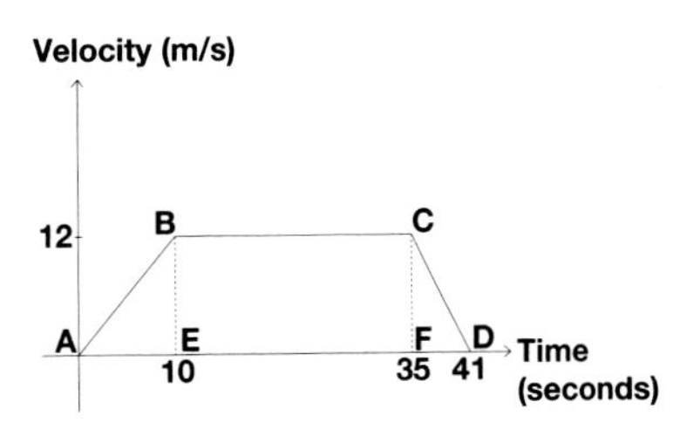

Figure 7.2

An examination of section AB of the graph prompts the question: in this case what does the gradient of AB represent?

$$\text{The gradient of AB} = \frac{12\,\text{m/s}}{10\,\text{s}} = 1.2\,\text{m/s}^2$$

The quantity $1.2\,\text{m/s}^2$ (i.e. 1.2 m per second per second) is a measure of acceleration.

The acceleration throughout the first 10 seconds is taken to be a constant in keeping with the information that the body increases its speed 'steadily'.

An acceleration of $1.2\,\text{m/s}^2$ implies that the body accelerates by 1.2 metres per second per second; in other words at $t = 0$, $v = 0$; at $t = 1$, $v = 1.2\,\text{m/s}$; at $t = 2$, $v = 2.4\,\text{m/s}$; at $t = 3$, $v = 3.6\,\text{m/s}$ and so on until at $t = 10$, $v = 12\,\text{m/s}$.

Clearly this type of uniform increase will be represented by a straight line and it has already been demonstrated that the gradient of AB represents the acceleration: the rate at which the velocity is changing.

For the section BC of the graph, the velocity remains constant at 12 m/s and so no acceleration occurs; this is in keeping with BC having zero gradient. Finally the body decelerates uniformly to execute the path represented by CD from a speed of 12 m/s at C to rest at D.

$$\text{Now the gradient of CD} = \frac{-12\,\text{m/s}}{6\,\text{s}} = -2\,\text{m/s}^2$$

Negative accelerations are termed 'retardations'.

It is instructive to consider the total distance travelled by the body on the path from A to D. First consider the section from A to B. The average velocity for this part of the motion is simply $\frac{1}{2}(0 + 12) = 6\,\text{m/s}$. It follows that the body travelled for 10 seconds covering, on average, 6 metres each second. Clearly the distance travelled is $10 \times 6 = 60\,\text{m}$. It is of interest to note that 60 is the area of the triangle ABE.

Similarly for the section C to D, the average velocity is 6 m/s (the body is moving at 12 m/s at the outset and 0 m/s at the conclusion). In moving from C to D the body takes 6 seconds and hence the body moved for 6 seconds covering, on average, 6 m each second. The distance travelled is 36 m and the reader should note that this is the area of the triangle CFD.

Finally the area of the rectangle BCFE is 25×12, numerically equal to the distance travelled by the body in 25 seconds maintaining a steady speed of 12 m/s.

It follows that the total distance travelled during journeys where motion takes place with uniform acceleration may be found by calculating the area under the velocity–time graph. For this body the total distance travelled is $60 + 36 + 300 = 396\,\text{m}$.

EXAMPLE 1

The velocity–time graph of Figure 7.3 represents the motion of a body which accelerates uniformly for 10 seconds, moves with constant velocity for 14 seconds and then decelerates uniformly for 8 seconds.

(i) Write down the initial and final velocities.
(ii) Find the acceleration and deceleration.
(iii) Calculate the total distance travelled.

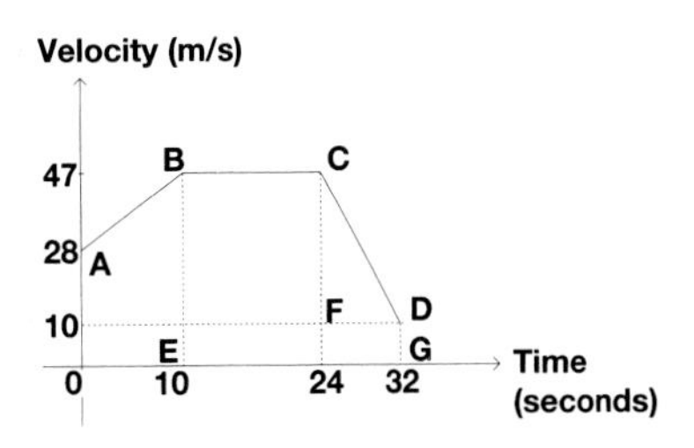

Figure 7.3

Solution 1

(i) The initial velocity is 28 m/s and the final velocity is 10 m/s.

(ii) Acceleration = gradient of AB

$$= \frac{47 - 28}{10} = 1.9\,\text{m/s}^2$$

Furthermore, the gradient of CD $= \dfrac{10 - 47}{8} = -\,4.625\,\text{m/s}^2$

Hence the acceleration is $1.9\,\text{m/s}^2$ and the deceleration is $4.625\,\text{m/s}^2$.

(iii) The area of OABE $= \frac{1}{2}(28 + 47)10 = 375$ using the formula for the area of a trapezium

and the area of BCFE $= 47 \times 14 = 658$

Also the area of CDGF $= \frac{1}{2}(47 + 10)8 = 228$

Therefore the total distance travelled is

$$375 + 658 + 228 = 1261\,\text{m} = 1.261\,\text{km}$$

EXAMPLE 2

An object accelerates uniformly from an initial velocity of 27 m/s at $3\,\text{m/s}^2$ for 5 seconds. The object then moves with the speed acquired in the acceleration phase for a further 19 seconds before decelerating uniformly. If the object travels 1.402 km in a total of 40 seconds, find the steady speed, the deceleration and the final velocity.

Solution 2

Figure 7.4 summarises the information supplied. The time for the deceleration phase has been deduced by subtracting the times for the first two phases from the total time.

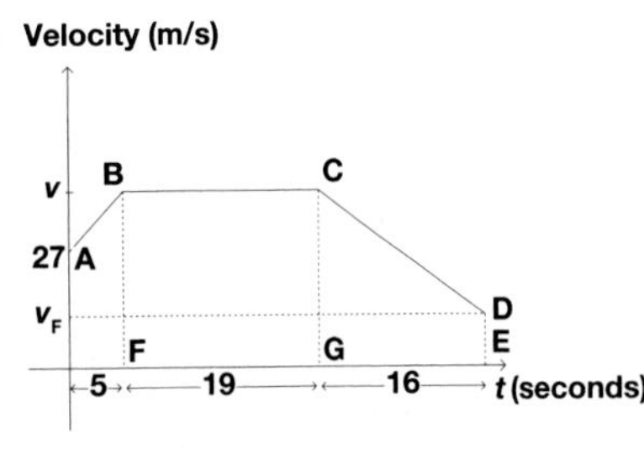

Figure 7.4

Now the gradient of the line AB is 3 (the acceleration is $3\,\text{m/s}^2$) and so the equation of AB is

$$v = 3t + 27$$

Therefore $t = 5$ gives $v = 42\,\text{m/s}$ and so the steady speed during the phase BC is 42 m/s.

The area under this velocity–time graph is

$$\tfrac{1}{2}(27 + 42)5 + 42 \times 19 + \tfrac{1}{2}(42 + v_F)16 = 1402$$

where v_F symbolises the final velocity and 1.402 km = 1402 m.

Hence

$$172.5 + 798 + 336 + 8v_F = 1402$$
$$\Rightarrow \quad v_F = 12\,\text{m/s}$$

It follows that the gradient of CD $= \dfrac{12 - 42}{16} = -1.875\,\text{m/s}^2$.

The deceleration is therefore $1.875\,\text{m/s}^2$.

EXAMPLE 3

A boy, jogging at a constant speed of 2 m/s along a straight road, leaves the city centre five minutes before a cyclist. The cyclist accelerates from rest at $0.5\,\text{m/s}^2$ for 20 seconds until she reaches a maximum speed which is maintained until she overtakes the boy. She then decelerates uniformly, coming to rest 40 seconds after passing the boy. Find

(i) the maximum speed of the cyclist,
(ii) the distance from the centre of the city at which the cyclist passes the boy,
(iii) the deceleration of the cyclist,
(iv) the distance between the boy and the cyclist when the cyclist comes to rest.

Solution 3

Let t be the time (in seconds) when the cyclist overtakes the boy. At the point of overtaking both will be equally displaced from the centre of the city. Figure 7.5 summarises the information offered in the example.

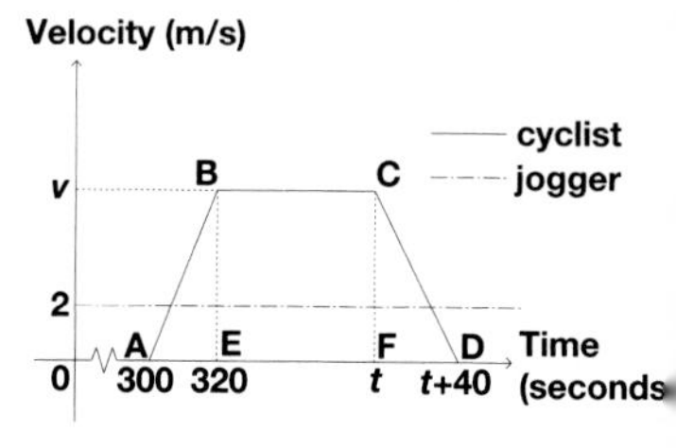

Figure 7.5

(i) Since the acceleration is $0.5\,\text{m/s}^2$ then the gradient of AB is 0.5. Furthermore, A is the point (300,0) and therefore the equation of AB is

$$v = 0.5(t - 300)$$
$$\text{or} \quad v = 0.5t - 150$$

In particular, if $t = 320$, $v = 10\,\text{m/s}$

The maximum speed is therefore 10 m/s.

(ii) Given t to be the time at which the cyclist overtakes the boy, then equating areas gives

$$2t = \tfrac{1}{2}(20)(10) + 10(t - 320)$$
$$\text{or} \quad 2t = 100 + 10t - 3200$$
$$\text{or} \quad t = 387.5$$

and $s = 2t$ gives the required distance to be 775 m.

(iii) The gradient of CD is $\dfrac{0-10}{40}$ i.e. $-0.25\,\text{m/s}^2$.

(iv) The total distance travelled by the cyclist is

$$\tfrac{1}{2}(20)(10) + 10(67.5) + \tfrac{1}{2}(40)(10) = 975\,\text{m}$$

The total distance travelled by the boy when $t = 427.5$ seconds is

$$427.5 \times 2 \quad \text{or} \quad 855\,\text{m}$$

Therefore the required separation is 120 m.

EXERCISE 7.1

1. In parts (i) to (v) of Figure 7.6 the velocity–time graph represents the motion of an object which accelerates uniformly for a period of time, then moves with a constant velocity and finally decelerates. In each case write down the initial velocity and the final velocity, calculate the acceleration and deceleration and find the total distance travelled by the object.

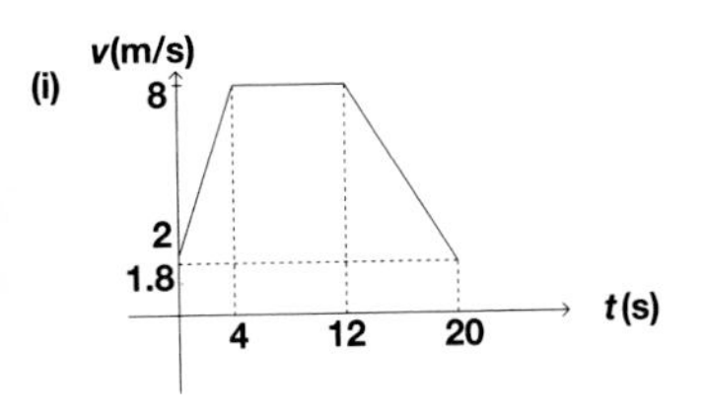

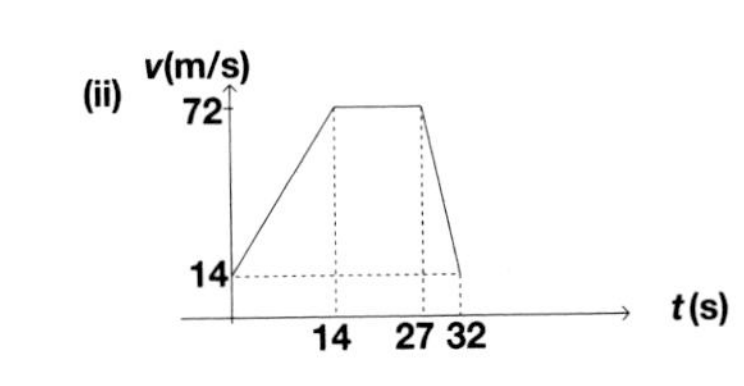

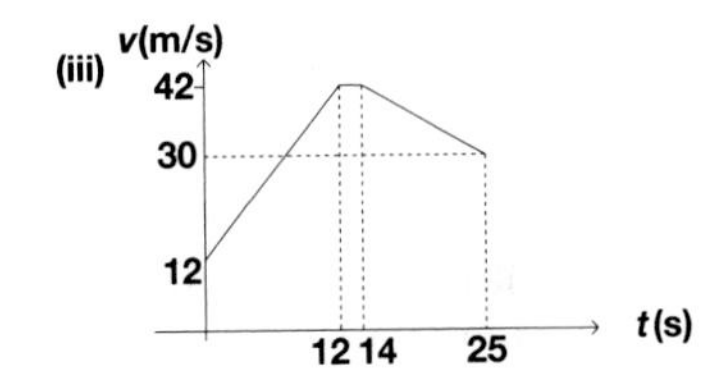

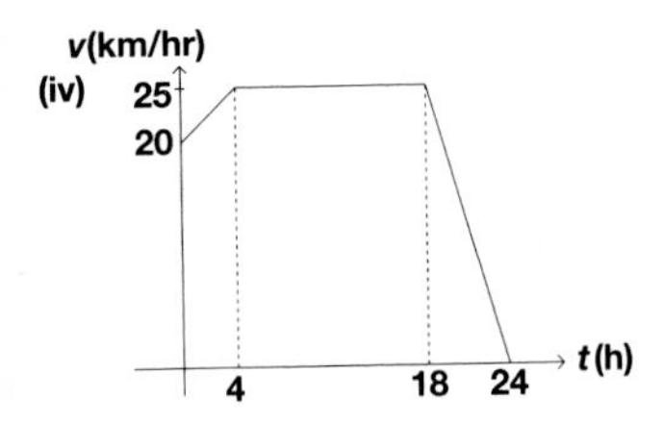

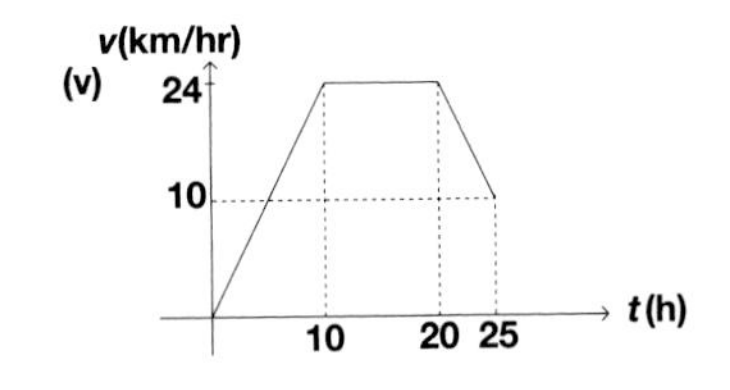

Figure 7.6

2. (i) An object accelerates uniformly from an initial velocity of 30 m/s, at $2.25\,\text{m/s}^2$ for 8 seconds. The object then moves with the maximum velocity acquired in the acceleration phase for a further 34 seconds before decelerating uniformly. The object travels 2.522 km in a total of 59 seconds. Find the maximum velocity, the deceleration and the final velocity.

(ii) An object accelerates uniformly from an initial velocity of 17 m/s, at $5\,\text{m/s}^2$ for 7 seconds. The object then moves with the maximum velocity acquired in the acceleration phase for a further 17 seconds before decelerating uniformly. The object travels 1524.5 m in a total of 38 seconds. Find the maximum velocity, the deceleration and the final velocity.

(iii) An object accelerates uniformly from an initial velocity of 24 m/s, at $9.357\,\text{m/s}^2$ for 14 seconds. The object then moves with the maximum velocity acquired in the acceleration phase for a further 43 seconds before decelerating uniformly. The object travels 14.965 km in a total time of 2 minutes 24 seconds. Find the maximum velocity, the deceleration and the final velocity.

(iv) An object accelerates uniformly from an initial velocity of 20 m/s, at $3\frac{1}{3}$ m/s^2 for 24 seconds. The object then moves with the maximum velocity acquired in the acceleration phase for a further 1 minute before decelerating uniformly. The object travels 9.96 km in a total time of 2 minutes. Find the maximum velocity, the deceleration and the final velocity.

(v) An object accelerates uniformly from an initial velocity of 24 m/s, at $2\frac{1}{6}$ m/s^2 for 12 seconds. The object then moves with the maximum velocity acquired in the acceleration phase for a further 14 seconds before decelerating uniformly. The object travels 1.774 km in a total of 40 seconds. Find the maximum velocity, the deceleration and the final velocity.

3. A body starts from rest, accelerates uniformly at 4 m/s^2 for 2 seconds, then decelerates uniformly to rest. If the total distance travelled is 20 m, what is the deceleration?

4. A body starts at rest, travels 8 m with constant acceleration, then 16 m at constant speed, and finally 4 m at a constant deceleration, finishing at rest. Given that the total time taken is 10 seconds, find the acceleration and deceleration.

5. Two bodies start together and move along the same straight line. If one moves with a constant speed of 6 m/s while the other starts from rest and accelerates at 4 m/s^2, how long will it be before they are together again?

6. A girl leaves a shop, walking at a steady speed of 1 m/s, and 6 seconds later her brother sets off from the same shop in pursuit, starting at rest and accelerating at $\frac{2}{3}$ m/s^2. How far do they go before they are together?

7. A car starts from rest and moves along a straight road. During the first 5 seconds of motion it accelerates uniformly to a speed of 10 m/s. The car continues with this velocity for 5 seconds and then decelerates uniformly, coming to rest after a further 3 seconds.

(i) Display this information on a velocity–time graph.
(ii) Calculate the total distance travelled.

(N. Ireland Additional Mathematics, 1989)

8. A train departs from station A and accelerates uniformly for 3 minutes reaching a speed of 90 km per hour. It continues at this speed for 10 minutes and it is then retarded uniformly for a further 5 minutes before coming to rest at Station B.

Find

(i) the distance between the stations,
(ii) the time taken to cover the distance between A and a signal box midway between A and B.

(N. Ireland Additional Mathematics, 1991)

THE UNIFORM ACCELERATION FORMULAE

A body accelerates uniformly from an initial velocity of u to a final velocity of v in t units of time. If the symbol a is used for acceleration then it is clear from the velocity–time graph of Figure 7.7 that

$$a = \frac{v - u}{t}$$

(Recall that acceleration is associated with the gradient of the velocity–time graph.)

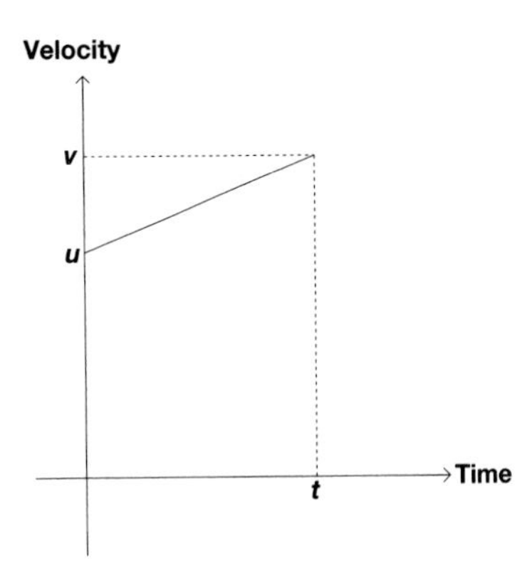

Figure 7.7

Furthermore, the distance travelled in accelerating from u to v is given by the area under the curve and so

$$s = \tfrac{1}{2}(u + v)t$$

Now $a = \dfrac{v - u}{t}$

may be rearranged

$$\boldsymbol{v = u + at} \qquad (1)$$

Inserting v from equation (1) in

$$s = \tfrac{1}{2}(u + v)t$$

gives $s = \tfrac{1}{2}(u + u + at)t$

or $\boldsymbol{s = ut + \tfrac{1}{2}at^2}$ (2)

Now equation (1) may be rearranged

$$t = \frac{v - u}{a}$$

Inserting $t = \dfrac{v - u}{a}$ in $s = \tfrac{1}{2}(u + v)t$ yields

$$s = \tfrac{1}{2}(u + v)\left(\frac{v - u}{a}\right)$$

or $2as = v^2 - u^2$

or $\boldsymbol{v^2 = u^2 + 2as}$ (3)

Equations (1) to (3) are referred to as the **uniform acceleration formulae** and are of fundamental importance in the analysis of motion with uniform acceleration. Their power will be illustrated in a series of examples.

EXAMPLE 4

A car travels at 25 km/h for 3 minutes, then at 72 km/h for 8 minutes and finally at 18 km/h for 2 minutes. Find the average speed for the journey.

Solution 4

The distance travelled in the first three minutes is given by $25 \times \dfrac{3}{60}$ km since the car covers 25 km in one hour.

Performing similar computations for the other two stages gives the total distance travelled to be

$$25 \times \frac{3}{60} + 72 \times \frac{8}{60} + 18 \times \frac{2}{60} = 11.45\,\text{km}$$

The total time taken is $(3 + 8 + 2) = 13$ minutes and so it follows that the average speed is

$$\frac{11.45}{\left(\frac{13}{60}\right)}\,\text{km/h} = 52.85\,\text{km/h}$$

EXAMPLE 5

A body with initial velocity 6 m/s moves along a straight line with constant acceleration, travelling 620 m in 38 s. Calculate the final velocity and the acceleration.

Solution 5

Here $u = 6\,\text{m/s}$, $s = 620\,\text{m}$ and $t = 38\,\text{s}$.

Equation (2) becomes

$$620 = 6(38) + \tfrac{1}{2}(a)(38)^2$$

Therefore

$$a = 0.5429\,\text{m/s}^2$$

Also, equation (1) yields

$$v = 6 + (0.5429)(38) = 26.63\,\text{m/s}$$

EXAMPLE 6

A lorry moves from rest with constant acceleration $5.6\,\text{m/s}^2$. Find the speed and distance travelled by the lorry after 4 seconds.

Solution 6

Here $u = 0$, $a = 5.6\,\text{m/s}^2$ and $t = 4$.

Equation (1) implies

$$v = 0 + (5.6)(4) = 22.4\,\text{m/s}$$

Finally, equation (2) yields

$$s = 0(4) + \tfrac{1}{2}(5.6)(4)^2 = 44.8\,\text{m}$$

EXAMPLE 7

A piece of machinery slides on an inclined surface at a factory. It begins from rest at the top of the plane and its velocity after 2 seconds is noted to be 3.2 m/s. If it continues to move in this way calculate its velocity after 6.2 seconds from the beginning of the motion, the uniform acceleration it experiences and the distance travelled down the plane in 6.2 seconds.

Solution 7

Here $u = 0$, $t = 2$ s and $v = 3.2$ m/s.

The acceleration may be calculated through equation (1)

i.e. $3.2 = 0 + a(2)$ or $a = 1.6\,\text{m/s}^2$

Now substituting $t = 6.2$ in equation (1) gives

$$v = 0 + (1.6)(6.2) = 9.92\,\text{m/s}$$

and equation (2) gives

$$s = (0)(6.2) + \tfrac{1}{2}(1.6)(6.2)^2 = 30.75\,\text{m}.$$

Hence the velocity is 9.92 m/s after 6.2 seconds and it has travelled 30.75 m.

EXAMPLE 8

A car freewheels from rest down a hill, with uniform acceleration. Given that the car covers 81 m in 9 seconds, find its speed 2 seconds after motion commences.

Solution 8

Here $u = 0$, $s = 81$ m and $t = 9$ s.

Equation (2) gives

$$81 = 0(9) + \tfrac{1}{2}(a)(9)^2 \quad \text{or} \quad a = 2\,\text{m/s}^2$$

To calculate its speed after 2 seconds, equation (1) is used with $u = 0$, $a = 2$, $t = 2$ providing

$$v = 0 + 2(2) = 4\,\text{m/s}$$

EXAMPLE 9

A car's speed increases from 21 km/h to 72 km/h in covering 80 m. Calculate the total time taken to execute the 80 m and the distance travelled during the third second.

Solution 9

In order that the same units be used throughout, the 80 m is converted to kilometres and written as 0.08 km.

Hence $u = 21$ km/h, $v = 72$ km/h and $s = 0.08$ km.

Equation (3) becomes

$$72^2 = 21^2 + 2a(0.08)$$
$$\text{or} \quad a = 29\,643.75\,\text{km/h}^2$$

Now equation (1) gives

$$72 = 21 + (29\,643.75)t$$

Hence $t = 1.72 \times 10^{-3}$ hours or 6.192 seconds.

The third second is the time interval between $t = 2$ and $t = 3$.

Now the value of s at $t = 2$ may be found by equation (2).

$$s_2 = (21)(\tfrac{2}{3600}) + \tfrac{1}{2}(29\,643.75)(\tfrac{2}{3600})^2$$

$$\Rightarrow \quad s_2 = 0.016\,24\,\text{km}$$

Similarly $s_3 = (21)(\tfrac{3}{3600}) + \tfrac{1}{2}(29\,643.75)(\tfrac{3}{3600})^2$

$$\Rightarrow \quad s_3 = 0.027\,79\,\text{km}.$$

Therefore the distance travelled in the third second is

$$(0.027\,79 - 0.016\,24)\,\text{km or } 11.55\,\text{m}$$

EXAMPLE 10

A car which is moving at 50 km/h decelerates at 3 m/s^2 for 4 seconds. Calculate the final speed and the distance covered in this time.

Solution 10

In order that the same units may be used throughout, 50 km/h is converted to m/s as follows.

$$50\,\text{km/h is } 50\,000\,\text{m per } 3600 \text{ seconds or } \frac{50\,000}{3600}\,\text{m/s}$$

It follows that 50 km/h is 13.89 m/s.

Hence $u = 13.89\,\text{m/s}$, $a = -3\,\text{m/s}^2$, $t = 4$

Equation (1) gives

$$v = 13.89 + (-3)(4) = 1.89\,\text{m/s (or } 6.804\,\text{km/h)}$$

and equation (2) becomes

$$s = (13.89)(4) + \tfrac{1}{2}(-3)(4)^2 = 31.56\,\text{m}.$$

Therefore the final speed is 6.804 km/h and the distance covered is 31.56 m.

EXERCISE 7.2

1.–8. Repeat the questions of Exercise 7.1 using the uniform acceleration formulae given in equations (1), (2) and (3).

9. A car approaching traffic lights at 18 m/s pulls up in 27 m. Find the retardation and the time taken to pull up.

10. A train starts from rest in one station, accelerates uniformly at 0.25 m/s^2 for 2 minutes, then travels at a uniform speed for 3 minutes, and finally retards uniformly for 1 minute, coming to rest in the next station. What is the top speed of the train and how far apart are the stations?

11. A particle is moving at 20 m/s at point P and is accelerated at 3 m/s^2 until its velocity becomes 60 m/s. It is then decelerated at 4 m/s^2 until it comes to rest at Q. Find the distance PQ and the total time taken in travelling from P to Q.

12. A train is uniformly decelerated from 35 m/s to 21 m/s in 350 m. Calculate

(i) the deceleration,
(ii) the total time taken to come to rest from a velocity of 35 m/s under this deceleration.

13. A particle moves with uniform acceleration 0.5 m/s^2 in a horizontal line ABC. The speed of the particle at C is 80 m/s and the times from A to B and from B to C are 40 seconds and 30 seconds respectively. Calculate

(i) the speed of the particle at A,
(ii) the distance BC.

14. A car moving along a straight line accelerates from rest until it has travelled x m. It then moves for 50 seconds at a constant speed and travels a further x m. Finally, the car uniformly decelerates and comes to rest after travelling a further distance $\frac{x}{2}$ m. Calculate the total time for the journey.

15. Two bus stops are 570 m apart. A bus accelerates uniformly from rest at the first stop to a speed of V m/s. It maintains this speed for T s and then, with a uniform retardation, comes to rest at the second stop. Sketch the velocity–time diagram for the journey.

(i) Given that $T = 6$ and that the bus accelerates for 15 s and slows down for 30 s, find the value of V.

(ii) Given that $V = 10$, and that the magnitude of the acceleration and of the retardation are unchanged, find the value of T.

A car starts from rest at the first bus stop and accelerates uniformly to 30 m/s in 18 s. On attaining this speed it retards uniformly, coming to rest at the second bus stop. Find the retardation.

(U.C.L.E.S. Additional Mathematics, November 1991)

16. (a) The velocity–time diagram in Figure 7.8 shows the velocities at different times of a particle moving in a straight line. Calculate

(i) the acceleration during the first 20 s,
(ii) the distance travelled in the first 60 s,
(iii) the total distance travelled in the 80 s,
(iv) the displacement of the particle from the starting point after 80 s.

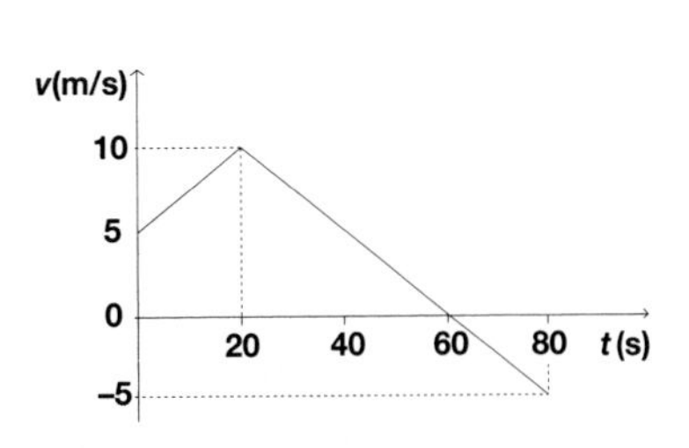

Figure 7.8

(*b*) Two particles, A and B, are 7 m apart on a smooth horizontal surface. Particle A is moving directly towards B with a speed of 2 m/s and an acceleration of $0.3\,\text{m/s}^2$. Particle B is moving in the same direction as A with a speed of 5 m/s and a deceleration of $0.2\,\text{m/s}^2$. Calculate the time taken before the particles collides.

(U.C.L.E.S. Additional Mathematics, June 1992)

17. (a) A particle starts from rest at a point A and moves in a straight line ABC with constant acceleration. At B its speed is 10 m/s and at C its speed is 15 m/s. Given that BC = 50 m, calculate

(i) the time taken to travel from B to C,
(ii) the acceleration of the particle,
(iii) the time taken to travel from A to B,
(iv) the distance AB.

(b) Figure 7.9 shows the velocity–time diagram for the motion of a particle which starts from rest and comes to instantaneous rest when $t = 30$ and $t = 100$. The greatest and least values of v are 12 and -3.6 respectively. Calculate the distance of the particle from its starting point when

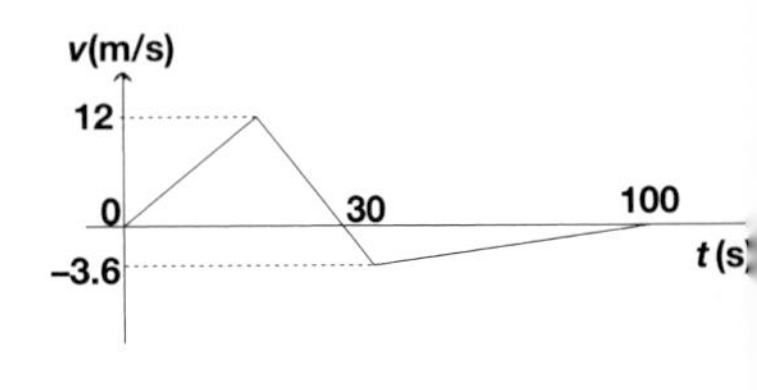

Figure 7.9

(i) $t = 30$,
(ii) $t = 100$.

Given also that $v = 12$ when $t = 20$, find

(iii) the value of t when $v = -3.6$,
(iv) the distance moved by the particle while its velocity is decreasing

(U.C.L.E.S. Additional Mathematics, June 1991)

VERTICAL MOTION UNDER GRAVITY

Despite the contents of many popular science books, little is known of the full range of experimental arrangements used by Galileo to confirm that all bodies (neglecting air resistance) fall to the earth with equal acceleration; the experiments where bodies of differing mass were dropped from the Leaning Tower of Pisa are almost certainly fanciful.

However, his 'inclined plane' experiments provided experimental verification that all objects near the surface of the Earth fall with a common acceleration, usually denoted $\mathbf{g}$.

In the United Kingdom $\mathbf{g = 9.8\,m/s^2}$ and this value of g will be used throughout this book although the reader is cautioned that a number of examination questions will be solved using the numerically **simpler approximation $g = 10\,m/s^2$**. The acceleration due to gravity varies with latitude, being $9.78\,m/s^2$ at the equator and $9.83\,m/s^2$ at the poles. On the Moon g is about one sixth of its value on Earth. While the solution of problems involving vertical motion under gravity merely involves the application of the methods of the previous section with a set equal to $9.8\,m/s^2$ in every case, the following examples will illustrate the importance of establishing a direction in which velocity, acceleration and displacement are positive. The convention that 'upwards is positive' leads to the gravitational acceleration always being written $a = -9.8\,m/s^2$ since gravity is directed downwards.

EXAMPLE 11

A toy is dropped from a cliff and strikes the water below after 6 seconds. Calculate the height of the cliff above the surface of the water and the speed with which the toy enters the water.

Solution 11

Since the toy is dropped the acceleration will be that of gravity.

Taking the origin as the edge of the cliff and 'upwards' as positive then

$$u = 0 \quad a = -9.8\,m/s^2 \quad t = 6$$

Therefore equation (1) gives

$$v = 0 + (-9.8)6$$
$$v = -58.8\,m/s$$

This result reveals that the toy strikes the water at 58.8 m/s moving downwards.

Now equation (2) yields

$$s = 0(6) + \tfrac{1}{2}(-9.8)(6)^2 = -176.4\,m$$

and this indicates that the point of entry of the toy to the water is 176.4 m below the cliff edge or that the cliff is 176.4 m above the water level.

EXAMPLE 12

A missile is thrown upwards at 42 m/s from the edge of a building of height 120 m. After what time and with what speed will the missile strike the ground below?

Solution 12

Taking the origin as the edge of the building and upwards as positive, the information given may be summarised thus

$$u = 42\,m/s \quad a = -9.8\,m/s^2 \quad s = -120$$

Now equation (3) gives

$$v^2 = (42)^2 + 2(-9.8)(-120)$$

Hence $v = \pm 64.16$ with the negative sign the obvious choice

i.e. $v = -64.16\,\text{m/s}$

Equation (1) implies

$$-64.16 = 42 + (-9.8)t \quad \text{or} \quad t = 10.83$$

Therefore the missile strikes the ground at 64.16 m/s and the time taken from the missile leaving the hand of the thrower until the impact with the ground is 10.83 seconds.

Alternatively

$$u = 42\,\text{m/s} \quad a = -9.8\,\text{m/s}^2 \quad s = -120$$

Now applying equation (2) directly gives

$$-120 = 42t + \tfrac{1}{2}(-9.8)t^2$$

or $4.9t^2 - 42t - 120 = 0$

Therefore $t = \dfrac{42 \pm \sqrt{((42)^2 + 4(4.9)(120))}}{9.8} = \dfrac{42 \pm 64.16}{9.8}$

and so $t = -2.261$ or 10.83 seconds

Clearly the negative time is 'unphysical' and so $t = 10.83\,\text{s}$.

Now equation (1) becomes

$$v = 42 + (-9.8)(10.83) = -64.13\,\text{m/s}$$

confirming that the missile strikes the ground with a downward-directed speed of 64.13 m/s after 10.83 seconds.

EXAMPLE 13

A fairground attraction consists of a gun which fires little missiles upwards at 100 m/s. Find the maximum height reached by the missiles, the time taken to reach that height and the times (measured from firing) at which the missile will be 200 m above the ground. Air resistance may be neglected.

Solution 13

Taking the gun as the origin and considering upwards as positive, the information given may be summarised

$$u = 100\,\text{m/s} \quad g = -9.8\,\text{m/s}^2.$$

Now at maximum height the upwards velocity is zero. If this were not the case then the missile would continue upwards and the term 'maximum height' would be a misnomer.

Now equation (1) becomes

$$0 = 100 + (-9.8)t$$

or $t = 10.2\,\text{s}$ is the time taken to reach the maximum height.

Attention now turns to the maximum height itself.

Equation (2) yields

$$s = 100(10.2) + \tfrac{1}{2}(-9.8)(10.2)^2 = 510.2\,\text{m}$$

Finally, when the missile is 200 m above the ground

$$s = ut + \tfrac{1}{2}at^2$$

becomes $200 = 100t + \tfrac{1}{2}(-9.8)t^2$

or $4.9t^2 - 100t + 200 = 0$

This equation has solution

$$t = \frac{100 \pm \sqrt{(100^2 - 4(4.9)(200))}}{9.8}$$

Accordingly $t = 18.16$ or 2.248 seconds

This result is interpreted physically by noting that the missile passes through the point 200 m above the ground level at $t = 2.248\,\text{s}$ on its path upwards to the maximum height and once again at $t = 18.16\,\text{s}$ on its return path from its maximum height to the ground.

EXERCISE 7.3

1. A particle is projected downwards at 40 m/s. How far has it travelled after 5 seconds? How far does it travel in the next 5 seconds?

2. An object is projected upwards at 245 m/s. Calculate its position after 10 seconds and find the total time that has elapsed when it has returned to that position again.

3. A building is 100 m high. A missile is projected upwards from the edge of the roof of the building with a velocity of 40 m/s. How long does it take to reach the ground below? What maximum height above ground level does it reach?

4. A cricket ball is thrown vertically upwards from ground level at 25 m/s. Taking $g = 10\,\text{m/s}^2$, find
   (i) the height to which the ball rises,
   (ii) the time taken to reach the greatest height,
   (iii) the speed and direction after 3 s,
   (iv) the height of the ball after 3 s,
   (v) the times when the ball is at height 20 m.

5. If the ball in question 4 was thrown up at 25 m/s but fell down a 70 m deep well instead of hitting the ground, find
   (i) the depth below ground after 6 seconds,
   (ii) the time (measured from the start) for the ball to strike the bottom of the well.

6. (Take g to be 10 m/s^2.)

(a) On a vertical television mast, whose height is 180 m, there are two platforms A and B. An object is dropped from rest from the top of the mast and passes A 2.4 seconds later. After a further t seconds it passes B, 16.2 m below A, at a speed of v m/s. Calculate

(i) t and v,
(ii) the time taken to reach the ground after passing B.

(b) A balloon is ascending at a uniform speed of 6 m/s. A small object is released from the balloon at a height of 95 m above the ground and moves freely under gravity. Find

(i) the greatest height above the ground attained by the object,
(ii) the speed with which the object hits the ground,
(iii) the time taken by the object to reach the ground from the moment of release.

After the release of the object the balloon accelerates uniformly upwards at 0.8 m/s^2. Find the height of the balloon above the ground when the object hits the ground

(U.C.L.E.S. Additional Mathematics, June 1991)

7. (Take g to be 10 m/s^2.)

A particle is dropped from rest from a height of H m and reaches the ground in 3 s. Find

(i) the speed with which it strikes the ground,
(ii) the value of H.

A second particle is projected vertically upwards with a speed of 15 m/s from a height of h m, and reaches the ground in 4 s. Find

(iii) the greatest height of this particle above the ground,
(iv) the value of h.

The second particle is projected upwards at the moment when the first particle is dropped. Find an expression, in terms of t, for the height of each particle above the ground after t s. Hence find the value of t when both particles are at the same height above the ground.

(U.C.L.E.S. Additional Mathematics, November 1991)

CHAPTER 8

Newton's laws

The French scientist **Charles Augustin de Coulomb** (1736–1806) trained as a military engineer before turning to physics and becoming Inspector-General of Public Instruction for Paris. He is responsible for the four laws of friction listed below. The SI unit of charge is named after him in recognition of his discovery that electric charges and magnetic poles attract and repel one another according to an 'inverse square law'. This fundamental physical law also explains the force of attraction between the planets; the Moon orbits the Earth because the Earth attracts the Moon with a gravitation field which obeys an inverse square law. We are surrounded by our own gravitation field which stretches to the end of the universe and we are all gravitationally attracted to one another.

INTRODUCTION

ewton's *Philosophiae Naturalis Principia Mathematica* (1st edn. 87) has had an enormous impact upon the way in which human ings see their world. The *Principia*, as it is often referred to by athematicians, presented three laws which govern the motion of objects. The laws still hold true today although they require odification before being applied to things very large (relativistic echanics) or things very small (quantum mechanics).

The **first law** is that every body remains stationary or in uniform motion in a straight line unless it is made to change direction by external forces.

The **second law** is that acceleration is proportional to the impressed force and acts along the same straight line.

The **third law** is that reaction is always equal and in the opposite direction to action.

hile students have little difficulty accepting that part of the first v which talks of a body remaining at rest unless acted upon by a force, they often struggle in coming to terms with the concept of uniform motion in the absence of a force in the direction of that motion. When studying the motion of a car travelling at constant speed on a motorway it is a common misconception to assume that the forward force of the engine exceeds the combined effect of the frictional force between the wheels and the road together with the wind resistance, both acting in the opposite direction. This conviction that the forward force must exceed the backward force in order for the car to move forward violates the first law of motion; the car is travelling with uniform speed and so there can be no net external force. A net external force of zero is achieved by balancing the total resistance to motion by the forward force of the engine; cars 'cruise' with the accelerator depressed sufficiently for the engine to deliver a forward force just equal to the total resistance to motion.

The second law is summarised in a single equation.

$$\boldsymbol{F = ma}$$

where F is the force, m the mass (the amount of matter in the body) and a the acceleration. In the SI system the units of force are newtons (N), mass is measured in kilograms (kg) and acceleration in m/s^2. The newton is defined to be the force required to accelerate a mass of 1 kg at $1\,m/s^2$. A special case of the $F = ma$ equation is where $a = g = 9.8\,m/s^2$, the acceleration due to gravity. In this instance F is assigned a special name, the **weight**. Therefore

$$\textbf{weight} = mg$$

The numerical value assigned to g throughout this book will be 9.8 although it will be convenient at times to use 10 for purposes of arithmetic simplicity.

Weight is an example of an 'action-at-a-distance' force. Action-at-a-distance forces are unlike 'contact' forces where the agency causing the acceleration is clearly visible. An example of a contact force is the force applied to a piano to move it from one side of a room to another. On the other hand, a ball falling from a tall building, while clearly accelerating, appears to be free of any force. However, the Earth is exerting a force of attraction upon the ball of magnitude equal to its mass multiplied by 9.8 and directed towards the Earth.

The orbit of the Moon about the Earth is a further example of action at a distance; there is no experimental evidence of the Moon and the Earth being connected by a length of string and yet the Moon maintains a circular path about the Earth. The Earth attracts the Moon and so holds it in its orbit with a force equal to the force with which the Moon attracts the Earth. This is an example of the interaction principle which is fundamentally a restatement of the third law: if a body A exerts a force on a body B, then B exerts on A a force of the same magnitude but in the opposite direction along the line joining the bodies. This principle will underpin the 'connected particle' examples considered later.

Finally in this section the force of **friction** is examined. Consider the following idealised situation. A pupil pushes the teacher's desk, laden with books, with a force of 20 N. The desk does not move. The pupil applies a force of 21 N – still to no avail. Finally the pupil musters a 22 N force and the desk moves. It is as though the desk were reacting to the pupil's efforts to move it; when the pupil pushes with 20 N the desk reacts with a frictional force of exactly that magnitude, and similarly for 21 N. However, there appears to be a limit to the ability of the desk to prevent motion and that limit would appear to lie somewhere between 21 and 22 N. Attempts to move the desk with forces less than this limit fail but the application of a force of, say, 25 N will cause the desk to move, resisting all the while with its limiting frictional force.

It can be confirmed experimentally that the limiting frictional force is proportional to the normal reaction between the desk and the floor. The normal reaction is the force acting upwards from the floor in order to balance the weight of the desk (if this force were absent the desk would plummet through the floor). Limiting friction acts at the moment when the desk is 'on the point of sliding'. Coulomb's fourth law of friction establishes the proportionality between the frictional force and normal reaction:

$$F_{\text{limiting}} = \mu N$$

where μ, the constant of proportionality, is called the **coefficient of friction**. Note that when motion begins the frictional force remains at the limiting value. **Coulomb's four laws of friction** are

- the frictional force between two surfaces opposes the relative motion,
- the magnitude of the frictional force is independent of the areas of the surfaces,
- the magnitude of the frictional force is independent of the velocity of one surface relative to the other,
- the magnitude of the limiting frictional force is proportional to the normal reaction i.e. $F = \mu R$, where μ is the coefficient of friction.

THE SECOND LAW

The following examples will serve to illustrate the power of the second law in describing the motion of objects under the influence of forces.

EXAMPLE 1

Figure 8.1 shows a body moving in the horizontal plane under the influence of two horizontal forces. State the magnitude of the force X.

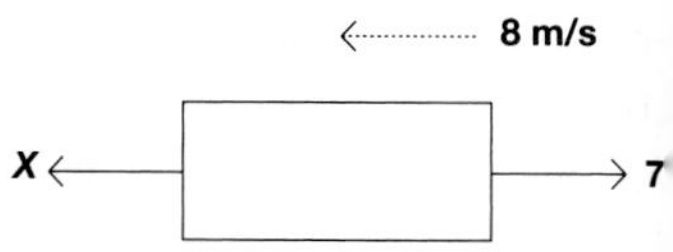

Figure 8.1

Solution 1

A body moving at constant speed is not accelerating and therefore there can be no net horizontal force in the direction of motion. It follows that the two forces acting on the body should balance. Therefore

$$X = 7\,\text{N}$$

EXAMPLE 2

Figure 8.2 illustrates a body of mass 10 kg moving in the horizontal plane with an acceleration of $2\,\text{m/s}^2$ under the action of two forces. Find Y.

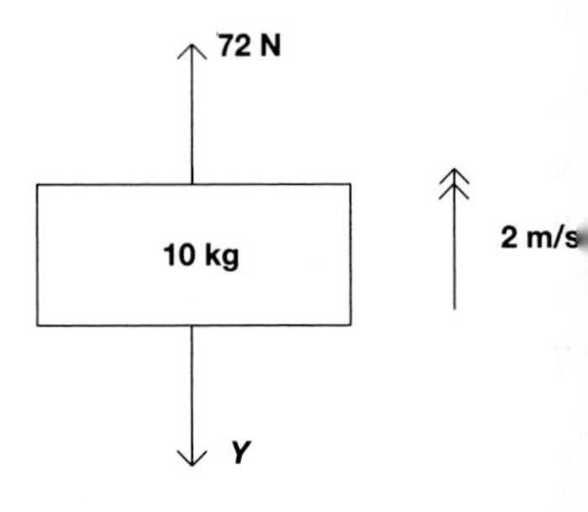

Figure 8.2

Solution 2

The net force in the direction of the acceleration is $72 - Y$. Using Newton's second law

$$72 - Y = 10(2)$$
$$\text{or} \quad Y = 52\,\text{N}$$

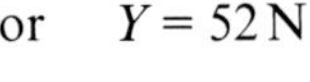

EXAMPLE 3

Figure 8.3 shows a body moving in the horizontal plane under the influence of a system of forces. Given that the body is moving at 4 m/s in the direction shown, find X and Y.

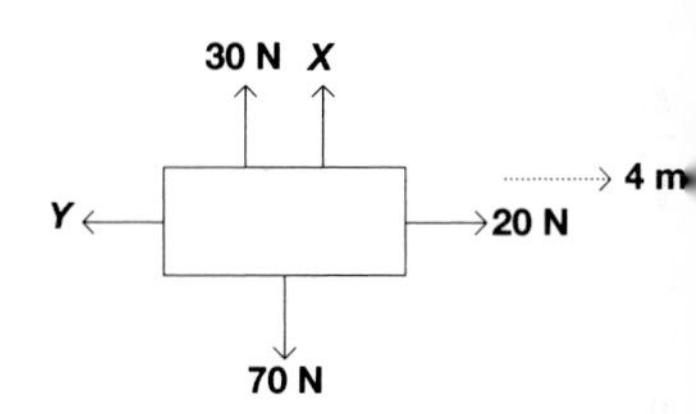

Figure 8.3

Solution 3

Since the body has no acceleration in the direction of the 30 N force and no acceleration (since it moves with constant speed) in the direction of the 20 N force then there can be no net force in either of these directions. Therefore

$$30 + X = 70 \quad \text{or} \quad X = 40\,\text{N}$$
$$\text{and} \quad 20 = Y \quad \text{or} \quad Y = 20\,\text{N}$$

EXAMPLE 4

Figure 8.4 shows a body of mass 4 kg moving in the horizontal plane under the action of a system of forces. Given that the body accelerates at $3\,\text{m/s}^2$ in the direction of the 58 N force, find X and Y.

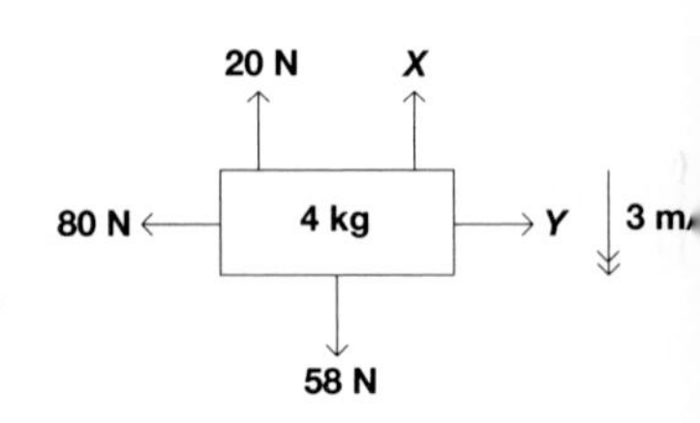

Figure 8.4

Solution 4

Since the body experiences no acceleration in the direction of the Y force

$$Y = 80\,\text{N}$$

The body experiences an acceleration of $3\,m/s^2$ in the direction of the net force

$58 - 20 - X$

Using $F = ma$

$$58 - 20 - X = 4(3)$$
$$\Rightarrow \quad X = 26\,N$$

EXAMPLE 5

A body of mass 4 kg travels vertically upwards under the action of a force of 80 N. Calculate its upward acceleration.

Solution 5

The forces acting are shown in Figure 8.5.

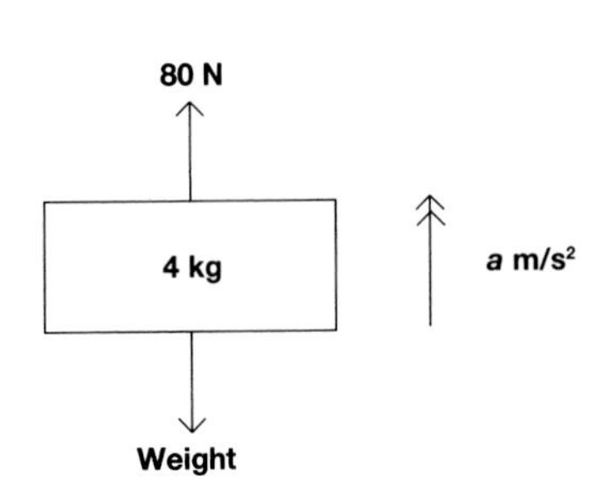

Figure 8.5

Note the presence of the weight of the body since it is moving vertically upwards. Now weight equals mass multiplied by the acceleration of gravity and so

$$\text{weight} = 4 \times 9.8\,N = 39.2\,N$$

The net force in the direction of the acceleration is therefore

$$80 - 39.2 = 40.8\,N$$

Using $F = ma$, therefore

$$40.8 = 4a$$
$$\text{or} \quad a = 10.2\,m/s^2$$

EXERCISE 8.1

1. Each diagram in Figure 8.6 represents an object under the action of a system of forces in the horizontal plane. Find the unknown forces A and B given that the object is at rest in every case.

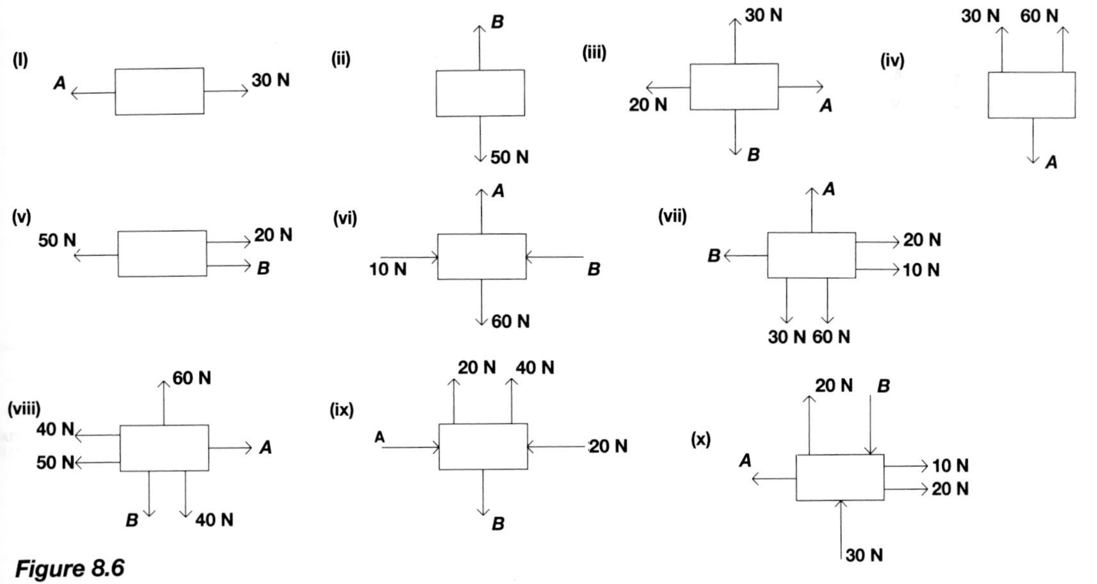

Figure 8.6

2. Each diagram in Figure 8.7 represents an object under the action of a system of forces in th horizontal plane. Find the unknown forces A and B given that the object moves with consta velocity in the direction indicated.

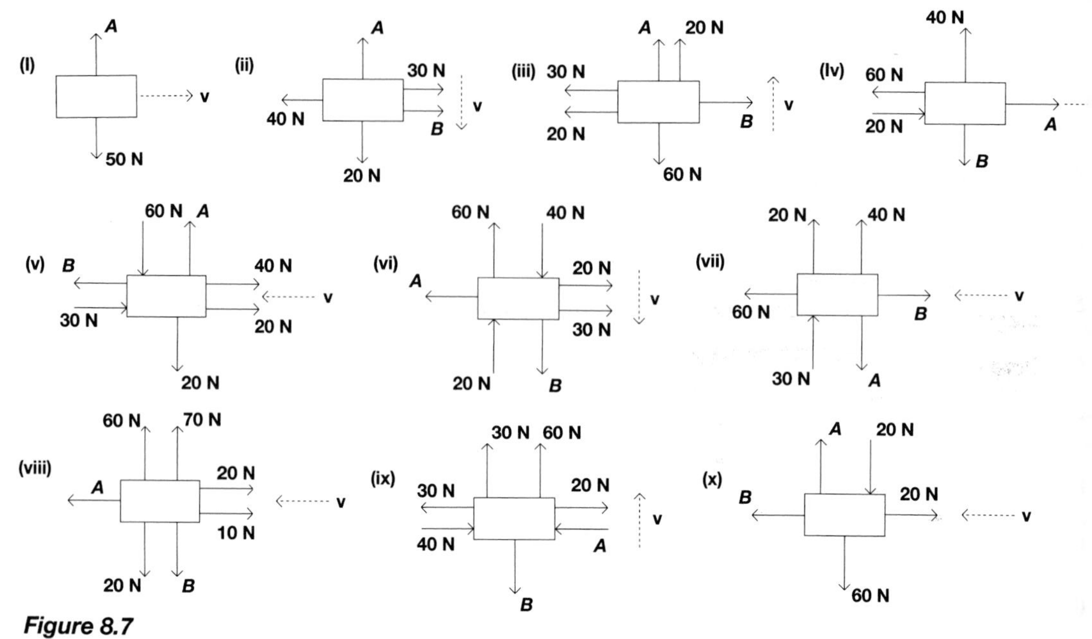

Figure 8.7

3. Each diagram in Figure 8.8 represents an object under the action of a system of forces in th horizontal plane. Find the unknown forces A and B given that the object accelerates horizontall in the direction shown.

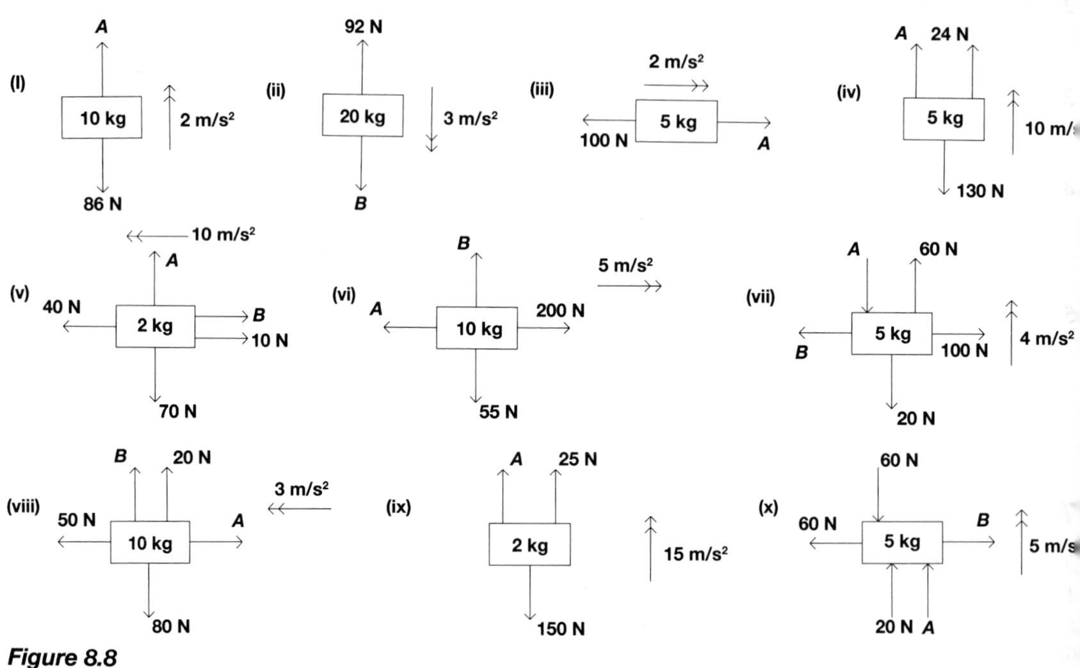

Figure 8.8

EXERCISE 8.2

1. An object experiences a force of 60 N. Given that the object's mass is 10 kg, what acceleration is the object given?

2. An object experiences a force of 48 N. Given that the object's mass is 12 kg, what acceleration is the object given?

3. An object experiences a force of 78 N. Given that the object's mass is 12 kg, what acceleration is the object given?

4. Find the magnitude of the force which will cause a mass of 8 kg to accelerate at 3 m/s^2.

5. Find the magnitude of the force which will cause a mass of 15 kg to accelerate at 8 m/s^2.

6. Find the magnitude of the force which will cause a mass of 10.5 kg to accelerate at 8 m/s^2.

7. A 30 N force causes a mass m kg to accelerate at 5 m/s^2. Find m.

8. A 56 N force causes a mass m kg to accelerate at 7 m/s^2. Find m.

9. A 75 N force causes a mass m kg to accelerate at 10 m/s^2. Find m.

10. Find the acceleration which results from a 2 N force acting upon an object of mass 50 g.

11. A vehicle of mass 750 kg accelerates at 3 m/s^2 along a level road with its engine exerting 1500 N. Find the resistance to motion.

12. A moped of mass 100 kg accelerates at 12 m/s^2 along a level road with its engine exerting 2000 N. Find the resistance to motion.

13. A van of mass 1500 kg accelerates along a level road with its engine exerting 7000 N. Find the acceleration given that the total resistance encountered can be represented by a force of 2500 N.

14. A limousine of mass 1200 kg moves along a level road against frictional resistances of 400 N and wind resistance of 200 N. Given that the forward force of the engine is 3000 N, find the rate at which the limousine is accelerating.

15. A car of mass 800 kg accelerates from rest to 30 m/s in 15 s. Find the forward force of the engine if this is to be accomplished against

    (a) no wind resistance,
    (b) wind resistance represented by a 500 N force.

The resolution principle

Consider the situation where a girl pulls a suitcase along the floor of an airport lounge with a force P. The suitcase is equipped with wheels and the floor is smooth. The situation is illustrated in Figure 8.9.

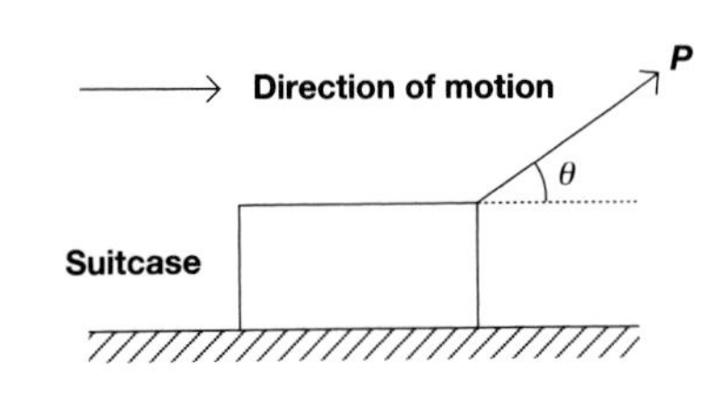

Figure 8.9

Now part of the force P is used to exert an upward force on the suitcase and hence not all of the force is being used to move the suitcase in the direction of motion. This may be remedied by applying the force P horizontally as in Figure 8.10. Here P is horizontal and none of it is used to apply an upward force on the suitcase.

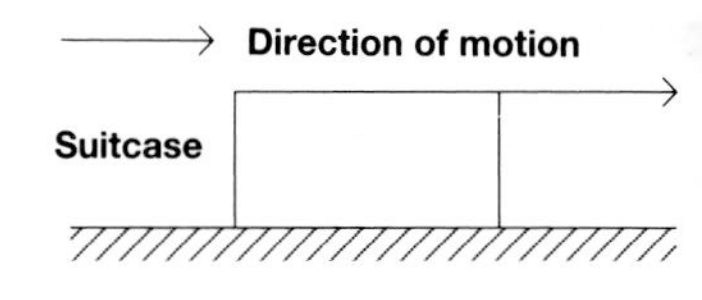

Figure 8.10

In Figure 8.9, therefore, only part of P is applied in the direction of motion while in Figure 8.10, all of P is in the direction of motion. The other 'extreme' is illustrated in Figure 8.11. Here none of P is exerted in the direction of motion – it is all applied as an upward force on the suitcase.

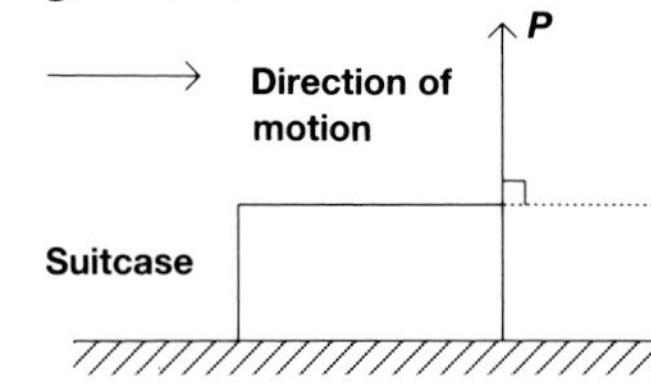

Figure 8.11

These three figures may be summarised in a single mathematical statement whose justification is clear from the trigonometry of Figure 8.12: the force in the direction of motion is given by $P\cos\theta$.

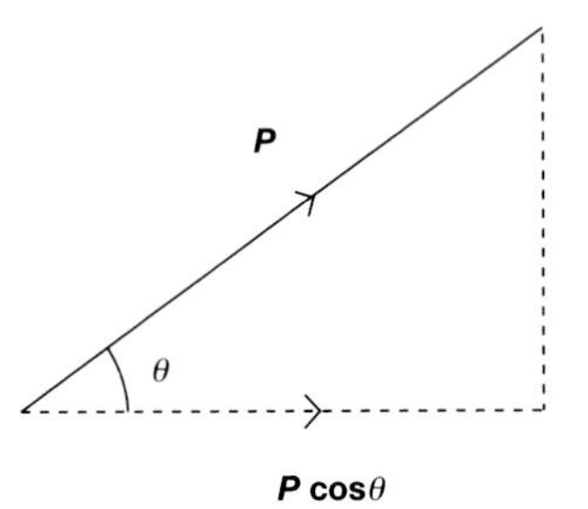

Figure 8.12

Figure 8.10 illustrates the situation where $\theta = 0°$ and here $P\cos\theta = P\cos 0° = P$ i.e. all of P is in the direction of motion.

Figure 8.11 illustrates the situation where $\theta = 90°$ and therefore $P\cos\theta = P\cos 90° = 0$ i.e. there is no force in the direction of motion. $P\cos\theta$ is known as the 'resolved part of P' in the direction of motion. What of the resolved part of P in a direction at right angles to the motion i.e. vertically upwards?

The same reasoning applies and therefore the resolved part of P in the vertical direction is

$$P\cos(90° - \theta)$$

since P makes an angle of $90° - \theta$ with the vertical direction as shown in Figure 8.13.

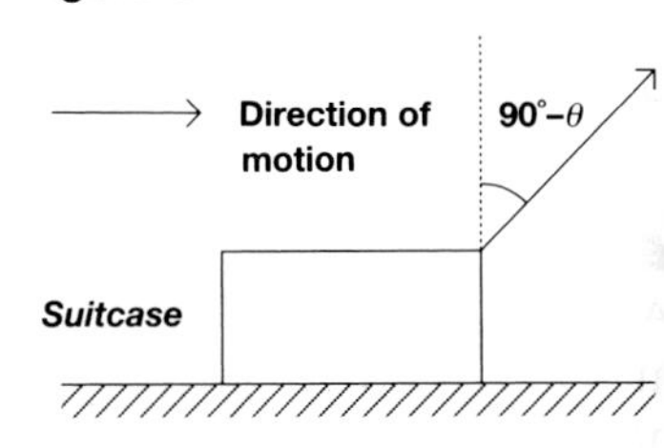

Figure 8.13

Now consider the triangle in Figure 8.14.

It is clear that

$$\cos(90° - \theta) = \frac{t}{s} = \sin\theta$$

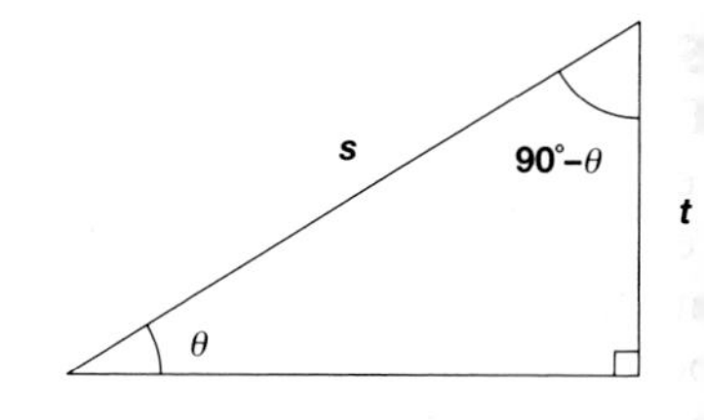

Figure 8.14

It follows that the resolved part of P in a direction at right angles to the direction of motion is $P\sin\theta$. In short, **where a force P makes an angle θ with the direction of motion, P may be resolved into a part $P\cos\theta$ along the direction of motion and a part $P\sin\theta$ at right angles to the direction of motion.** The examples which follow will demonstrate how the resolved parts of forces play their part in the resolution principle. The **resolution principle** states that **if a direction is chosen and the forces acting upon a body are resolved along that direction, the sum of the resolved parts equals the mass multiplied by the acceleration in that direction.** The resolution principle is merely the application of Newton's second law in the resolved direction.

EXAMPLE 6

A body of mass 4.8 kg accelerates along a rough surface at $1.8\,\text{m/s}^2$ under the action of a horizontal force of magnitude 20 N as shown in Figure 8.15. Find the coefficient of friction.

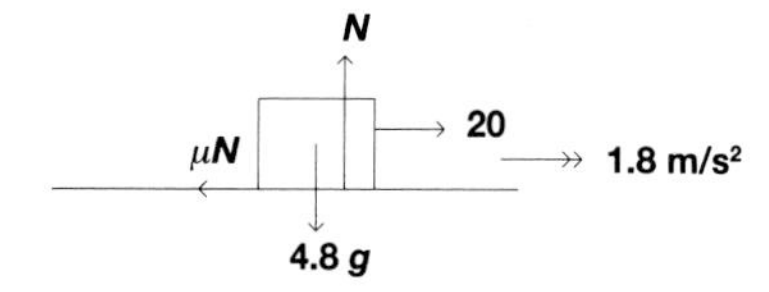

Figure 8.15

Solution 6

Figure 8.15 shows the weight of the body acting downwards and of magnitude $4.8g$ (recall that the weight, in newtons, is calculated by multiplying the mass, 4.8 kg, by $g = 9.8\,\text{m/s}^2$); N represents the normal (perpendicular to the surface) reaction of the surface on the body. Finally the friction exerts a force μN in the opposite direction to motion, where μ is the coefficient of friction.

Now, since there is no upward acceleration there can be no net force in this direction so

$$N = 4.8g$$

Applying $F = ma$ horizontally

$$20 - \mu N = (4.8)(1.8)$$

But $N = 4.8g$ and so

$$\begin{aligned} & 20 - \mu(4.8g) = (4.8)(1.8) \\ \Rightarrow\quad & 20 - 47.04\mu = 8.64 \\ \Rightarrow\quad & \mu = 0.2415 \end{aligned}$$

Hence the coefficient of friction is 0.2415.

EXAMPLE 7

A body of mass 5.4 kg accelerates along a rough horizontal surface at $2.8\,\text{m/s}^2$ under the action of a force of 41 N inclined at 58° to the horizontal. Find the coefficient of friction.

Solution 7

The forces on the body are illustrated in Figure 8.16. The resolution principle must be applied here – the rule is always to resolve forces either along or perpendicular to the direction of motion. The three forces N, $5.4g$ and μN are each parallel or perpendicular to the direction of motion, with the latter force being strictly anti-parallel to the direction of motion. The 41 N force has a resolved part $41\cos 58°$ along the direction of motion and $41\sin 58°$ at right-angles to it. Figure 8.16 may therefore be re-drawn as Figure 8.17.

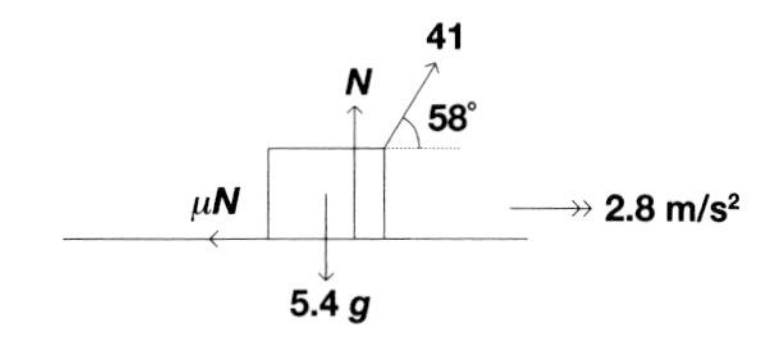

Figure 8.16

Once again, as there is no acceleration in the vertical direction, there can be no net force i.e.

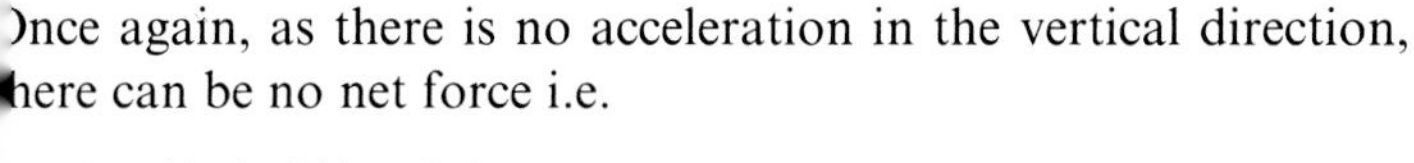

$$\begin{aligned} & N + 41\sin 58° = 5.4g \\ \Rightarrow\quad & N = 18.15 \text{ newtons} \end{aligned}$$

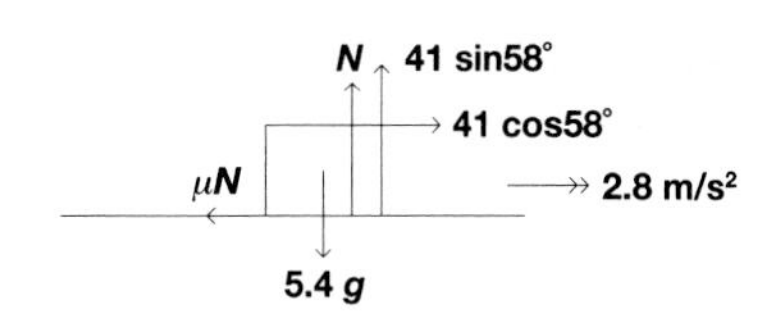

Figure 8.17

Applying $F = ma$ horizontally

$$41 \cos 58° - \mu N = (5.4)(2.8)$$

that is

$$21.73 - 18.15\mu = 15.12$$
$$\Rightarrow \quad \mu = 0.3642$$

The coefficient of friction is 0.3642.

EXAMPLE 8

A body of mass 5 kg accelerates along a rough horizontal surface at $1.7\,\text{m/s}^2$ under the action of a force of 55 N inclined at 42° to the horizontal and directed towards the floor. Find the coefficient of friction.

Solution 8

This information is summarised in Figure 8.18.

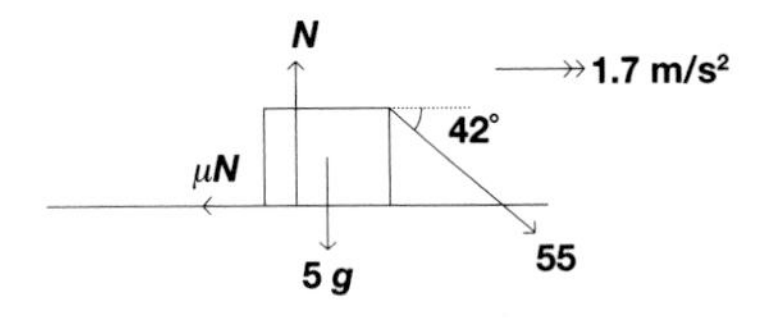

Figure 8.18

Once again the only force which is not resolved along or perpendicular to the direction of motion is the pulling force, 55 N. The 55 N force may be resolved into a part 55 cos 42° in the direction of motion and a part 55 sin 42° pressing the body into the surface as in Figure 8.19.

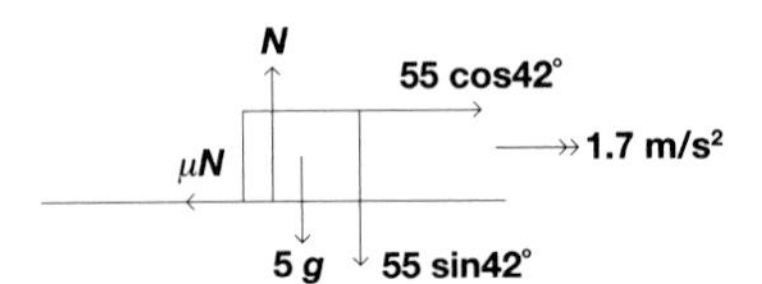

Figure 8.19

Again, since the body experiences no motion at right angles to the surface there can be no net force in this direction and so

$$N = 5g + 55 \sin 42°$$
$$\Rightarrow \quad N = 85.8 \text{ newtons}$$

Applying $F = ma$ horizontally

$$55 \cos 42° - \mu N = (5)(1.7)$$
$$\Rightarrow \quad 55 \cos 42° - \mu(85.8) = (5)(1.7)$$
$$\Rightarrow \quad \mu = 0.3773$$

The coefficient of friction is 0.3773.

EXERCISE 8.3

In all questions take $g = 9.8\,\text{m/s}^2$ unless otherwise instructed.

1. In parts (i) to (vi) of Figure 8.20 an object of mass 10 kg accelerates along a rough surface with an acceleration, as shown, under the action of a force which is horizontal in cases (i) to (iv) and inclined to the horizontal in cases (v) and (vi). R symbolises the normal reaction and F the frictional force. Calculate the coefficient of friction in each case.

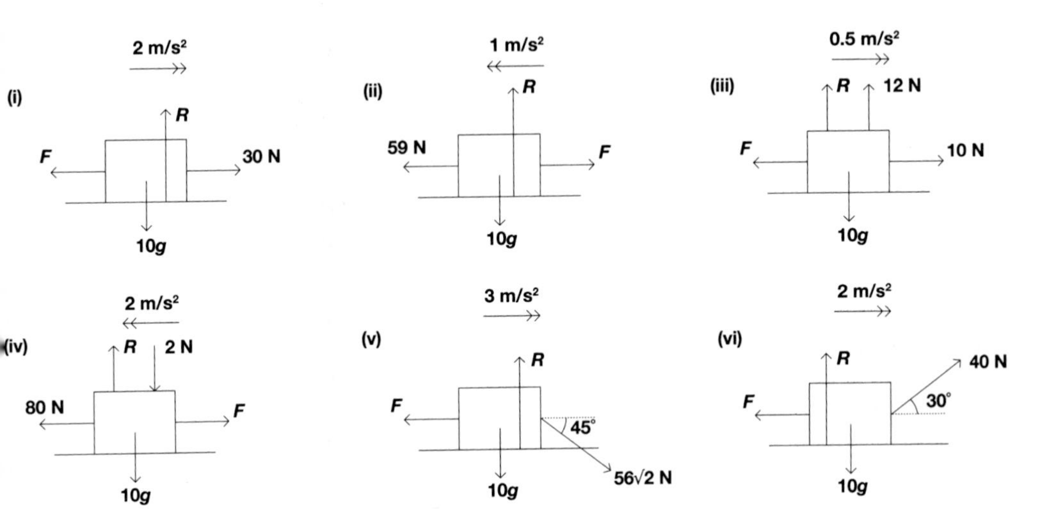

Figure 8.20

2. A body of mass 5 kg is at rest on a rough horizontal floor. When acted upon by a horizontal force of 14 N, the body is on the point of slipping and therefore friction is limiting. Find the coefficient of friction between the body and the floor.

3. A body of mass 15 kg is at rest on a rough horizontal floor, the coefficient of friction between body and floor being 0.1. A horizontal force of 10 N is applied to the body. Calculate the frictional force with which the body resists this force and state whether or not the body will move.

4. The coefficient of friction between a 12 kg mass and a rough horizontal floor is 0.2. Calculate the maximum frictional force which the mass can offer and find the acceleration of the mass when acted upon by a 25 N horizontal force.

5. The coefficient of friction between a 750 g mass and a horizontal floor is 0.05. Calculate the maximum frictional force which the mass can offer and find the acceleration of the mass when acted upon by a 5 N horizontal force.

6. A body of mass 750 g is at rest on a rough horizontal floor. When acted upon by a horizontal force 0.6 N, the body is on the point of slipping and therefore friction is limiting. Find the coefficient of friction between the body and the floor.

7. An object of mass 5 kg is at rest on a rough horizontal floor with the coefficient of frictio between floor and object being 0.2. Calculate the horizontal force which must be applied to th object in order to maintain it in motion with a constant acceleration of $2\,\text{m/s}^2$.

8. An object of mass 12 kg is at rest on a rough horizontal floor with the coefficient of frictio between the floor and the object being 0.5. Calculate the horizontal force which must be applie to the object in order to maintain it in motion with a constant acceleration of $1\,\text{m/s}^2$.

9. A horizontal force of 25 N, applied to an object of mass 10 kg at rest on a rough horizonta surface, accelerates the object along the surface at $2\,\text{m/s}^2$. Calculate the coefficient of frictio between object and surface.

10. A body of mass 5 kg slides with constant speed 5 m/s on a smooth surface before encountering rough patch of ground with $\mu = 0.1$. How far will the body slide on the rough patch befor coming to rest?

11. A body of mass 8 kg slides with constant speed 3 m/s on a smooth surface before encountering rough patch of ground with $\mu = 0.75$. How far will the body slide on the rough patch befor coming to rest?

12. Take $g = 10\,\text{m/s}^2$.

    A block of wood of mass 12 kg, initially at rest, is pulled with constant acceleration in a straigh line along a horizontal plane surface. After 30 seconds the block is moving with a velocity c 25 m/s.

    (i) Find the acceleration of the block.

    The block is moving under the action of a horizontal force of 50 N.

    (ii) Find the force opposing the motion of the block.

    It is assumed that the force opposing the motion is entirely due to friction between the block an the plane.

    (iii) Find the coefficient of friction.

    (N. Ireland Additional Mathematics, 1989

13. Take $g = 10\,\text{m/s}^2$.

    A block of mass 12 kg is pulled along a rough horizontal plane by a force of 100 N inclined at an angle 30° to the horizontal as shown in Figure 8.21. The coefficient of friction between the block and the plane is 0.2.

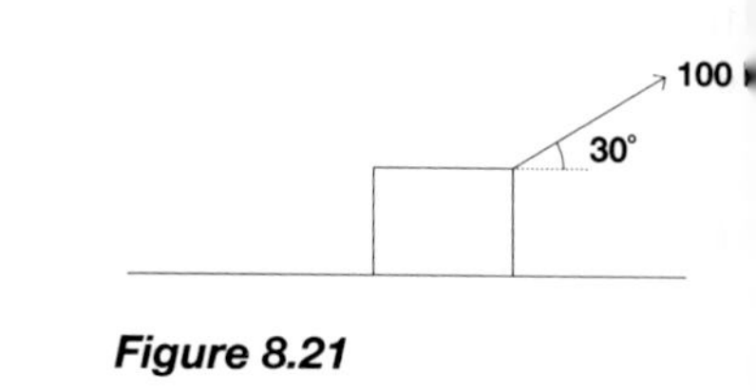

Figure 8.21

    (i) Copy Figure 8.21 and mark clearly on it all the forces acting on the block.

    (ii) Calculate the acceleration of the block.

    (N. Ireland Additional Mathematics, 1990

14. Take $g = 10\,\text{m/s}^2$.

A block of mass 5 kg rests on a horizontal table. The coefficient of friction between the block and table is 0.25. A light string is attached to the block and is held taut at 30° above the horizontal.

Draw a diagram showing clearly all the forces acting on the block.

(i) Express the normal reaction R in terms of T, the tension in the string.

(ii) Express the frictional force F in terms of T.

(iii) Calculate the tension in the string when the block is about to move.

(iv) If the tension in the string is increased to 25 N show that the block remains in contact with the table and calculate the acceleration of the block along the table.

(N. Ireland Additional Mathematics, 1987)

15. A force P newtons is acting on a body of mass 2 kg which remains at rest on a rough horizontal surface. The force P is directed downwards towards the body at an angle of 30° to the vertical.

Draw a neat diagram showing clearly all the forces acting on the body.

Express in terms of P

(i) the normal reaction R newtons on the body,
(ii) the frictional force F newtons on the body.

When P takes the value 146, the body is on the point of slipping. Calculate the coefficient of friction between the body and the surface.

Take $g = 9.8\,\text{m/s}^2$.

(N. Ireland Additional Mathematics, 1984)

The inclined plane

The inclined plane holds a special place in the history of the study of motion. Galileo investigated the concept of acceleration by timing the trajectories of small balls down an inclined plane using water dripping from a leaking tank as the timing device. The examples which follow illustrate how methods hitherto applied to the investigation of motion in the horizontal plane may be extended to the study of motion on an inclined plane.

EXAMPLE 9

A body of mass 5 kg is pulled down a smooth plane inclined at 27° to the horizontal by a force of 12 N as shown in Figure 8.22. Find the acceleration of the body and the normal reaction of the plane on the body.

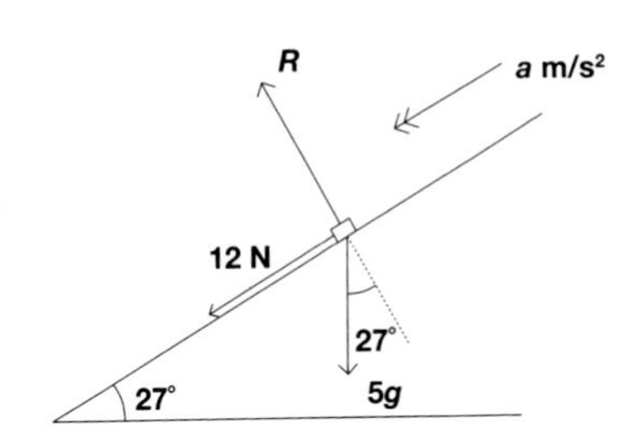

Figure 8.22

Solution 9

The normal reaction, R, and acceleration down the plane, a, are both shown in Figure 8.22. Resolving the weight along and at right angles to the direction of motion transforms Figure 8.22 into Figure 8.23.

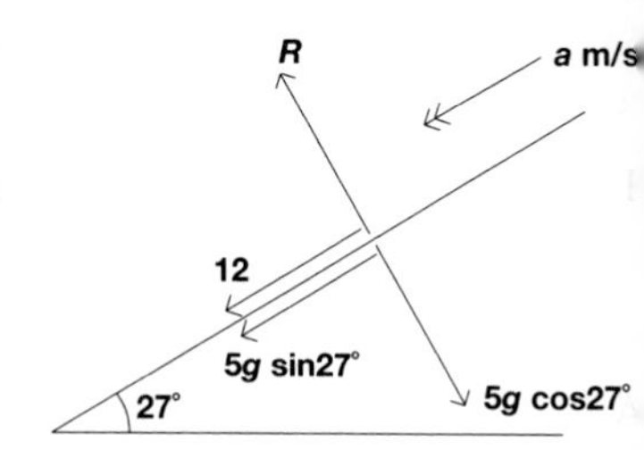

Figure 8.23

Since there is no acceleration at right angles to the plane, there can be no net force in this direction. It follows that

$$R = 5g \cos 27^\circ$$
$$\Rightarrow \quad R = 43.66\,\text{N}$$

and the application of Newton's second law to motion down the plane yields

$$12 + 5g \sin 27^\circ = 5a$$
i.e. $$a = 6.849\,\text{m/s}^2$$

EXAMPLE 10

A body of mass 4 kg rests on a rough plane inclined at 40° to the horizontal. When the body is pulled down the plane by a force of 38 N inclined at 21° to the plane, it experiences a constant frictional force of 3 N. Find the acceleration of the body down the plane and the normal reaction of the surface of the plane on the body.

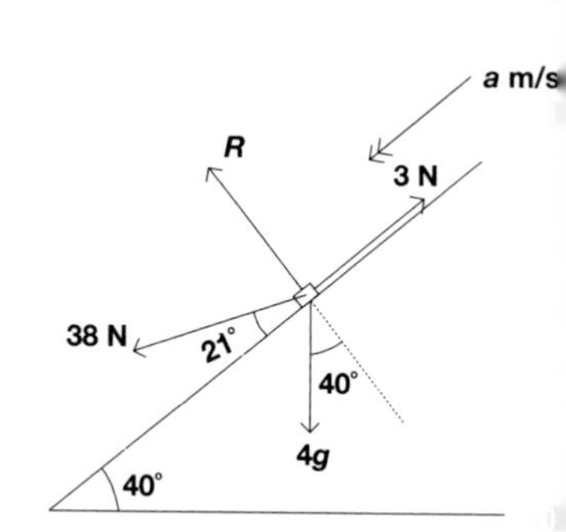

Figure 8.24

Solution 10

The situation described above is summarised in Figure 8.24.

The normal reaction, R, is at right angles to the direction of motion and the 3 N frictional force is anti-parallel to the direction of motion. However, the 38 N force and the weight must be resolved along and at right angles to the direction of motion. Carrying out this resolution gives the new force diagram of Figure 8.25.

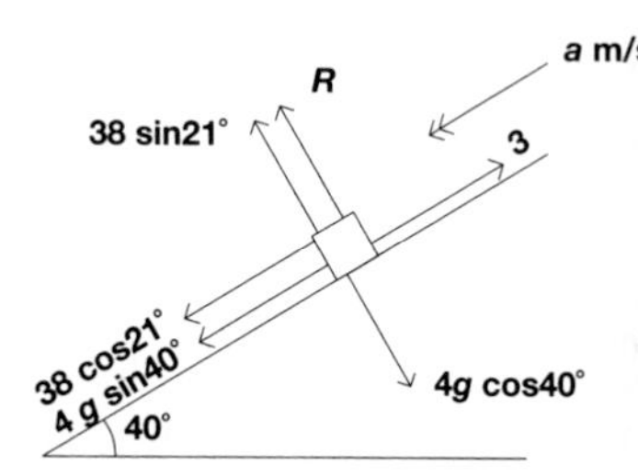

Figure 8.25

Since there is no acceleration at right angles to the plane there can be no net force in this direction.

$$R + 38 \sin 21^\circ = 4 \times 9.8 \cos 40^\circ$$

Applying Newton's second law along the plane

$$38 \cos 21^\circ + 4 \times 9.8 \sin 40^\circ - 3 = 4a$$

it follows that $R = 16.41\,\text{N}$ and $a = 14.42\,\text{m/s}^2$.

EXAMPLE 11

A body of mass 5 kg rests on a rough plane inclined at 30° to the horizontal. When the body is pulled up the plane by a force of 50 N inclined at 25° to the plane, it experiences a constant frictional resistance of 2 N. Find the acceleration of the body up the plane and the normal reaction of the surface of the plane on the body.

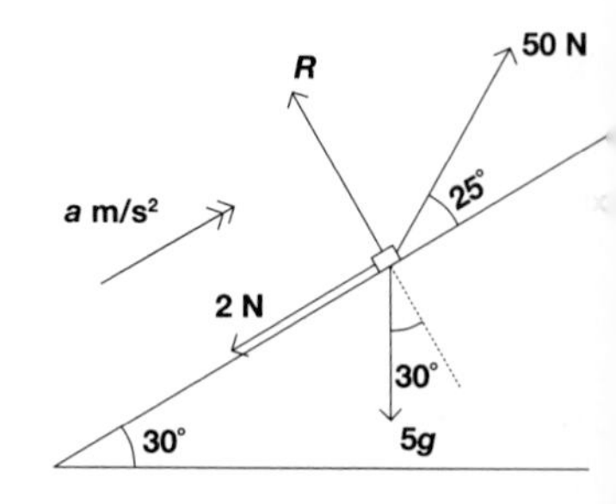

Figure 8.26

Solution 11

Figure 8.26 reflects the information given. Resolving the 50 N and

5g forces along and perpendicular to the direction of motion results in Figure 8.27.

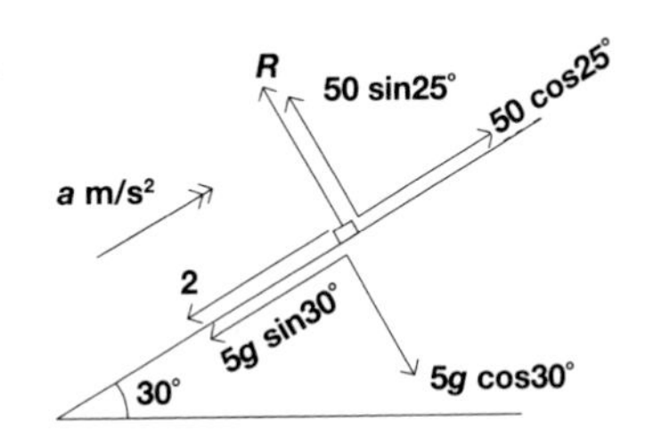

Figure 8.27

Therefore

$$R + 50 \sin 25° = 5g \cos 30°$$
$$\Rightarrow \quad R = 21.3\,\text{N}$$

Also $F = ma$ gives

$$50 \cos 25° - 2 - 5g \sin 30° = 5a$$
$$18.82 = 5a$$
$$\Rightarrow \quad a = 3.763\,\text{m/s}^2$$

EXERCISE 8.4

For each of the situations illustrated in Figure 8.28 find the normal reaction R and the acceleration a. Take $g = 9.8\,\text{m/s}^2$ throughout.

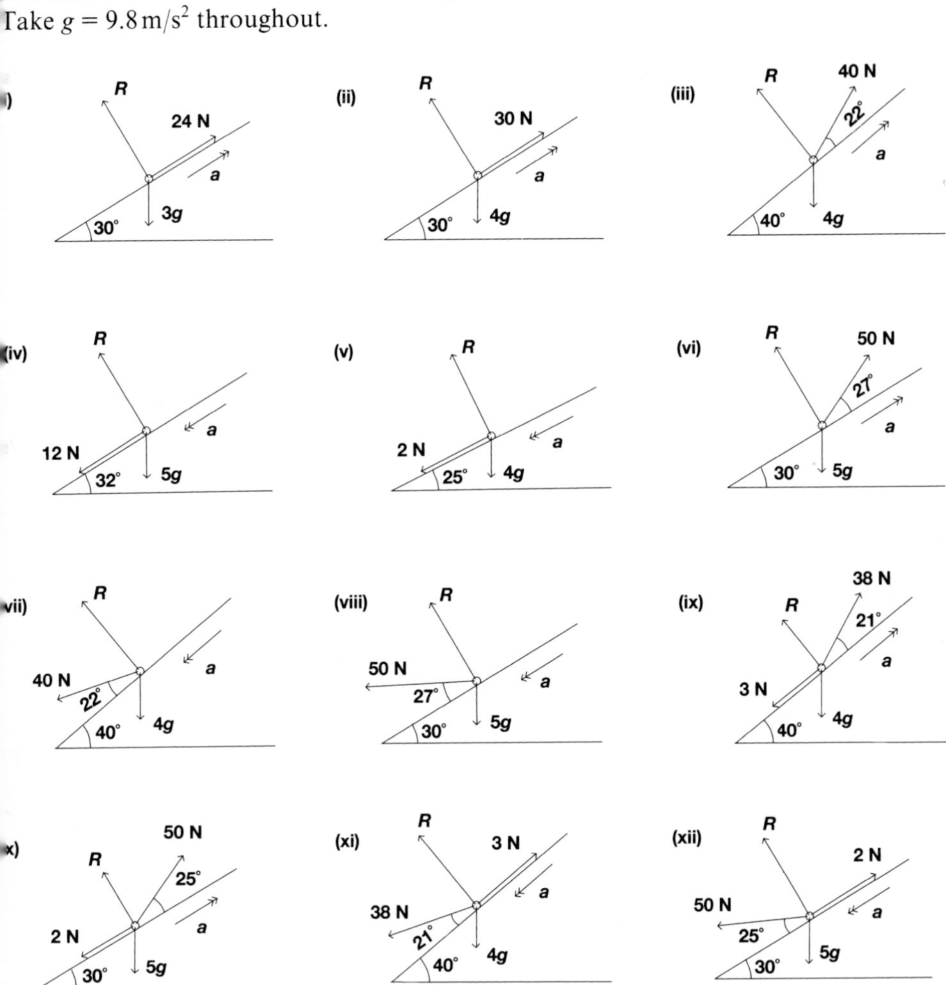

Figure 8.28

Friction on the inclined plane

EXAMPLE 12

A body of mass 7 kg rests on a rough plane inclined at 30° to the horizontal. The coefficient of friction between the body and the plane is 0.2 and a force X, applied in a direction up and parallel to the plane, just prevents the body from slipping down the plane. Find X and the normal reaction of the plane on the body.

Solution 12

Since the body is on the point of slipping down the plane, the frictional force will act up the plane in the direction of the force X as shown in Figure 8.29.

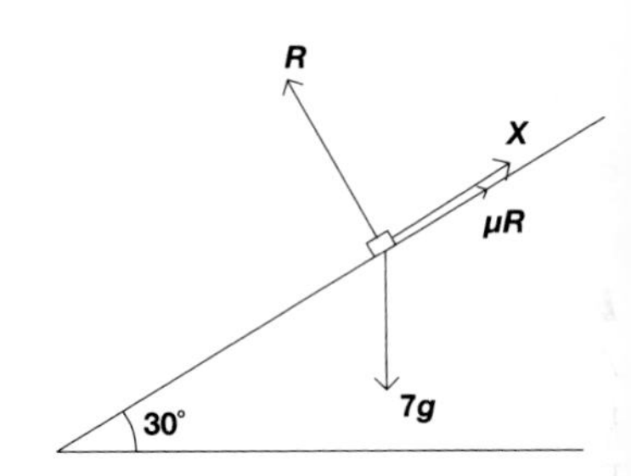

Figure 8.29

If R is the normal reaction, the frictional force is μR, with $\mu = 0.2$ in this case. Resolving the weight along and perpendicular to the plane transforms Figure 8.29 to Figure 8.30.

Since the body remains in equilibrium, there is no net force along or perpendicular to the plane. Hence, with reference to Figure 8.30

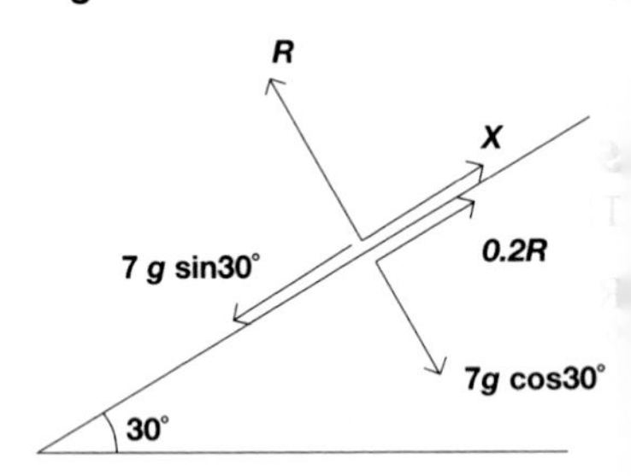

Figure 8.30

$$R = 7g\cos 30°$$
$$\text{and} \quad 7g\sin 30° = 0.2R + X$$

The first of these equation yields

$$R = 59.41\,\text{N}$$

Substituting this value in the second equation

$$7g\sin 30° = 0.2(59.41) + X$$
$$\text{or} \quad X = 22.42\,\text{N}$$

EXAMPLE 13

A body of mass 4.92 kg rests on a rough plane inclined at 29° to the horizontal. The coefficient of friction between the body and the plane is 0.14 and a force X, applied in a direction up the plane and inclined at 29.2° to it, just prevents the body from slipping down the plane. Find X and the normal reaction.

Solution 13

If the body is about to slip down the plane then friction will be directed upwards along the surface of the plane. Figure 8.31 is the appropriate force diagram.

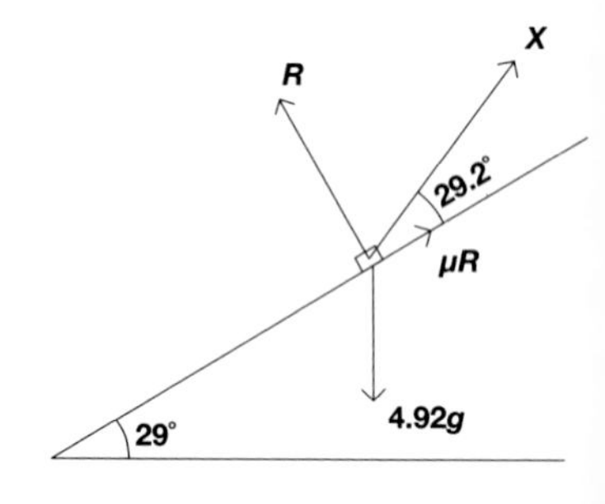

Figure 8.31

Resolving the X and $4.92g$ force along and at right angles to the direction of motion, transforms Figure 8.31 to Figure 8.32.

Since no acceleration occurs there can be no net force at right angles or parallel to the plane.

Hence

$$R + X \sin 29.2° = 4.92g \cos 29°$$
and $$X \cos 29.2° + \mu R = 4.92g \sin 29°$$

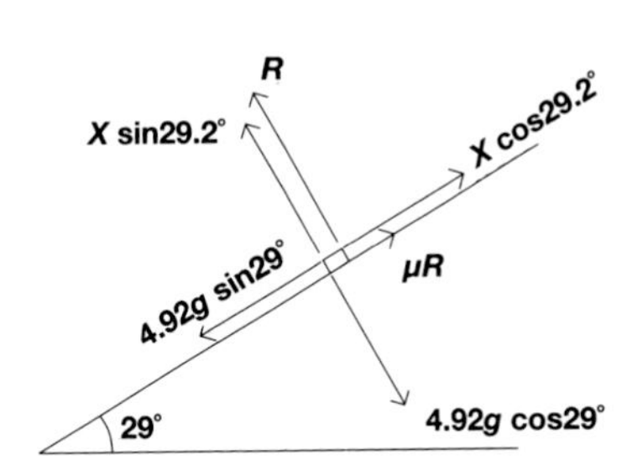

Figure 8.32

Inserting R, found by rearranging the first equation, in the second equation

$$X \cos 29.2° + 0.14(4.92g \cos 29° - X \sin 29.2°) = 4.92g \sin 29°$$
$$\Rightarrow \quad 0.8729\,X + 5.904 - 0.0683\,X = 23.38$$
$$\Rightarrow \quad 0.8046\,X = 17.48$$

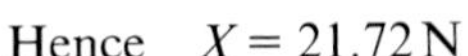

Hence $X = 21.72\,\text{N}$
and $R = 4.92g \cos 29° - 21.72 \sin 29.2°$
$\Rightarrow$ $R = 31.57\,\text{N}$

EXAMPLE 14

An object of mass 2.08 kg rests on a rough plane inclined at 19.8° to the horizontal. A force of 14.98 N, applied at 45.2° to the plane, pulls the object up the plane with acceleration $0.992\,\text{m/s}^2$. Find μ, the coefficient of friction between object and plane.

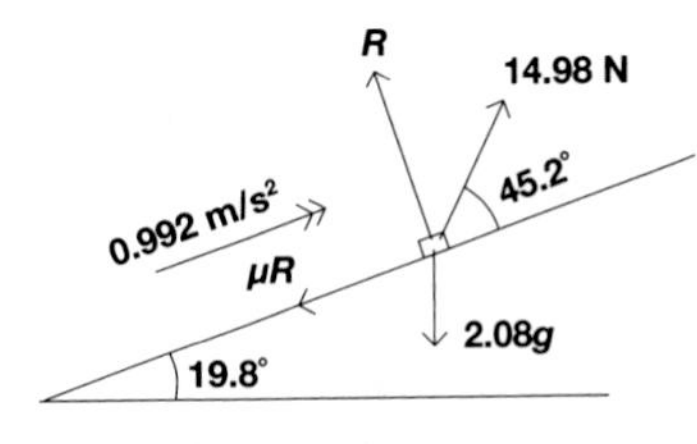

Figure 8.33

Solution 14

The force diagram in this case is illustrated in Figure 8.33.

Resolving the weight and the 14.98 N force yields Figure 8.34.

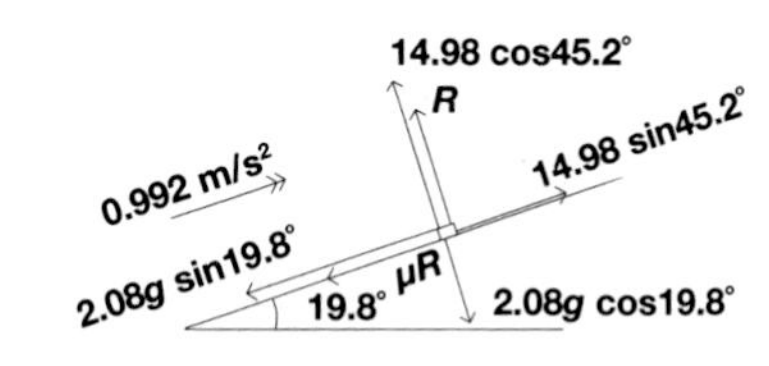

Figure 8.34

Therefore $$R + 14.98 \sin 45.2° = 2.08g \cos 19.8°$$
and $$14.98 \cos 45.2° - 2.08g \sin 19.8° - \mu R = 2.08(0.992)$$

It follows that $R = 8.55\,\text{N}$ (from the first equation)
therefore $\mu = 0.1856$ (from the second).

EXERCISE 8.5

Take $g = 9.8\,\text{m/s}^2$ throughout this exercise unless otherwise instructed.

1. The angle of inclination and coefficient of friction of eight planes are given below. In each case find the limiting friction for a body of mass 5 kg placed on the plane and state whether the body will accelerate down the plane or remain at rest upon it.

(i) angle of inclination 10° $\mu = 0.5$
(ii) angle of inclination 40° $\mu = 0.5$
(iii) angle of inclination 15° $\mu = 0.5$
(iv) angle of inclination 50° $\mu = 0.5$
(v) angle of inclination 10° $\mu = 0.25$
(vi) angle of inclination 15° $\mu = 0.25$
(vii) angle of inclination 30° $\mu = 0.25$
(viii) angle of inclination 8° $\mu = 0.25$

2. A rough plane is inclined at 30° to the horizontal. A body of mass 750 g is placed on the plane such that the coefficient of friction between the body and the plane is 0.5. Calculate the limiting frictional force and decide if the body will slide.

3. A rough plane is inclined at 20° to the horizontal. A mass of 10 kg is placed on the plane. The mass is in limiting equilibrium when a force of 30 N parallel to and up the plane is applied. Find μ.

4. A rough plane is inclined at 20° to the horizontal. A mass of 3 kg is placed on the plane. When a force of 30 N is applied to the mass, parallel to and up the plane, the mass accelerates up the plane at $2\,\text{m/s}^2$. Find μ.

5. A rough plane is inclined at 15° to the horizontal and is 3.2 m long. The coefficient of friction between a body of mass 0.5 kg and the plane is 0.1. If the body is released from the top of the plane find its speed upon reaching the bottom.

6. A rough plane is inclined at 25° to the horizontal and is very long. An object of mass 5 kg is released from the top of the plane and after 1.5 s has reached a speed of 3.5 m/s. Find μ, the coefficient of friction between object and plane.

7. A body of mass 5 kg is placed on a rough plane ($\mu = \frac{1}{7}$) inclined at 30.2° to the horizontal. A force of X just prevents the body from slipping down the plane. The X force is directed up the plane at 36° to it as illustrated in Figure 8.35. Copy and complete the force diagram in Figure 8.35 and calculate X.

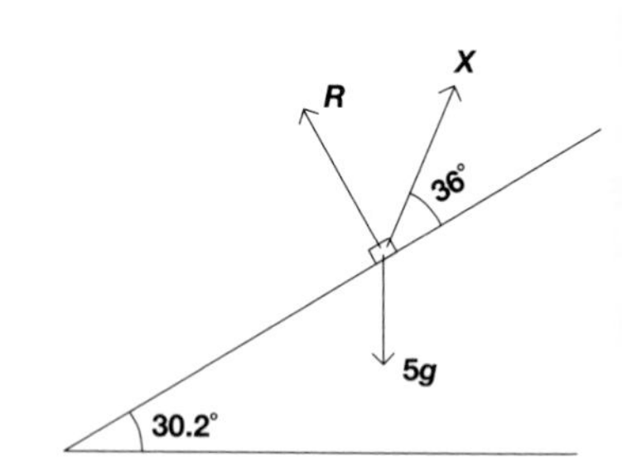

Figure 8.35

8. A body of mass 3 kg is placed on a rough plane ($\mu = 0.34$) inclined at 30.8° to the horizontal. A force of X at 18.7° to the plane and directed up the plane acts upon the body in such a way that it is just about to move up the plane. Copy and complete the force diagram in Figure 8.36 and find X.

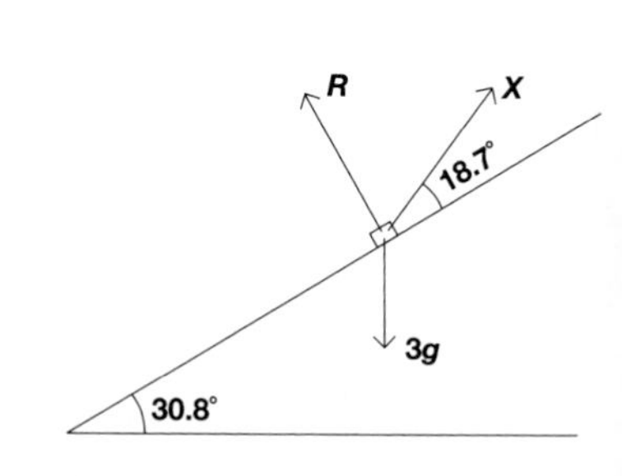

Figure 8.36

9. A body of mass 2 kg is placed on a rough plane inclined at 20.4° to the horizontal. A force of 15 N is applied to the body at 45° to the plane and acting up the plane, causing the body to accelerate up the plane at $1.01\,\text{m/s}^2$. Find μ, the coefficient of friction between body and plane.

10. A body of mass 3.02 kg is placed on a rough ($\mu = 0.332$) plane inclined at 30.2° to the horizontal. A force of X acting up the plane and at 17.8° to it is applied to the body such that it is just about to move up the plane. Find X and the normal reaction.

11. Take g to be $10\,\text{m/s}^2$.

An engine of mass 50 tonnes moves along a railway track. The frictional resistance to the motion of the engine is constant and of magnitude 50 N per tonne. When the engine travels along a straight section of the horizontal track the force generated by the engine is 15 kN.

(i) Calculate the acceleration, $f\,\text{m/s}^2$, of the engine.

The engine ascends a hill inclined at an angle α to the horizontal where $\sin\alpha = \frac{1}{100}$. The acceleration is now $\dfrac{f}{2}$ and the frictional resistance remains at 50 N per tonne.

(ii) Calculate the force, in kN, generated by the engine.

(iii) Determine the force, in kN, which the engine must generate when climbing the hill to ensure that it is travelling with constant velocity.

(N. Ireland Additional Mathematics, 1989)

12. Take g to be $10\,\text{m/s}^2$.

A body of mass 10 kg is on the point of sliding down a rough plane inclined at 30° to the horizontal, as shown in Figure 8.37.

(i) Copy Figure 8.37 and mark clearly on it all the forces acting on the body.
(ii) Find the coefficient of friction between the body and the inclined plane.
(iii) If the body is being pulled *up* the plane with constant velocity by a force F acting parallel to the plane, find the magnitude of F.
(iv) Determine the force parallel to the plane required to pull the body *down* the plane with an acceleration of $0.5\,\text{m/s}^2$.

(N. Ireland Additional Mathematics, 1990)

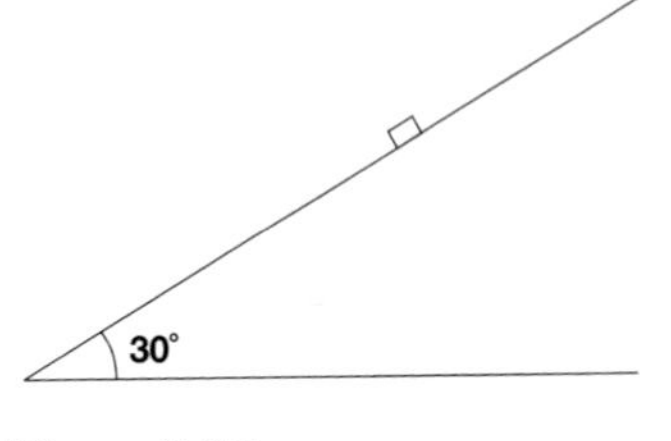

Figure 8.37

13. Take g to be $10\,\text{m/s}^2$.

(a) A body of mass 20 kg rests on a rough plane inclined to the horizontal at an angle of 35° as shown in Figure 8.38.

The body is on the point of slipping down the plane.

(i) Copy Figure 8.38 and mark clearly on it all the forces acting on the body.

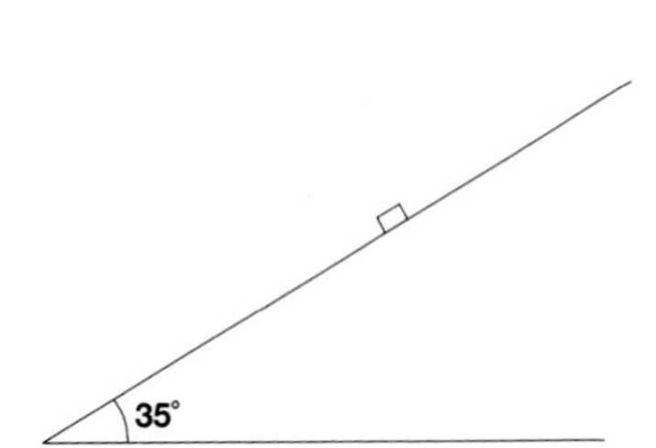

Figure 8.38

Calculate
(ii) the normal reaction of the plane on the body,
(iii) the coefficient of friction between the body and the plane.

(b) The inclination of the plane is now reduced to 30° as shown in Figure 8.39. Find the force P, which, acting at 20° to the plane, will cause the body to move up the plane with constant velocity.

(N. Ireland Additional Mathematics, 1991)

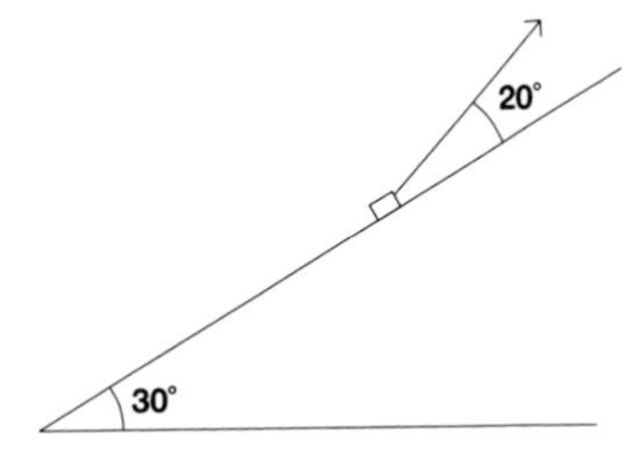

Figure 8.39

CONNECTED PARTICLES

Newton's third law plays an important role in the analysis of the motion of objects which are connected together. The third law requires that a body cannot exert a force upon another body without feeling the effects of that force in the opposite sense. If a trailer is being pulled along by a car (see Figure 8.40) and the trailer experiences a force X, pulling it in the direction of motion, then the car experiences a pull X in the direction opposite to the direction of motion, this pair of forces being transmitted via the device connecting car and trailer.

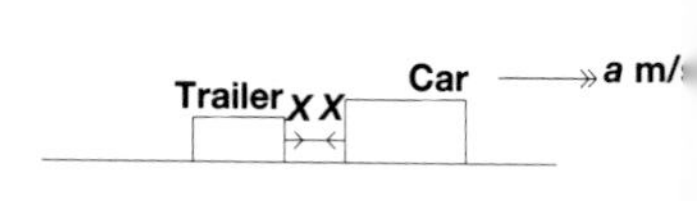

Figure 8.40

These equal and opposite forces which exist in all links are fundamental to connected particle problems. The examples which follow illustrate how such problems are solved.

EXAMPLE 15

A large model train, illustrated in Figure 8.41, consists of an engine of mass 8 kg, capable of exerting 60 N and three carriages of mass 1 kg, 2 kg and 4 kg. Calculate the acceleration of the train and the tensions T_1, T_2 and T_3 in the couplings.

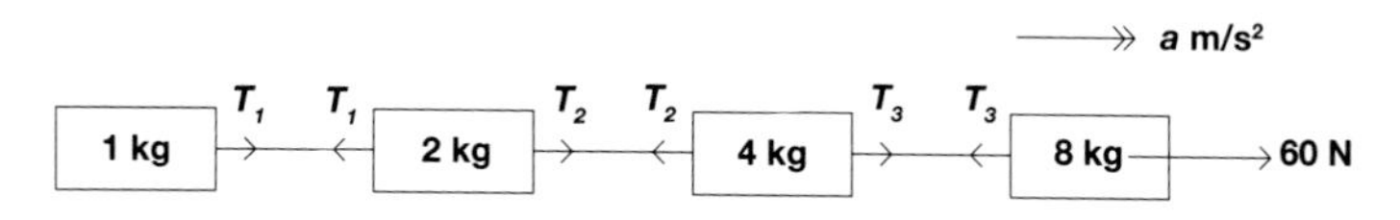

Figure 8.41

Solution 15

If each component is isolated the four diagrams of Figure 8.42 result.

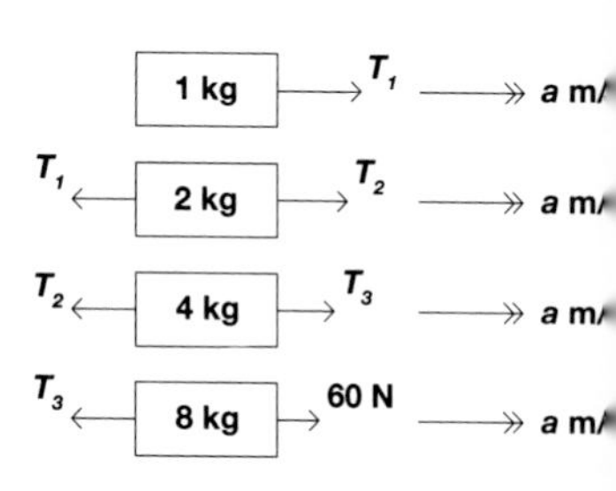

Figure 8.42

Since the parts are **connected** they all accelerate at the same rate, a m/s^2.

Applying Newton's second law to each diagram in Figure 8.42

$$T_1 = 1a$$
$$T_2 - T_1 = 2a$$
$$T_3 - T_2 = 4a$$
$$60 - T_3 = 8a$$

If these equations are added, then

$$T_1 + T_2 - T_1 + T_3 - T_2 + 60 - T_3 = 1a + 2a + 4a + 8a$$

or $\quad 60 = 15a$

i.e. $\quad a = 4$

Substituting $a = 4$ in each of the above equations in turn yields

$$T_1 = 4\,\text{N}$$
$$T_2 - 4\,\text{N} = 2 \times 4\,\text{N} \quad \Rightarrow \quad T_2 = 12\,\text{N}$$
$$T_3 - 12\,\text{N} = 4 \times 4\,\text{N} \quad \Rightarrow \quad T_3 = 28\,\text{N}$$

Hence the common acceleration is $4\,\text{m/s}^2$ with the couplings experiencing tensions of 4, 12 and 28 N.

EXAMPLE 16

Masses of 7 kg and 4 kg are connected by a light inextensible string which passes over a light, smooth, fixed pulley. Find the acceleration of the masses and the tension in the string.

Solution 16

The situation is illustrated in Figure 8.43.

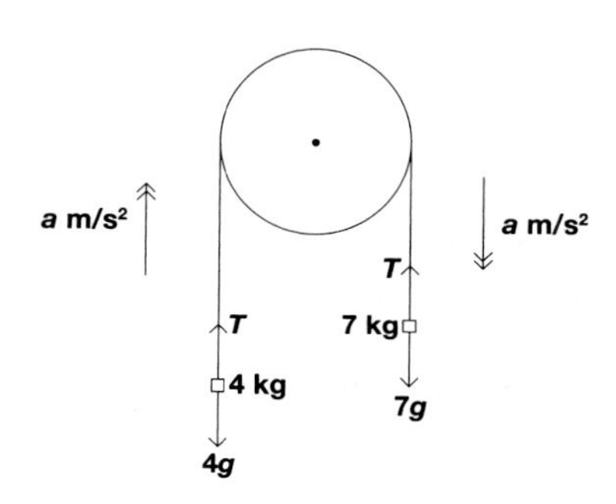

Figure 8.43

It may be verified experimentally that, provided the pulley is smooth and light, the tension is the same throughout the string. Clearly that end of the string bearing the 7 kg mass will move down with acceleration, $a\,\text{m/s}^2$, and the 4 kg end will move up at the same rate since they are connected.

Applying Newton's second law to each mass in turn

7 kg mass: $7g - T = 7a$
4 kg mass: $T - 4g = 4a$

Adding these equations yields $3g = 11a$ or $a = 2.673\,\text{m/s}^2$, and substituting this value of a, in either of the two equations above, gives $T = 49.89\,\text{N}$.

EXAMPLE 17

Figure 8.44 represents two bodies of mass 12 kg and 3 kg connected by a light inextensible string passing over a smooth light pulley. The bodies move on the smooth sloping faces of a wedge as shown. Find the acceleration, $a\,\text{m/s}^2$, as illustrated and the tension in the string.

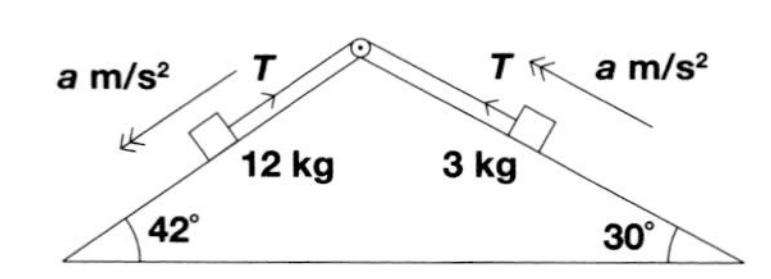

Figure 8.44

Solution 17

Completing the force diagram by including the normal reactions and the resolved components of the weight along and at right angles to the direction of motion results in Figure 8.45.

Figure 8.45

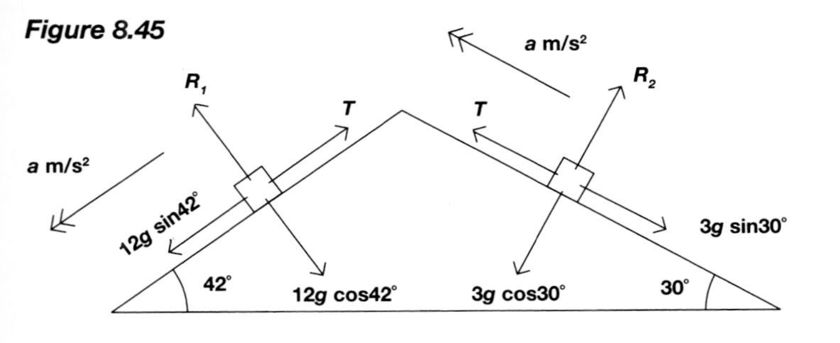

Applying Newton's second law along the sloping faces of the wedge

$$12g \sin 42° - T = 12a$$
$$T - 3g \sin 30° = 3a$$

and adding these equations leads to

$$12g \sin 42° - 3g \sin 30° = 15a$$

Hence $a = 4.266\,\text{m/s}^2$ and $T = 27.5\,\text{N}$.

EXAMPLE 18

Figure 8.46 represents three bodies of mass 2 kg, 10 kg and 24 kg connected by a light inextensible string. The surfaces on which they move are smooth; the 2 kg mass hangs freely, the 10 kg mass moves on a horizontal surface and the 24 kg mass moves down a plane inclined at 32° to the horizontal. The tension in that portion of the string connecting the 2 kg and 10 kg bodies is T_1, and that between the 10 kg and 24 kg bodies is T_2. Find the common acceleration, a, and these two tensions.

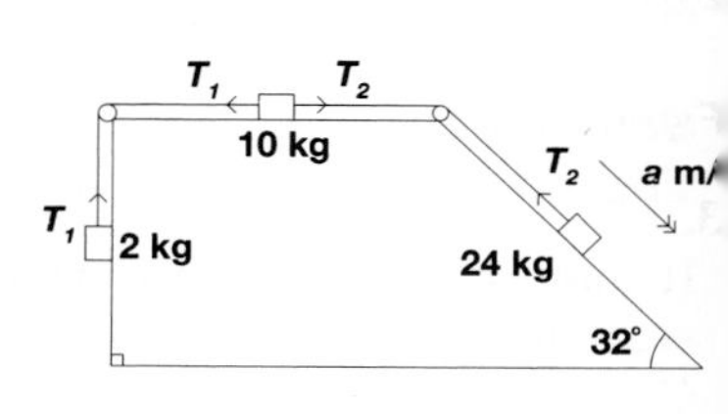

Figure 8.46

Solution 18

The application of Newton's second law gives

$$\begin{aligned} &2\,\text{kg mass:} && T_1 - 2g = 2a \\ &10\,\text{kg mass:} && T_2 - T_1 = 10a \\ &24\,\text{kg mass:} && 24g \sin 32° - T_2 = 24a \end{aligned}$$

Adding these equations

$$24g \sin 32° - 2g = 36a$$
$$\Rightarrow \quad a = 2.918\,\text{m/s}^2$$

It follows that

$$T_1 = 25.44\,\text{N} \quad \text{and} \quad T_2 = 54.62\,\text{N}$$

EXERCISE 8.6

Take $g = 9.8\,\text{m/s}^2$ throughout this exercise unless otherwise instructed.

1. For each of the diagrams in Figure 8.47 find the common acceleration and the tension in the couplings.

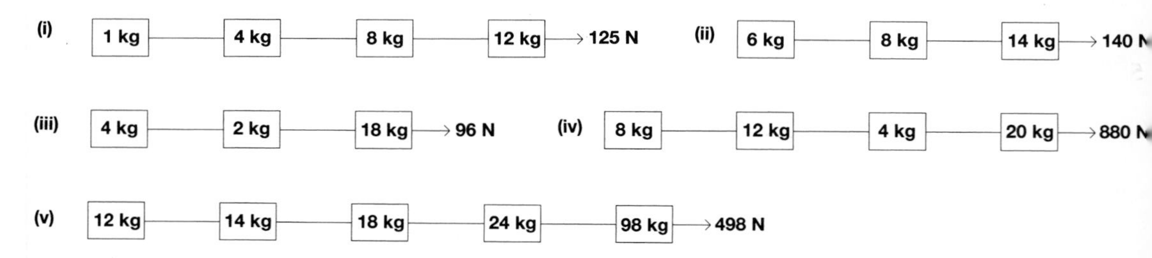

Figure 8.47

2. For each of the diagrams in Figure 8.48 find the common acceleration and the tension in the string.

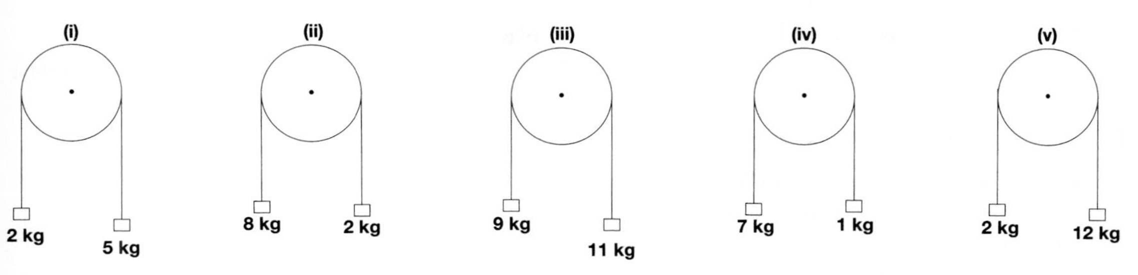

Figure 8.48

3. Calculate the common acceleration and the tensions in the five connected motions illustrated in Figure 8.49. Assume all surfaces to be smooth.

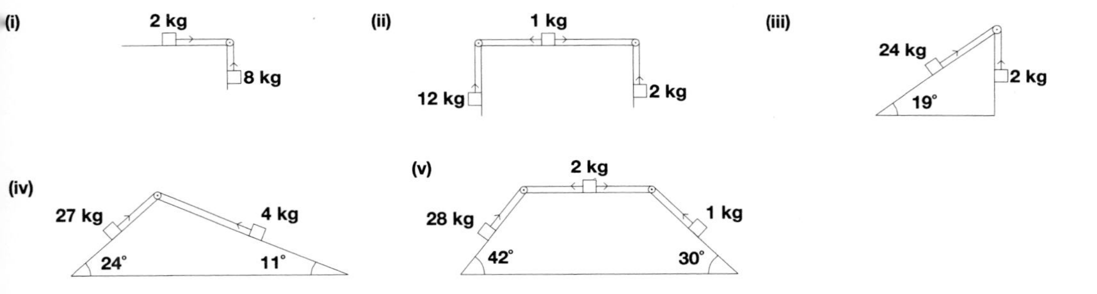

Figure 8.49

4. Two weights P and Q of mass 0.8 kg and 0.6 kg respectively, are connected by a light inextensible string which passes over a smooth fixed pulley as shown in Figure 8.50.

   The weights are released from rest with the string taut and the hanging parts vertical. Taking g to be $10\,\text{m/s}^2$, find

   (i) the acceleration of P,
   (ii) the tension in the string.

   (N. Ireland Additional Mathematics, 1989)

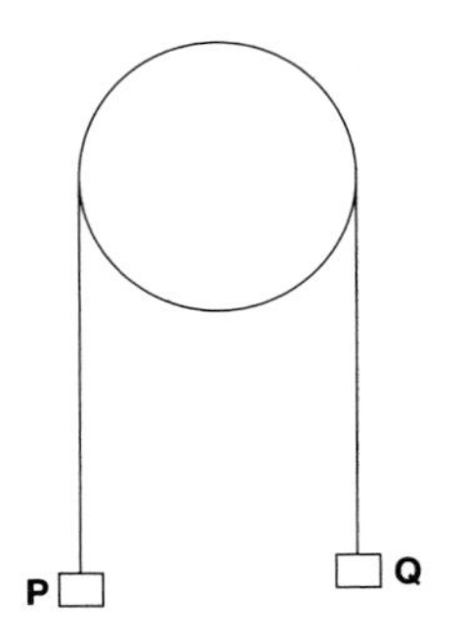

Figure 8.50

5. Take g to be $10\,\text{m/s}^2$.

   Two blocks P and Q of mass 7 kg and 5 kg respectively are connected by a light inextensible rope passing over a smooth fixed pulley. The blocks are held so that both parts of the string are vertical and P is 12 m above the ground.

   The blocks are released from rest. Assuming Q does not hit the pulley, calculate

   (i) the acceleration of Q,
   (ii) the speed of P when it hits the ground.

When P hits the ground the string will becomes slack and Q will initially continue to rise. Calculate

(iii) the additional distance by which Q will rise after P hits the ground,
(iv) the time that elapses between P hitting the ground and the string becoming taut again.

(N. Ireland Additional Mathematics, 1990)

6. Take g to be $10\,\text{m/s}^2$.

(*a*) Figure 8.51 shows two particles, of mass M kg and 5 kg, connected by a light inextensible string passing over a smooth fixed peg. The system is released from rest. Given that $M < 5$, and that the acceleration of each particle is $2.5\,\text{m/s}^2$, calculate

(i) the tension in the string,
(ii) the value of M.

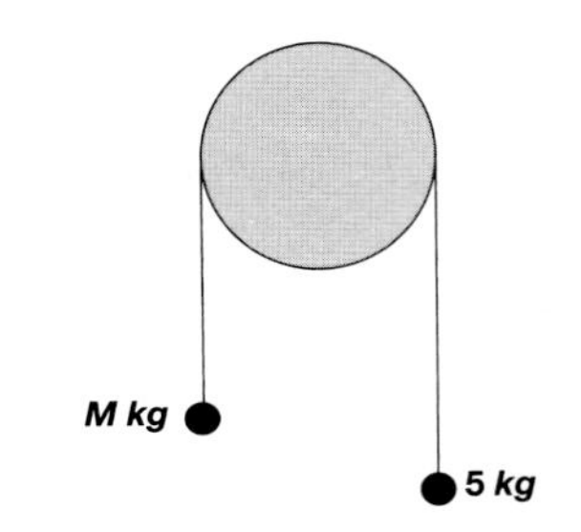

Figure 8.51

(*b*) A particle, of mass 3 kg, is projected up a line of greatest slope of a rough plane inclined at an angle α to the horizontal, where $\tan\alpha = \frac{5}{12}$. Given that the speed of projection is 8 m/s and that the coefficient of friction between the particle and the plane is $\frac{1}{8}$, calculate

(i) the retardation of the particle,
(ii) the distance the particle moves up the plane before coming to instantaneous rest,
(iii) the acceleration of the particle as it slides down the plane in the subsequent motion.

(U.C.L.E.S. Additional Mathematics, November 1991)

7. Table g to be $10\,\text{m/s}^2$.

(a) Figure 8.52 shows a particle P of mass 4 kg connected by a light string to a fixed point O and by another light string to a particle Q of mass 5 kg. Both particles are hanging at rest. Find the tension in

(i) the lower string,
(ii) the upper string.

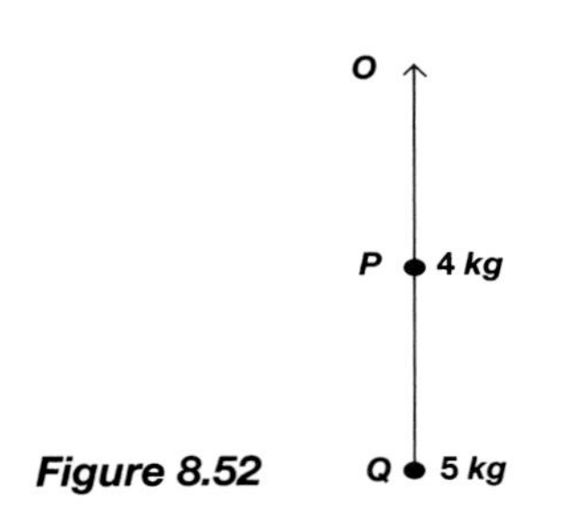

Figure 8.52

(b) Figure 8.53 shows a particle A of mass 0.3 kg held at rest on a smooth slope inclined at an angle of 30° to the horizontal. Particle A is connected by a light inextensible string passing over a smooth pulley at the bottom of the slope to a particle B, of mass 0.2 kg, hanging freely at rest. A is 1.54 m from the pulley and B is 0.56 m above the floor. The system is released from rest. Find

(i) the acceleration of each particle and the tension in the string while the string remains taut,
(ii) the speed with which B hits the floor.

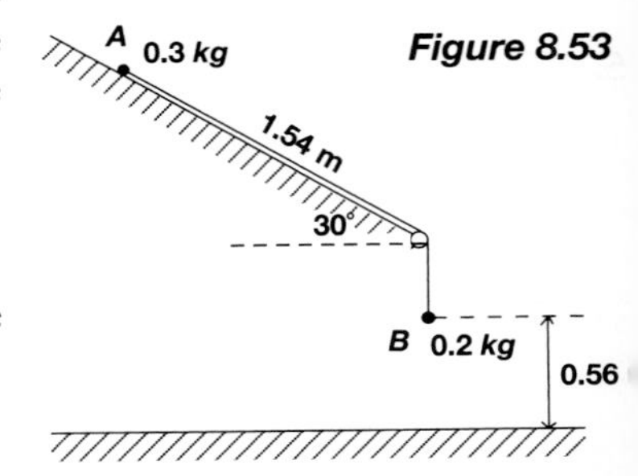

Figure 8.53

Assuming that B is brought to rest when it hits the floor, find the speed with which A reaches the pulley.

(U.C.L.E.S. Additional Mathematics, June 1992)

8. Take g to be $10\,\text{m/s}$.

(a) A particle of mass 4 kg lies on a rough horizontal plane and is acted on by an upwards force of P N at an angle α to the vertical, where $\cos\alpha = 0.8$. The coefficient of friction between the particle and the plane is 0.75.

(i) Find the value of P when the particle is about to move.

(ii) In the case where $P = 30$, find the acceleration of the particle.

(b) Figure 8.54 shows two particles, A and B, connected by a light inextensible string passing over a smooth peg. The particle A, mass 8 kg, is held on a smooth fixed plane inclined at an angle θ to the horizontal, where $\sin\theta = 0.75$. The particle B, mass 12 kg, hangs freely at a height of 2 m above the horizontal plane through A. The system is released from rest. Assuming particle A does not reach the pulley, calculate

(i) the acceleration of the particles while both are in motion,

(ii) the velocity with which B strikes the horizontal plane.

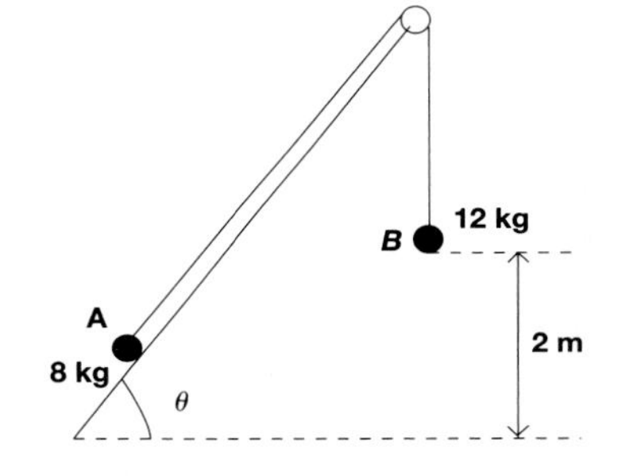

Figure 8.54

The particle A comes to instantaneous rest at a point on the inclined plane. Calculate the distance that A has then travelled up the plane.

(U.C.L.E.S. Additional Mathematics, June 1991)

CHAPTER 9

Kinematics

Albert Einstein (1879–1955) extended kinematics (the study of motion in space and time) to situations in which the speed of motion approached that of light. He was born in Ulm and struggled academically at school, particularly with mathematics. His poor grades did little to help him secure a post as a teacher and he took up a position as a patent examiner. However, these early academic deficiencies were short lived and, in 1909, he became a junior professor at Zurich University, with promotion to full professor following three years later. In 1913 he secured a directorship of the Kaiser Wilhelm Institute in Berlin. Fortunately for civilisation, he was in California when Hitler came to power and he remained in the USA, continuing his research at Princeton. Einstein received the Nobel Prize in 1921 for his 1905 discovery that light consisted of 'particles' but he will be remembered for his work on special relativity (the study of motion near the speed of light) and general relativity (the study of gravity).

It is small wonder that Einstein's work met with derision in some quarters when one considers that in the early 1900s he was suggesting that light was heavy and bends under gravity, that the Earth is plummeting down through the universe, that gravity warps space and that high speed travel can help retard the ageing process.

Furthermore, the equations of a paper published in 1915 allow for the possibility of a 'black hole' from which nothing can escape. In 1978 experimental evidence was presented for a huge black hole at the heart of the M87 galaxy, providing confirmation of an aspect of Einstein's work which many eminent thinkers deemed more appropriate to the realms of science fiction.

The Nazis referred contemptuously to relativity as the 'Jewish' theory constructed from 'botched-up theories consisting of some ancient knowledge and a few arbitrary additions', and the Vatican endorsed the view that relativity might result in 'universal doubt about God and his creation'. In an address at the Sorbonne in Paris, Einstein responded to his critics:

If my theory of relativity is proven successful, Germany will claim me as a German and France will declare that I am a citizen of the world. Should my theory prove untrue, France will say I am a German and Germany will declare that I am a Jew.

INTRODUCTION

At time t seconds a particle P is s metres from a fixed point O as shown in Figure 9.1 where $s = 4t^2 - 2t + 7$.

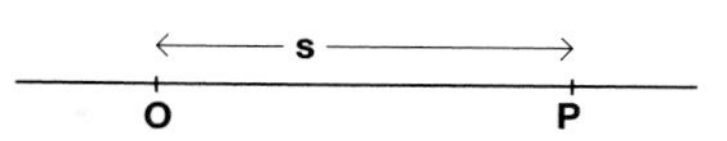

Figure 9.1

Suppose the velocity of the particle were required at a particular time. The small distance, δs, travelled in a small time δt (from time $= t$ to time $= t + \delta t$) can be calculated and the result divided by δt to produce the average speed during that interval. By making δt as small as possible, the interval $(t, t + \delta t)$ 'closes in' around t and the average speed over the interval becomes the instantaneous velocity at time t.

At time t: $\quad s = 4t^2 - 2t + 7$
At time $t + \delta t$: $\quad s = 4(t + \delta t)^2 - 2(t + \delta t) + 7$

The distance travelled in the time interval from time $= t$ to time $= t + \delta t$ is given by

$$\begin{aligned}\delta s &= 4(t + \delta t)^2 - 2(t + \delta t) + 7 - (4t^2 - 2t + 7)\\ &= 4t^2 + 8t\delta t + 4\delta t^2 - 2t - 2\delta t + 7 - 4t^2 + 2t - 7\\ &= 8t\delta t + 4\delta t^2 - 2\delta t\end{aligned}$$

$$\frac{\delta s}{\delta t} = 8t + 4\delta t - 2$$

Making the time interval as small as possible ($\delta t \to 0$) gives

$$\lim_{\delta t \to 0} \frac{\delta s}{\delta t} = 8t - 2$$

Using equation (2) of Chapter 6 this becomes

$$\frac{ds}{dt} = 8t - 2$$

It can be seen that the instantaneous velocity is found by simply differentiating the displacement expression with respect to time.

$$\mathbf{v = \frac{ds}{dt}} \qquad (1)$$

This is a general result often stated as 'velocity is the rate of change of displacement'.

Now acceleration is defined to be the rate of change of velocity.

$$\mathbf{a = \frac{dv}{dt}} \qquad (2)$$

Since acceleration has the units of velocity divided by time, it is measured in metres per second per second or metres per second squared (m/s^2).

It has been demonstrated in Chapter 6 that integration is the reverse of differentiation. Equation (1) may therefore be 'reversed' as

$$s = \int v\,dt \qquad (3)$$

and equation (2) as

$$v = \int a\,dt \qquad (4)$$

KINEMATICS

EXAMPLE 1

A particle moves along a straight line with its displacement from a fixed point O in the line given by $s = 2t^3 - 15t^2 + 36t + 4$ metres at time t seconds.

(i) Sketch the velocity–time graph.
(ii) Calculate the times at which the particle is at rest.
(iii) Calculate the velocity at $t = 1$.
(iv) Calculate the acceleration at $t = 1$.

Solution 1

(i) $$s = 2t^3 - 15t^2 + 36t + 4$$

Equation (1) yields

$$v = 6t^2 - 30t + 36$$

In sketching the curve of velocity against time it is instructive to determine the points at which the curve cuts the axes and to establish the location and nature of any stationary point(s). Now the curve intersects the velocity axis when $t = 0$ i.e. at $v = 36$.

It intersects the time axis when $v = 0$

i.e. $6t^2 - 30t + 36 = 0$
$\Rightarrow \quad t^2 - 5t + 6 = 0$
$\Rightarrow \quad (t - 2)(t - 3) = 0$
$\Rightarrow \quad t = 2$ and $t = 3$.

The calculation of the turning point now follows.

$$\frac{dv}{dt} = 12t - 30$$

and $$\frac{d^2v}{dt^2} = 12$$

Since $\frac{d^2v}{dt^2} > 0$ the velocity curve has a minimum turning point when

$$12t - 30 = 0$$
i.e. $\quad t = 2.5$
Here $\quad v = 6(2.5)^2 - 30(2.5) + 36 = -1.5\,\text{m/s}$

The velocity–time graph is shown in Figure 9.2.

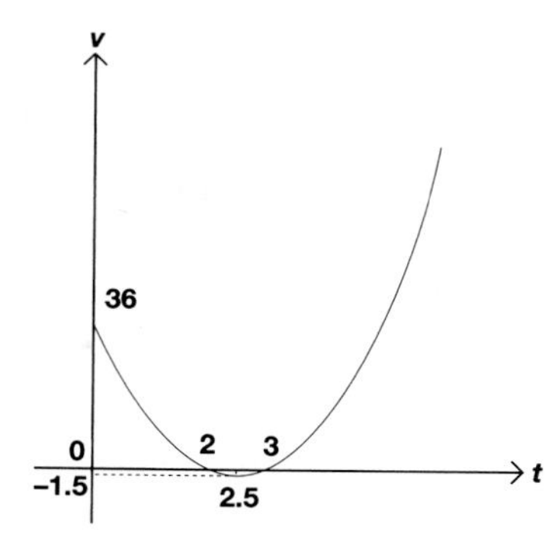

Figure 9.2

(ii) The particle is at rest when its velocity is zero. From the graph the times at which the particle is at rest are $t = 2$ and $t = 3$.

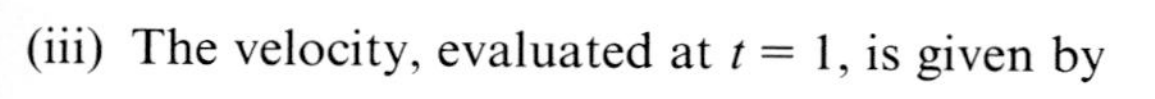

(iii) The velocity, evaluated at $t = 1$, is given by

$$6(1)^2 - 30(1) + 36 = 12\,\text{m/s}.$$

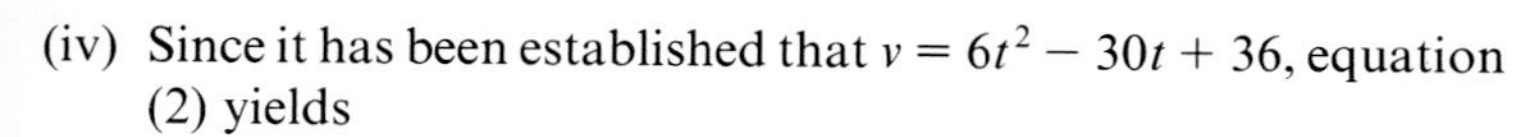

(iv) Since it has been established that $v = 6t^2 - 30t + 36$, equation (2) yields

$$a = 12t - 30$$

Thus the acceleration at $t = 1$ is $-18\,\text{m/s}^2$.

EXERCISE 9.1

All the questions in this exercise involve motion in a straight line. In each question the displacement, s, of a particle is given as a function of time. All distances are measured in metres and all times in seconds.

1. Given $s = 3t^3 - 2t^2 + 7t - 2$

(i) sketch the velocity–time graph,
(ii) calculate the times at which the particle is at rest,
(iii) calculate the velocity at $t = 1$,
(iv) calculate the acceleration at $t = 4$.

2. Given $s = 2t^2 - 4t + 7$

(i) sketch the velocity–time graph,
(ii) calculate the time at which the particle is at rest,
(iii) calculate the velocity at $t = 4$,
(iv) calculate the acceleration at $t = 0$.

3. Given $s = t^3 - 8t^2 - 2$

(i) sketch the velocity–time graph,
(ii) calculate the times at which the particle is at rest,
(iii) calculate the velocity at $t = 2$,
(iv) calculate the acceleration at $t = 1$.

4. Given $s = 4t^2 - 8t + 2$

(i) sketch the velocity–time graph,
(ii) calculate the time at which the particle is at rest,
(iii) calculate the velocity at $t = 12$,
(iv) calculate the acceleration at $t = 7$.

5. Given $s = 8t^2 - 2t + 5$

(i) sketch the velocity–time graph,
(ii) calculate the time at which the particle is at rest,
(iii) calculate the velocity at $t = 2$,
(iv) calculate the acceleration at $t = 2$.

6. Given $s = 7t^3 - 2t^2 - 5t + 2$

(i) sketch the velocity–time graph,
(ii) calculate the time at which the particle is at rest,
(iii) calculate the velocity at $t = 4$,
(iv) calculate the acceleration at $t = 5$.

7. Given $s = 8t^2 - 5t + 1$

(i) sketch the velocity–time graph,
(ii) calculate the time at which the particle is at rest,
(iii) calculate the velocity at $t = 2$,
(iv) calculate the acceleration at $t = 2$.

8. Given $s = -2t^3 + 11t - 8$

(i) sketch the velocity–time graph,
(ii) calculate the time at which the particle is at rest,
(iii) calculate the velocity at $t = 1$,
(iv) calculate the acceleration at $t = 4$.

9. Given $s = 2t + 5$

(i) sketch the velocity–time graph,
(ii) calculate the time at which the particle comes to rest,
(iii) calculate the velocity at $t = 2$,
(iv) calculate the acceleration at $t = 4$.

10. Given $s = 4t$

(i) sketch the velocity–time graph,
(ii) calculate the time at which the particle comes to rest,
(iii) calculate the velocity at $t = 1$,
(iv) calculate the acceleration at $t = 12$.

EXAMPLE 2

The velocity of a particle t seconds after leaving a fixed point is given by $v = 2t^2 - 3t + 1$ m/s. The particle is constrained to move in a straight line and the fixed point from which the body starts is 4 metres to the right of point O.

(i) Find an expression for s, the distance of the particle from O at time t.
(ii) Determine the direction of motion of the particle during the time interval $0 \le t \le 2$ adopting the convention that positive velocity is directed to the right.
(iii) Determine the total distance travelled in the time interval $0 \le t \le 2$.

Solution 2

(i) From equation (3)

$$s = \int (2t^2 - 3t + 1)\,dt$$
$$s = \tfrac{2}{3}t^3 - \tfrac{3}{2}t^2 + t + c$$
$s = 4$ when $t = 0$ and so
$$4 = \tfrac{2}{3}(0)^3 - \tfrac{3}{2}(0)^2 + 0 + c$$
i.e. $c = 4$

s may therefore be rewritten

$$s = \tfrac{2}{3}t^3 - \tfrac{3}{2}t^2 + t + 4$$

(ii) In order to establish the direction of motion in any given time interval, it is instructive to sketch the velocity–time graph.

Since $v = 2t^2 - 3t + 1$, then $v = 1$ when $t = 0$ i.e. the velocity curve intersects the v axis at $v = 1$.

At what times does the curve interesect the t axis?

Applying the quadratic formula to $2t^2 - 3t + 1 = 0$ results in the roots $t = 1$ or $t = \frac{1}{2}$ i.e. the velocity curve intersects the t axis at $t = 1$ and $t = \frac{1}{2}$.

A knowledge of the position and nature of the turning point of the velocity–time curve will assist in establishing the shape of the curve. To this end the derivatives

$$\frac{dv}{dt} = 4t - 3$$

$$\frac{d^2v}{dt^2} = 4$$

reveal a minimum turning point $\left(\frac{d^2v}{dt^2} > 0\right)$ where $4t - 3 = 0$.

The velocity curve therefore has a minimum turning point at $t = \frac{3}{4}$.

Here $\quad v = 2(\frac{3}{4})^2 - 3(\frac{3}{4}) + 1 = -0.125$

and the graph in Figure 9.3 may be constructed.

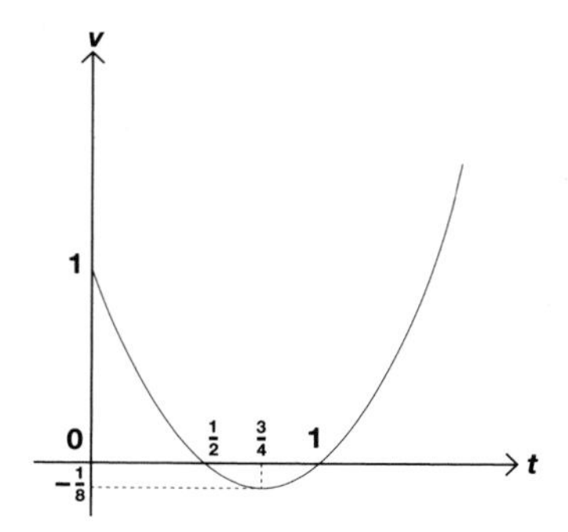

Figure 9.3

Bearing in mind that regions in which the curve appears above the time axis represent motion with positive velocity and regions where the curve lies below the time axis represent motion with negative velocity, inspection of the curve reveals that, for the time interval $0 \leq t < \frac{1}{2}$, the velocity is positive, for the time interval $\frac{1}{2} < t < 1$ the velocity is negative, and for the time interval $1 < t \leq 2$ the velocity is once again positive.

This may be visualised

in the time interval	$0 \leq t < \frac{1}{2}$	the particle is moving to the right
in the time interval	$\frac{1}{2} < t < 1$	the particle is moving to the left
in the time interval	$1 < t \leq 2$	the particle is moving to the right

(iii) The important time points to consider are the starting time, the times at which the particle is at rest and the completion time, namely

$$t = 0 \quad t = \tfrac{1}{2} \quad t = 1 \quad t = 2$$

Now since

$s = \frac{2}{3}t^3 - \frac{3}{2}t^2 + t + 4$ it follows

at $\quad t = 0 \Rightarrow s = 4$

$t = \frac{1}{2} \Rightarrow s = 4\frac{5}{24}$

$t = 1 \Rightarrow s = 4\frac{1}{6}$

$t = 2 \Rightarrow s = 5\frac{1}{3}$

Figure 9.4 serves as an aid to visualising the motion. In this figure the particle moves to the right (for a period of one half-second), to the left (for a further half-second) and then moves right again (for one second).

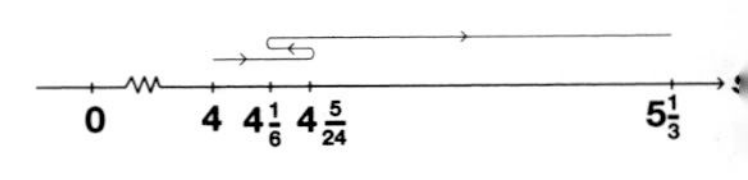

Figure 9.4

Using Figure 9.4, the total distance travelled is thus

$$\tfrac{5}{24} + \tfrac{1}{24} + (5\tfrac{1}{3} - 4\tfrac{1}{6}) = 1\tfrac{5}{12}\,\text{m}$$

EXERCISE 9.2

For questions 1–4 determine the direction of motion using the convention that positive velocity is interpreted as motion to the right. In each case the displacement is given from some fixed point on the path.

1. $s = -\frac{1}{3}t^3 + \frac{5}{2}t^2 - 4t + 2$

2. $s = t^3 - 9t^2 + 24t + 3$

3. $s = -t^3 + 18t^2 - 96t + 12$

4. $s = t^3 - 21t^2 + 135t + 15$

For questions 5–9 determine the direction of motion and the total distance travelled in the time interval specified. In each case the velocity is given as a function of time. The initial displacement from some fixed point on the path is also supplied.

5. Time interval: $0 \le t \le 6$
$v = 2t^2 - 14t + 24$
and $s = 3$ when $t = 0$.

6. Time interval: $0 \le t \le 3$
$v = -4t^2 + 12t - 8$
and $s = 2$ when $t = 0$.

7. Time interval: $0 \le t \le 2$
$v = -3t^2 + 4t - 1$
and $s = 4$ when $t = 0$.

8. Time interval: $0 \le t \le 6$
$v = t^2 - 7t + 10$
and $s = 7$ when $t = 0$.

9. Time interval: $0 \le t \le 5$
$v = -t^2 + 4t - 3$
and $s = 7$ when $t = 0$.

EXAMPLE 3

A particle moves along a straight line OAB.

The acceleration, $a\,\text{m/s}^2$, at time t seconds after passing O is given by

$$a = t - 5$$

The particle passes through O with velocity 12 m/s. Find

(i) the velocity of the particle 2 seconds after passing through O,
(ii) the distance of the particle from O after 4 seconds.

The particle is instantaneously at rest, first at B and then at A.

(iii) Calculate the distance AB.
(iv) Sketch the graph of the velocity of the particle against time for the interval $0 \le t \le 10$.

Describe, briefly, the path of the particle during this time interval.

(N. Ireland Additional Mathematics, 1991)

Solution 3

(i) $a = t - 5$

Equation (4) $\Rightarrow$ $v = \int (t-5)\,\mathrm{d}t$

$\Rightarrow$ $v = \frac{1}{2}t^2 - 5t + c$

But $v = 12$ when $t = 0$ and so

$12 = c$

The velocity may therefore be written

$v = \frac{1}{2}t^2 - 5t + 12$

and v at $t = 2$ is

$\frac{1}{2}(2)^2 - 5(2) + 12 = 4$

Accordingly, the velocity after 2 seconds is 4 m/s.

(ii) Using equation (3)

$s = \int (\frac{1}{2}t^2 - 5t + 12)\,\mathrm{d}t$
$s = \frac{1}{6}t^3 - \frac{5}{2}t^2 + 12t + c$

At $t = 0$, $s = 0$, since the particle is initially at O and s represents distance from O. Hence $c = 0$

$\Rightarrow$ $s = \frac{1}{6}t^3 - \frac{5}{2}t^2 + 12t$

It follows that the value of s at $t = 4$ is

$\frac{1}{6}(4)^3 - \frac{5}{2}(4)^2 + 12(4) = 18\frac{2}{3}$

The particle is $18\frac{2}{3}$ m from O after 4 seconds.

(iii) The particle is at rest when its velocity is zero, therefore

$\frac{1}{2}t^2 - 5t + 12 = 0$
$\Rightarrow$ $t^2 - 10t + 24 = 0$
$\Rightarrow$ $(t-6)(t-4) = 0$
i.e. $t = 4$ or $t = 6$

Now since s is $\frac{1}{6}(4)^3 - \frac{5}{2}(4)^2 + 12(4) = 18\frac{2}{3}$ m when $t = 4$
and s is $\frac{1}{6}(6)^3 - \frac{5}{2}(6)^2 + 12(6) = 18$ m when $t = 6$,
the required distance is $18\frac{2}{3} - 18 = \frac{2}{3}$ m.

(iv) $v = \frac{1}{2}t^2 - 5t + 12$ intersects the v axis when $t = 0$ i.e. at $v = 12$, and intersects the t axis when $v = 0$ i.e. when $\frac{1}{2}t^2 - 5t + 12 = 0$.

Applying the quadratic formula to the latter equation yields $t = 4$ and $t = 6$.

Furthermore, differentiation of $v = \frac{1}{2}t^2 - 5t + 12$ with respect to time leads to

$$\frac{\mathrm{d}v}{\mathrm{d}t} = t - 5$$

$$\frac{d^2v}{dt^2} = 1$$

Therefore v has a minimum turning point at $t = 5$.

At $t = 5$

$$v = \tfrac{1}{2}(5)^2 - 5(5) + 12 = -\tfrac{1}{2}\text{m/s}$$

The velocity–time curve is given in Figure 9.5.

A close examination of Figure 9.5 reveals that the particle moves to the right for the first 4 seconds, to the left for the next 2 seconds and thereafter to the right. This motion is illustrated in Figure 9.6.

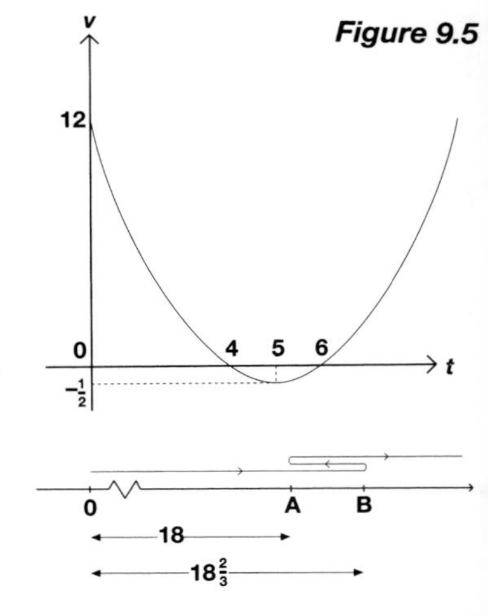

Figure 9.5

Figure 9.6

EXAMPLE 4

A particle moves in a straight line and its velocity, v m/s, at time t seconds is given by

$$v = 2t^2 - 7t + 3$$

When $t = 0$, the particle is moving through a fixed point O in the line.

Find

(i) the velocity with which the particle is moving initially, through O,
(ii) the time at which the acceleration of the particle is zero,
(iii) the times at which the particle is instantaneously at rest,
(iv) the displacement of the particle from O when it is first instantaneously at rest,
(v) the total distance travelled by the particle during the first second of its motion.

(N. Ireland Additional Mathematics, 1989)

Solution 4

(i) At $t = 0$, $v = 3$ m/s

(ii)
$$v = 2t^2 - 7t + 3$$
$$\Rightarrow \quad a = 4t - 7$$
$$4t - 7 = 0$$
$$\Rightarrow \quad t = 1\tfrac{3}{4}$$

Accordingly, the acceleration is zero after 1.75 seconds.

(iii) $2t^2 - 7t + 3 = 0$ gives $t = \frac{1}{2}$ or $t = 3$.

(iv) $s = \int v\,dt$ yields

$$s = \int(2t^2 - 7t + 3)\,dt = \tfrac{2}{3}t^3 - \tfrac{7}{2}t^2 + 3t + c$$

Therefore $s = \frac{2}{3}t^3 - \frac{7}{2}t^2 + 3t$ since $s = 0$ at $t = 0$.

Thus s at $t = \frac{1}{2}$ is

$$\tfrac{2}{3}(\tfrac{1}{2})^3 - \tfrac{7}{2}(\tfrac{1}{2})^2 + 3(\tfrac{1}{2}) = \tfrac{17}{24}$$

or the displacement is $\frac{17}{24}$m at $t = \frac{1}{2}$.

(v) $$v = 2t^2 - 7t + 3$$

results in the velocity–time graph of Figure 9.7.

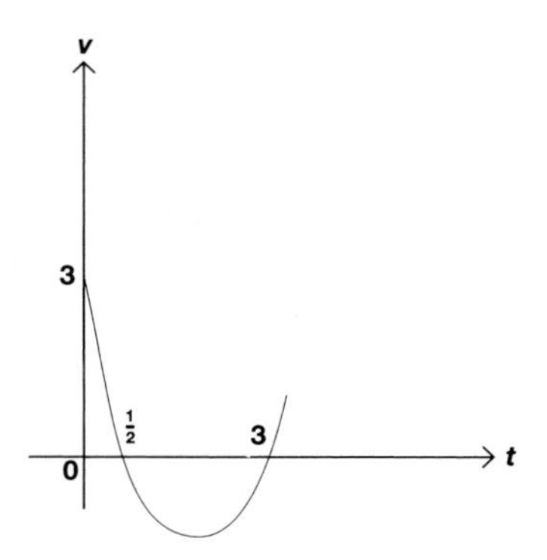

Figure 9.7

An examination of Figure 9.7 reveals that the particle is moving right for the first half-second of motion and subsequently to the left for the second half-second.

Now s is $\frac{2}{3}(1)^3 - \frac{7}{2}(1)^2 + 3(1) = \frac{1}{6}$m when $t = 1$.

The path is illustrated in Figure 9.8.

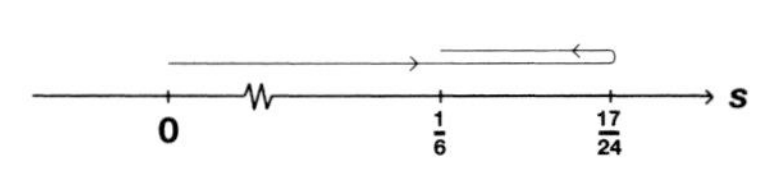

Figure 9.8

The total distance travelled is hence

$$\tfrac{17}{24} + (\tfrac{17}{24} - \tfrac{1}{6}) = 1\tfrac{1}{4}\text{m}$$

and therefore the particle travels $1\frac{1}{4}$m in the first second of motion.

EXAMPLE 5

(a) A particle moves along a straight line OX so that t seconds after leaving O its distance (in metres) from O is given by $t^2 - 4t$.

(i) Find the time at which the particle is at rest.
(ii) Determine the acceleration of the particle.
(iii) Calculate the time at which the particle returns to O.
(iv) Find the total distance travelled by the particle during the first 6 seconds.

(b) Another particle moves along a different straight line OY so that t seconds after leaving O, its distance (in metres) from O is $At^3 + Bt^2$ where A and B are constants.

When $t = 2$ the particle is 0.5 m from O and is moving with a velocity of 3 m/s.

Find the values of A and B.

(N. Ireland Additional Mathematics, 1988)

Solution 5

(a) (i) Applying equation (1) to $s = t^2 - 4t$ gives

$$v = 2t - 4$$

The particle is therefore at rest after 2 seconds.

(ii) $$a = \frac{dv}{dt} = 2\,\text{m/s}^2$$

(iii) s represents distance from O. It follows that $s = 0$ will have solutions which indicate the times at which the particle is at O.

$$s = 0 \quad \Rightarrow \quad t^2 - 4t = 0$$
$$\text{or} \qquad t(t-4) = 0$$
$$\text{i.e.} \qquad t = 0 \text{ or } t = 4.$$

The particle returns to O at $t = 4$.

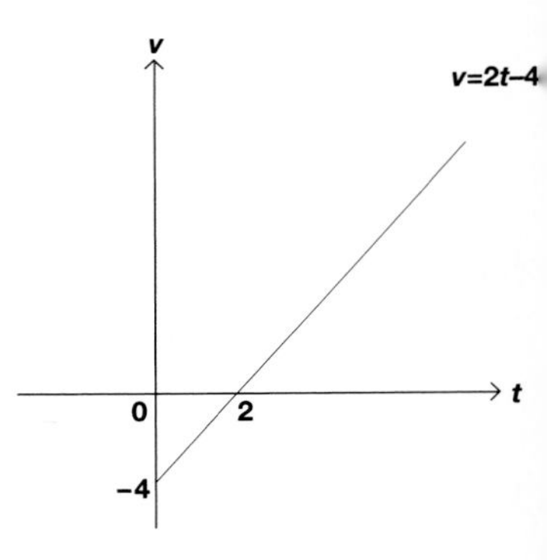

Figure 9.9

(iv) Consider the velocity–time graph for $v = 2t - 4$, sketched in Figure 9.9.

The particle is moving left for the first 2 seconds of motion and to the right thereafter.

The distance travelled at $t = 2$ is

$$s = (2)^2 - 4(2) = -4\,\text{m}$$

and the distance travelled at $t = 6$ is

$$s = (6)^2 - 4(6) = 12\,\text{m}$$

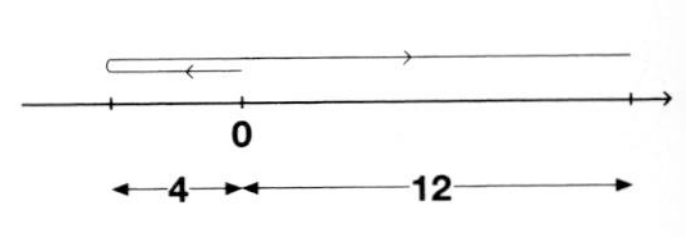

Figure 9.10

This is illustrated in Figure 9.10. A glance at Figure 9.10 reveals the total distance travelled to be 20 m.

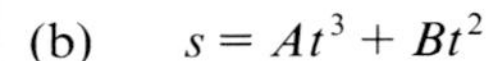

(b) $s = At^3 + Bt^2$

Given that $s = 0.5$ at $t = 2$, it follows that

$$0.5 = 8A + 4B$$

Differential calculus yields

$$v = 3At^2 + 2Bt$$

and since $v = 3$ at $t = 2$, it follows that

$$3 = 3A(2)^2 + 2B(2) \quad \text{i.e.} \quad 3 = 12A + 4B$$

Solving simultaneously for A and B yields

$$A = \tfrac{5}{8} \text{ and } B = -\tfrac{9}{8}.$$

EXAMPLE 6

A particle moves in a straight line through a fixed point O. Its velocity in metres per second, t seconds after passing through O, is given by $v = 3t^2 - 10t + 3$. If the particle comes to instantaneous rest at the point A, calculate

(i) the time taken to reach A for the first time,
(ii) the distance OA,
(iii) the velocity of the particle when its acceleration is zero,
(iv) the total distance travelled from O to the point at which the particle is instantaneously at rest for the second time.

Sketch the displacement–time graph for the motion of the particle from $t = 0$ to $t = 4$.

Solution 6

(i) $v = 3t^2 - 10t + 3 = 0$ at point A since the particle comes to rest at A.

This quadratic equation has roots $t = 3$ and $t = \frac{1}{3}$, and so the particle reaches A for the first time after $\frac{1}{3}$ second.

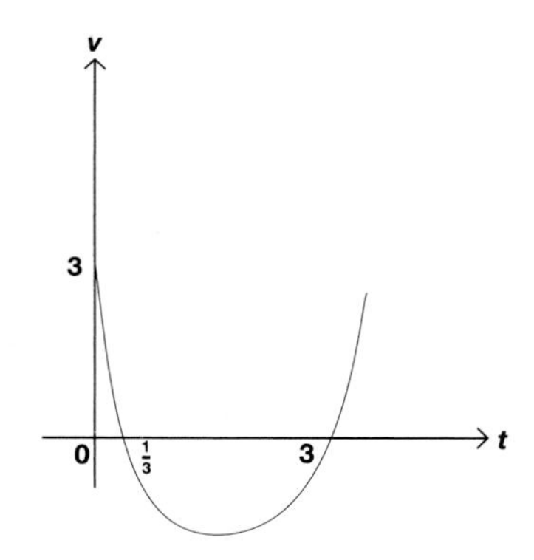

Figure 9.11

(ii) $s = \int v\,dt$ gives

$$s = \int(3t^2 - 10t + 3)dt = t^3 - 5t^2 + 3t + c$$

At $t = 0$, $s = 0$ (the particle is initially at O) and so $c = 0$.

Accordingly, since $s = t^3 - 5t^2 + 3t$, it follows that s is $\frac{13}{27}$ when $t = \frac{1}{3}$ or OA $= \frac{13}{27}$m.

(iii) $a = \dfrac{dv}{dt} \Rightarrow a = 6t - 10$

Hence $a = 0 \Rightarrow t = \frac{10}{6}$

Now v is $-\frac{16}{3}$ when $t = \frac{10}{6}$ and the velocity may be concluded to be $-\frac{16}{3}$m/s when the acceleration is zero.

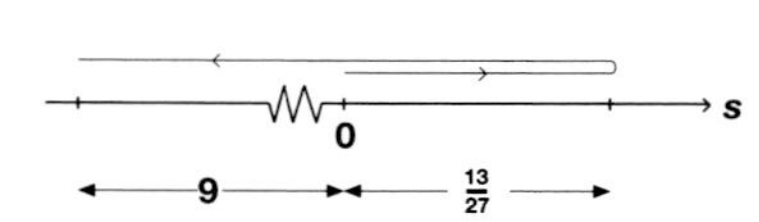

Figure 9.12

(iv) Using methods already established in previous examples, $v = 3t^2 - 10t + 3$ may be graphed as in Figure 9.11.

In order to calculate the total distance travelled, it is useful to calculate the displacement at the two 'rest' points i.e. s is $\frac{13}{27}$ when $t = \frac{1}{3}$ and s is -9 when $t = 3$.

Now the particle moves right for $0 \le t < \frac{1}{3}$ and then left during the time interval $\frac{1}{3} < t \le 3$. The path travelled may be illustrated as in Figure 9.12.

The total distance travelled is $\frac{13}{27} + \frac{13}{27} + 9$m and so the total distance travelled from O to the second rest point is $9\frac{26}{27}$m.

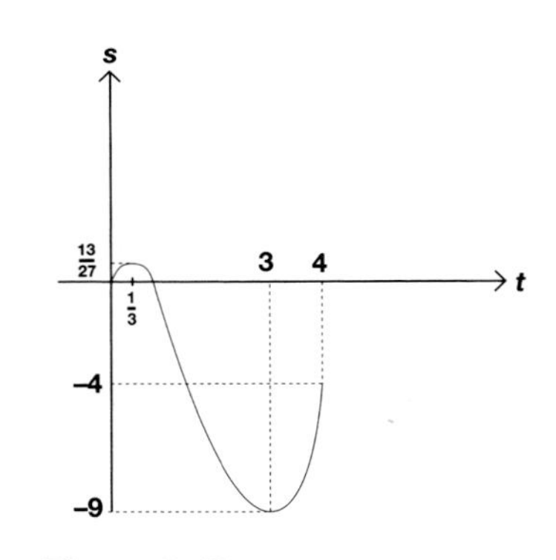

Figure 9.13

(v) The earlier parts of this solution involved calculations of s, the displacement, at $t = 0$, $t = \frac{1}{3}$, $t = 3$ and, since $s = -4$ when $t = 4$, the graph of s as a function of time is as shown in Figure 9.13.

EXERCISE 9.3

1. A particle, initially at a point O, moves in a straight line so that its displacement s metres from O after t seconds is given by $s = 2t^3 - 21t^2 + 60t$.

(i) By finding the velocity v in terms of t, determine the times at which the particle is momentarily at rest.

(ii) State the initial velocity of the particle.

(iii) Calculate the acceleration of the particle after 2 seconds.

(N. Ireland Additional Mathematics, 1990)

2. A particle P moves in a straight line through a fixed point O. When P first passes through O, its velocity is 12 m/s and t seconds later its acceleration is $(6t - 15)\,\text{m/s}^2$.

(i) Find the velocity of P at time t.

(ii) Calculate the times t_1 and t_2 $(t_1 < t_2)$ when P is instantaneously at rest.

(iii) Calculate the displacements s_1 and s_2 of P from O at times t_1 and t_2 respectively, and explain the significance of the negative value of s_2.

(iv) Find the total distance travelled by P from time $t = 0$ until time $t = 6$.

(v) Calculate the times after $t = 0$ at which P again passes through O, giving your answers correct to three significant figures.

(vi) Find the greatest velocity of P between times $t = 0$ and $t = 6$.

(N. Ireland Additional Mathematics, 1985)

3. A particle moves in a straight line through a fixed point O and t seconds after passing through O its acceleration is $a = (6t - k)\,\text{m/s}^2$ where k is a positive constant.

(i) Given that the velocity of the particle at O is b m/s, find an expression for its velocity in terms of k, b and t.

(ii) Hence find an expression for its displacement from O at time t, in terms of k, b and t, given that the particle is at O at time $t = 0$.

(iii) The displacements of the particle from O at times $t = 2$ and $t = 3$ are 2 m and 12 m respectively. Find k and b.

(iv) Calculate the distance travelled by the particle while its velocity is negative.

(N. Ireland Additional Mathematics, 1983)

4. In Figure 9.14, AB and DC are two smooth parallel wires; ABCD forms a rectangle lying in the horizontal plane.

Initially, a bead P is projected from A with speed 10 m/s and constant acceleration 1 m/s^2 towards B.

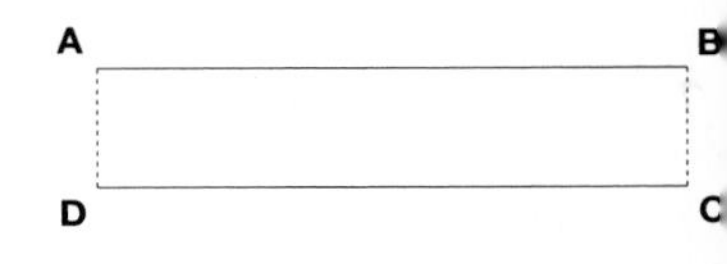

Figure 9.14

At the same instant a bead Q is projected from D with speed 12 m/s and acceleration $(4 - 2t)\,\text{m/s}^2$ towards C, where t is the time, in seconds, from the instant of projection.

(i) Find the speed of each particle after 3 seconds.

(ii) Determine the time, t, when Q is instantaneously at rest.

(iii) If Q just comes to rest at the point C, find the velocity with which P arrives at the point B.

(iv) Find the distance of P from A when P overtakes Q. Give your answer correct to three significant figures.

(N. Ireland Additional Mathematics, 1987)

5. A particle moves in a straight line through a fixed point O and t seconds after passing through O its velocity is $v = t^2 - 5t + 6$ m/s.

Find
(i) the times t_1 and t_2 between which the velocity of the particle is negative,
(ii) the displacements of the particle from O at times t_1 and t_2,
(iii) the total distance travelled by the particle from time $t = 0$ until time $t = 4$,
(iv) the times in the range $0 \leq t \leq 4$ at which the velocity of the particle takes

(a) its minimum value,
(b) its maximum value.

(N. Ireland Additional Mathematics, 1984)

6. A particle moves in a straight line and its acceleration at time t seconds is $(6t - 15)$ m/s^2. When $t = 0$, the particle is at a point B whose displacement from a fixed point A is 9 m, and the particle is moving with velocity 12 m/s in the direction AB.

(i) Find the velocity of the particle, and its displacement from A, at time t.

(ii) Calculate the times t_1 and t_2 ($t_1 < t_2$) between which the velocity of the particle is in the direction BA.

(iii) Calculate the displacements of the particle from A at times t_1 and t_2.

(iv) Find the greatest speed of the particle

(a) in the direction AB,
(b) in the direction BA,
between times $t = 0$ and $t = 6$.

(v) Sketch the displacement–time graph for the motion of the particle from $t = 0$ to $t = 6$, measuring displacement from A.

(N. Ireland Additional Mathematics, 1986)

7. A particle moves in a straight line through a fixed point O such that its velocity v m/s, t seconds after passing through O, is given by $v = 3t^2 - 8t + 4$. If the particle first comes to instantaneous rest at the point A, calculate

(i) the velocity of the particle when it initially passes through O,
(ii) the value of t when the particle first reaches A,
(iii) the distance OA,
(iv) the distance of the particle from O when it is again instantaneously at rest.

Sketch the displacement–time graph for the motion of the particle, for the first 3 seconds of motion after passing through O.

8. A and B are two points 30 m apart, on a horizontal plane. A particle moves from rest at A, in a straight line towards B, with a variable acceleration of $(2t - 1)\,\text{m/s}^2$, where t is the time in seconds for which the particle has been in motion. At the same instant a second particle is projected from B in a straight line towards A with an initial velocity of 2 m/s and moves with a constant acceleration of $3\,\text{m/s}^2$.

(i) What is the speed of each particle after 2 seconds?

(ii) What distance apart are the particles after 3 seconds?

9. A particle moves in a straight line and its velocity, v m/s, at time t seconds is given by

$$v = t^3 - 3t^2 + 2t$$

When $t = 0$, the particle is instantaneously at rest at a fixed point A in the line.

Find
(i) the subsequent times at which the particle is instantaneously at rest,
(ii) the times at which the acceleration of the particle is zero,
(iii) the time at which the acceleration of the particle is minimum,
(iv) the displacements of the particle from A at the times at which the particle is instantaneously at rest,
(v) the total distance moved by the particle from $t = 0$ to $t = 3$.

Sketch the velocity–time graph for values of t between -1 and 3.

10. A particle starts from a point O and moves in a straight line so that its displacement, s m, from O, t seconds after leaving O, is given by $s = t(t - 6)^2$. Obtain an expression for the velocity of the particle in terms of t. Hence determine the value of t when the particle first comes to instantaneous rest and find the acceleration at this instant.

The particle is next at O when $t = T$. Find
(i) the value of T,
(ii) the distance traveled from $t = 0$ to $t = T$.

(U.C.L.E.S. Additional Mathematics, November 1991)

11. A particle X moves along a horizontal straight line so that its displacement, s m, from a fixed point O, t seconds after motion has begun, is given by $s = 28 + 4t - 5t^2 - t^3$. Obtain expressions, in terms of t, for the velocity and acceleration of X, and state the initial velocity and the initial acceleration of X.

A second particle Y moves along the same horizontal straight line as X and starts from O at the same instant that X begins to move. The initial velocity of Y is 2 m/s and its acceleration, a m/s, t seconds after motion has begun, is given by $a = 2 - 6t$. Find the value of t at the instant when X and Y collide and determine whether or not X and Y are travelling in the same direction at this instant.

(U.C.L.E.S. Additional Mathematics, June 1991)

CHAPTER 10

Describing straight line motion in 2D space

The Irish mathematician **Sir William Rowan Hamilton** (1805–1865) discovered quaternions, a new algebra in which the commutative law for multiplication (i.e. $ab = ba$) did not apply. Hamilton's work on quaternions together with similar but independent research by Grassmann in Germany prompted Josiah Willard Gibbs (1839–1903), at Yale, to publish a number of articles on 'vectors'. In 1881 Gibbs published his *Vector Analysis* but many of the concepts presented in this work had already been addressed in Hamilton's earlier quaternion work.

Hamilton's parents died when he was young and he was despatched to his uncle at Trim when he was three years old. Hamilton was a brilliant linguist who had mastered thirteen languages by the age of thirteen years, including Sanskrit, Persian, Arabic, Chaldee and Syriac. Despite being a poor poet, he counted Wordsworth and Coleridge among his friends.

At twenty two, and prior to graduating from Trinity College, Dublin, he was appointed to the chair of astronomy at that university. He became Astronomer Royal of Ireland and director of the Dunsink Observatory before being knighted at the age of thirty. His obsessional interest in quaternions caused Hamilton to neglect his responsibilities to Irish astronomy; indeed, he appointed three members of his family to key positions at Dunsink so that he might be free to pursue his mathematical ideas.

After fifteen years of struggling with the quaternion concept, the possibility of relaxing the standard algebraic requirement that multiplication should be commutative suddenly occurred to him while walking his semi-invalid wife along Royal Canal in Dublin. In triumph he carved the quaternion combination rules on Brougham Bridge with a small knife. He wrote two large texts in the years which followed this flash of inspiration: *Lectures on quaternions* appeared in 1853 and *Elements of quaternions* was published in 1865, following Hamilton's death.

While quaternions feature little in contemporary mainstream mathematics, Hamilton's important contribution to the development of new algebras lives on in modern vector analysis.

ELEMENTARY VECTOR ALGEBRA IN TWO DIMENSIONS

All of the motion examined to this point has taken place in one dimension. In order to extend the investigation to two dimensions an appeal must be made to vector algebra. Quantities such as mass, volume, distance, speed and temperature have size but no associated direction. Such quantities are called **scalars** and obey the laws of elementary algebra. A second category of variables, such as velocity, force and acceleration, have **magnitude and direction**. The elements of this category are known as **vectors** and have properties which cannot be accommodated by elementary algebra; vectors are manipulated using vector algebra. Each vector may be represented by a straight line the length of which reflects the magnitude of the vector and the orientation of which indicates the direction of the vector.

This chapter will be concerned with two vector types: the **directed-line segment** and the **position vector** as illustrated in Figure 10.1. These two are distinguished in that the points denoting the extremities of the directed-line segment may be chosen with complete freedom while all position vectors emanate from the origin.

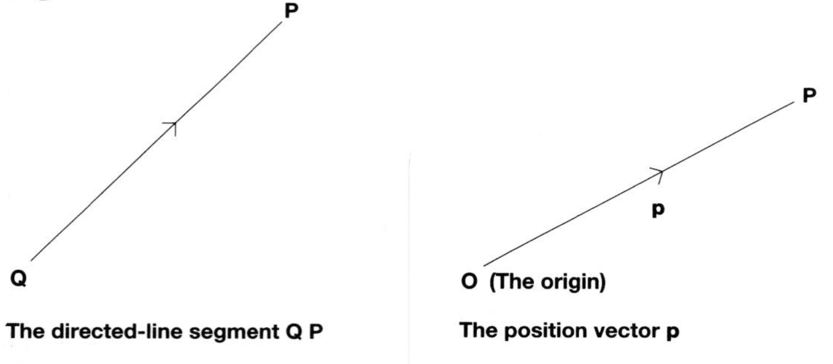

Figure 10.1

The distinction between directed-line segments and position vectors is sometimes made by stressing that while **QP** expresses the position of P relative to any point Q, **p** always expresses the position of P relative to the origin.

In this chapter vectors will be distinguished from other quantities by the use of bold print, as above. In handwritten mathematics attention may be drawn to the fact that **PQ** is a vector by writing PQ with an arrow above it. In the case of position vectors, **a** may be identified as a vector by underlining the letter a.

The length or magnitude of a vector is its **modulus**, the modulus of **a** being denoted $|\mathbf{a}|$. A suitable context for illustrating the role of the modulus symbol is in establishing the mathematical link between such fundamental dynamical measures as speed and velocity, distance and displacement. The velocity of a car may be measured as 70 km/h northwards – note that both magnitude and direction must be specified. The speed of the car is 70 km/h, that is, the speed is the magnitude (or modulus) of the velocity.

In general

$$\text{speed} = |\mathbf{v}|$$

which translates as '**speed is the modulus of velocity**'. Similarly

$$\text{distance} = |\mathbf{s}|$$

or '**distance is the modulus of displacement**'.

Finally, a vector of modulus unity is known as a **unit vector**.

The point was made above that scalars obey the rules of elementary algebra whereas vectors conform to the rules of vector algebra. The contrast between these two algebras may be seen at once in the manner in which vectors are added. Consider the sum of vectors **PS** and **SL**. The rules of vector algebra require that the sum of **PS** and **SL** be represented by the third side of the triangle PSL in which **PS** and **SL** form the other two sides – see Figure 10.2.

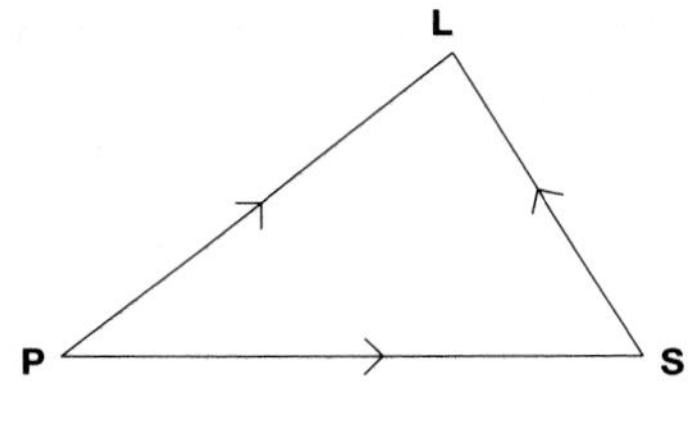

Figure 10.2

According to the triangle rule of vector addition

$$\mathbf{PS} + \mathbf{SL} = \mathbf{PL}$$

Mathematicians speak of the 'tail' of **SL** being coincident with the 'nose' of **PS**, hence the frequent reference to the above addition procedure as 'nose to tail' addition.

Equal vectors are represented by parallel lines in the same sense and of equal length while the **negative** of a vector has the same magnitude as the vector itself but is directed in the opposite sense; therefore

$$-\mathbf{SL} = \mathbf{LS}$$

This rule facilitates the subtraction of vectors where

PS − **LS**	
= **PS** + **SL**	(since − **LS** = **SL**)
= **PL**	(applying the triangle law of addition to Figure 10.3).

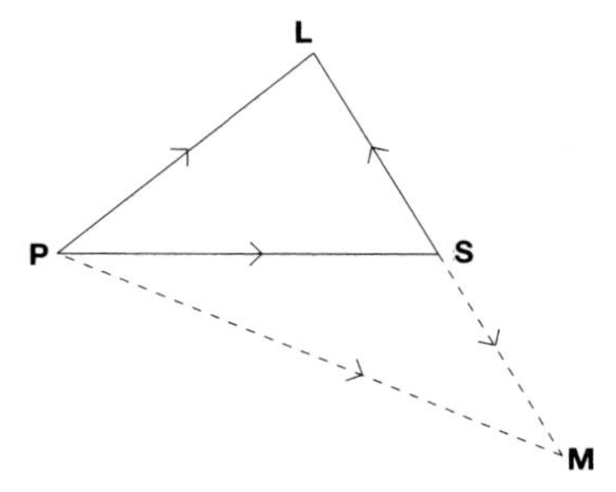

Figure 10.3

Furthermore

PS − **SL**	
= **PS** + **LS**	(since − **SL** = **LS**)
= **PS** + **SM**	(since **LS** = **SM** being in the same direction and equal in length)
= **PM**	(applying the triangle law of addition to Figure 10.3)

In vector algebra if α is a positive number then $\alpha\mathbf{m}$ is a vector in the same sense as **m** but with α times its magnitude.

The **scalar** (or **dot**) product of two vectors **p** and **q**, inclined at an angle θ to each other, is denoted **p.q** and defined through the equation

$$\mathbf{p}.\mathbf{q} = |\mathbf{p}|\,|\mathbf{q}|\cos\theta \qquad (1)$$

An obvious consequence of this definition is that vectors which are perpendicular to one another will have zero scalar product ($\cos 90° = 0$).

Consider now the point $P(x, y)$ in the two-dimensional Cartesian plane of Figure 10.4.

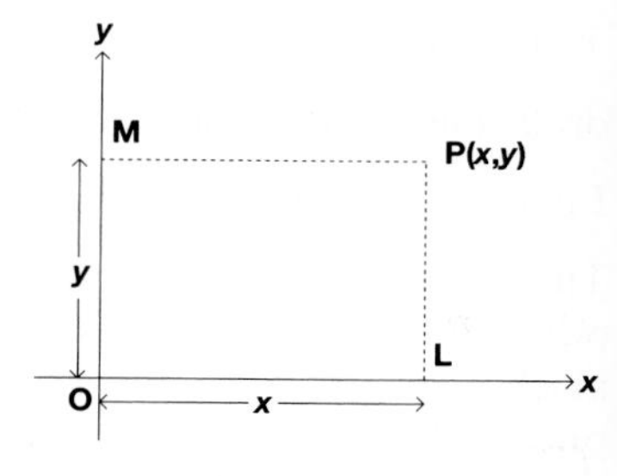

Figure 10.4

The symbols **i** and **j** are, by convention, reserved for the unit vectors along the x and y axes respectively. In compliance with the rules of vector algebra, the vector $x\mathbf{i}$ denotes a vector of magnitude x times the magnitude of **i** and in the sense of **i**. But, as the magnitude of **i** is 1 (being a unit vector), the vector $x\mathbf{i}$ represents a vector of magnitude x in the **i** sense. This is just the vector **OL** in Figure 10.4.

Hence $\quad \mathbf{OL} = x\mathbf{i}$
Similarly $\quad \mathbf{OM} = \mathbf{LP} = y\mathbf{j}$

Using the method of addition of vectors, detailed above, where the sum of two vectors is represented by the third side of a triangle, the other two sides of which are the vectors to be added (positioned 'nose to tail') then

$$\mathbf{OP} = \mathbf{OL} + \mathbf{LP}$$

or $\quad \mathbf{p} = x\mathbf{i} + y\mathbf{j} \qquad (2)$

The vector **p** is said to be written in 'component' or 'Cartesian' form. It follows that, for example, L(4,7) has an associated position vector $\mathbf{l} = 4\mathbf{i} + 7\mathbf{j}$.

By Pythagoras' theorem in Figure 10.4

$$|\mathbf{p}| = \sqrt{(x^2 + y^2)} \qquad (3)$$

and this is the general definition for the modulus of a vector expressed in component form.

The **unit vector** associated with the vector **a** is denoted $\hat{\mathbf{a}}$ where

$$\hat{\mathbf{a}} = \frac{\mathbf{a}}{|\mathbf{a}|} \qquad (4)$$

i.e. the unit vector associated with a given vector is found by dividing that vector by its modulus.

For vectors $\mathbf{l} = a\mathbf{i} + b\mathbf{j}$ and $\mathbf{m} = c\mathbf{i} + d\mathbf{j}$ the scalar product (see above) becomes

$$\mathbf{l.m} = (a\mathbf{i} + b\mathbf{j}).(c\mathbf{i} + d\mathbf{j}) = ac\mathbf{i.i} + ad\mathbf{i.j} + bc\mathbf{j.i} + bd\mathbf{j.j} \qquad (5)$$

Using the definition of scalar product in equation (1)

$$\mathbf{i.i} = |\mathbf{i}|\,|\mathbf{i}| \cos 0° = 1 \times 1 \times 1 = 1$$

and

$$\mathbf{i.j} = |\mathbf{i}|\,|\mathbf{j}| \cos 90° = 1 \times 1 \times 0 = 0$$

since **i** and **j** are unit vectors perpendicular to one another.

Equation (5) may therefore be further simplified

$$\mathbf{l.m} = ac + bd \qquad (6)$$

The examples which now follow will help provide more concrete illustrations of the above abstract theory.

EXAMPLE 1

A triangle ABC has vertices A(1,4), B(3,1) and C(5,8).

(i) Write down the vectors **a**, **b** and **c**.
(ii) Find in terms of **i** and **j** the vectors **AB**, **BC** and **CA**.
(iii) Calculate $|\mathbf{a}|$, $|\mathbf{AB}|$ and $|\mathbf{CA}|$.
(iv) Find $\hat{\mathbf{a}}$.
(v) Calculate **a.b**, **b.c** and **AC.BC**.

Solution 1

(i) $\mathbf{a} = \mathbf{i} + 4\mathbf{j} \quad \mathbf{b} = 3\mathbf{i} + \mathbf{j} \quad \mathbf{c} = 5\mathbf{i} + 8\mathbf{j}$

(ii) Applying the triangle rule of vector addition to Figure 10.5

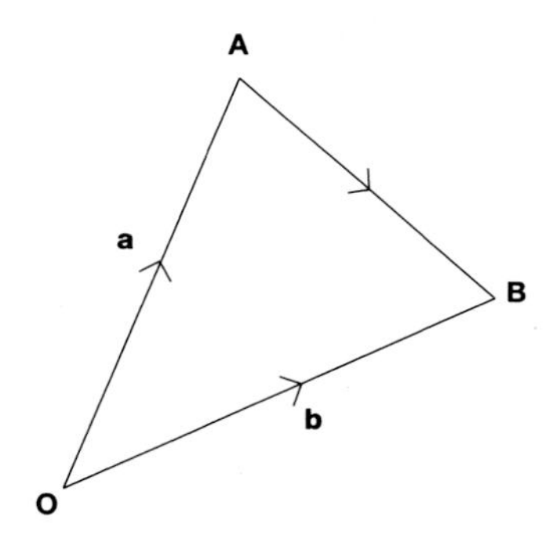

Figure 10.5

$$\mathbf{b} = \mathbf{a} + \mathbf{AB}$$

or $\mathbf{AB} = \mathbf{b} - \mathbf{a}$

Therefore $\mathbf{AB} = 3\mathbf{i} + \mathbf{j} - (\mathbf{i} + 4\mathbf{j}) = 2\mathbf{i} - 3\mathbf{j}$

Similarly $\mathbf{BC} = \mathbf{c} - \mathbf{b} = 5\mathbf{i} + 8\mathbf{j} - (3\mathbf{i} + \mathbf{j}) = 2\mathbf{i} + 7\mathbf{j}$

and $\mathbf{CA} = \mathbf{a} - \mathbf{c} = \mathbf{i} + 4\mathbf{j} - (5\mathbf{i} + 8\mathbf{j}) = -4\mathbf{i} - 4\mathbf{j}$

(iii) $|\mathbf{a}| = \sqrt{(1^2 + 4^2)} = \sqrt{17}$ (using equation (3))

$|\mathbf{AB}| = \sqrt{(2^2 + (-3)^2)} = \sqrt{13}$

and $|\mathbf{CA}| = \sqrt{((-4)^2 + (-4)^2)} = 4\sqrt{2}$

(iv) Equation (4) governs the calculation of $\hat{\mathbf{a}}$.

$$\hat{\mathbf{a}} = \frac{\mathbf{a}}{|\mathbf{a}|} = \frac{(\mathbf{i} + 4\mathbf{j})}{\sqrt{17}} = \left(\frac{1}{\sqrt{17}}\right)\mathbf{i} + \left(\frac{4}{\sqrt{17}}\right)\mathbf{j}$$

(v) It has already been established (equation (6)) that for

$$\mathbf{l} = a\mathbf{i} + b\mathbf{j} \quad \mathbf{m} = c\mathbf{i} + d\mathbf{j}$$

then $\mathbf{l.m} = ac + bd$.

It follows that

$$\mathbf{a.b} = (\mathbf{i} + 4\mathbf{j}).(3\mathbf{i} + \mathbf{j}) = 1 \times 3 + 4 \times 1 = 7$$
$$\mathbf{b.c} = (3\mathbf{i} + \mathbf{j}).(5\mathbf{i} + 8\mathbf{j}) = 3 \times 5 + 1 \times 8 = 23$$
$$\mathbf{AC.BC} = (-\mathbf{CA}).\mathbf{BC}$$
$$= (4\mathbf{i} + 4\mathbf{j}).(2\mathbf{i} + 7\mathbf{j}) \quad \text{(see section (ii) of this solution for } \mathbf{CA} \text{ and } \mathbf{BC}\text{)}$$
$$= 8 + 28 = 36$$

EXAMPLE 2

The three vectors **a**, **b** and **c** are represented by the directed line segments **AB**, **BC** and **CA** respectively.

(i) Show with the aid of a diagram that $\mathbf{a} + \mathbf{b} + \mathbf{c} = \mathbf{0}$.

Hence if

$$\mathbf{a} = (3x + y)\mathbf{i} + (2x - y)\mathbf{j}$$
$$\mathbf{b} = (y - 2x)\mathbf{i} + 4y\mathbf{j}$$
$$\mathbf{c} = 3\mathbf{i} + 4\mathbf{j}$$

(ii) find the values of x and y and
(iii) find the constant k so that $\mathbf{a.b} = k|\mathbf{c}|$ where $|\mathbf{c}|$ denotes the modulus of **c**.

(N. Ireland Additional Mathematics, 1980 (part question))

Solution 2

(i) From Figure 10.6 and the triangle rule of vector addition

$$-\mathbf{c} = \mathbf{a} + \mathbf{b}$$
$$\text{or} \quad \mathbf{a} + \mathbf{b} + \mathbf{c} = \mathbf{0}$$

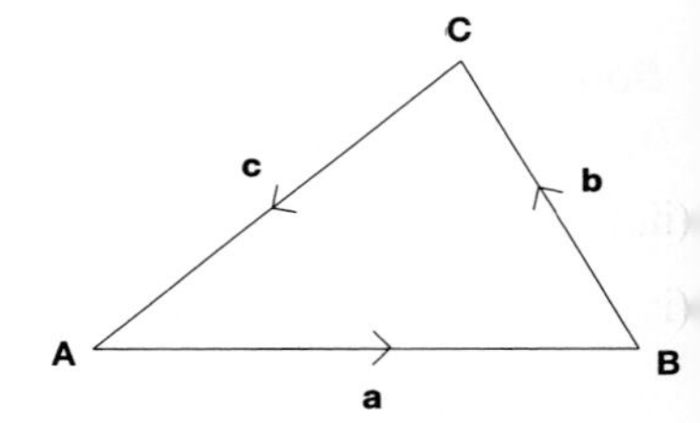

Figure 10.6

(ii) Since $\mathbf{a} + \mathbf{b} + \mathbf{c} = \mathbf{0}$

then $(3x + y)\mathbf{i} + (2x - y)\mathbf{j} + (y - 2x)\mathbf{i} + 4y\mathbf{j} + 3\mathbf{i} + 4\mathbf{j} = \mathbf{0}$

$\Rightarrow \quad (x + 2y + 3)\mathbf{i} + (2x + 3y + 4)\mathbf{j} = \mathbf{0}$

It follows that

$$x + 2y = -3 \quad \text{and} \quad 2x + 3y = -4$$

Solving these equation simultaneously gives $x = 1$, $y = -2$.

(iii) Substituting the values for x and y found in part (ii) allows **a**, **b** and **c** to be rewritten

$$\mathbf{a} = \mathbf{i} + 4\mathbf{j} \quad \mathbf{b} = -4\mathbf{i} - 8\mathbf{j} \quad \mathbf{c} = 3\mathbf{i} + 4\mathbf{j}$$

Hence $\quad \mathbf{a.b} = 1 \times -4 + 4 \times -8 = -36$
and $\quad |\mathbf{c}| = \sqrt{(3^2 + 4^2)} = 5$

Therefore $\mathbf{a.b} = k|\mathbf{c}|$ implies

$$-36 = 5k \text{ or } k = -7.2$$

EXAMPLE 3

$$\mathbf{a} = 2\mathbf{i} + \mathbf{j} \quad \mathbf{b} = -3\mathbf{i} + 2\mathbf{j} \quad \mathbf{u} = x\mathbf{i} + y\mathbf{j}$$

are three vectors in component form such that $\mathbf{a}.\mathbf{u} = 2$ and $\mathbf{b}.\mathbf{u} = -17$.

(i) Calculate the values of x and y.
(ii) Find in component form the unit vector which is parallel to **u**.

The vector $\mathbf{e} = 4\mathbf{a} + 5\mathbf{b}$.

(iii) Express **e** in terms of **i** and **j**.
(iv) Show that **a** is perpendicular to **e**.

(N. Ireland Additional Mathematics, 1981 (part question))

Solution 3

(i)
$$\mathbf{a}.\mathbf{u} = (2\mathbf{i} + \mathbf{j}).(x\mathbf{i} + y\mathbf{j}) = 2x + y$$
$$\mathbf{b}.\mathbf{u} = (-3\mathbf{i} + 2\mathbf{j}).(x\mathbf{i} + y\mathbf{j}) = -3x + 2y$$

The information offered in the example then leads to the simultaneous equations

$$2x + y = 2$$
$$-3x + 2y = -17$$

These solve as $x = 3$, $y = -4$ and hence $\mathbf{u} = 3\mathbf{i} - 4\mathbf{j}$.

(ii) The calculation of $\hat{\mathbf{u}}$ requires the modulus of **u**.

Now $|\mathbf{u}| = \sqrt{(3^2 + (-4)^2)} = 5$

Therefore $\hat{\mathbf{u}} = \dfrac{\mathbf{u}}{|\mathbf{u}|} = \dfrac{(3\mathbf{i} - 4\mathbf{j})}{5}$ or $\hat{\mathbf{u}} = 0.6\mathbf{i} - 0.8\mathbf{j}$

(iii) $\mathbf{e} = 4(2\mathbf{i} + \mathbf{j}) + 5(-3\mathbf{i} + 2\mathbf{j}) = -7\mathbf{i} + 14\mathbf{j}$

(iv) As has already been pointed out, the definition of scalar product as $\mathbf{a}.\mathbf{b} = |\mathbf{a}||\mathbf{b}| \cos\theta$ provides a means of 'testing' for perpendicular vectors ($\theta = 90°$). Since perpendicular vectors will have $\mathbf{a}.\mathbf{b} = 0$, the scalar product may be used to discover such vector pairs. In this case

$$\mathbf{a}.\mathbf{e} = (2\mathbf{i} + \mathbf{j}).(-7\mathbf{i} + 14\mathbf{j}) = -14 + 14 = 0$$

and so **a** is perpendicular to **e**.

EXAMPLE 4

$\mathbf{a} = 2\mathbf{i} - 3\mathbf{j}$, $\mathbf{b} = 3\mathbf{i} + \mathbf{j}$ and $\mathbf{c} = 3\mathbf{i} + 2\mathbf{j}$ are three vectors in component form.

(i) Calculate in component form the vectors
$\mathbf{a} - \mathbf{c}$ and $\mathbf{b} - 2\mathbf{c} + \mathbf{a}$.
(ii) Show, using the above vectors, that $\mathbf{b}.(\mathbf{c} + \mathbf{a}) = \mathbf{b}.\mathbf{c} + \mathbf{b}.\mathbf{a}$.
(iii) Find which two of the vectors **a**, **b** and **c** are at right angles to one another.

(iv) Find a unit vector perpendicular to **a**.

(N. Ireland Additional Mathematics, 1982 (part question))

Solution 4

(i) $\mathbf{a} - \mathbf{c} = (2\mathbf{i} - 3\mathbf{j}) - (3\mathbf{i} + 2\mathbf{j}) = -\mathbf{i} - 5\mathbf{j}$

$\mathbf{b} - 2\mathbf{c} + \mathbf{a} = (3\mathbf{i} + \mathbf{j}) - 2(3\mathbf{i} + 2\mathbf{j}) + 2\mathbf{i} - 3\mathbf{j} = -\mathbf{i} - 6\mathbf{j}$

(ii) $\mathbf{b}.(\mathbf{c} + \mathbf{a}) = (3\mathbf{i} + \mathbf{j}).(5\mathbf{i} - \mathbf{j}) = 15 - 1 = 14$

$\mathbf{b}.\mathbf{c} = (3\mathbf{i} + \mathbf{j}).(3\mathbf{i} + 2\mathbf{j}) = 11$

$\mathbf{b}.\mathbf{a} = (3\mathbf{i} + \mathbf{j}).(2\mathbf{i} - 3\mathbf{j}) = 6 - 3 = 3$

and since $14 = 11 + 3$

then $\mathbf{b}.(\mathbf{c} + \mathbf{a}) = \mathbf{b}.\mathbf{c} + \mathbf{b}.\mathbf{a}$

(iii) $\mathbf{a}.\mathbf{c} = (2\mathbf{i} - 3\mathbf{j}).(3\mathbf{i} + 2\mathbf{j}) = 6 - 6 = 0$

and, since it has already been established that $\mathbf{b}.\mathbf{c} = 11$ and $\mathbf{b}.\mathbf{a} = 3$, then **a** and **c** are the only perpendicular pair of vectors.

(iv) Let $\mathbf{u} = x\mathbf{i} + y\mathbf{j}$ be any vector perpendicular to **a**.

Obviously **u** must satisfy $\mathbf{u}.\mathbf{a} = 0$ and so

$(x\mathbf{i} + y\mathbf{j}).(2\mathbf{i} - 3\mathbf{j}) = 0$

$\Rightarrow \quad 2x - 3y = 0$

It follows that any vector $\mathbf{u} = x\mathbf{i} + y\mathbf{j}$, such that $2x = 3y$, will be perpendicular to **a**.

Let $y = 2$, say, and so $x = 3$ i.e. $\mathbf{u} = 3\mathbf{i} + 2\mathbf{j}$.

Therefore $\hat{\mathbf{u}} = \dfrac{\mathbf{u}}{|\mathbf{u}|} = \dfrac{(3\mathbf{i} + 2\mathbf{j})}{\sqrt{(3^2 + 2^2)}}$

or $\hat{\mathbf{u}} = \left(\dfrac{3}{\sqrt{13}}\right)\mathbf{i} + \left(\dfrac{2}{\sqrt{13}}\right)\mathbf{j}$

EXERCISE 10.1

The first four questions of this exercise refer to the points

A(3,5), B(2,4), D(1,7), E(2,5), F(1,10), H(−1,3), I(1,−4), M(2,−7), N(4,−9), and S(2,0).

1. For each point above

(i) write the corresponding position vector,
(ii) calculate the modulus of the position vector,
(iii) find the corresponding unit vector.

2. Calculate the following scalar products with reference to the points listed above.

(i) **a.b** (ii) **b.d** (iii) **d.e** (iv) **e.f** (v) **h.i** (vi) **m.n** (vii) **n.s** (viii) **b.m** (ix) **m.s** (x) **a.s**

3. Express these in terms of **i** and **j**.

   (i) **AB** (ii) **AD** (iii) **BD** (iv) **BE** (v) **AF** (vi) **HI** (vii) **IM** (viii) **NS** (ix) **SH** (x) **AS**

4. Find the scalar products.

   (i) **AB.BD** (ii) **AH.BN** (iii) **DE.EN** (iv) **EF.HM** (v) **NS.AB**

   (vi) **HS.SN** (vii) **MD.EN** (viii) **FS.AH** (ix) **IS.BD** (x) **HI.IM**

5. Calculate the scalar product for each vector pair below and identify any vector pairs which are mutually perpendicular.

   (i) $2\mathbf{i} + \mathbf{j}, 4\mathbf{i} - \mathbf{j}$ (ii) $\mathbf{i} + 2\mathbf{j}, 4\mathbf{i} + 3\mathbf{j}$ (iii) $\mathbf{i} - 2\mathbf{j}, 2\mathbf{i} + \mathbf{j}$ (iv) $3\mathbf{i} - \mathbf{j}, \mathbf{i} - \mathbf{j}$

   (v) $\mathbf{i}, 3\mathbf{i} + \mathbf{j}$ (vi) $\mathbf{j}, 2\mathbf{i} + \mathbf{j}$ (vii) $\mathbf{i}, 4\mathbf{i} - \mathbf{j}$ (viii) $2\mathbf{i} + \mathbf{j}, \mathbf{i}$

   (ix) $4\mathbf{j}, 3\mathbf{i} - \mathbf{j}$ (x) $-3\mathbf{j}, -4\mathbf{i} + \mathbf{j}$

6. $\mathbf{a} = x\mathbf{i} + y\mathbf{j}$, $\mathbf{b} = (2 + x)\mathbf{i} + (y - 4)\mathbf{j}$ and $\mathbf{c} = 13\mathbf{i} + 4\mathbf{j}$ are three vectors in component form.

   If $\mathbf{a} + 2\mathbf{b} = \mathbf{c}$ find

   (i) the values of x and y,
   (ii) a unit vector parallel to **a**,
   (iii) the scalar product $\mathbf{a}.(\mathbf{b} + \mathbf{c})$.

   (N. Ireland Additional Mathematics (part question))

7. $\mathbf{a} = 3\mathbf{i} + \mathbf{j}$, $\mathbf{b} = 2\mathbf{i} - 3\mathbf{j}$ and $\mathbf{c} = 3\mathbf{i} + 2\mathbf{j}$ are three vectors in component form.

   (i) Calculate, in component form, the vectors $\mathbf{b} + \mathbf{c}$ and $2\mathbf{a} - 3\mathbf{b}$.
   (ii) Show, using the above vectors, that $\mathbf{a}.(\mathbf{b} + \mathbf{c}) = \mathbf{a}.\mathbf{b} + \mathbf{a}.\mathbf{c}$.
   (iii) Show that **b** and **c** are at right angles to each other.

   (N. Ireland Additional Mathematics (part question))

8. $\mathbf{a} = 3\mathbf{i} + 2\mathbf{j}$, $\mathbf{b} = -6\mathbf{i} - \mathbf{j}$ and $\mathbf{c} = 9\mathbf{j}$ are three vectors in component form.

   Find in the simplest form possible

   (i) $\mathbf{a} + 2\mathbf{b} + \mathbf{c}$,
   (ii) the modulus of $2\mathbf{a} - \mathbf{b}$,
   (iii) a unit vector parallel to $\mathbf{b} + \mathbf{c}$,
   (iv) the scalar product $(\mathbf{a} - \mathbf{b}).(\mathbf{a} + \mathbf{b})$.

   (N. Ireland Additional Mathematics (part question))

9. $\mathbf{a} = 2x\mathbf{i} + 3y\mathbf{j}$, $\mathbf{b} = (x + 4y)\mathbf{i} + (2x + y)\mathbf{j}$ and $\mathbf{c} = 3\mathbf{i} + 4\mathbf{j}$ are three vectors in component form.

   If $\mathbf{a} + \mathbf{b} = 2\mathbf{c}$ find

   (i) the values of x and y,
   (ii) a unit of vector parallel to $\mathbf{a} + \mathbf{b} - \mathbf{c}$.

   (N. Ireland Additional Mathematics (part question))

USING VECTORS TO DESCRIBE MOTION IN THE PLANE

Newton's second law and the constant acceleration formulae have been used in earlier chapters to describe motion in a straight line – motion in a single dimension. The second law and constant acceleration formulae may be used directly to investigate motion in two dimensions by converting each law/formula to its vector analogue. Since the only scalars involved are time and mass

$F = ma$	becomes	$\mathbf{F} = m\mathbf{a}$
$v = u + at$	becomes	$\mathbf{v} = \mathbf{u} + \mathbf{a}t$
$s = ut + \frac{1}{2}at^2$	becomes	$\mathbf{s} = \mathbf{u}t + \frac{1}{2}\mathbf{a}t^2$
$v^2 = u^2 + 2as$	becomes	$\mathbf{v.v} = \mathbf{u.u} + 2\mathbf{a.s}$

Motion in two dimensions may therefore be analysed using the equations encountered in previous chapters by converting these to vector form – a simple conversion rule might be to convert every variable (with the exception of mass and time) in a given equation to its vector equivalent. The examples below will serve to demonstrate that two-dimensional straight line motion may be analysed with the same 'ease' as one-dimensional motion.

EXAMPLE 5

Initially a particle P of mass 3 kg is at rest at the point D whose displacement relative to the origin O is $\mathbf{OD} = (\mathbf{i} + 2\mathbf{j})$ metres.

If the force $\mathbf{F} = (9\mathbf{i} + 3\mathbf{j})$ newtons acts on P, find the displacement **OE** of P from O after 2 seconds.

(N. Ireland Additional Mathematics, 1980 (part question))

Solution 5

Applying $\mathbf{F} = m\mathbf{a}$ where $\mathbf{F} = 9\mathbf{i} + 3\mathbf{j}$ and $m = 3$

$$9\mathbf{i} + 3\mathbf{j} = 3\mathbf{a}$$
or $\mathbf{a} = 3\mathbf{i} + \mathbf{j}$

In order to find the displacement which takes place in the 2 seconds $\mathbf{s} = \mathbf{u}t + \frac{1}{2}\mathbf{a}t^2$ is applied, where $\mathbf{u} = 0\mathbf{i} + 0\mathbf{j}$ (the particle starts from rest).

$$\mathbf{s} = (0\mathbf{i} + 0\mathbf{j})(2) + \tfrac{1}{2}(3\mathbf{i} + \mathbf{j})(2)^2$$
or $\mathbf{s} = 6\mathbf{i} + 2\mathbf{j}$

Figure 10.7 illustrates this displacement from the point D.

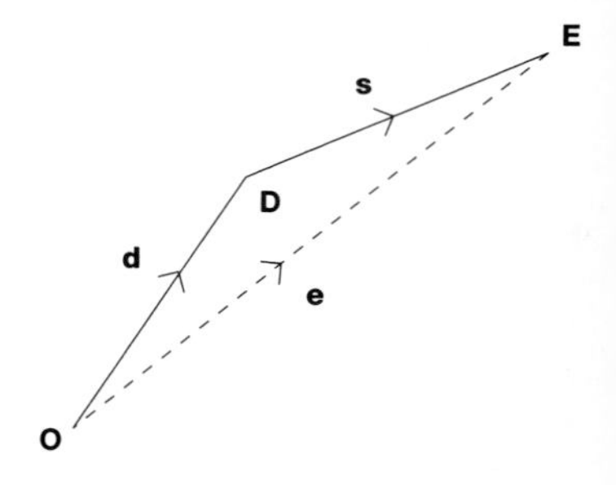

Figure 10.7

Applying the triangle rule for vector addition

$$\mathbf{e} = \mathbf{d} + \mathbf{s}$$
or $\mathbf{e} = \mathbf{i} + 2\mathbf{j} + 6\mathbf{i} + 2\mathbf{j}$

hence $\mathbf{e} = 7\mathbf{i} + 4\mathbf{j}$

i.e. $\mathbf{OE} = 7\mathbf{i} + 4\mathbf{j}$

EXAMPLE 6

A particle P of mass 2 kg is at rest at the origin O. The forces $\mathbf{F} = (4\mathbf{i} + 3\mathbf{j})$ newtons and $\mathbf{G} = (2\mathbf{i} + \mathbf{j})$ newtons act simultaneously on P.

Find the displacement of P from O after 4 seconds.

(N. Ireland Additional Mathematics, 1981 (part question))

Solution 6

The total force on P is

$$4\mathbf{i} + 3\mathbf{j} + 2\mathbf{i} + \mathbf{j} = 6\mathbf{i} + 4\mathbf{j}$$

Therefore $\mathbf{F} = m\mathbf{a}$ becomes $6\mathbf{i} + 4\mathbf{j} = 2\mathbf{a}$ or $\mathbf{a} = 3\mathbf{i} + 2\mathbf{j}$.

Now, since the body starts from rest at the origin, its displacement from O after 4 seconds is given by

$$\mathbf{s} = (0\mathbf{i} + 0\mathbf{j})(4) + \tfrac{1}{2}(3\mathbf{i} + 2\mathbf{j})(4)^2$$

or $$\mathbf{s} = 24\mathbf{i} + 16\mathbf{j}$$

EXAMPLE 7

Initially a particle P of mass 2 kg is at rest at the point A whose displacement relative to the origin O is $\mathbf{OA} = (2\mathbf{i} - \mathbf{j})$ metres. If the force $\mathbf{F} = (6\mathbf{i} - 7\mathbf{j})$ newtons acts on P, find the displacement $\mathbf{OB}$ of P from O after 3 seconds.

(N. Ireland Additional Mathematics, 1982 (part question))

Solution 7

Applying $\mathbf{F} = m\mathbf{a}$ in this case yields

$$6\mathbf{i} - 7\mathbf{j} = 2\mathbf{a}$$

or $$\mathbf{a} = 3\mathbf{i} - 3.5\mathbf{j}$$

The displacement from the initial position is therefore

$$\mathbf{s} = (0\mathbf{i} + 0\mathbf{j})(3) + \tfrac{1}{2}(3\mathbf{i} - 3.5\mathbf{j})(3)^2$$

or $$\mathbf{s} = 13.5\mathbf{i} - 15.75\mathbf{j}$$

The displacement $\mathbf{b}$ is the initial displacement plus the displacement which occurred in the three seconds.

Therefore

$$\mathbf{b} = 2\mathbf{i} - \mathbf{j} + 13.5\mathbf{i} - 15.75\mathbf{j}$$

or $$\mathbf{OB} = 15.5\mathbf{i} - 16.75\mathbf{j}$$

EXAMPLE 8

A particle of mass 5 kg is acted upon by a constant force of $15\mathbf{i} + 20\mathbf{j}$. If, after 8 seconds, its velocity is $24\mathbf{i} + 20\mathbf{j}$, find its initial velocity.

Solution 8

Applying $\mathbf{F} = m\mathbf{a}$ leads to $\mathbf{a} = 3\mathbf{i} + 4\mathbf{j}$.

Here, $\mathbf{v} = \mathbf{u} + \mathbf{a}t$ is the appropriate constant acceleration equation.
Inserting the relevant values of $\mathbf{v}$, $\mathbf{a}$ and t

$$24\mathbf{i} + 20\mathbf{j} = \mathbf{u} + (3\mathbf{i} + 4\mathbf{j})8$$
$$\Rightarrow \quad 24\mathbf{i} + 20\mathbf{j} = \mathbf{u} + 24\mathbf{i} + 32\mathbf{j}$$

Therefore

$$\mathbf{u} = -12\mathbf{j}$$

EXERCISE 10.2

1. (i) A particle of mass 3 kg starts with velocity $2\mathbf{i} + \mathbf{j}$ (m/s) at a point P where $\mathbf{p} = 4\mathbf{i} + \mathbf{j}$ (m). It is acted upon by a force $\mathbf{F} = 6\mathbf{i} + 9\mathbf{j}$ (N) and, after 2 seconds, arrives at Q. Find

(a) the velocity at Q and,
(b) the vector $\mathbf{q}$.

(ii) A particle of mass 9 kg starts with velocity $2\mathbf{i} + \mathbf{j}$ (m/s) at a point P where $\mathbf{p} = 2\mathbf{i} + 4\mathbf{j}$ (m). It is acted upon by a force $\mathbf{F} = 27\mathbf{i} + 54\mathbf{j}$ (N) and, after 2 seconds, arrives at Q. Find

(a) the velocity at Q and,
(b) the vector $\mathbf{q}$.

(iii) A particle of mass 4 kg starts with velocity $3\mathbf{i} - \mathbf{j}$ (m/s) at a point P where $\mathbf{p} = 3\mathbf{i} + 4\mathbf{j}$ (m). It is acted upon by a force $\mathbf{F} = 8\mathbf{i} + 12\mathbf{j}$ (N) and, after 4 seconds, arrives at Q. Find

(a) the velocity at Q and,
(b) the vector $\mathbf{q}$.

(iv) A particle of mass 5 kg starts with velocity $3\mathbf{i} + \mathbf{j}$ (m/s) at a point P where $\mathbf{p} = 2\mathbf{i} + 4\mathbf{j}$ (m). It is acted upon by a force $\mathbf{F} = 5\mathbf{i} + 10\mathbf{j}$ (N) and, after 6 seconds, arrives at Q. Find

(a) the velocity at Q and,
(b) the vector $\mathbf{q}$.

(v) A particle of mass 6 kg starts with velocity $-2\mathbf{i} + \mathbf{j}$ (m/s) at a point P where $\mathbf{p} = 3\mathbf{i} + 8\mathbf{j}$ (m). It is acted upon by a force $\mathbf{F} = -6\mathbf{i} + 18\mathbf{j}$ (N) and, after 3 seconds, arrives at Q. Find

(a) the velocity at Q and,
(b) the vector $\mathbf{q}$.

(vi) A particle of mass 4 kg starts with velocity $\mathbf{i} + \mathbf{j}$ (m/s) at a point P where $\mathbf{p} = 3\mathbf{i} - 2\mathbf{j}$ (m). It is acted upon by a force $\mathbf{F} = 6\mathbf{i} + 8\mathbf{j}$ (N) and, after 5 seconds, arrives at Q. Find

(a) the velocity at Q and,
(b) the vector $\mathbf{q}$.

(vii) A particle of mass 7 kg starts with velocity $\mathbf{i} + 2\mathbf{j}$ (m/s) at a point P where $\mathbf{p} = 3\mathbf{i}$ (m). It is acted upon by a force $\mathbf{F} = -14\mathbf{i} - 21\mathbf{j}$ (N) and, after 4 seconds, arrives at Q. Find

(a) the velocity at Q and,
(b) the vector $\mathbf{q}$.

(viii) A particle of mass 5 kg starts with velocity $\mathbf{j}$ (m/s) at a point P where $\mathbf{p} = 2\mathbf{i} - \mathbf{j}$ (m). It is acted upon by a force $\mathbf{F} = 10\mathbf{i} + 15\mathbf{j}$ (N) and, after 6 seconds, arrives at Q. Find

(a) the velocity at Q and,
(b) the vector $\mathbf{q}$.

(ix) A particle of mass 6 kg starts with velocity $-\mathbf{i}$ (m/s) at a point P where $\mathbf{p} = 4\mathbf{i}$ (m). It is acted upon by a force $\mathbf{F} = 12\mathbf{i} - 24\mathbf{j}$ (N) and, after 2 seconds, arrives at Q. Find

(a) the velocity at Q and,
(b) the vector $\mathbf{q}$.

(x) A particle of mass 11 kg starts with velocity $-\mathbf{i} - \mathbf{j}$ (m/s) at a point P where $\mathbf{p} = 2\mathbf{i} - \mathbf{j}$ (m). It is acted upon by a force $\mathbf{F} = 33\mathbf{i} - 44\mathbf{j}$ (N) and, after 4 seconds, arrives at Q. Find

(a) the velocity at Q and,
(b) the vector $\mathbf{q}$.

2. A particle starts from the origin O of perpendicular axes, with initial velocity $3\mathbf{i} - \mathbf{j}$ (m/s), and accelerates uniformly for 4 seconds, when its new position is P, and its new velocity is $-9\mathbf{i} + 7\mathbf{j}$ (m/s). Find in component vector form

(i) the acceleration of the particle,
(ii) the displacement vector **OP**.

(N. Ireland Additional Mathematics (part question))

3. (i) A particle, of mass 3 kg, acted upon by the three forces $2p\mathbf{i} - 3q\mathbf{j}$ (N), $q\mathbf{i} + 2p\mathbf{j}$ (N) and $-7\mathbf{i} + 5\mathbf{j}$ (N) is in equilibrium. Find p and q.

(ii) If the force $2p\mathbf{i} - 3q\mathbf{j}$ (N) ceases to act, find, in component form, the resulting acceleration of the particle.

(N. Ireland Additional Mathematics (part question))

4. A body of mass 20 kg, initially at rest at O, is acted on by two forces $\mathbf{P} = 3\mathbf{i} + \mathbf{j}$ (N) and $\mathbf{Q} = x\mathbf{i} + y\mathbf{j}$ (N) and the resulting acceleration of the body is $\frac{1}{10}(2\mathbf{i} + 5\mathbf{j})$ (m/s^2). Find

(i) the values of x and y,
(ii) the displacement vector **OA** where A is the position of the body after 5 seconds of motion.

(N. Ireland Additional Mathematics (part question))

5. A body of mass 5 kg is initially at rest at the point A whose displacement vector **OA** is $-2\mathbf{i} + 3\mathbf{j}$. The body is acted on by two forces $\mathbf{P} = 7\mathbf{i} + \mathbf{j}$ (newtons) and $\mathbf{Q} = -3\mathbf{i} - 2\mathbf{j}$ (newtons).

If the body is at the point B after 2 seconds of motion find the displacement vector **OB** and show that the body is moving in a direction parallel to the unit vector $\frac{1}{\sqrt{17}}(4\mathbf{i} - \mathbf{j})$.

(N. Ireland Additional Mathematics (part question))

CHAPTER 11

The equilibrium of particles and rigid bodies

The first mathematician to study bodies in equilibrium was **Archimedes** (287–212 BC) although he is better remembered for his book *On Floating Bodies* than for his treatise *On the Equilibrium of Planes*. Archimedes was born in Syracuse, a Greek settlement south-east of Sicily. He discovered the principle of buoyancy in the bath from which he ran naked into the street shouting 'Eureka!, Eureka!' (I have discovered it). He derived the formulae for the surface area and volume of the sphere stating these to be four times the area of a great circle (i.e. $4\pi r^2$) and two thirds of the volume of the smallest cylinder to enclose the sphere (i.e. $\frac{2}{3} \times \pi r^2(2r) = \frac{4}{3}\pi r^3$). Archimedes also left us some interesting approximations. To quote but two of these

$$3\tfrac{10}{71} < \pi < 3\tfrac{1}{7} \quad \text{and} \quad \tfrac{265}{153} < \sqrt{3} < \tfrac{1351}{780}$$

In his famous *Sand-Reckoner Axiom* he illustrates that the amount of sand in the world is finite.

Archimedes' knowledge of levers was put to good use in the defence of Syracuse against the Romans. He invented many large-scale weapons which were used to hurl boulders out to sea at incoming ships. When the Romans entered the city one soldier found Archimedes deep in thought over a mathematical problem. Archimedes' plea for a short stay of execution until he solved the problem fell on deaf ears.

Archimedes and Newton together are the inventors of that field of science now known as mathematical physics.

INTRODUCTION

Since this chapter addresses the equilibrium of particles and rigid bodies, it is instructive to begin with a definition of the terms 'particle', 'rigid body' and 'equilibrium'.

A **particle** is a quantity of matter so small that the distances between its extremities are negligible. A **rigid body** is a collection of particles forming a solid mass in which the distance between any two particles is not negligible and remains constant and **equilibrium** is achieved when the rigid body or particle does not move when acted upon by two or more forces.

This chapter considers the resultant of two forces and the technique for resolving a single force into its components. It then explores the equilibrium of a particle under the action of a number of forces before illustrating the application of Lami's theorem to problems in which a particle is in equilibrium under the action of three forces. Finally, it details how the conditions for equilibrium of a rigid body can be exploited to discover a variety of unknown forces acting on that body.

The acceleration of gravity, g, will be taken to be 9.8 m/s^2 unless otherwise specified.

RESULTANT FORCES AND COMPONENTS

A force F is said to be the **resultant** of forces F_1, F_2 and F_3, for example, if F acting alone can produce exactly the same effect as F_1, F_2 and F_3 acting together. A simple illustration is possible in the special case of parallel or anti-parallel forces. In Figure 11.1 the 7 N force is the resultant of the 3 N and 4 N forces.

Figure 11.1

In Figure 11.2 the 7 N force is the resultant of the 10 N and 3 N forces.

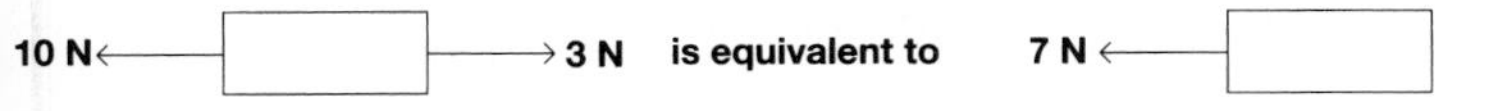

Figure 11.2

The resultant of two coplanar forces at right angles to one another is represented by the diagonal of the rectangle whose sides represent the two forces. Therefore in Figure 11.3 the force F is the resultant of the pair of forces F_1, F_2.

Figure 11.3

The resultant of the forces F_1 and F_2 has magnitude F, where $F = \sqrt{(F_1^2 + F_2^2)}$, the units of F being newtons when the units of F_1 and F_2 are newtons. However, a glance at Figure 11.3 will reveal that the resultant force is not uniquely defined until some measure of its direction is added. Notice that F makes an angle θ with the direction of F_2 where

$$\tan\theta = \frac{F_1}{F_2}$$

In summary then the resultant of F_1 and F_2 is a force of magnitude $\sqrt{(F_1^2 + F_2^2)}$ at θ to the direction of F_2 where $\tan\theta = \dfrac{F_1}{F_2}$.

EXAMPLE 1

Find the resultant of the coplanar forces 7 N and 2 N in Figure 11.4.

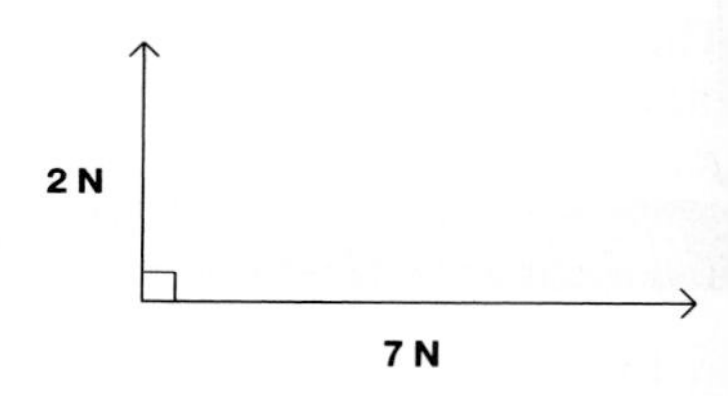

Figure 11.4

Solution 1

Using the reasoning outlined above this pair of forces may be combined as in Figure 11.5 where

$$F = \sqrt{(2^2 + 7^2)} = 7.280\,\text{N}$$

and $\tan\theta = \frac{2}{7}$ or $\theta = 15.95°$

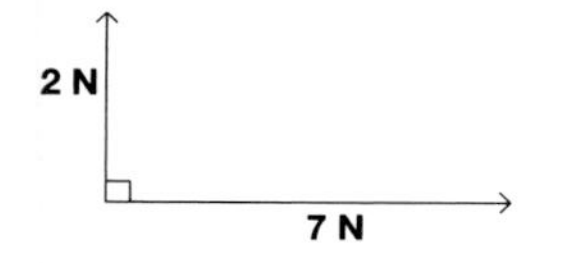

is equivalent to

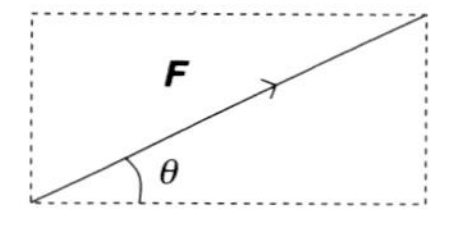

Figure 11.5

The resultant has magnitude 7.280 N and is directed at 15.95° to the direction of the 7 N force.

EXAMPLE 2

Find the resultant of the coplanar forces 4.2 N and 1.7 N in Figure 11.6.

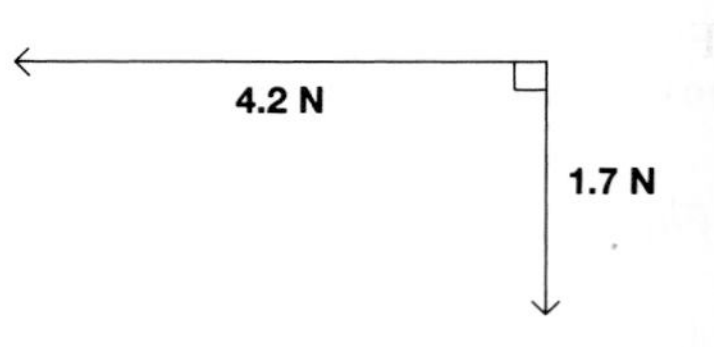

Figure 11.6

Solution 2

Using the theory above, the pair of forces may be combined as in Figure 11.7 where

$$F = \sqrt{(4.2^2 + 1.7^2)} = 4.531\,\text{N}$$

and $\tan\theta = \dfrac{1.7}{4.2}$ or $\theta = 22.04°$

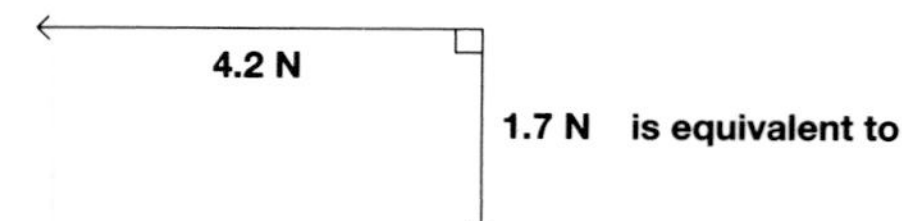

is equivalent to

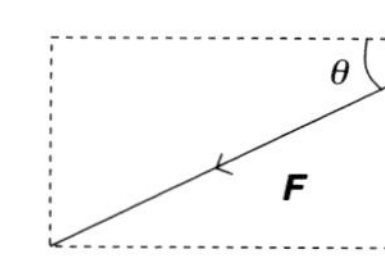

Figure 11.7

Hence the resultant has magnitude 4.531 N and is directed at 22.04° to the direction of the 4.2 N force.

Just as two forces at right angles may be combined to produce a single force (the resultant), then any force may be **resolved** into two mutually perpendicular forces, called the **components** of the force.

Therefore, by reversing the process of calculating the resultant, a single force F at θ to the horizontal may be split into two mutually perpendicular forces F_1 and F_2 which together have the same effect as the F force. This is illustrated in Figure 11.8.

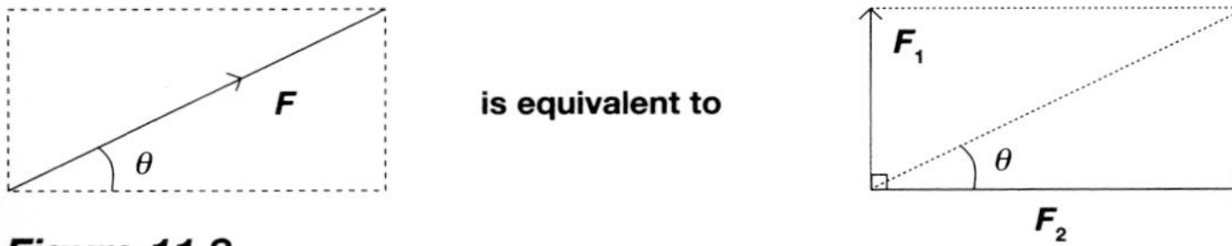

Figure 11.8

The application of elementary trigonometry to Figure 11.8 gives

$$\boldsymbol{F_2 = F\cos\theta}$$
$$\boldsymbol{F_1 = F\sin\theta}$$

$F\cos\theta$ and $F\sin\theta$ (measured in newtons when F is measured in newtons) are said to be the components or 'resolutes' of F in the horizontal and vertical direction respectively. The process of splitting F into its component parts is referred to as **resolving *F* horizontally and vertically**.

EXAMPLE 3

A force of 7 N lies in the first quadrant and makes an angle of 27° with the x axis. Find the components of this force in the direction of the x and y axes.

Solution 3

Figure 11.9 shows the 7 N force resolved along the x and y axes.

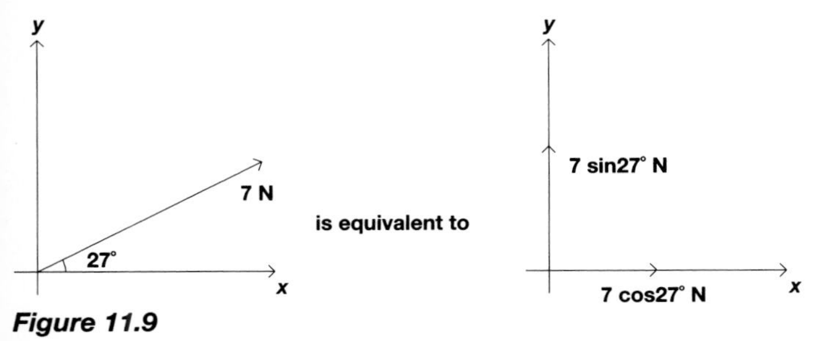

Figure 11.9

Since $7\sin 27° = 3.18\,\text{N}$
and $7\cos 27° = 6.24\,\text{N}$

it follows that the 7 N force is equivalent to a force of 6.24 N along the x axis together with a force of 3.18 N along the y axis.

PARTICLE EQUILIBRIUM UNDER THE ACTION OF MANY FORCES

The techniques presented above for combining two forces may be extended to problems involving more than two forces as illustrated in examples 4–9.

EXAMPLE 4

The system of 3 forces on the left of Figure 11.10 is equivalent to the single resultant F at angle ψ to the horizontal, shown on the right of Figure 11.10. Find F and ψ.

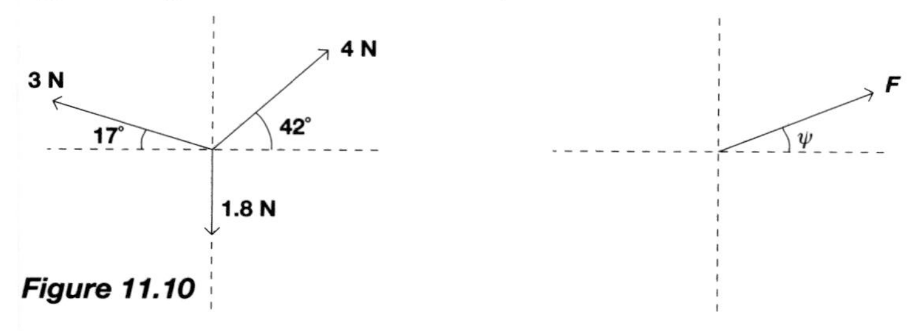

Figure 11.10

Solution 4

The three forces may each be resolved into a horizontal and a vertical component. For example, the 4 N force has horizontal and vertical components of $4\cos 42°$ N and $4\sin 42°$ N respectively. Clearly, the 1.8 N force has no horizontal component. The net horizontal component to the right is

$$4\cos 42° - 3\cos 17° = 0.1037\,\text{N}$$

and the net vertical force, directed upwards is

$$3\sin 17° + 4\sin 42° - 1.8 = 1.754\,\text{N}$$

The diagram on the left of Figure 11.10 may therefore be replaced by that on the left of Figure 11.11.

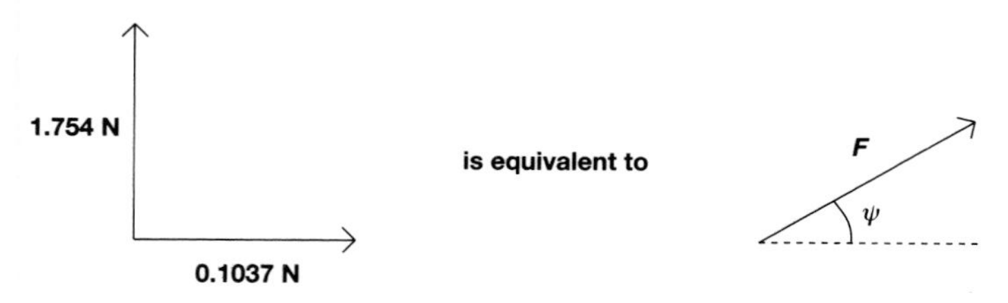

Figure 11.11

Hence $F = \sqrt{(1.754^2 + 0.1037^2)}$
or $F = 1.757\,\text{N}$

and $\tan\psi = \dfrac{1.754}{0.1037}$

i.e. $\psi = 86.61°$

EXAMPLE 5

The system of 5 forces on the left of Figure 11.12 can be shown to be equivalent to the single resultant force F at angle ψ to the horizontal, shown on the right of the figure. Find F and ψ.

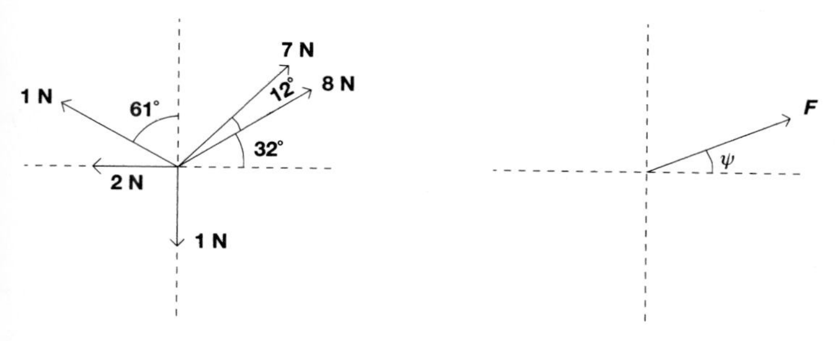

Figure 11.12

Solution 5

Each force may be resolved into a horizontal component and a vertical component i.e. the 8 N force has components $8\cos 32°$ N and $8\sin 32°$ N in the horizontal and vertical directions respectively. By combining these components it may be shown that the net horizontal component directed to the right becomes

$$8\cos 32° + 7\cos 44° - 1\sin 61° - 2 = 8.945\,\text{N}$$

and the net vertical force, directed upwards, is

$$8\sin 32° + 7\sin 44° + 1\cos 61° - 1 = 8.587\,\text{N}$$

The diagram on the left of Figure 11.12 may therefore be replaced by the diagram on the left of Figure 11.13.

Figure 11.13

It follows that

$$F = \sqrt{(8.587^2 + 8.945^2)}$$

or $$F = 12.40\,\text{N}$$

and $$\tan\psi = \frac{8.587}{8.945}$$

i.e. $$\psi = 43.83°$$

EXAMPLE 6

A particle of mass 4 kg is attached to the lower end of a light inextensible string. The upper end is fixed to a wall. A horizontal force of P N is applied to the free end of the string so that the string makes an angle of θ with the downward vertical and experiences a tension of 200 N. If the 4 kg mass rests in equilibrium, find P and θ.

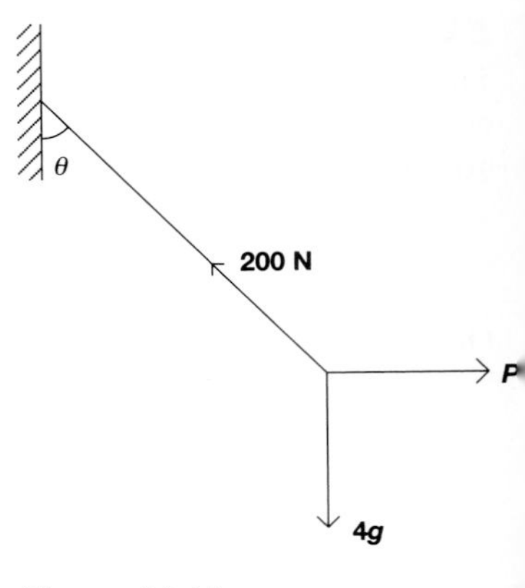

Figure 11.14

Solution 6

Figure 11.14 shows the forces acting on the particle. The weight (acting downwards) is calculated by multiplying the mass by $g = 9.8\,\text{m/s}^2$ and the particle experiences a pull of 200 N due to being on the end of the string.

The P force is horizontal and the $4g$ force vertical. For ease of calculation the tension is replaced by its components as shown in Figure 11.15.

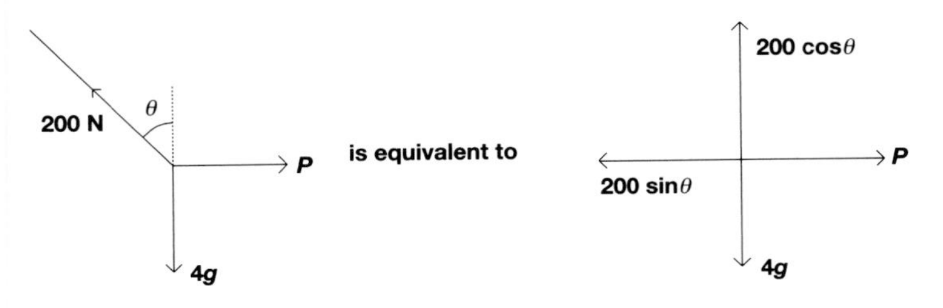

Figure 11.15

In order to comply with the definition of equilibrium offered on page 229, all forces are required to 'balance'; if this were not the case a net force might exist in, say, the horizontal direction, resulting in movement and a breakdown of equilibrium.

Therefore $200\cos\theta = 4g$
and $200\sin\theta = P$

From the first of these equations

$$\cos\theta = \frac{4 \times 9.8}{200}$$

$$\Rightarrow \qquad \theta = 78.70°$$

Inserting this value of θ in the second equation yields

$$P = 200\sin 78.70°$$

$$\Rightarrow \qquad P = 196.1\,\text{N}$$

EXAMPLE 7

A particle of mass m kg is attached to the lower end of a light inextensible string, the upper end of which is fixed to a wall. A horizontal force of 40 N is applied to the free end of the string so that the string makes an angle of θ with the downward vertical and experiences a tension of 90 N. If the particle rests in equilibrium, find θ and m.

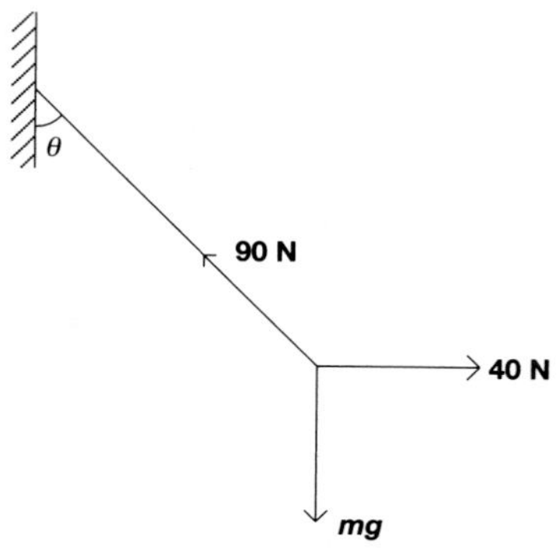

Figure 11.16

Solution 7

Figure 11.16 shows the forces acting on the particle. Since the mass of the particle is m, its weight is mg newtons.

Replacing the 90 N force by its components, transforms the force diagram on the left of Figure 11.17 to that on the right.

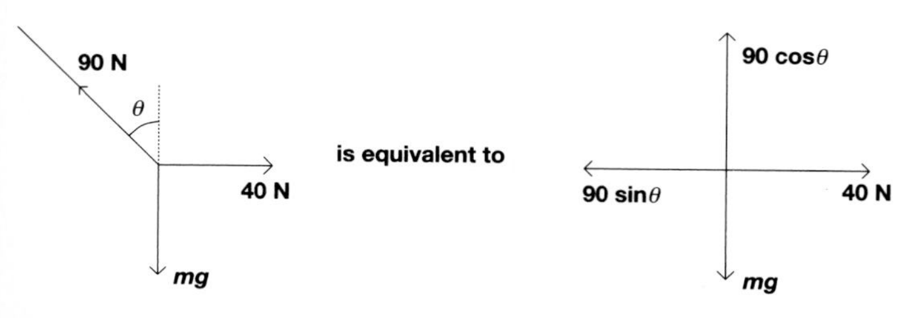

Figure 11.17

Since the particle is in equilibrium

$$90\cos\theta = mg$$
$$\text{and} \quad 90\sin\theta = 40$$

The second of these equations gives

$$\sin\theta = \frac{40}{90}$$

$$\text{or} \qquad \theta = 26.39°$$

The first equation therefore becomes

$$90\cos 26.39° = m \times 9.8$$
$$\text{or} \qquad m = 8.227\,\text{kg}$$

EXAMPLE 8

A particle of mass 5 kg is attached to the lower end of a light inextensible string. The upper end is fixed to a wall. A horizontal force of 40 N is applied to the free end of the string so that the string makes an angle θ with the downward vertical and experiences a tension T N. If the 5 kg mass rests in equilibrium, find θ and T.

Solution 8

The situation is illustrated in Figure 11.18.

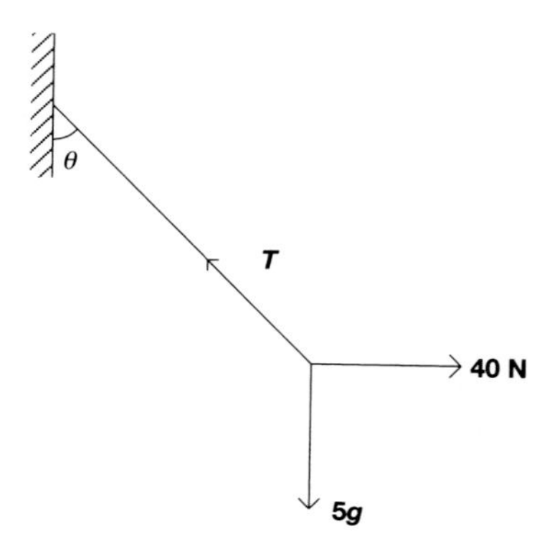

Figure 11.18

Once again, resolving the tension horizontally and vertically leads to

$$T\sin\theta = 40$$
$$\text{and} \quad T\cos\theta = 5 \times 9.8 \quad \text{respectively.}$$

'Dividing' the first equation by the second removes T since

$$\frac{T\sin\theta}{T\cos\theta} = \frac{40}{5 \times 9.8}$$

or $\quad \tan\theta = 0.8163$

$\Rightarrow \quad \theta = 39.23°$

Inserting this θ value in $T\sin\theta = 40$ yields

$T\sin(39.23°) = 40$

$\Rightarrow \quad T = 63.25\,\text{N}$

EXAMPLE 9

Figure 11.19 shows a body of mass 5 kg supported by two inextensible strings, the other ends of which are attached to two fixed points P and Q in a ceiling. The 5 kg mass rests in equilibrium with one string experiencing a tension T N and inclined at 30° to the horizontal and the other experiencing a force of S N and inclined at 45° to the horizontal. Find T and S.

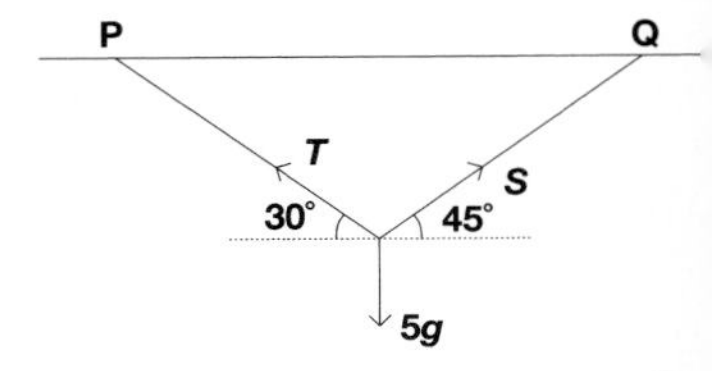

Figure 11.19

Solution 9

Resolving T and S into their horizontal and vertical components, the forces acting on the particle are as illustrated in Figure 11.20.

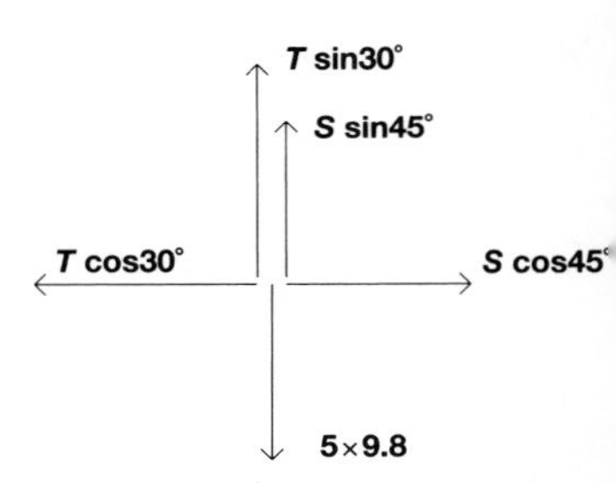

Figure 11.20

Since the particle is in equilibrium it follows that

$$T\cos 30° = S\cos 45°$$

and $\quad T\sin 30° + S\sin 45° = 49$

The first of these equations may be rearranged

$$T = \left(\frac{\cos 45°}{\cos 30°}\right) S$$

or $\quad T = 0.8165S$

and substituting this in the second equation gives

$(0.8165S)\sin 30° + S\sin 45° = 49$

$\Rightarrow \quad 0.4083S + 0.7071S = 49$

i.e. $\quad S = 43.93\,\text{N}$

But $\quad T = 0.8165S$

hence $\quad T = 35.87\,\text{N}$

EXERCISE 11.1

Take $g = 9.8\,\text{m/s}^2$ throughout.

1. Demonstrate that the system of forces on the left is equivalent to the single force on the right in parts (a) to (d) of Figure 11.21.

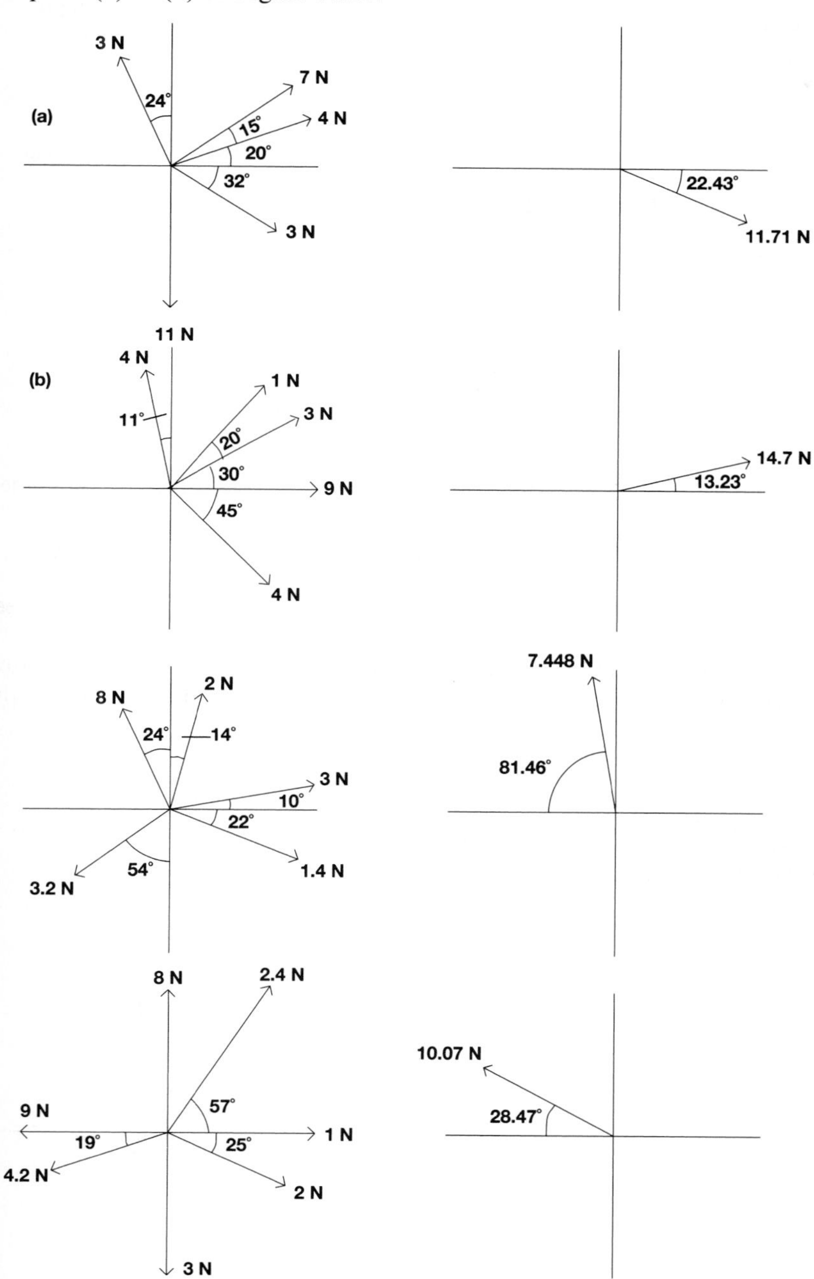

Figure 11.21

2. Find θ and P in parts (a) to (d) of Figure 11.22. Each diagram illustrates a particle in equilibrium on the end of a string of given tension attached to a wall. The particle is pulled horizontally with a force P until the string becomes inclined at θ to the vertical.

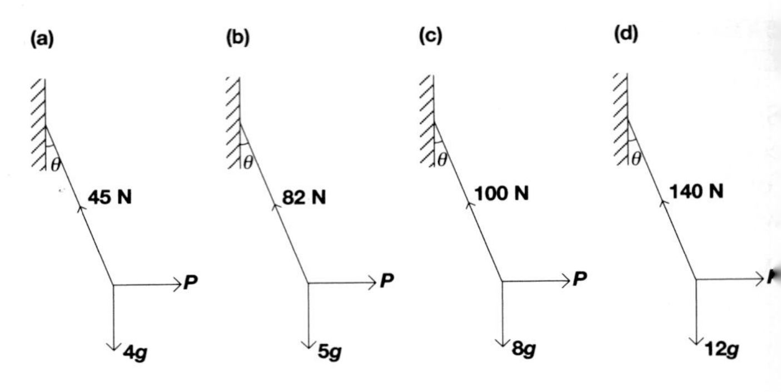

Figure 11.22

3. Find θ and m in parts (a) to (d) of Figure 11.23. Each diagram illustrates a particle of mass m in equilibrium on the end of a string of given tension attached to a wall. The particle is pulled horizontally until the string becomes inclined at θ to the vertical.

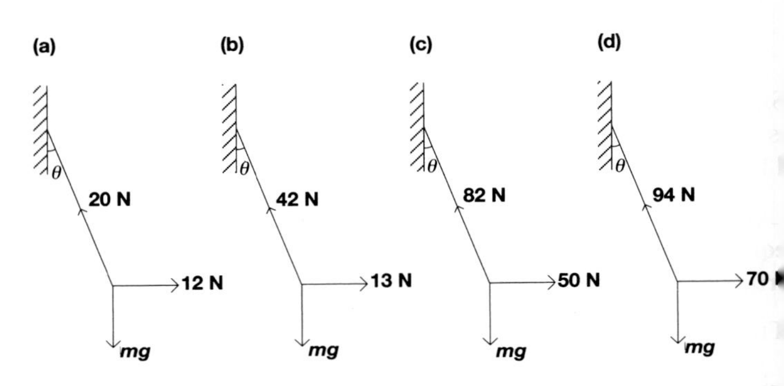

Figure 11.23

4. Find T_1 and T_2 in parts (a) to (d) of Figure 11.24. Each diagram illustrates a particle in equilibrium supported by two inextensible strings inclined at given angles to the vertical. In each case the particle is in equilibrium.

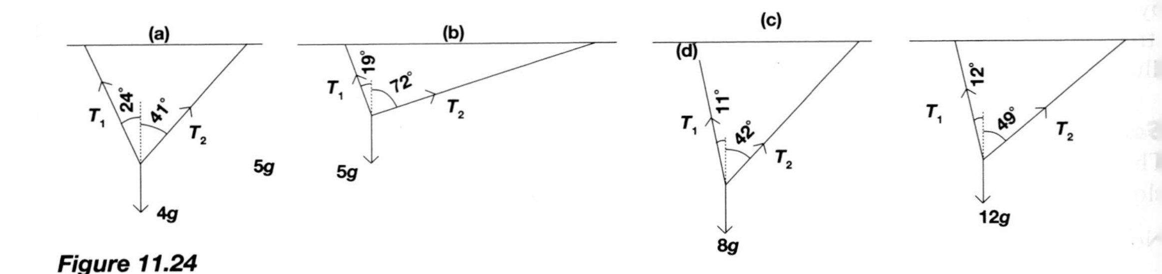

Figure 11.24

LAMI'S THEOREM AND ITS APPLICATIONS

Suppose the forces F_1, F_2 and F_3 act at a point and are in equilibrium. Lami's theorem states that the magnitude of each force is proportional to the size of the angle between the other two, with the same constant of proportionality in each case.

With reference to Figure 11.25 then

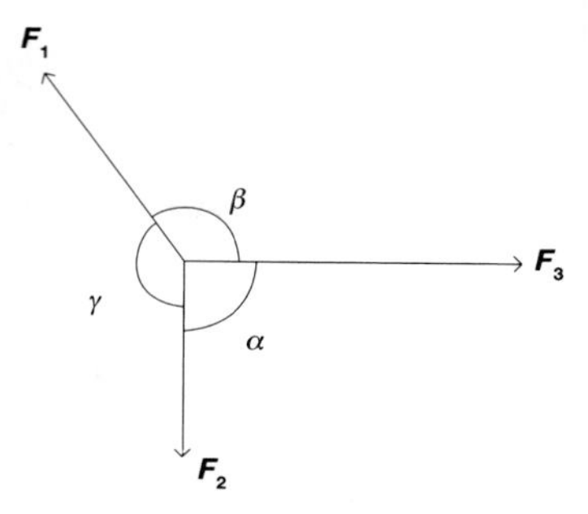

Figure 11.25

$$F_1 = k \sin \alpha$$
$$F_2 = k \sin \beta$$
$$F_3 = k \sin \gamma$$

$$\text{or} \quad k = \frac{F_1}{\sin \alpha} = \frac{F_2}{\sin \beta} = \frac{F_3}{\sin \gamma}$$

Since the sine of an angle is equal to the sine of its supplement, this is simply a statement of the sine rule for the triangle in Figure 11.26.

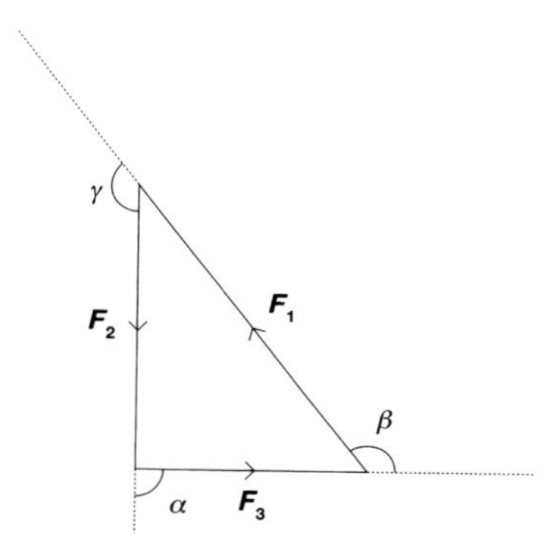

Figure 11.26

Lami's theorem is often stated: **if three forces acting at a point are in equilibrium then these forces may be represented by the sides of a triangle with the forces taken in order round the triangle**.

This triangle is frequently referred to as the **triangle of forces.**

Lami's theorem also applies to rigid bodies (see page 228) under the action of three forces. The use of the theorem is best illustrated through an example.

EXAMPLE 10

A beam of negligible mass which supports a mass of 5 kg at some point along its length, rests horizontally in equilibrium supported by two strings. If the tensions in the strings are T_1 and T_2 N and the strings are inclined at 30° and 40° respectively to the horizontal (as illustrated in Figure 11.27), find T_1 and T_2.

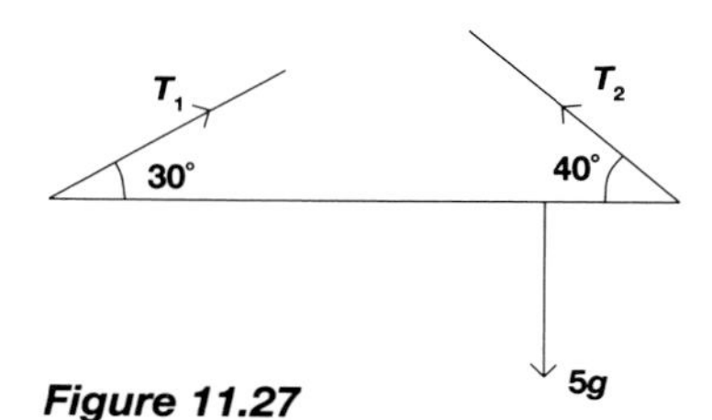

Figure 11.27

Solution 10

The three forces T_1, T_2 and $5g$ may be represented as the sides of a closed triangle as shown in Figure 11.28.

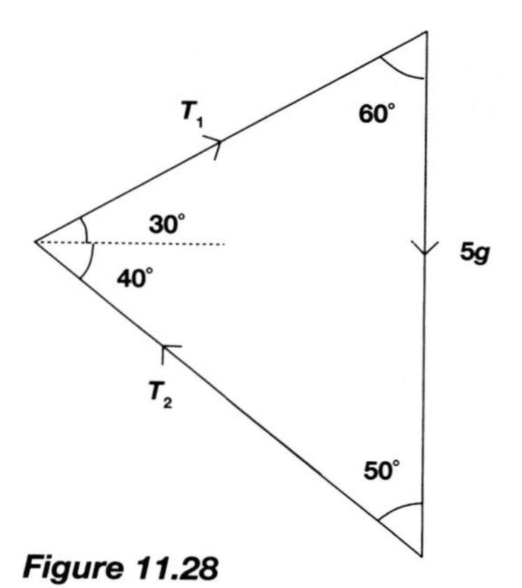

Figure 11.28

Now $5g = 5 \times 9.8 = 49\,\text{N}$

and, using sine rule in Figure 11.28

$$\frac{49}{\sin 70°} = \frac{T_2}{\sin 60°} = \frac{T_1}{\sin 50°}$$

Hence the two tensions may be computed by 'splitting' this equation and pursuing two parallel calculations.

$$\frac{49}{\sin 70°} = \frac{T_2}{\sin 60°} \qquad \frac{49}{\sin 70°} = \frac{T_1}{\sin 50°}$$

$$\frac{49}{0.9397} = \frac{T_2}{0.8660} \qquad \frac{49}{0.9397} = \frac{T_1}{0.7660}$$

$$T_2 = \frac{49 \times 0.8660}{0.9397} \qquad T_1 = \frac{49 \times 0.7660}{0.9397}$$

Hence $T_2 = 45.16$ N and $T_1 = 39.94$ N.

EXERCISE 11.2

Solve the following using the triangle of forces (in conjunction with sine and/or cosine rule), taking g to be 10 m/s^2 throughout.

1. A particle of mass 0.1 kg hangs on a string. What is the inclination of the string to the vertical when the particle is held aside by a force of 0.6 N directed at 20° above the horizontal? What is the tension in the string?
2. A particle has mass 125 g and hangs on a string. The particle is pulled aside by a force of 0.75 N so that the string makes an angle of 30° with the vertical. In what direction is the force applied? There are two possible answers – find the two directions and also the tension in each case.
3. A particle of mass 1×10^{-4} kg is supported by two strands of thread. If the tensions in the threads are 1.8×10^{-3} N and 1.9×10^{-3} N, find the angles which the threads make with the vertical.
4. An oil platform is towed at a constant speed by the horizontal pulls of the cables from two liners. If the tensions in the cables are 5×10^4 N and 7×10^4 N and the resistance to motion is 1.1×10^5 N, find the angles which the cables make with the direction of motion.
5. A particle of mass 4 kg is suspended by two strings of lengths 70 and 240 mm respectively, which are attached to two points at the same level and 250 mm apart. Find the tensions in the strings.
6. The tensions in two cables which together support a mass of 210 kg are 1260 N and 1680 N. Find the angles which these cables make with the vertical.
7. A machine component of mass 68 kg is slung by two chains 15 m and 8 m long from points 17 m apart on the roof of a loading bay. Find the tensions in the chains.

THE MOMENT OF A FORCE AND THE EQUILIBRIUM OF A RIGID BODY UNDER A NUMBER OF FORCES

A full understanding of the equilibrium of a rigid body is impossible without an appreciation of the concept of the moment of a force.

The moment of a force F about a point P is found by multiplying the magnitude of the force by the perpendicular distance from P to the line of action of the force. The SI unit of moment is therefore the newton metre (Nm). In Figure 11.29 the moment of F about P is therefore Fl.

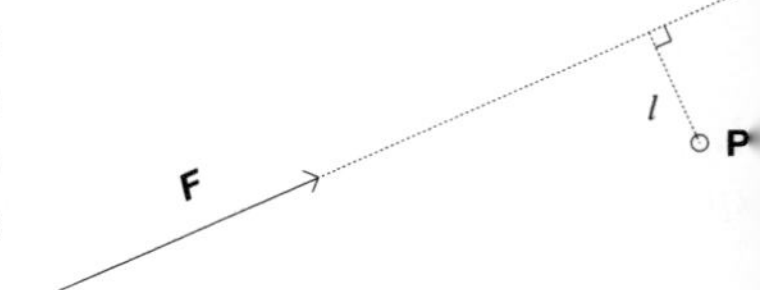

Figure 11.29

The special case where P lies on the line of action of F is illustrated in Figure 11.30. In this case the distance from P to the line of action of F is zero and therefore the moment of F about P is zero.

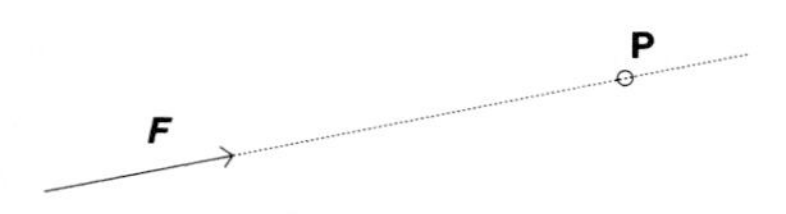

Figure 11.30

It is customary to give the 'sense' of the moment. In Figure 11.29 the moment of F about P is stated as Fl clockwise since F is directed such as to have a clockwise turning effect at P. The concept of the moment of a force plays a vital role in the definition of equilibrium.

The two conditions sufficient to ensure the equilibrium of a rigid body subject to several forces are as follows.

(a) **The sum of the components of the forces in any two directions must be zero**; and
(b) **The sum of the moments of the forces about any point must be zero**.

The examples which follow all concern situations in which a rigid body is known to be in equilibrium. Each example demonstrates how the two equilibrium conditions may be exploited to discover a variety of unknown forces. However, before proceeding to the examples, it is instructive to digress briefly to consider the precise mathematical definitions of three words: 'uniform', 'smooth' and 'light'. These three words have considerable mathematical implications when used as descriptors in the examples below. Their mathematical interpretation borrows much from their everyday usage while, as one might expect, imposing a rigour not to be found in the dictionary.

In all examples **uniform** implies that the weight of the body acts at its 'centre of mass' (the mid-point of the rod in the case of a uniform rod). Bodies resting against **smooth** walls experience no friction and so the only force on the point of the body in contact with the wall is the reaction perpendicular (or normal) to the wall.

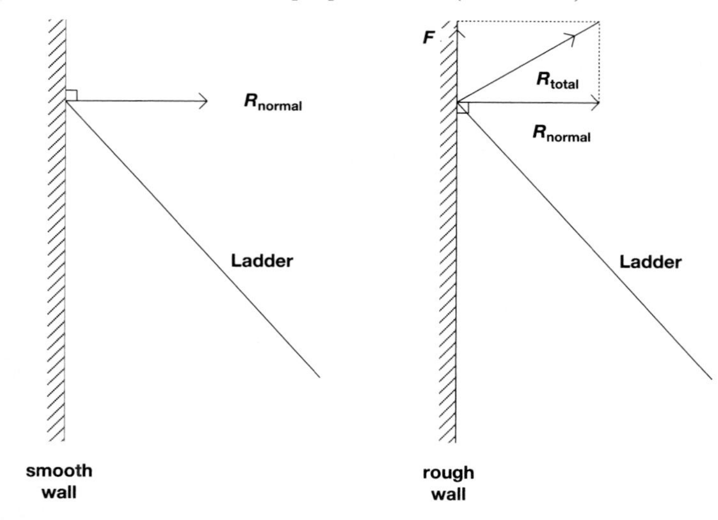

Figure 11.31

Hence, when a body rests against a smooth wall the reaction is at right angles to the wall but when the wall is not smooth, the reaction is the resultant of a perpendicular reaction and friction. In such cases the total reaction is not normal, as illustrated in Figure 11.31.

A **light** body is understood to have zero mass.

EXAMPLE 11

A uniform rod AB of mass 6 kg and length 4 m is supported at end A and at a point 0.7 m from end B as shown in Figure 11.32. Find the reaction at each support.

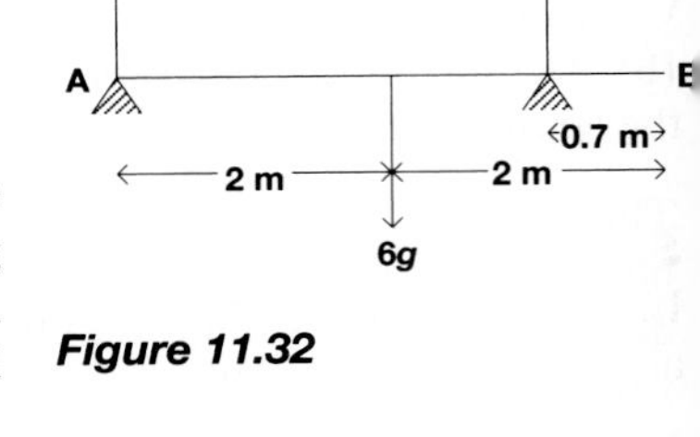

Figure 11.32

Solution 11

The weight acts at the mid-point of the rod and has a clockwise turning effect about A. The force S has an anticlockwise turning effect about A. Finally, the reaction R has no turning effect at A since it passes through A. Therefore, equating clockwise and anticlockwise moments at A

$$6g(2) = S(3.3)$$

where each side of this equation is made up of a force multiplied by the perpendicular distance to that force from the point about which moments are being calculated. This equating of clockwise and anticlockwise moments ensure condition (b) above is satisfied.

Hence $S = 35.64\,\text{N}$

Now, in order that condition (a) be complied with

$$R + S = 6 \times 9.8$$
$$\Rightarrow \quad R = 23.16\,\text{N}$$

EXAMPLE 12

A uniform bar AB of length 6.8 m has mass 8 kg and is supported at two points, one 0.7 m from A and the other 0.9 m from B. A body of mass 40 kg is moved gradually from end A towards B. How far is the body from end B when the bar first begins to tilt about the support nearer B?

Solution 12

This situation is illustrated in Figure 11.33 where the zero reaction at support E is a consequence of the tilting taking place at support C i.e. the bar is just losing contact with support E.

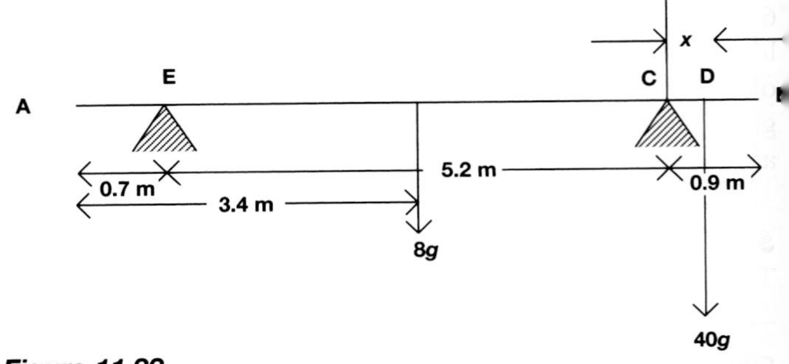

Figure 11.33

The body has been moved from A to the point D, x m from C. Taking moments about C and equating clockwise and anticlockwise moments

$$8g(2.5) = 40g(x) + 0(5.2)$$
$$\Rightarrow \quad x = 0.5\,\text{m}.$$

It follows that the body has been moved 0.4 m from B when tilting of the bar occurs.

EXAMPLE 13

Where must a mass of 80 kg be hung on a uniform rod of mass 10 kg and length l so that a child at one end provides one third of the support provided by an adult at the other?

Solution 13

In Figure 11.34 the child provides an upward force of P while the adult provides a $3P$ force.

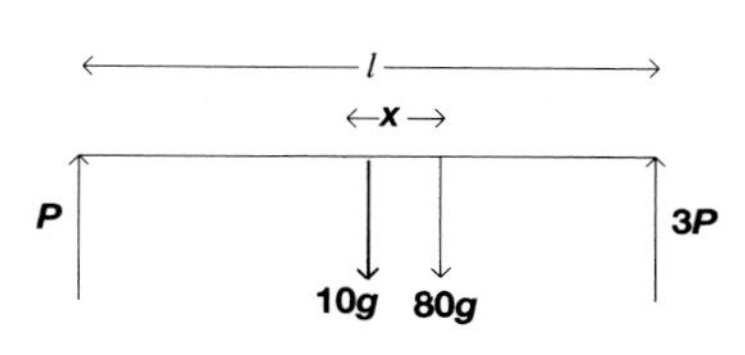

Figure 11.34

Equating vertical forces

$$P + 3P = 90g$$
$$\text{or} \quad P = 22.5g$$

Taking moments about the end supported by the adult with the 80 kg mass positioned a distance x from the centre

$$\frac{l}{2}(10g) + \left(\frac{l}{2} - x\right)80g = l(22.5g)$$

$$\Rightarrow \quad 5gl + 40gl - 80xg = 22.5gl$$
$$\text{i.e.} \quad 22.5gl = 80xg$$
$$\text{or} \quad x = 0.2813\,l$$

The body should therefore be positioned $0.7813\,l$ from the child's end.

EXAMPLE 14

A light rod rests in equilibrium against a smooth vertical wall with its lower end on rough ground as shown in Figure 11.35. The rod is 6 m long and has a mass of 2 kg attached at a point 2 m from its lowest point and a further mass of 9 kg attached 1 m from its point of contact with the wall. Find the normal reactions at the wall and ground and the size of the frictional force given that the rod makes an angle of 26° with the wall.

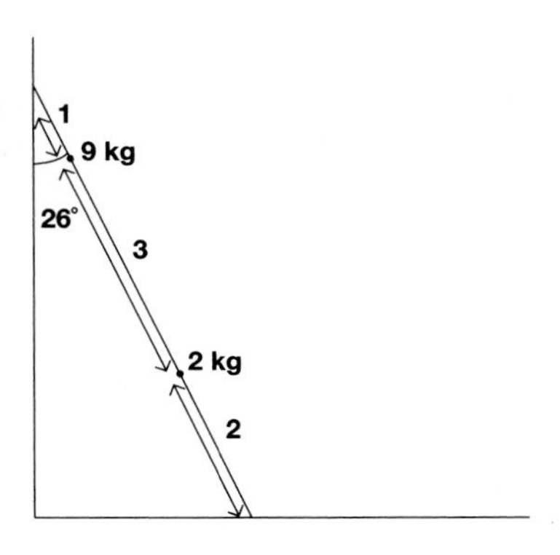

Figure 11.35

Solution 14

The forces on the rod are shown in Figure 11.36.

The normal reaction at the wall is R and at the ground is S. The frictional force acts in the direction which opposes motion and, since the rod is likely to slip outwards (away from the wall), then the frictional force F acts towards the wall. If moments are taken about B then F and S have no moment as B lies on the line of action of both forces. An examination of the turning effects of the remaining forces reveals that R tends to turn the rod in a clockwise direction about B while the $9g$ and $2g$ forces will turn the rod in the anticlockwise direction. At the beginning of this section two equilibrium conditions were cited. Condition (b) requires that the anticlockwise moments equal the clockwise moments. The implications of condition (b) are now investigated.

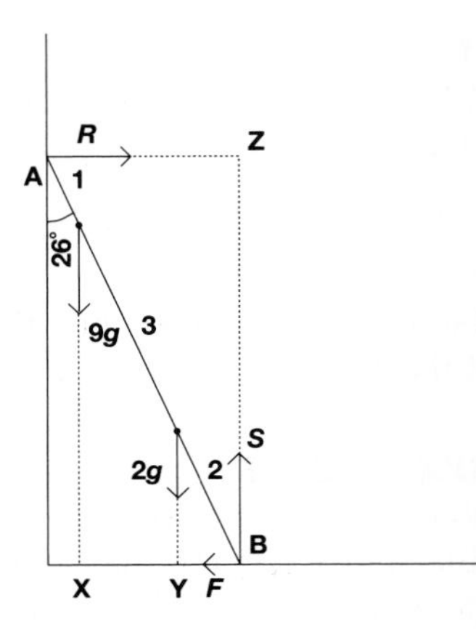

Figure 11.36

Moment is defined as force multiplied by perpendicular distance (from the point about which moments are being taken) to the line of action of the force.

It follows that the clockwise moment is

$$R \times \mathrm{BZ}$$
or $R(6\cos 26°)$.

The anticlockwise moments are

$$9g \times \mathrm{BX} + 2g \times \mathrm{BY}$$
or $9g(5\sin 26°) + 2g(2\sin 26°)$

Equating clockwise and anticlockwise moments yields

$$R(6\cos 26°) = 9g(5\sin 26°) + 2g(2\sin 26°)$$
or $R(5.393) = 193.3 + 17.18$
i.e. $R = 39.03\,\mathrm{N}$.

Condition (a) for equilibrium requires that the complete set of horizontal and the complete set of vertical forces each 'balance' in order that the sum of components in these two directions be zero.

Hence, with reference to Figure 11.36

$R = F$ (horizontal)
and $S = 9g + 2g$ (vertical)
But $R = 39.03\,\mathrm{N}$ and so $F = 39.03\,\mathrm{N}$
Also $S = 11g = 107.8\,\mathrm{N}$

In summary, the normal reaction at the wall is 39.03 N, the normal reaction at the ground is 107.8 N and the frictional force is 39.03 N.

EXAMPLE 15

A uniform rod AB of mass 5 kg is freely hinged at A to a vertical wall. A force P maintains equilibrium by supporting the other end of the rod. This force makes an angle of 20° with the rod and the rod itself makes 45° with the wall. The reaction at the hinge is shown in the Figure 11.37 and is inclined at ϕ to the upward vertical. Find P, the reaction at the hinge and ϕ.

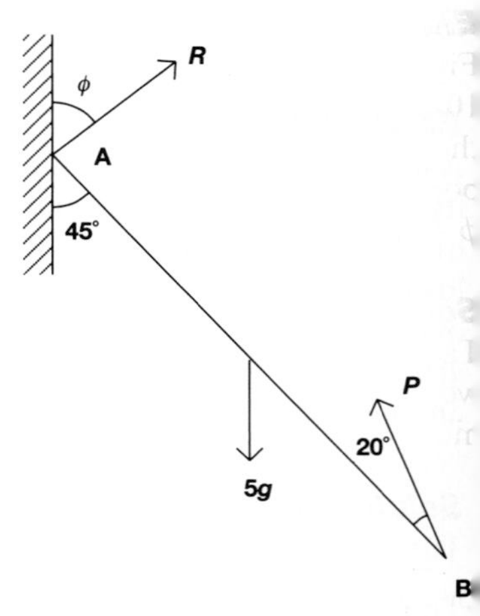

Figure 11.37

Solution 15

The forces in the problem are illustrated in Figure 11.38.

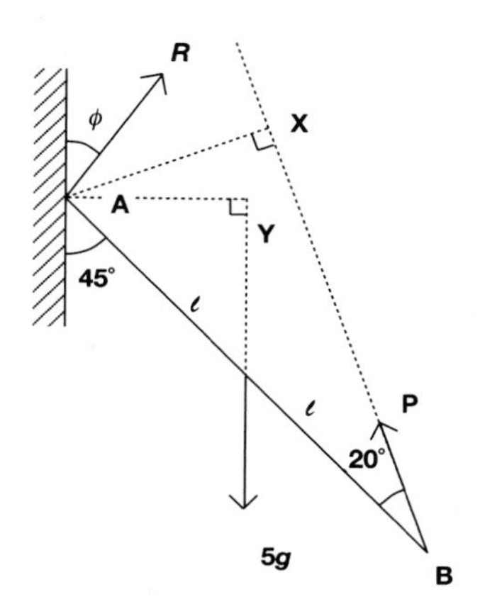

Figure 11.38

Let the length of the rod be $2l$ metres.

Taking moments about A, R has no moment since its line of action passes through A. P will rotate the rod anticlockwise about A and the $5g$ force will rotate it clockwise.

$$\begin{aligned} \text{Hence} \quad P(AX) &= 5g(AY) \\ \text{i.e.} \quad P(2l\sin 20°) &= 5g(l\sin 45°) \\ \Rightarrow \quad 2P\sin 20° &= 5g\sin 45° \\ \Rightarrow \quad P &= 50.65\,\text{N} \end{aligned}$$

In order to balance the horizontal and vertical components of the forces in this problem separately it will be necessary to resolve the R and P forces into components in these two directions. The $5g$ force is already vertical. Figure 11.39 illustrates the complete set of components.

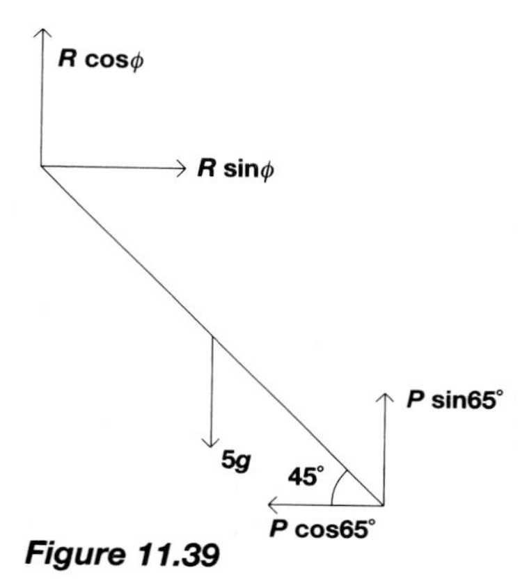

Figure 11.39

$$\begin{aligned} \text{Hence} \quad R\sin\phi &= P\cos 65° \\ \text{and} \quad R\cos\phi + P\sin 65° &= 5g \\ \Rightarrow \quad R\sin\phi &= 21.41 \\ \text{and} \quad R\cos\phi &= 3.096 \end{aligned}$$

'Dividing' this pair of equations

$$\frac{R\sin\phi}{R\cos\phi} = \frac{21.41}{3.096}$$

$$\begin{aligned} \Rightarrow \quad \tan\phi &= 6.915 \\ \text{hence} \quad \phi &= 81.77° \\ \text{i.e.} \quad R\sin 81.77° &= 21.41 \\ \Rightarrow \quad R &= 21.63\,\text{N} \end{aligned}$$

EXAMPLE 16

Figure 11.40 shows a uniform bar AB of length 36 m and mass 10 kg, freely hinged at A, resting on a smooth support C at 45° to the vertical. AC is 21 m and a body of mass 5 kg is suspended from a point D on the bar where BD = 9 m. Calculate S and R and prove $\phi = 45°$.

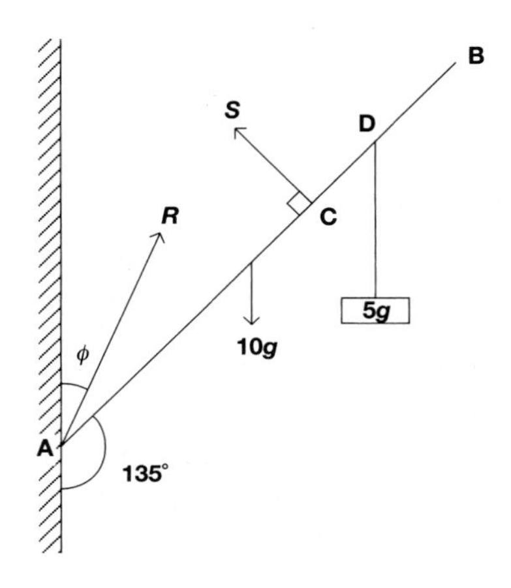

Figure 11.40

Solution 16

Taking moments about A, S has an anticlockwise moment and the weight of the bar and the 5 kg weight together have clockwise moments. Accordingly

$$21S = (18\sin 45°)10g + (27\sin 45°)5g$$

Therefore $S = 103.9\,\text{N}$

Consider Figure 11.41 in which the horizontal and vertical components of all forces are illustrated.

For a net horizontal force of zero

$$R\sin\phi = S\cos 45°$$

and for a net vertical force of zero

$$R\cos\phi + S\sin 45° = 15g$$

Since $S = 103.9\,\text{N}$ it follows that

$$R\sin\phi = 73.47$$

and $\quad R\cos\phi = 73.53$

Thus $\quad \dfrac{R\sin\phi}{R\cos\phi} = \dfrac{73.47}{73.53}$

$\Rightarrow \quad \tan\phi = 0.9992$

i.e. $\quad \phi = 45°$ (approx.)

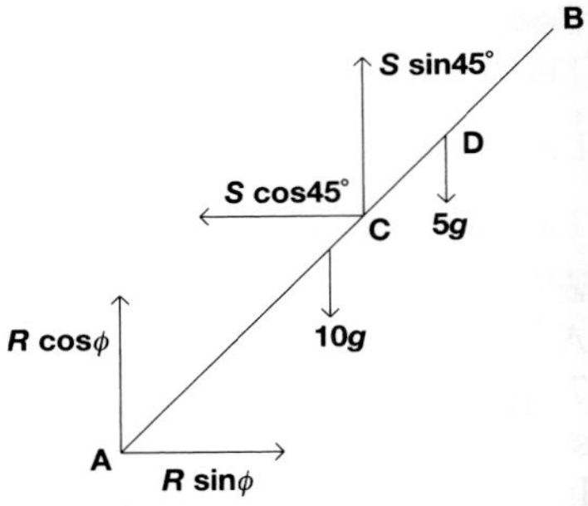

Figure 11.41

Substituting for ϕ in the equation $R\sin\phi = 73.47$ yields

$$R\sin 45° = 73.47\,\text{N}$$

$\Rightarrow \quad R = 103.9\,\text{N}.$

EXAMPLE 17

A light rod of length 6 m is freely hinged at A to a vertical wall as illustrated in Figure 11.42. A mass of 3.8 kg is attached at a point 2 m from A and a further mass of 5.2 kg at a point 4 m from A. A force P is applied at the other end of the rod (at 48.2° with the rod) in order to maintain equilibrium. The rod makes 39.4° with the horizontal and R is the force of reaction at the hinge, making an angle ϕ with the upward vertical. Find the forces P and R and the angle ϕ.

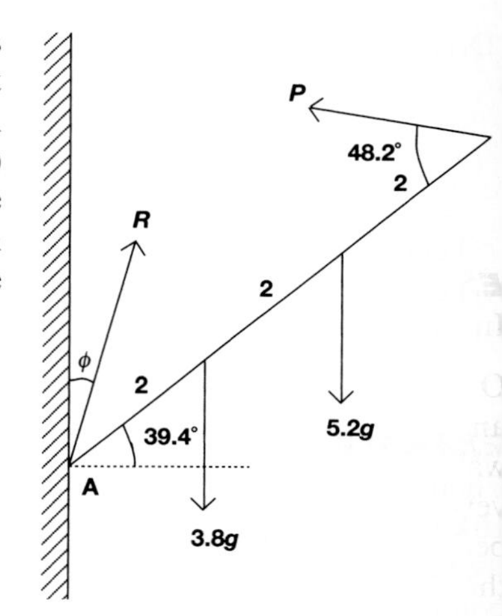

Figure 11.42

Solution 17

Equating clockwise and anticlockwise moments about A

$$3.8g(2\cos 39.4°) + 5.2g(4\cos 39.4°) = P(6\sin 48.2°)$$

$$\Rightarrow \quad P = 48.08\,\text{N}$$

Equating horizontal and vertical components yields

$$R\sin\phi = 48.08\cos 8.8°$$

and $\quad 48.08\sin 8.8° + R\cos\phi = 3.8g + 5.2g$

(since P makes an angle of 8.8° (i.e. 48.2° − 39.4°) with the horizontal).

Accordingly $\quad R\sin\phi = 47.51$

and $\quad R\cos\phi = 80.84$

Therefore $\quad \dfrac{R\sin\phi}{R\cos\phi} = \dfrac{47.51}{80.84}$

$\Rightarrow \quad \tan\phi = 0.5877$

i.e. $\quad \phi = 30.44°$

So $\quad R\sin 30.44° = 47.51$

$\Rightarrow \quad R = 93.78\,\text{N}$

EXAMPLE 18

A light uniform rod AB of length 8 m has masses of 13.8 kg and 7.4 kg attached at distances of 2 m and 5 m from A respectively. It is supported in equilibrium, inclined at 21° to the horizontal, by a light inextensible cable joining the end B to the wall vertically above A such that this cable makes an angle of 38° with the rod. Figure 11.43 illustrates this information. The reaction at A is inclined at angle ϕ to the wall. Find P (the tension in the cable), the reaction R and the angle ϕ.

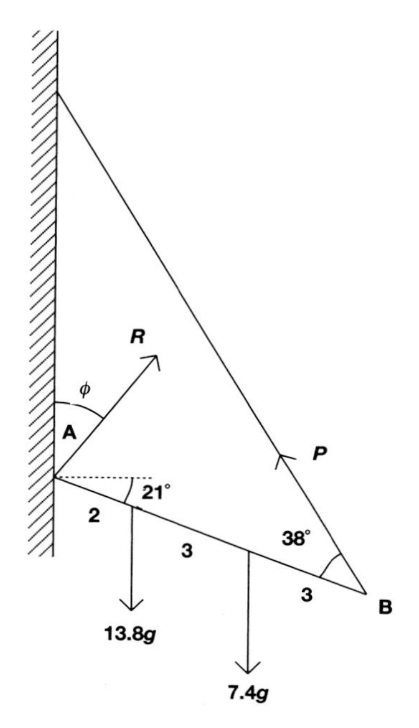

Figure 11.43

Solution 18

Equating anticlockwise and clockwise moments about A

$$P(8\sin 38°) = (2\cos 21°)13.8g + (5\cos 21°)7.4g$$
$$\Rightarrow \quad P = 120\,\text{N}$$

Equating horizontal and vertical components yields

$$R\sin\phi = 120\cos 59°$$
and $$R\cos\phi + 120\sin 59° = 13.8g + 7.4g$$
i.e $$R\sin\phi = 61.8$$
and $$R\cos\phi = 104.9$$
hence $$R = 121.8\,\text{N} \text{ and } \phi = 30.5°$$

EXAMPLE 19

In this question take $g = 10\,\text{m/s}^2$.

O is the centre of mass of a non-uniform ladder XY of mass 25 kg and length 8 m. The ladder rests with X against a smooth vertical wall and with Y on rough horizontal ground. The ladder is in a vertical plane perpendicular to the wall. The coefficient of friction between the ladder and the ground is 0.2, XY is inclined at 60° to the horizontal and the ladder is on the point of slipping down the wall. Copy Figure 11.44 and show clearly on your copy all forces which act on the ladder.

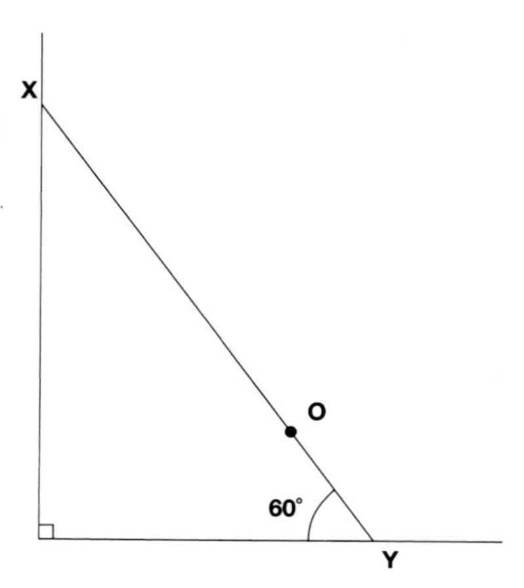

Figure 11.44

Calculate

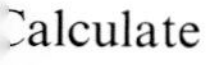

(i) the horizontal and vertical components of the reaction at Y,
(ii) the reaction at X,
(iii) the distance of the centre of mass of the ladder from Y.
(iv) Determine the horizontal force which should be applied at Y so that a man of mass 120 kg can ascend to a point Z on the ladder, where YZ equals 6 m without the ladder slipping.

(N. Ireland Additional Mathematics, 1987)

Solution 19

Since the centre of mass of the non-uniform ladder is at O, the weight acts at O as shown in Figure 11.45.

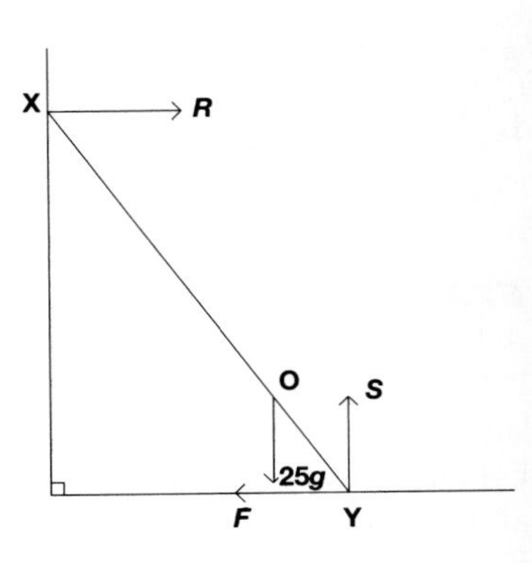

Figure 11.45

(i) Equating horizontal and vertical components

$$R = F$$
$$S = 25g = 250\,\text{N}$$

The concept of limiting friction was introduced in Chapter 8. Using the principles detailed there, together with the knowledge that the ladder is on the point of slipping

$$F = \mu S = \tfrac{1}{5}(S) = 50\,\text{N}$$

It follows that the horizontal and vertical components of the reaction at Y are 50 N and 250 N respectively.

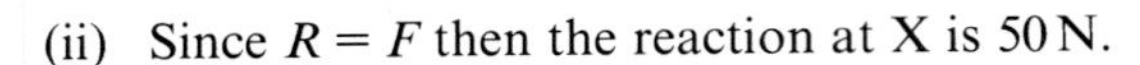

(ii) Since $R = F$ then the reaction at X is 50 N.

(iii) Taking moments about Y

$$R(8\sin 60°) = 250(\text{OY}\cos 60°)$$
i.e. $$50(8\sin 60°) = 250(\text{OY}\cos 60°)$$
$$\Rightarrow \quad \text{OY} = 2.771\,\text{m}$$

(iv) Consider the new force diagram in Figure 11.46.

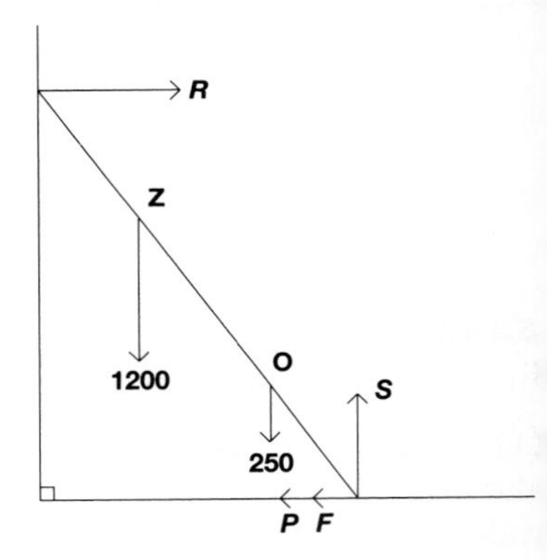

Figure 11.46

Let the new horizontal force be denoted P.

Equating horizontal and vertical forces

$$R = P + F$$
and $$S = 1200 + 250 = 1450$$

Once again, taking moments about Y

$$R(8\sin 60°) = 1200(6\cos 60°) + 250(2.771\cos 60°)$$
$$\Rightarrow \quad R(6.928) = 1200(3) + 250(1.386)$$
i.e. $$R = 570\,\text{N}$$
Therefore $$570 = P + F$$
But $$F = \mu S$$

and so $$F = \frac{1450}{5} = 290\,\text{N}$$

Hence, since P and F must sum to 570 N, $P = 280\,\text{N}$.

EXERCISE 11.3

In this exercise take $g = 9.8\,\text{m/s}^2$ unless otherwise instructed.

1. A light rod of length 1.4 m is supported at its ends. Masses of 4 kg and 7 kg are located at 0.2 m and 0.9 m respectively from one end. Calculate the reaction at each support.

2. A light rod AB of length 1 m has a mass of 8 kg attached at a point 0.2 m from A and a mass of 2 kg at B. The rod is supported at A and at a point 0.1 m from B. Find the reaction at each support.

3. A uniform beam AD of length 1.4 m and of mass 8 kg has a 2 kg mass attached at D. The beam is supported at two points, one 0.2 m from A and the other 0.48 m from D. Find the reaction at each support.

4. A uniform rod AD of length 56 cm and mass 5.8 kg rests horizontally on supports at B and C where AB = 8 cm and CD = 27 cm. Find

(a) the reaction at each support,
(b) the downward force at A which will cause the rod to tilt about the support at B.

5. Two painters of mass 50 kg and 70 kg stand on a uniform horizontal plank ABCDEF of mass 40 kg and length 8 m. B, C, D and E are respectively 1 m, 2 m, 5 m and 6 m from A. The 50 kg painter stands at B, the 70 kg painter at D and the plank is supported at C and E. Find the reaction at each support.

6. A uniform plank AB is 5.2 m long and is pivoted at a point 2 m from A. A girl of mass 60 kg sits on the plank at a point 1 m from A. If the plank is in equilibrium, calculate the mass of the plank.

7. Jane and Ruth are big game hunters. They return from a safari carrying a 98 kg lion on a uniform pole of length 2 m and mass 12 kg. End A of the pole is on Jane's shoulder with end B on Ruth's and the lion is tied to the pole 0.92 m from A. Find the force exerted on Ruth's shoulder by the pole.

8. A uniform metre rule of mass 0.1 kg is suspended horizontally by two vertical threads A and B attached to the rule at points 0.18 m and 0.28 m from the ends. Find the distance from the centre of the rule at which a 0.2 kg mass should be suspended such as to

(a) make thread A slack,
(b) make thread B slack.

9. A uniform beam of mass 20 kg and length l supports a 45 kg mass at a point which is $0.25l$ from the centre of the beam. Find the magnitude of the forces acting at the ends of the beam which will support it in equilibrium.

10. A uniform 10 kg pipe of length l is to be used as a balance. If masses of 20 kg and 50 kg are to be supported in equilibrium at the extremes of the pipe, find the position of the point at which the pipe is balanced.

11. In the diagrams in Figure 11.47, a uniform bar AB of mass 15 kg and length 6 m is freely hinge to a wall at A. The bar is held in equilibrium by a force P and the horizontal and vertic components, X and Y respectively, of the reaction at the hinge are shown. Find the magnitud of the forces P, X and Y.

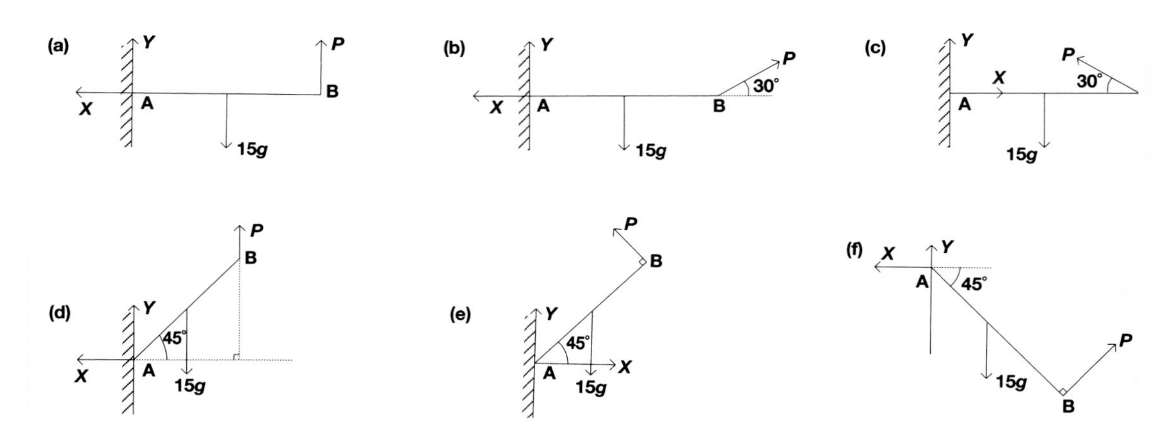

Figure 11.47

12. In the diagrams in Figure 11.48, a uniform bar AB of mass 10 kg and length 5 m is freely hinge to a wall at A. The bar is held in equilibrium by a force P. The reaction at the hinge is R and th line of action of this force makes an angle θ with the wall. Find the magnitudes of the forces and R and the size of the angle θ.

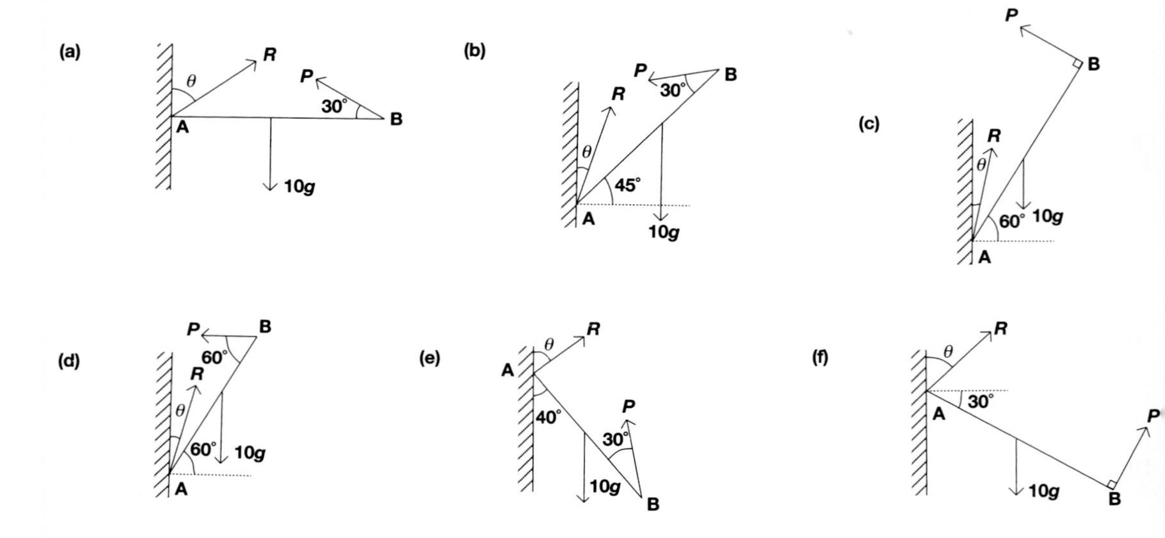

Figure 11.48

13. In the diagrams in Figure 11.49, a uniform bar AB of mass 6 kg has end A resting on a rough horizontal floor of coefficient of friction μ. The bar is held in equilibrium by a string attached to B. T is the tension in the string, F is the frictional force at A and R is the normal reaction at A. Find the magnitude of T, F and R and the value of μ given that the bar is on the point of slipping.

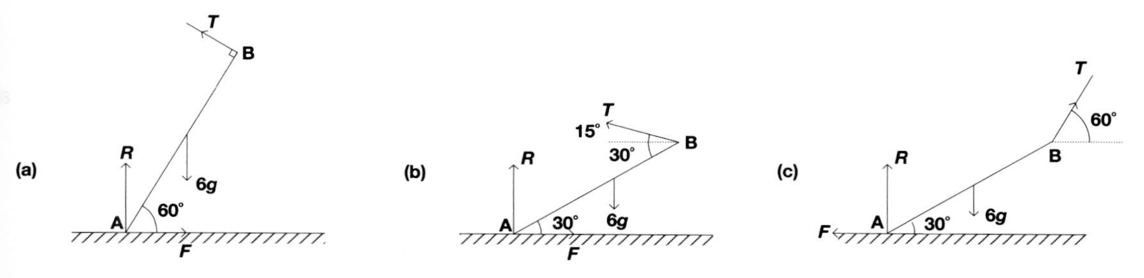

Figure 11.49

14. In the diagrams in Figure 11.50, a uniform ladder AB of mass 15 kg and length 4 m rests against a smooth vertical wall at A and on a smooth horizontal floor at B. A light horizontal string with one end attached to B and the other end attached to the wall holds the ladder in equilibrium. R is the normal reaction at the floor, S is the normal reaction at the wall and T is the tension in the string. Find R, S and T.

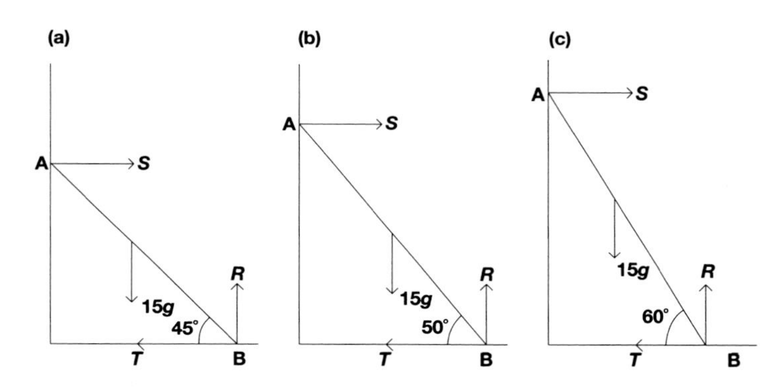

Figure 11.50

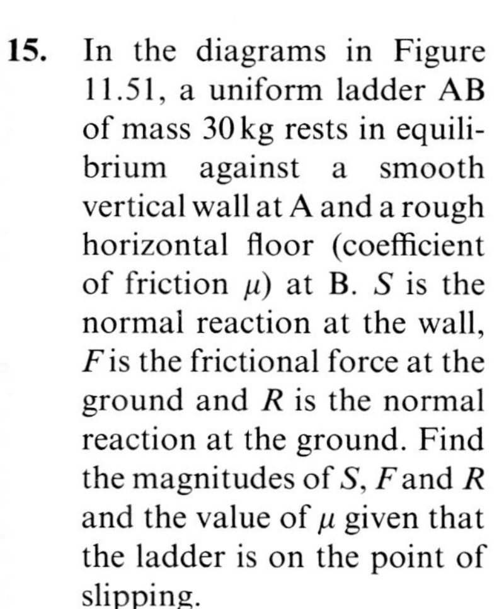

15. In the diagrams in Figure 11.51, a uniform ladder AB of mass 30 kg rests in equilibrium against a smooth vertical wall at A and a rough horizontal floor (coefficient of friction μ) at B. S is the normal reaction at the wall, F is the frictional force at the ground and R is the normal reaction at the ground. Find the magnitudes of S, F and R and the value of μ given that the ladder is on the point of slipping.

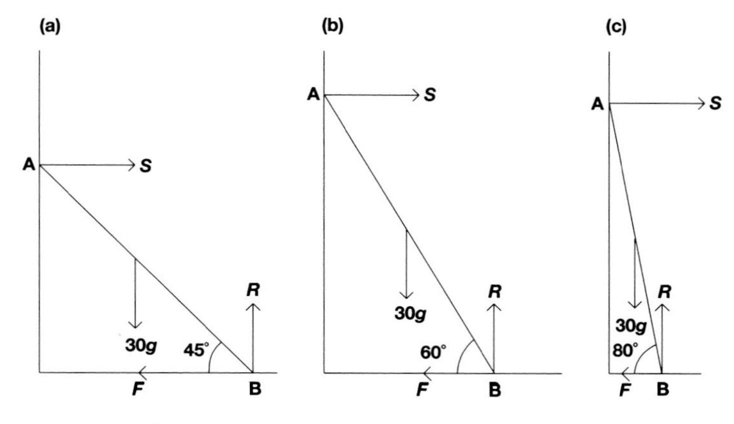

Figure 11.51

16. In the diagrams in Figure 11.52, a light ladder has a mass (or masses) attached and rests in equilibrium against a smooth wall with its lowest point in contact with rough ground. The numbers along the ladder in each diagram represent lengths so that, for example, in part (i) the ladder is 8 m long and has masses of 3 kg and 8 kg attached at points 2 m and 4 m respectively from that end in contact with the ground. In each case find R, S and F.

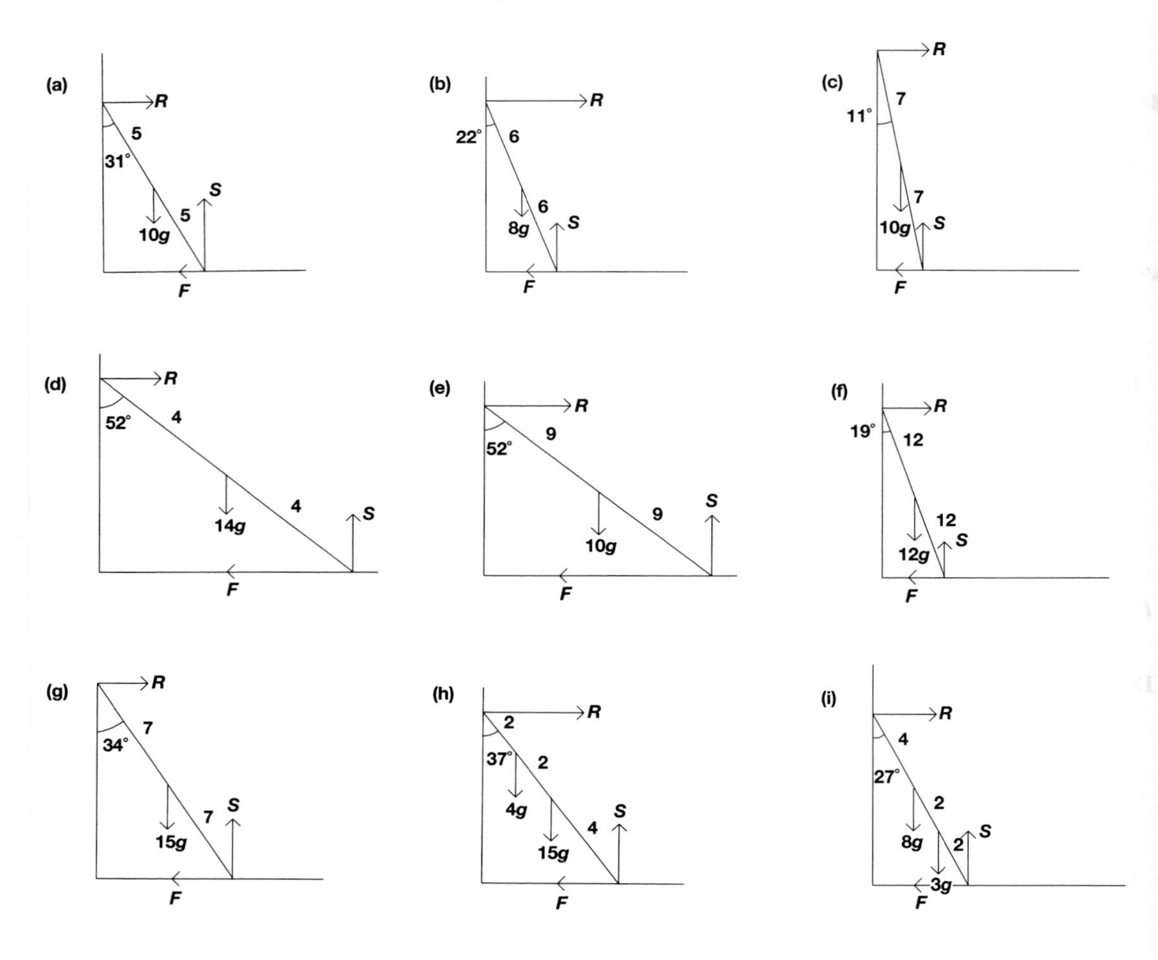

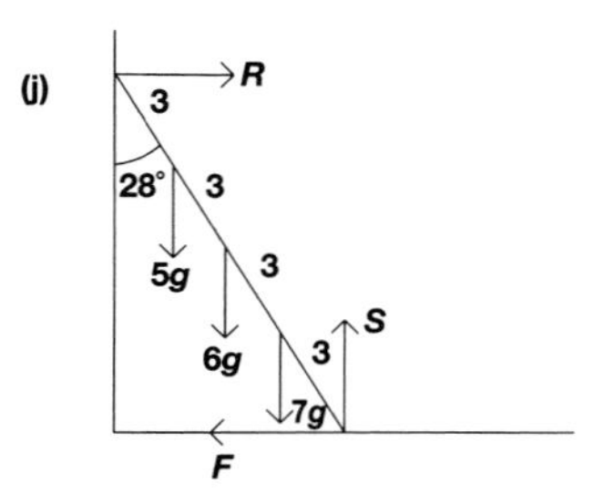

Figure 11.52

17. In the diagrams in Figure 11.53, a light rod has a mass (or masses) attached and is freely hinged at one end. The other end is acted upon by a force P such that the rod is maintained in equilibrium. The numbers along the rod in each diagram represent lengths so that, for example, in part (e) the rod is 8 m long and has masses of 14 kg and 7 kg at points 2 m and 5 m respectively from the end of the rod in contact with the wall. In each case find P, R and ϕ.

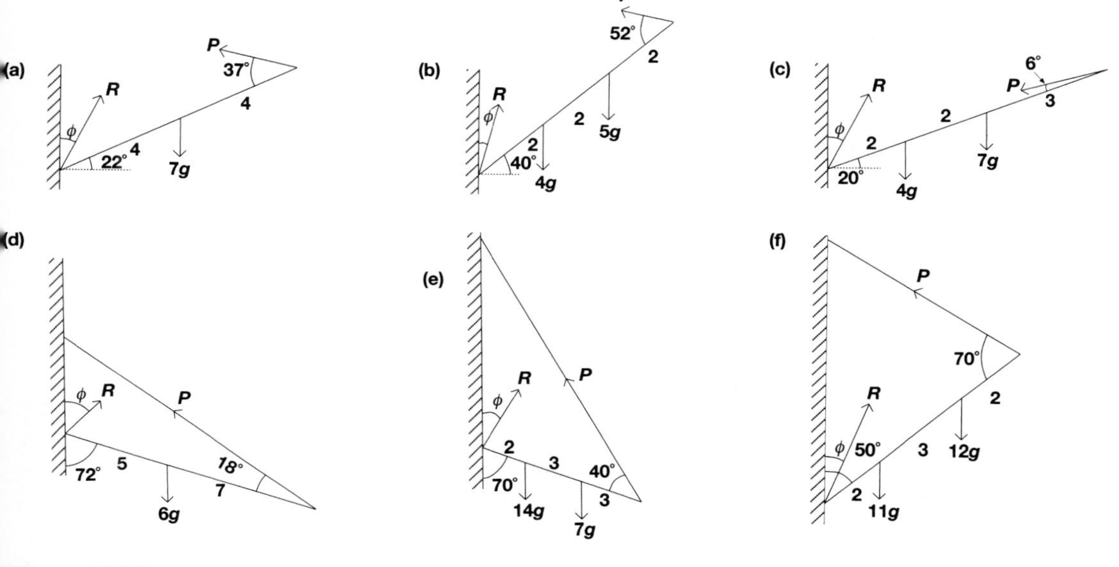

Figure 11.53

18. (a) A uniform plank of mass 20 kg and length 4 m rests horizontally on two supports which are positioned 1 m from each end of the plank. A man of mass 100 kg climbs on to the plank at its centre and walks slowly towards one end. Find how close he can get to the end without tilting the plank.

(b) AB is a uniform rod of mass 10 kg and length 3 m which is hinged at A to a vertical wall. The end B is attached by a light inextensible rope to a point C on the wall 6 m vertically above A. The system is at rest with angle ABC = 90°, as shown in Figure 11.54.

Draw a neat diagram to show clearly the direction and line of action of each of the forces acting on the rod AB.

Calculate
(i) the angle CAB,
(ii) the tension in the rope,
(iii) the magnitude and direction of the reaction on the rod at A.

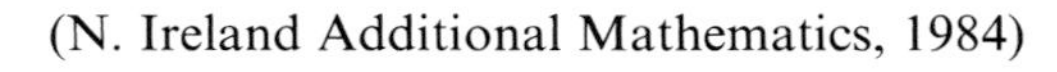
(N. Ireland Additional Mathematics, 1984)

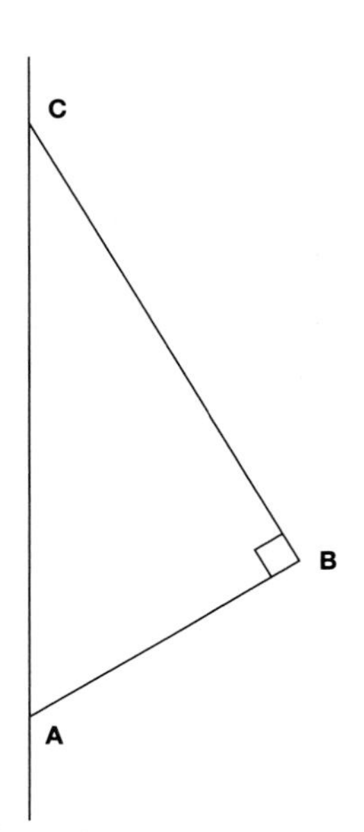

Figure 11.54

19. In this question take g to be $10\,\text{m/s}^2$.

A light inextensible string is attached to two points A and B in a horizontal ceiling. Two masses, 4 kg and 12 kg, are securely fixed to the string at P and Q respectively as shown in Figure 11.55. The angles ABQ and BQP are 60° and 90° respectively.

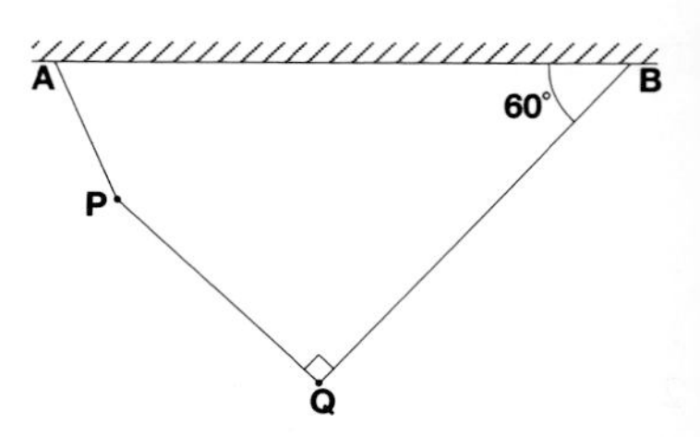

Figure 11.55

Draw separately two clearly labelled diagrams to show the forces acting on the particle at P and the particle at Q.

Calculate
(i) the tension in QB,
(ii) the tension in PQ.

Hence find
(iii) the tension in PA,
(iv) the angles BAP and APQ.

(N. Ireland Additional Mathematics, 1986)

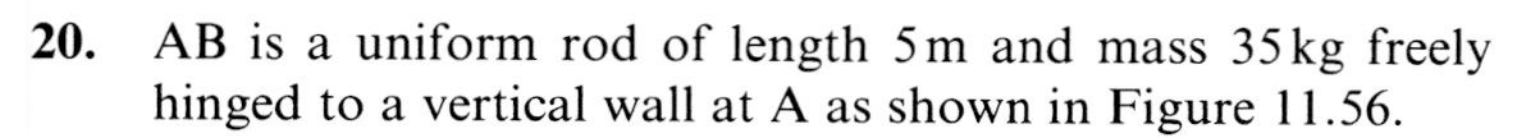

20. AB is a uniform rod of length 5 m and mass 35 kg freely hinged to a vertical wall at A as shown in Figure 11.56.

The rod is held in a horizontal position by a light rope BC, C being a point on the wall vertically above A such that angle ABC = 40°.

A load M of mass 100 kg is suspended from the rod at a point D, 3 m from A.

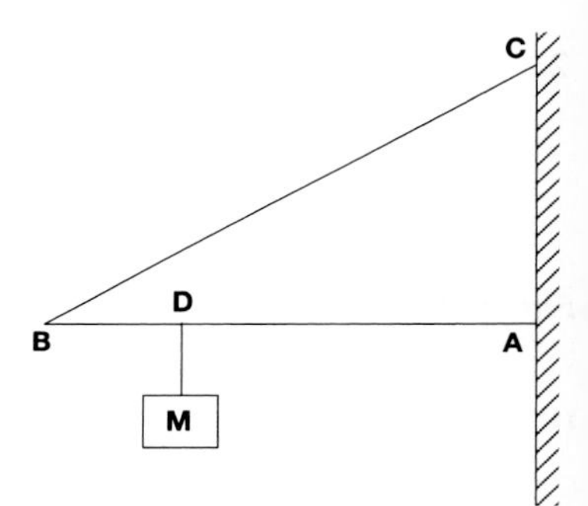

Figure 11.56

Draw a neat diagram showing clearly all of the forces acting on the rod AB.

Calculate
(i) the tension in the rope,
(ii) the magnitude and direction of the reaction on the rod at A.

If the rope BC breaks when its tension exceeds $200g$ N, find the greatest mass which can be added to M, if it is still suspended from the same point D on the rod, without causing the rope to break.

(N. Ireland Additional Mathematics, 1985)

21. A uniform rod AB of length 12 m and mass 50 kg is pivoted at a point C in the rod, 3 m from A. A mass of 200 kg is suspended from A.

A force, F newtons, applied to the rod at B in a direction perpendicular to the rod keeps the rod in equilibrium with A lower than B and AB inclined at 60° to the horizontal as shown in Figure 11.57.

Figure 11.57

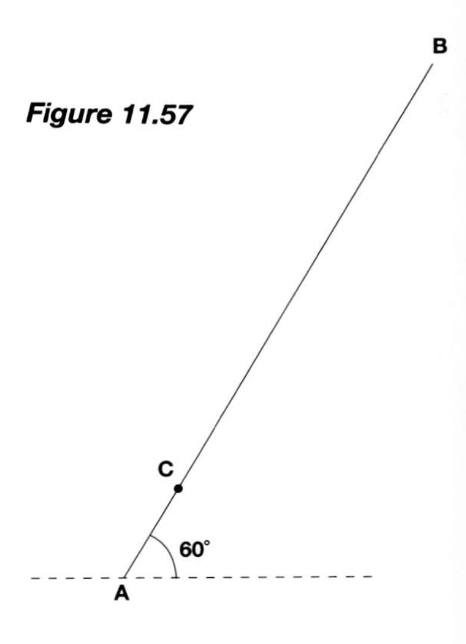

Calculate

(i) the value of F,

(ii) the magnitude and direction of the reaction on the rod at the pivot.

(N. Ireland Additional Mathematics)

22. The ends of a light inextensible string are attached to two points A and B in a horizontal ceiling. Two masses 10 kg and 5 kg are securely fixed to the string at P and Q respectively and the system is held in equilibrium in the position shown in Figure 11.58 by a horizontal force F of 20 N acting on the mass at P. P is vertically below A and PQ is inclined at 45° to the horizontal.

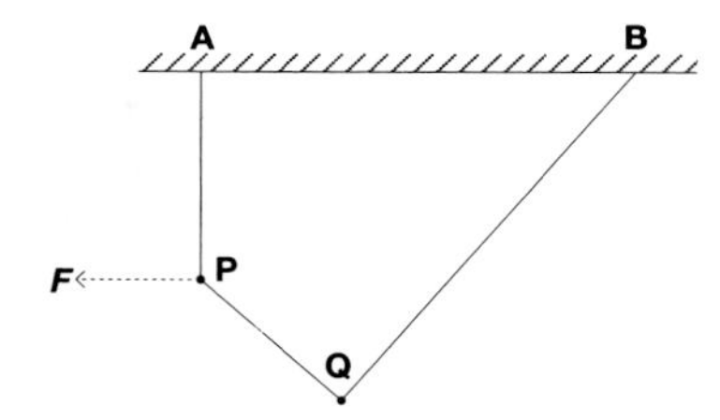

Figure 11.58

Calculate

(i) the tension in the string PQ,

(ii) the tension in the string PA.

Hence find the tension in the string QB and the angle ABQ.

(N. Ireland Additional Mathematics)

23. In this question take g to be $10\,\text{m/s}^2$.

A uniform beam PQ of length 5 m and mass 100 kg is freely hinged at P to a vertical wall and is kept in a horizontal position by a light inextensible rope attached to Q and to a point C on the wall 2 m vertically above P.

(i) Draw a neat diagram showing clearly the direction and lines of action of all forces acting on PQ.

Calculate

(ii) the tension in the rope,

(iii) the magnitude and direction of the force at P.

(iv) If the rope can withstand a tension of 2000 N before breaking, calculate the maximum load which can be safely attached to the beam at Q.

(N. Ireland Additional Mathematics, 1988)

24. In this question take g to be $10\,\text{m/s}^2$.

(a) A uniform plank AD of mass 10 kg and length 3 m rests horizontally on two supports B and C such that AB = 0.5 m and CD = 1.0 m.

Determine, in newtons, the reaction at each of the supports B and C.

(b) A non-uniform plank PS of length 3 m and mass M kg rests horizontally on two supports Q and R where PQ = 0.5 m and RS = 1.0 m. The centre of gravity of the plank is x metres from P. If a child of mass 20 kg stands on the plank at either end, the plank is on the point of toppling.

(i) Use the principle of moments to obtain two equations connecting M and x. Hence find M and x.

(ii) Calculate the distance from P at which the child must stand for the reaction at Q to be twice the reaction at R.

(N. Ireland Additional Mathematics, 1988)

25. In this question take g to be $10\,m/s^2$.

Figure 11.59 shows a non-uniform ladder PQ of length 6 m and mass 20 kg. It stands on a smooth horizontal floor and leans against a smooth vertical wall. The foot, Q, of the ladder is at a distance of 2 m from the wall. The ladder is held in equilibrium by a light inextensible rope TR where QT = 1 m and TR is parallel to the floor.

When a man of mass 80 kg stands on the ladder at S where PS = 1.5 m, the tension in the rope is 278 N. O, P, Q, R, S and T all lie in a vertical plane at right angles to the wall.

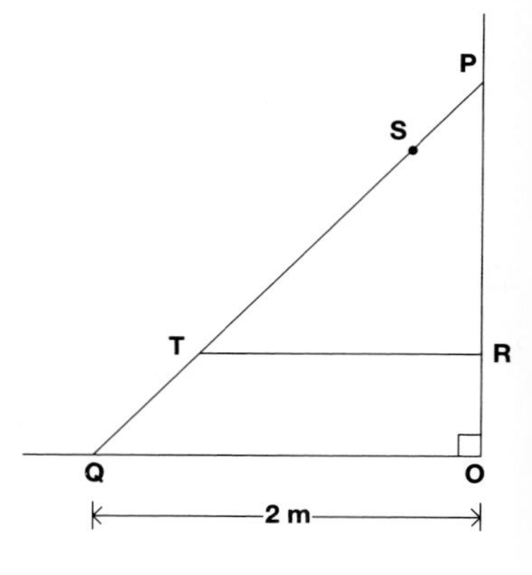

Figure 11.59

(i) Copy Figure 11.59 and mark clearly on it all the forces acting on the ladder PQ.

Calculate

(ii) the magnitude of the force exerted on the ladder at P by the wall,
(iii) the magnitude of the force exerted on the floor at Q by the ladder.
(iv) Find the distance of the centre of gravity of the ladder from Q.

(N. Ireland Additional Mathematics, 1989)

26. In this question take g to be $10\,m/s^2$.

A uniform rod AB of mass 0.2 kg and length 4 m is held in a horizontal position by two light inextensible vertical strings as shown in Figure 11.60 where AC = 1 m and DB = 0.5 m.

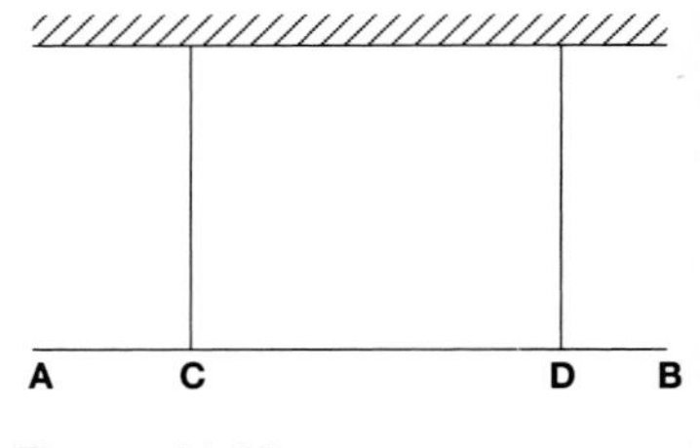

Figure 11.60

(i) Calculate the tension in each string.

(ii) Either string will break if the tension in it exceeds 20 N. Determine the maximum load which can be placed on the rod 1 m from B.

(N. Ireland Additional Mathematics, 1990)

27. In this question take g to be $10\,m/s^2$.

Figure 11.61 shows a uniform beam AB of length 2 m and mass 10 kg, freely hinged at A. It is held in a horizontal position by a light rope BC where C is 1 m vertically above A.

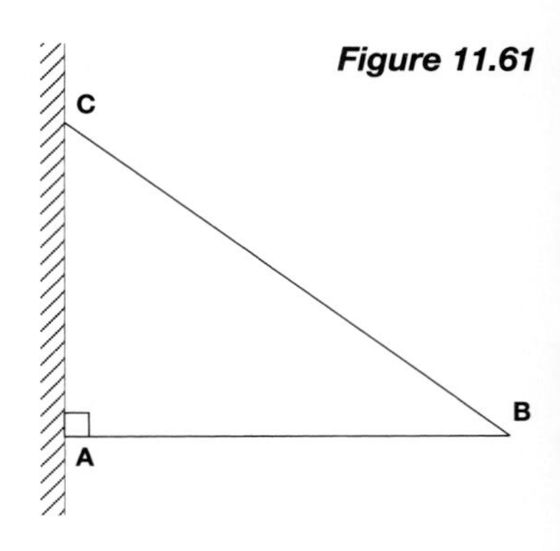

Figure 11.61

(i) Copy Figure 11.61 and mark clearly on it all the forces acting on the beam.

(ii) Find the tension in the rope.

(N. Ireland Additional Mathematics, 1991)

28. In this question take g to be $10\,\text{m/s}^2$.

(a) A uniform plank PS of mass 20 kg and length 5 m rests horizontally on two supports Q and R such that PQ = 1 m, and RS = 2 m. Determine, in newtons, the reaction at the supports Q and R.

(b) A non-uniform plank AD of length 5 m and mass 100 kg rests horizontally on two supports B and C where AB = 1.5 m and CD = 3 m. The centre of mass of the plank is x m from A.

If a boy of mass 50 kg stands on the plank 4.5 m from A, the plank is on the point of toppling.

(i) Find x.
(ii) Calculate how far along the plank from A the boy must stand for the reaction at B to be $\frac{1}{4}$ of the reaction at C.

(N. Ireland Additional Mathematics, 1991)

29. [Take g to be $10\,\text{m/s}^2$.]

(a) A particle of mass 2 kg rests on a rough plane inclined at an angle θ to the horizontal where $\tan\theta = \frac{3}{4}$. The particle is just prevented from sliding down the plane by a force of 8 N acting directly up the plane. Find the coefficient of friction between the particle and the plane.

When the force of 8 N is increased to Q N, the particle is just about to move up the plane. Find the value of Q.

(b) The resultant of forces X N and 12 N acting at right angles to each other is equal in magnitude to the resultant of forces 8 N and 7 N acting at 60° to each other. Find the value of X.

(U.C.L.E.S. Additional Mathematics, June 1992)

30. (a) Find the magnitude of the resultant of the forces 3 N, 4 N and 5 N shown in Figure 11.62.

Find also the angle which the resultant makes with the x axis.

(b) The resultant of the two forces, 4 N and 4 N, shown in Figure 11.63, is a force R N, and the resultant of the other two forces, 5 N and P N, is a force Q N. Given that R is perpendicular to Q, find the values of R, P and Q.

(U.C.L.E.S. Additional Mathematics, November 1991)

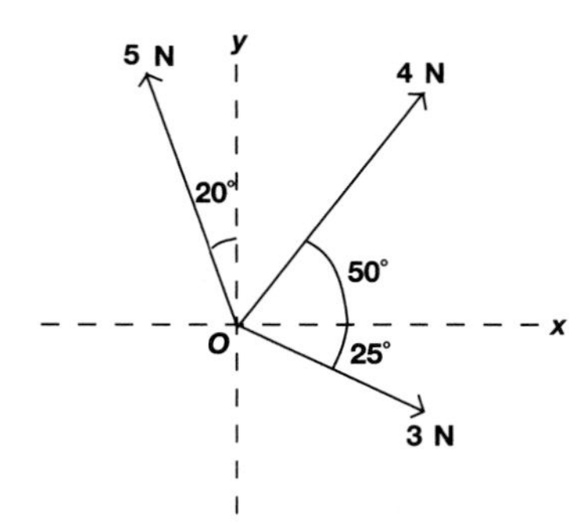

Figure 11.62

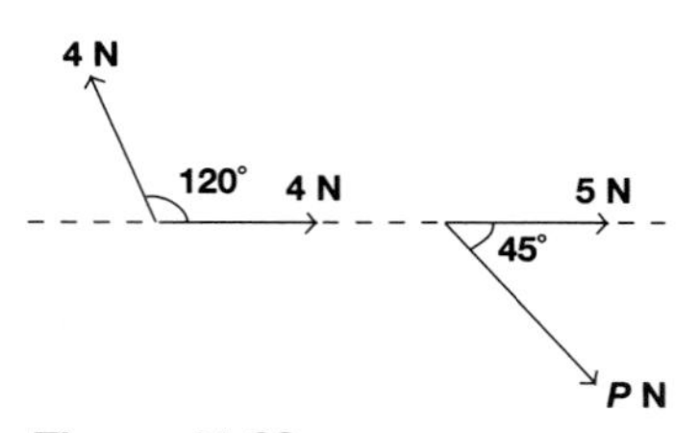

Figure 11.63

31. [Take g to be $10\,\text{m/s}^2$.]

(a) Forces 10 N and 5 N act along OA and OB respectively as shown in Figure 11.64, where angle AOB is 60°. The resultant of these two forces makes an angle $\theta°$ with OA. Find the value of θ.

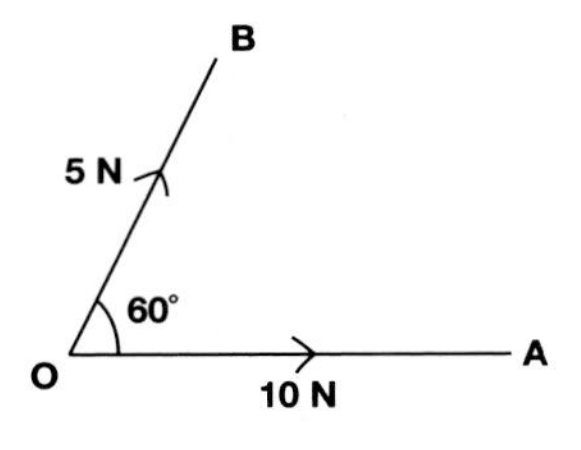

Figure 11.64

(b) The resultant of the three forces shown in Figure 11.65 has a magnitude of R N and acts along OX. Find the value of

(i) P,
(ii) R.

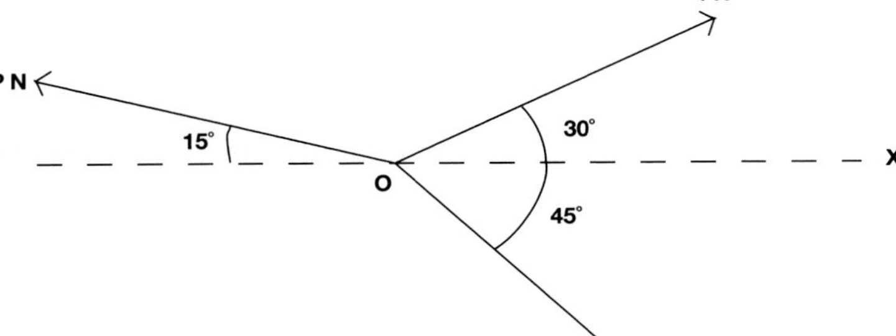

Figure 11.65

(c) Figure 11.66 shows a particle at A of mass 0.8 kg suspended by strings attached to fixed points at B and C. Given that AB and AC are inclined to the vertical at 45° and 60° respectively, find the tension in

(i) AB,
(ii) AC.

(U.C.L.E.S. Additional Mathematics, June 1991)

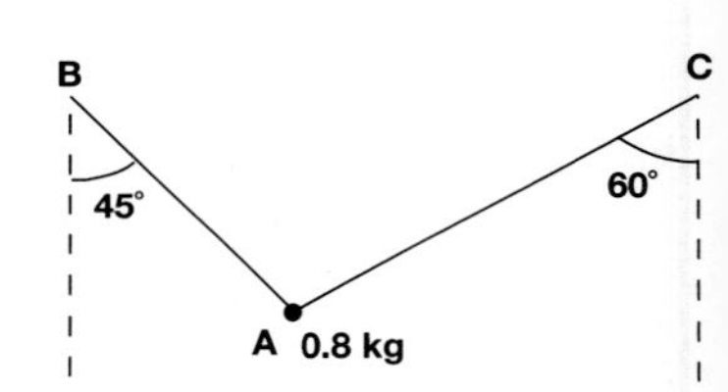

Figure 11.66

CHAPTER **12**

Classifying and illustrating statistical data

The English mathematician **Sir Francis Galton** (1822–1911) was born near Birmingham and, despite receiving the Gold Medal of the Royal Geographical Society and being elected to a Fellowship of the Royal Society, is regarded as a mathematical 'lightweight'. Karl Pearson described Galton as the 'master builder' of modern statistical theory, more for his obsession with counting and measuring than for his standing as a great mathematical thinker.

His early education was unhappy and his research output later in life was often halted through breakdowns which Galton referred to as periods of 'sprained brain'. Two of his experiments illustrate his passion for measurement.

He analysed 'fidget rates', drawing the rather obvious conclusions that children fidget a lot, middle-aged people are moderate fidgets while elderly philosophers 'will sometimes remain rigid for minutes together'. He classified passing girls into three categories: 'attractive, indifferent and repulsive', unobtrusively building a database by piecing paper secreted in his pocket, with a pin. An analysis of the distribution of pin pricks allowed him to compile his 'Beauty Map' of the British Isles.

He had a macabre fascination for designing suffocation devices which he tested upon himself with near disastrous consequences on a number of occasions, and almost drowned himself in an invention which allowed him to read underwater in the bath. The following extract from his 1869 book *Hereditary Genius* concerns 'idiots' and provides a valuable insight into the primitive nature of Galton's early statistical thinking, couched in the insensitive language of the day.

... more than forty per cent have become capable of the ordinary transactions of life, under friendly control; of understanding moral and social abstractions, and of working like two-thirds of a man. And, lastly, that from twenty-five to thirty per cent come nearer and nearer to the standard of manhood, till some of them will defy the scrutiny of good judges, when compared with ordinary young men and women. In the order next above idiots and imbeciles are a large number of milder cases scattered among private families and kept out of sight, the existence of whom is, however, well known to relatives and friends; they are too silly to take part in general society, but are easily amused with some trivial, harmless occupation.

INTRODUCTION

A central aim of statistics is to summarise data; the statistician condenses large sets of numbers into a small number of groups. He or she may then present these groupings as a diagram. The statistician therefore summarises the data in such a way that the main features are easily recognisable.

VARIABLES

An apple, for example, can be described by a number of variables associated with it; its taste, its weight, its colour, its volume, the number of pips it contains and so on. **Variables to which numbers cannot be assigned are known as qualitative variables**; taste and colour are examples of such variables.

Quantitative variables may be subdivided into discrete and continuous categories. **Discrete variables are capable of being described by whole numbers**; the number of pips in the apple, the number of desks in the classroom are instances of discrete variables.

Continuous variables, on the other hand, have associations with measurement – the apple has mass 83.24 g or the volume of the apple is 42.27 cm^3 are instances of the use of continuous variables. To use such numbers to describe discrete variables would clearly be absurd; consider statements such as 'the apple has 83.24 pips' or 'there are 42.27 desks in our classroom'. Continuous variables can take any value in a given range. For example, a pencil of length 12 cm (correct to the nearest cm) could have a length anywhere in the range $11.5\,\text{cm} \leq \text{length} < 12.5\,\text{cm}$.

CLASSIFICATION OF DATA

A large quantity of data may be collected in a relatively small number of **class intervals** using a variety of notations, three of which will be used in this book. Consider firstly the four intervals

2–7   8–12   13–17   18–20

In assigning numbers to these intervals

3 would be inserted in interval 1
9 would be inserted in interval 2
14 would be inserted in interval 3
19 would be inserted in interval 4
7 would be inserted in interval 1
12 would be inserted in interval 2
17 would be inserted in interval 3

A second notation used to specify class intervals seems at first sight ambiguous; it would appear that certain numbers may belong to two intervals simultaneously. An illustration of this second means of writing class intervals may be had from consideration of the intervals

2–8   8–12   12–17   17–20

There does indeed appear to be a difficulty here – surely 17 could be allocated to the third and/or fourth intervals. This problem disappears when one appreciates that statisticians interpret these intervals to be

2–7.999...   8–11.999...   12–16.999...   17–19.999...

Therefore

3 would be inserted in interval 1
9 would be inserted in interval 2
14 would be inserted in interval 3
19 would be inserted in interval 4
7 would be inserted in interval 1
12 would be inserted in interval 3
17 would be inserted in interval 4

The final notation is merely another means of writing the second notation i.e.

2–8   8–12   12–17   17–20

is often written

2–   8–   12–   17–20

Data are assigned to class intervals in **group frequency tables** as illustrated in the following example.

EXAMPLE 1

The distance in km (rounded to the nearest km) travelled by each of 30 spectators at a soccer match is

19	34	10	47	25	20	27	38	14	25
33	5	41	16	26	6	22	13	39	31
15	22	20	5	26	13	32	25	29	15

Using this data, construct a group frequency table with class intervals

(i) 5–9   10–14   15–19   ...   45–49
(ii) 5–9   9–13   13–17   ...   45–49
(iii) 5–   10–   15–   ...   40–49

Solution 1

Using simple tally methods and the interval notation guidance above, the following groupings emerge.

(i) ***Table 12.1***

Distance travelled by spectator (km)	Frequency
5–9	3
10–14	4
15–19	4
20–24	4
25–29	7
30–34	4
35–39	2
40–44	1
45–49	1

(ii) ***Table 12.2***

Distance travelled by spectator (km)	Frequency
5–9	3
9–13	1
13–17	6
17–21	3
21–25	2
25–29	6
29–33	3
33–37	2
37–41	2
41–45	1
45–49	1

(iii) ***Table 12.3***

Distance travelled by spectator (km)	Frequency
5–	3
10–	4
15–	4
20–	4
25–	7
30–	4
35–	2
40–49	2

EXERCISE 12.1

1. Group the following numbers using intervals 15–17, 18–20, 21–23, etc.

16	33	36	24	32	26	31	32
24	18	22	29	36	21	32	27
32	24	15	23	24	18	24	18
29	19	24	22	21	35	21	19

2. Listed below are the shoe sizes of 30 pupils. Group the data using the intervals 1–2, 3–4, 5–6, etc.

1	3	9	8	4	6	9	2	5	6
6	2	4	4	8	3	2	1	5	6
3	2	5	7	9	9	7	2	6	4

3. Listed below are the heights of 50 pupils to the nearest cm. Group the data using intervals 140–144, 145–149, etc.

162	148	152	143	149	156	158	162	168	160
142	153	156	157	161	162	158	146	147	149
141	140	150	162	160	168	162	150	156	153
142	149	162	153	148	146	150	152	160	163
141	140	163	162	158	156	160	157	158	149

4. Listed below are the ages of members of a tennis club. Group the data using intervals 15–19, 20–24, etc.

15	26	23	42	47	32	36	20	24	26
19	24	32	36	35	42	46	49	26	35
26	32	30	19	20	26	27	34	36	42
26	32	32	36	19	18	17	16	26	27

5. Listed below are the wages of part-time workers (in £s) in a supermarket. Group the data using intervals 0–9, 10–19, etc.

9	9	16	25	32	68	64	59
18	26	32	37	49	64	83	72
25	25	26	19	20	31	46	59
81	62	75	72	19	26	34	52

6. The following are the heights of various shrubs, measured to the nearest cm. Classify the data using groups 30–33, 33–36, 36–39, etc.

31	37	42	45	38	39	46	32
38	40	49	36	39	42	41	41
40	30	31	32	36	38	42	46
41	32	34	35	41	46	42	41

7. The following are the heights (in cm) of 30 schoolchildren. Classify the data using groups 140–145, 145–150, 150–155, etc.

142	146	153	162	160	159	146	150	155	165
141	163	162	149	153	150	156	146	151	162
145	155	162	161	160	140	146	156	141	162

8. The following are the weights of 40 men (to the nearest kg). Classify the data using groups 45–50, 50–55, 55–60, etc.

49	56	63	62	80	69	72	69
52	55	49	57	81	71	75	80
63	65	82	68	82	70	65	79
81	62	71	72	76	75	62	61

9. The following are the weekly wages (to the nearest £) earned by 30 bank part-time workers. Classify the data using groups 100–110, 110–120, 120–130, etc.

112	126	163	162	100	140
114	145	162	158	105	119
116	160	149	136	126	129
121	120	140	102	125	136
134	145	120	100	130	140

10. The following are the ages of people attending a leisure centre during a particular period of the day. Classify the data using groups 10–20, 20–30, etc.

16	36	28	37	32	10	11	30
24	35	19	28	40	20	12	29
32	24	32	26	59	34	16	26
60	26	46	45	60	61	19	24

11. The following are the delays (in minutes) on several train journeys. Group the data according to 0–, 5–, 10–, etc.

5	12	19	8	16	20	18	19	31	2
6	24	20	3	14	32	29	24	32	3
10	36	40	6	9	34	36	22	20	1

12. The following are the ages of people attending a conference. Group the data using intervals 10–, 20–, 30–, 40–, etc.

19	24	30	39	42	18	40
22	20	18	15	16	21	23
29	30	39	20	15	17	41
36	32	29	26	24	32	35
42	35	36	29	28	20	19

13. The following are the heights of 25 pupils in cm to the nearest cm. Group the data using intervals 140–, 145–, 150–, etc.

142	156	162	149	151
159	162	160	147	146
152	158	149	140	145
152	155	165	164	163
160	162	152	147	144

14. The following are the lengths of a number of lines to the nearest cm. Group the data using intervals 30–, 34–, 38–, etc.

32	41	46	38	32	37
36	41	49	51	42	36
38	39	40	42	45	49
32	36	41	47	52	47

15. The following are the times (in minutes to the nearest minute) taken for a number of motorists to cover a particular journey. Classify the data using groups 10–, 12–, 14–, 16–, etc.

10	15	20	22	16	18	19
18	19	21	24	25	32	17
18	26	24	30	32	31	19
18	24	31	20	16	12	18
25	26	22	18	21	13	25

TERMINOLOGY

Consider Table 12.4, the frequency table for the mass (given to the nearest kg) of 25 army recruits.

Table 12.4

Mass (kg)	Frequency, f
53–55	1
56–58	2
59–61	4
62–64	9
65–67	9

The third class interval, 59–61, is said to have **upper class limit** 61 and **lower class limit** 59.

Furthermore, since masses are assigned to intervals after rounding to the nearest kg, then the second interval actually includes recruits with masses in the range 55.5 kg to 58.5 kg since any mass within this range will be rounded and assigned to the second class interval. In statistical terminology 55.5 is said to be the **lower class boundary** of the second class and 58.5 the **upper class boundary** of this class. It is important to note that, had the class intervals been written 53–55, 55–58, 58–61, 61–64, 64–67 then the upper class limit and upper class boundary are indistinguishable as are the lower class limit and lower class boundary. A convention often adhered to by statisticians when computing boundaries is to calculate the upper class boundary of class m by calculating the average of the upper class limit of m and the lower class limit of class $m + 1$. Similarly the lower class boundary of m is calculated by finding the average of the lower class limit of m and the upper class limit of $m - 1$.

The **class mark** or **mid-mark** of a class interval is the average of its class limits. Hence the class mark for the fifth class in the recruits data is $\frac{1}{2}(65 + 67) = 66$.

The **class width** or **size** or **strength** is the difference between the upper class and lower class boundaries of the class. Therefore the size of the first class in the recruits data is $55.5–52.5 = 3$.

Finally, the **relative frequency** is defined to be the actual frequency divided by the total frequency. It follows that the relative frequency of the second class is $\frac{2}{25} = 0.08$ and the relative frequency of the fourth class is $\frac{9}{25} = 0.36$.

EXERCISE 12.2

1. Table 12.5 is a frequency distribution of the shoe sizes of 30 pupils.

Table 12.5

Size	1–2	3–4	5–6	7–8	9–10
Frequency	4	6	12	8	1

Calculate

(i) the upper limit of the third interval,
(ii) the lower boundary of the fourth interval,
(iii) the class mark of the fifth interval,
(iv) the size of the second interval,
(v) the relative frequency of the second interval.

2. Table 12.6 is a frequency distribution showing the heights, in cm to the nearest cm, of 100 pupils.

Table 12.6

Height (cm)	140–144	145–149	150–154	155–159	160–164
Frequency	12	24	21	32	11

Calculate

(i) the lower limit of the second interval
(ii) the upper boundary of the fifth interval,
(iii) the class mark of the third interval,
(iv) the size of the fourth interval,
(v) the relative frequency of the first interval.

3. The weights, in kg to the nearest kg, of 50 women are distributed as in Table 12.7.

Table 12.7

Weight (kg)	45–49	50–54	55–59	60–64	65–69	70–74
Frequency	2	5	13	16	5	9

Calculate

(i) the lower limit of the third interval,
(ii) the lower boundary of the third interval,
(iii) the class mark of the fifth interval,
(iv) the size of the sixth interval,
(v) the relative frequency of the fourth interval.

4. The percentage score of 100 pupils in an examination was distributed as shown in Table 12.8.

Table 12.8

%	0–9	10–19	20–29	30–39	40–49	50–59	60–69	70–79	80–89	90–99
Frequency	0	2	3	6	22	25	13	12	11	6

Calculate

(i) the upper limit of the first interval,
(ii) the lower boundary of the tenth interval,
(iii) the class mark of the fourth interval,
(iv) the size of the second interval,
(v) the relative frequency of the sixth interval.

5. The number of typing errors made by several students had the following frequency distribution.

Table 12.9

Number of errors	0–2	3–5	6–8	9–11	12–14	15–17	18–20
Frequency	4	2	6	3	4	5	1

Calculate

(i) the lower limit of the fourth interval,
(ii) the upper boundary of the fifth interval,
(iii) the class mark of the sixth interval,
(iv) the size of the first interval,
(v) the relative frequency of the sixth interval.

6. The delay, in minutes to the nearest minute, on a train journey over a particular month had the following frequency distribution.

Table 12.10

Delay (min)	0–4	5–9	10–14	15–19	20–24	25–29
Frequency	1	6	2	10	5	6

Calculate

(i) the lower limit of the fourth interval,
(ii) the lower boundary of the fourth interval,
(iii) the class mark of the third interval,
(iv) the size of the sixth interval,
(v) the relative frequency of the fifth interval.

7. The lifetime (to the nearest hour) of each of 80 lightbulbs was noted in Table 12.11.

Table 12.11

Life (hours)	600–649	650–699	700–749	750–799
Frequency	12	24	32	12

Calculate

(i) the upper limit of the second interval,
(ii) the upper boundary of the second interval,
(iii) the class mark of the second interval,
(iv) the size of the second interval,
(v) the relative frequency of the second interval.

8. The time, in minutes to the nearest minute, taken by pupils to walk to school on a particular da had the following frequency distribution.

Table 12.12

Time (min)	0–3	4–7	8–11	12–15	16–19	20–23	24–27
Frequency	18	13	16	15	10	9	19

Calculate

(i) the lower limit of the third interval,
(ii) the upper boundary of the seventh interval,
(iii) the class mark of the first interval,
(iv) the size of the fourth interval,
(v) the relative frequency of the fourth interval.

9. The age distribution of factory workers is shown in Table 12.13.

Table 12.13

Age	16–19	20–23	24–27	28–31	32–35
Frequency	8	5	12	13	12

Calculate

(i) the lower limit of the first interval,
(ii) the upper boundary of the fifth interval,
(iii) the class mark of the second interval,
(iv) the size of the third interval,
(v) the relative frequency of the fifth interval.

10. The daily take home pay (to the nearest £) of a number of factory workers is shown in Table 12.14.

Table 12.14

Wage (£)	15–24	25–34	35–44	45–54	55–64
Frequency	10	13	27	32	18

Calculate

(i) the upper limit of the fifth interval,
(ii) the lower boundary of the second interval,
(iii) the class mark of the first interval,
(iv) the size of the second interval,
(v) the relative frequency of the first interval.

ILLUSTRATING STATISTICAL DATA

The pie chart

Newspapers often present data in the form of a pie chart. Broadly speaking, a pie chart is a circular illustration where the sectors of the circle represent the principal groupings of the distribution. The following example illustrates the construction of such charts.

Table 12.15 gives the number of pupils, in a class of 30, entering for foundation, intermediate or high level GCSE mathematics.

Table 12.15

Level	Foundation	Intermediate	High
Number of pupils	4	24	2

A circle represents the complete class (hence the term pie chart) while sectors of the circle represent the foundation, intermediate and high candidates. Each group subtends an angle at the centre in proportion to its size.

Therefore the sector angle in the case of foundation students is

$$\frac{4}{30} \times 360° = 48°$$

And for intermediate entries

$$\frac{24}{30} \times 360° = 288°$$

And finally for high level entries

$$\frac{2}{30} \times 360° = 24°$$

Figure 12.1 summarises these calculations.

It is customary to omit the angles from the final pie chart (they are included in the answers to Exercise 12.3 in order that the reader may check his or her calculations) and to present Figure 12.1 as Figure 12.2.

24° High
48° Foundation
288° Intermediate

Figure 12.1

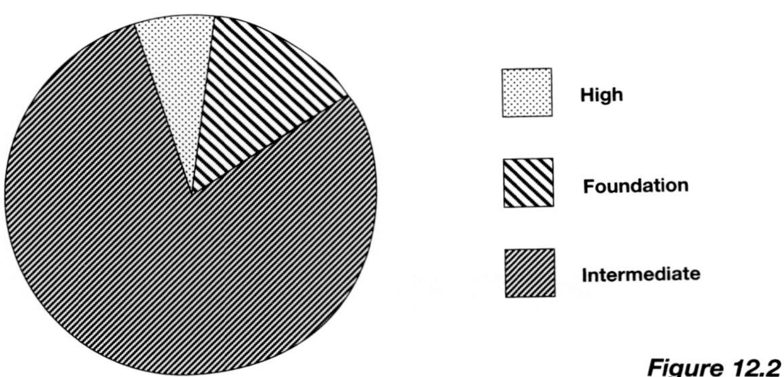

Figure 12.2

Pie charts are particularly appropriate when the intention is to demonstrate how subgroups combine to form a whole with sufficient precision that the relative contribution of each subgroup to the whole is clear at a glance. Furthermore, pie charts are rarely used when a large number of subgroups are involved. A glance at Figure 12.2 is sufficient to inform the reader that the majority of pupils in the class are entered for the intermediate papers, a smaller number for the foundation papers, while about twice as many pupils are entered for foundation level as for high level.

EXERCISE 12.3

1. The voting returns in a particular area during an election were

Labour	2472
Conservative	10421
Liberal Democrat	5107

Illustrate this information on a pie chart.

2. The numbers of pupils studying particula modern languages in a school is

French	56
German	25
Spanish	9

Illustrate this information on a pie chart

3. The numbers of vehicles crossing a toll bridge on a given day were recorded as

cars	9876
lorries	1043
buses	576
motorbikes	505

Illustrate this information on a pie chart.

4. In a survey on holiday resorts it was found that, of those who travelled abroad, the numbers travelling to particular resorts were

Spain	1036
France	1592
Germany	965
Greece	1207

Illustrate this information on a pie chart

5. The numbers of fans who spectated at their teams' home games on a particular Saturday were

Arsenal	28 746
Nottingham Forest	48 231
Stoke	14967
Luton	16056

Illustrate this information on a pie chart.

6. The numbers of people who attended a concert in one week were

Monday	197
Tuesday	205
Wednesday	163
Thursday	215
Friday	256
Saturday	254
Sunday	210

Illustrate these data on a pie chart.

7. On average Ken spent £36 each day in the following way.

£4	on entertainment
£6	on petrol
£6	on savings
£5	on clothes
£10	on rent
£5	on food

Illustrate his expenditure on a pie chart.

8. In a music survey the results of 360 people asked about their favourite type of music were

rave	96
rock	145
classical	24
indie	95

Illustrate this information on a pie chart.

9. In a health survey the numbers of under 5s who suffered from childhood diseases were

measles	47
mumps	28
chicken-pox	45

Illustrate this information on a pie chart.

10. The information below shows the current careers of last year's graduates of a particular university.

nursing	164
teaching	237
industry	248
accounting	123
dentistry	72
medicine	56

Illustrate this on a pie chart.

The histogram

The most important means of illustrating statistical data is the histogram. Unlike the bar chart where the heights of the bars are proportional to the frequencies of the classes, **the histogram is designed such that the area of a given rectangle is proportional to the frequency of the class it represents**. Since, in general, the height of the rectangle in the histogram is not proportional to the frequency of the class it represents (the exception being the histogram representing a distribution where all class widths are equal), then it is quite wrong to label the vertical axis 'frequency'. The vertical axis of an unequal-interval histogram should be labelled **frequency-density** to alert the unsuspecting reader to the perils of misinterpreting the vertical scale. The construction of a histogram will be illustrated in example 2.

EXAMPLE 2

Construct the histogram appropriate to the following frequency distribution of lengths of metal rods (measured to the nearest cm) used in the manufacture of small metal cages.

Table 12.16

Length of rod (cm)	1–3	4–6	7–9	10–14	15–19	20–26	27–29	30–39
Frequency	3	8	12	16	18	14	10	4

Solution 2

The lower and upper class boundaries of, say, the first class are 0.5 and 3.5 respectively and the size of this class is therefore $3.5–0.5 = 3$. Furthermore, if frequency density is defined as frequency divided by class size, Table 12.17 may be constructed.

Table 12.17

Class interval	Frequency	Boundaries	Size	Frequency density
1–3	3	0.5–3.5	3	1
4–6	8	3.5–6.5	3	2.67
7–9	12	6.5–9.5	3	4
10–14	16	9.5–14.5	5	3.2
15–19	18	14.5–19.5	5	3.6
20–26	14	19.5–26.5	7	2
27–29	10	26.5–29.5	3	3.33
30–39	4	29.5–39.5	10	0.4

By plotting class boundary on the horizontal axis and frequency density on the vertical axis a histogram appropriate to the above distribution may be constructed as in Figure 12.3.

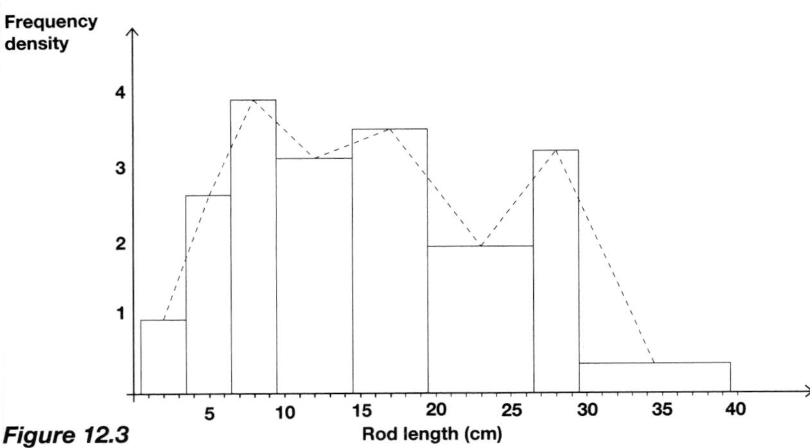

Figure 12.3

It is a straightforward exercise to check that the area of each rectangle represents the frequency of the interval on whose boundaries it 'rests'.

For example, in the case of the rectangle third from the left in Figure 12.3

$$\text{area of rectangle} = 3 \times 4 = 12 \quad \text{the frequency of the third class}$$

The series of straight lines joining the mid-points of the tops of the rectangles is known as a **frequency polygon**. When the intervals are of equal width, it is the convention to extend the extremes of the frequency polygon to the mid-point of an imaginary first and last interval as in Figure 12.5.

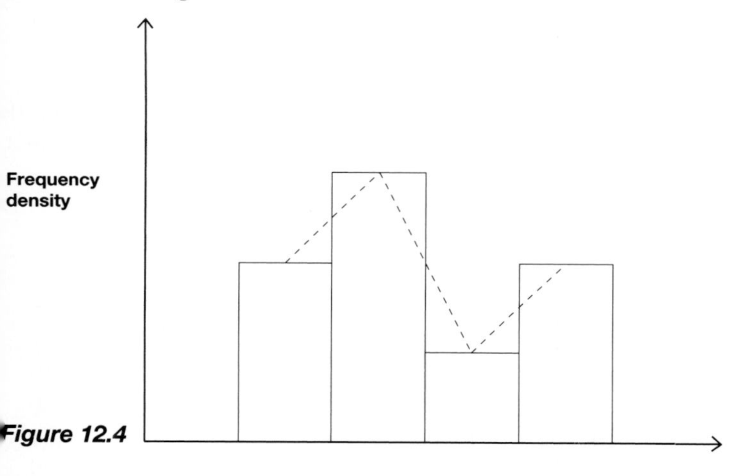

Figure 12.4

becomes

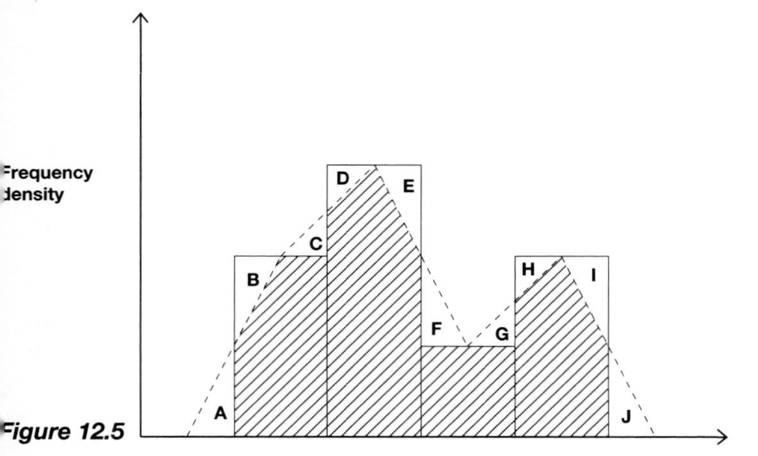

Figure 12.5

The reason for this convention becomes clear when the area of the histogram in Figure 12.5 is written as

$$\text{area of the histogram} = \text{shaded area} + B + D + E + H + I$$

But this equation may be re-expressed

area of the histogram = shaded area + A + C + F + G + J
or area of the histogram = area under the frequency polygon

The area under the frequency polygon therefore represents the total frequency in the special case where the class intervals are all of equal width.

Exercise 12.4 offers the reader the opportunity to consolidate his or her understanding of the sequence of steps involved in drawing a histogram. There will be many opportunities in later chapters to construct histograms in order to illustrate data presented in real-life contexts.

EXERCISE 12.4

Draw a histogram for each of the distributions given in Tables 12.18–26 and construct th corresponding frequency polygon. The column headed 'f' represents the frequencies appropriate t the classes given.

1.

Class	f
1–4	2
5–8	4
9–15	7
16–18	10
19–22	18
23–27	12
28–37	6
38–40	3

Table 12.18

2.

Class	f
1–5	3
6–10	5
11–19	8
20–26	10
27–39	14
40–52	17
53–59	12
60–64	4

Table 12.19

3.

Class	f
3–8	7
9–19	12
20–25	18
26–31	19
32–38	30
39–49	21
50–52	20
53–59	12
60–61	8

Table 12.20

4.

Class	f
4–12	3
13–21	6
22–30	7
31–35	14
36–38	20
39–40	27
41–57	21
58–61	15
62–70	10
71–79	5
80–86	4

Table 12.21

5.

Class	f
3–6	2
7–12	4
13–15	3
16–20	9
21–23	6
24–26	6
27–29	3
30–35	5

Table 12.22

6.

Class	f
1–6	12
7–10	13
11–13	12
14–17	12
18–20	10
21–23	9
24–25	7

Table 12.23

7.

Class	f
4–9	3
10–15	12
16–20	15
21–23	4
24–25	11
26–31	10
32–35	10
36–40	11

Table 12.24

8.

Class	f
10–12	9
13–18	12
19–21	3
22–24	4
25–27	15
28–32	16
33–35	15
36–40	10

Table 12.25

9.

Class	f
1–5	16
6–10	20
11–15	22
16–20	13
21–30	14
31–40	24
41–45	16
46–50	8

Table 12.26

Chapter 13

Summary measures

John Graunt, a London haberdasher born in 1620, is credited with founding that branch of mathematics now known as statistics. He compiled the first statistical database of births and deaths for 1604–1661 and published an analysis of the causes of death in a diminutive volume with the rather grand title *Natural and Political Observations made upon the Bills of Mortality*. Graunt's *Observations* summarised a wealth of data, gathered from parish clerks, in a series of tables augmented by brief commentaries, crude by the standards of modern inferential statistics.

Charles II proposed Graunt for membership of the Royal Society and he was raised to fellowship of that prestigious Society in 1662. The following extract from his writing illustrates the timelessness of some social issues.

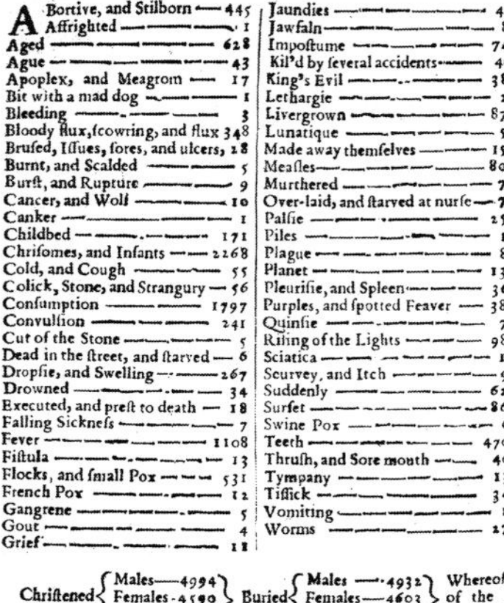

The Diſeaſes, and Caſualties this year being 1632.

Abortive, and Stilborn	445	Jaundies	43
Affrighted	1	Jawfaln	8
Aged	628	Impoſtume	74
Ague	43	Kil'd by ſeveral accidents	46
Apoplex, and Meagrom	17	King's Evil	38
Bit with a mad dog	1	Lethargie	2
Bleeding	3	Livergrown	87
Bloody flux, ſcowring, and flux	348	Lunatique	5
Bruſed, Iſſues, ſores, and ulcers,	28	Made away themſelves	15
Burnt, and Scalded	5	Meaſles	80
Burſt, and Rupture	9	Murthered	7
Cancer, and Wolf	10	Over-laid, and ſtarved at nurſe	7
Canker	1	Palſie	25
Childbed	171	Piles	1
Chriſomes, and Infants	2268	Plague	8
Cold, and Cough	55	Planet	13
Colick, Stone, and Strangury	56	Pleuriſie, and Spleen	36
Conſumption	1797	Purples, and ſpotted Feaver	38
Convulſion	241	Quinſie	7
Cut of the Stone	5	Riſing of the Lights	98
Dead in the ſtreet, and ſtarved	6	Sciatica	1
Dropſie, and Swelling	267	Scurvey, and Itch	9
Drowned	34	Suddenly	62
Executed, and preſt to death	18	Surfet	86
Falling Sickneſs	7	Swine Pox	6
Fever	1108	Teeth	470
Fiſtula	13	Thruſh, and Sore mouth	40
Flocks, and ſmall Pox	531	Tympany	13
French Pox	12	Tiſſick	34
Gangrene	5	Vomiting	1
Gout	4	Worms	27
Grief	11		

Chriſtened { Males—4994, Females—4590, In all—9584 } Buried { Males—4932, Females—4603, In all—9535 } Whereof, of the Plague 8

Increaſed in the Burials in the 122 Pariſhes, and at the Peſthouſe this year 993
Decreaſed of the Plague in the 122 Pariſhes, and at the Peſthouſe this year. 266

C 7 In

The Observation, which I shall add hereunto, is, That the vast numbers of Beggars, *swarming up and down this City, do all live, and seem to be most of them healthy and strong; whereupon I make this Question, Whether, since they all live by Begging, that is, without any kind of labour; it were not better for the State to keep them, even although they earned nothing; that so they might live regularly, and not in that Debauchery, as many Beggars do; and that they might be cured of their bodily Impotencies, or taught to work, &c. each according to his condition, and capacity; or by being employed in some work (not better undone) might be accustomed, and fitted for labour.*

INTRODUCTION

The most frequently encountered summary measures in statistics are measures of central tendency (mean, mode, median) and measures of dispersion (range, semi-interquartile range, mean absolute deviation from the mean, standard deviation). While the former may be viewed as indicating where the data is located, the latter provide some measure of the extent to which the data is spread.

MEASURES OF CENTRAL TENDENCY FOR SMALL NUMERICAL DATA SETS

The mean

The arithmetic mean of a number of observations is simply the sum of those observations divided by the number of observations. In statistics this is written

$$\textbf{mean } (\bar{x}) = \frac{\Sigma x}{n}$$

where sigma (Σ) merely symbolises 'add up'; so that the formula reads 'add up the x values and divide by the number of these'.

EXAMPLE 1

In a shooting competition seven competitors scored as follows.

7 3 5 9 2 6 6

Calculate the mean score.

Solution 1

$$\bar{x} = \frac{\Sigma x}{n}$$

mean $= \frac{1}{7}(7 + 3 + 5 + 9 + 2 + 6 + 6)$

mean $= 5.43$

The median

The median of a set of observations is the middle observation when the data is ranked. In the event of the number of observations being even, the average of the two central observations is computed.

EXAMPLE 2

The hourly rates of pay for five people in a variety of occupations are

£32 £24 £12 £15 £18

Calculate the median hourly rate.

Solution 2

The numbers are arranged in increasing order

12 15 18 24 32

and the central number chosen.

It follows that the median is £18 per hour.

EXAMPLE 3

Twelve applicants for a secretarial job declared their typing speeds in words per minute to be

36 58 47 51 42 43 41 40 42 43 48 45

Calculate the median typing speed.

Solution 3

Arranging the numbers in increasing order

36 40 41 42 42 43 43 45 47 48 51 58

the central pair is identified as 43, 43.

The median is therefore $\frac{1}{2}(43 + 43) = 43$.

It follows that the median is 43 words per minute.

The mode

The mode of a set of observations is that observation which occurs most frequently. The mode may not exist and, in the event that it does, it may not be unique.

EXAMPLE 4

Compute the mode for each of these three data sets.

(i) 2, 5, 2, 7, 9, 4, 8
(ii) 5, 2, 7, 4, 8, 3
(iii) 19, 2, 17, 4, −2, 17, 3, 4, 5

Solution 4

(i) The mode is 2 since this number occurs twice while the remaining numbers only occur once.

(ii) This data has no mode – all numbers appear once.

(iii) This data is bimodal – 17 and 4 both occur twice.

It is of interest to note, in respect of the mean, that **if a set of numbers $a_1, a_2, a_3, \ldots a_n$, has mean $\bar{a}$ then the set $ma_1 + c, ma_2 + c, ma_3 + c, \ldots ma_n + c$, has mean $m\bar{a} + c$.**

EXERCISE 13.1

1. For the following list of numbers calculate the mean, the mode and the median.

1 6 3 6 4 3 5 6 9 8 5 4

2. Calculate the mean, mode and median for the following set of numbers.

2.4 1.9 2.4 3.6 1.8 2.0 4.1

3.9 2.6 2.7 4.9 4.0 3.8 3.3

3. Below are the shoe sizes of 15 pupils. Calculate the mean, mode and median.

1	4	6	10	9	5	6	5	2	2	3	7	4	5	6

4. Listed below are the heights of several children in centimetres. Calculate the mean, mode and median heights.

158	149	162	160	159	140
153	161	154	153	160	145

5. In a class of 30 pupils, the marks in a mathematics examination were

62	47	64	80	50	39
49	52	63	82	35	42
35	63	51	84	49	65
82	72	41	44	54	83
96	71	94	62	63	86

Find the mean, mode and median scores.

6. In 40 matches a football team scored the following numbers of goals.

2	3	0	3	4	1	1	1	2	3
3	2	4	5	3	6	0	0	2	1
1	1	3	2	1	4	3	2	1	6
5	2	4	1	1	2	1	2	2	3

Find the mean, mode and median number of goals per match.

7. The weights of 16 people in kilograms have been listed below. Find the mean, mode and median weights.

60	62	48	52
75	64	60	66
70	56	57	59
63	67	64	49

8. In successive rounds a golfer took the following numbers of strokes.

93	62	79	80	62	68	75

Calculate the mean, mode and median number of strokes.

9. In a class each pupil was asked how many children there were in their family. The following results were obtained.

2	4	6	4	5	4	6	8
1	2	1	1	9	5	3	4
3	4	2	4	2	8	7	
3	5	3	3	3	4	4	

Find the mean, mode and median number of children per family.

10. The numbers of vehicles crossing a toll bridge for 30 successive days were

168	183	193	173	215	192
206	196	204	178	226	145
159	184	223	193	189	163
268	172	234	182	184	194
152	156	165	146	162	200

What is the mean, mode and median number of vehicles per day?

11. Using the mean calculated in question 1, write down the mean for the following data.

2 12 6 12 8 6 10 12 18 16 10 8

12. Using the mean calculated in question 2, write down the mean for the following data.

5.4	4.9	5.4	6.6	4.8	5.0	7.1
6.9	5.6	5.7	7.9	7.0	6.8	6.3

13. Using the mean calculated in question 3, write down the mean for the following data.

5 11 15 23 21 13 15 13 7 7 9 17 11 13 15

14. Using the mean calculated in question 8, write down the mean for the following data.

179 86 137 140 86 104 125

THE ADVANTAGES AND DISADVANTAGES OF THE THREE MEASURES OF CENTRAL TENDENCY

The mean

While the mean is simple to calculate, well understood and makes use of all the data, it suffers from being sensitive to extreme values; the data should first be inspected for 'rogue' values or 'outliers' before the mean is calculated.

The median

The median is not sensitive to extreme values and has the additional advantage of being one of the observed values in cases where the number of observations is odd.

However, for large data sets, it is tedious to calculate by hand since ordering of the data is time consuming.

The mode

The mode has the advantage of being easily understood but the disadvantage that, in many cases, it may not even exist.

THE MEAN OF GROUPED DATA

The method used in estimating the mean for a set of grouped data is illustrated in the next example.

EXAMPLE 5

The time in months (measured to the nearest month) between the onset of a particular illness and its recurrence was recorded for 80 patients. The following frequency distribution was obtained.

Table 13.1

Time (months)	4–6	7–9	10–12	13–15	16–18	19–21	22–24
Frequency, f	6	15	24	19	10	4	2

Estimate the mean of the distribution.

Solution 5

The mean is calculated using the formula

$$\bar{x} = \frac{\Sigma fx}{\Sigma f}$$

where x is the mid-point of each class interval.

The first stage in the use of the formula is to construct Table 13.2.

Table 13.2

Time (months)	Frequency, f	Mid-mark, x	fx
4–6	6	5	30
7–9	15	8	120
10–12	24	11	264
13–15	19	14	266
16–18	10	17	170
19–21	4	20	80
22–24	2	23	46
	$\Sigma f = 80$		$\Sigma fx = 976$

The third column is constructed by finding the mid-mark of each class interval, for example, the third entry is the average of 10 and 12. Each entry in the final column is the product of the corresponding entries in columns two and three.

Adding the second column gives $\Sigma f = 80$ and adding the final column gives $\Sigma fx = 976$

$$\text{Hence } \bar{x} = \frac{\Sigma fx}{\Sigma f} = \frac{976}{80} = 12.2 \text{ months}$$

EXAMPLE 6

Table 13.3 shows the frequencies with which words of various 'lengths' appear in the first paragraph on a page of a particular book.

Table 13.3

Number of letters	1–3	4–6	7–9	10–12	13–15
Number of words	35	33	34	13	5

Estimate the mean length of these words.

Solution 6

The table in this case becomes

Table 13.4

Number of letters	f	x	fx
1–3	35	2	70
4–6	33	5	165
7–9	34	8	272
10–12	13	11	143
13–15	5	14	70
	$\Sigma f = 120$		$\Sigma fx = 72($

Once again the reader is reminded that column 3 represents the class marks of the intervals in column 1 and the entries in column 4 are generated from the product of the corresponding entries in columns 2 and 3. Applying the formula for the mean

$$\bar{x} = \frac{\Sigma fx}{\Sigma f} = \frac{720}{120} = 6 \text{ letters}$$

The coding method

The coding method is a technique for reducing the amount of tedious computation involved in calculating the mean from a frequency distribution. With the advent of the electronic calculator the coding method effectively becomes redundant but, as many examinations require candidates to show the development of their answers, the coding method can still contribute much to the presentation and accuracy of solutions to questions in public examinations.

The mean in the coding method is given by

$$\bar{x} = A + c\frac{\Sigma fu}{\Sigma f}$$

where c is usually the class size (where all classes have a common size), A is an 'assumed' mean and $u = \frac{(x - A)}{c}$. This terminology will become clear in the example which follows.

EXAMPLE 7

Repeat example 6 using coding to estimate the mean.

Solution 7

The first three columns of Table 13.5 are identical to those of Table 13.4.

Table 13.5

Number of letters	f	x	$x - A$	$u = \frac{(x - A)}{c}$	fu
1–3	35	2	− 6	− 2	− 70
4–6	33	5	− 3	− 1	− 33
7–9	34	8←A	0	0	0
10–12	13	11	3	1	13
13–15	5	14	6	2	10

$\Sigma f = 120$ $\Sigma fu = -80$

In order to generate column 4 a value of x near the 'centre' of column 3 is chosen.

Here $A = 8$. A is subtracted from each entry in column 3 to generate column 4. Now c, the class width, is constant and equal to 3 (for example in the third class $c = 9.5–6.5 = 3$).

Column 5 is now generated by dividing each column 4 entry by c. Finally the entries of the final column are the products of the corresponding entries in columns 2 and 5.

Using the coding formula therefore

$$\bar{x} = A + c\frac{\Sigma fu}{\Sigma f}$$

becomes $\bar{x} = 8 + 3\left(\frac{-80}{120}\right) = 6$ as before

The concept of coding will be returned to when standard deviation is considered later.

Grouping error

It would be remiss to leave this section on the mean of grouped data without some reference to the fact that the mean of such data is merely an approximation, albeit a good one in many instances, to the number that would have resulted had the data not been grouped. This idea is now developed.

Consider the following data.

138	164	150	132	144	125	149	157
146	158	140	147	136	148	152	144
168	126	138	176	163	119	154	165
146	173	142	147	135	153	140	135
161	145	135	142	150	156	145	128

Should this data be grouped into classes 118–126, 127–135 etc. the frequency table results in Table 13.6.

Table 13.6

Class	f	x
118–126	3	122
127–135	5	131
136–144	9	140
145–153	12	149
154–162	5	158
163–171	4	167
172–180	2	176

The mid-marks listed in column three have been calculated in the normal way. In grouping the data in this way the three numbers in the original data grouped in the first class interval, i.e. 125, 126, 119, are effectively replaced by 122, 122, 122. Clearly grouping results in a loss of information and it is this that gives rise to the 'grouping error'. The magnitude of that error depends upon how closely the original data set approximates to the data set

122	122	122	131	131	131	131	131
140	140	140	140	140	140	140	140
140	149	149	149	149	149	149	149
149	149	149	149	149	158	158	158
158	158	167	167	167	167	176	176

constructed from the frequency table. Grouping error may be reduced by arranging intervals to be small and to have mid-marks which coincide as much as possible with the original data.

EXERCISE 13.2

1. The table below shows the scores of 40 pupils in an English examination. Estimate the mean score from the grouped data.

Table 13.7

Score	Frequency
1–5	2
6–10	3
11–15	1
16–20	2
21–25	10
26–30	12
31–35	2
36–40	5
41–45	2
46–50	1

2. The age distribution of workers in a large company is as follows.

Table 13.8

Age	Frequency
16–20	10
21–25	10
26–30	13
31–35	12
36–40	15
41–45	22
46–50	10
51–55	12
56–60	8

3. The estimated heights (to the nearest cm) of 30 pupils are as follows.

Table 13.9

Height	Frequency
140–144	1
145–149	2
150–154	3
155–159	7
160–164	15
165–169	1
170–174	1

Estimate the mean height in cm.

4. The daily earnings (to the nearest £) of part-time employees in a supermarket are shown in Table 13.10.

Table 13.10

Wage (£)	Frequency
10–12	5
13–15	6
16–18	6
19–21	10
22–24	12
25–27	7
28–30	4

Estimate the mean daily wage.

5. Table 13.11 shows the number of errors made by each of 60 pupils in a typing exercise. Estimate the mean number of errors made by a pupil.

Table 13.11

Errors	Frequency
1–5	13
6–10	19
11–15	4
16–20	5
21–25	10
26–30	2
31–35	3
36–40	4

6. Table 13.12 shows the number of days missed by each of 80 pupils during a year.

Table 13.12

Number of days	Frequency
0–4	29
5–9	32
10–14	5
15–19	4
20–24	3
25–30	7

Estimate the mean number of days missed by a pupil.

7. Table 13.13 shows the lifetime (to the nearest hour) of a batch of light bulbs produced in a factory. Estimate the mean lifetime of a bulb.

Table 13.13

Lifetime (h)	Frequency
0–9	16
10–19	15
20–29	26
30–39	32
40–49	30
50–59	25
60–69	12

8. The distribution of pupils' weights (to the nearest kg) in a class of 30 senior pupils is shown in Table 13.14. Estimate the mean weight.

Table 13.14

Weight (kg)	Frequency
60–62	2
63–65	4
66–68	4
69–71	12
72–74	6
75–77	1
78–80	1

9. The students staying in a hostel are grouped by age as follows.

Table 13.15

Age	Frequency
15–19	16
20–24	13
25–29	19
30–34	12

Estimate the mean age.

10. The length of time (to the nearest minute) spent by patients in a waiting room is given in Table 13.16.

Table 13.16

Time (min)	Frequency
26–35	32
36–45	29
46–55	10
56–65	9

Estimate the mean waiting time.

THE MEDIAN OF GROUPED DATA

Cumulative frequency curves

Consider the following information.

A company, having problems with absenteeism, chooses a sample of 64 employees and compiles a table of the numbers of days for which these employees failed to turn up for work last year.

Table 13.17

Number of days absent from work	Frequency, f
16–20	7
21–25	17
26–30	14
31–35	12
36–40	8
41–45	5
46–50	1

It is interesting to note that 7 workers were absent for less than 20.5 days (those workers absent for 16 to 20 days inclusive), 24 workers were absent for less than 25.5 days (the 7 workers in the 16–20 day category plus those absent for 21 to 25 days).

Continuing this process the following **cumulative frequency** table may be constructed.

Table 13.18

Number of days absent from work (less than . . .)	Number of workers
20.5	7
25.5	24
30.5	38
35.5	50
40.5	58
45.5	63
50.5	64

The numbers in the left-hand column are the upper class boundaries for each class interval in the original data and the right-hand column is known as the cumulative frequency. The data in Table 13.18 may be graphed with upper class boundary on the horizontal axis and cumulative frequency on the vertical to give the **cumulative frequency curve** or **ogive** of Figure 13.1.

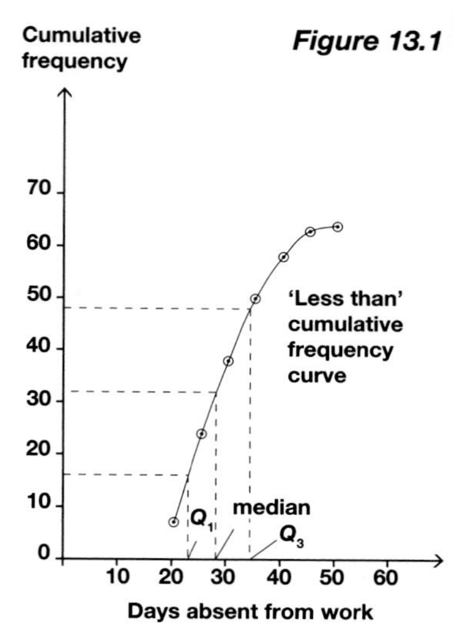

Figure 13.1

Alternatively the data may be presented using the lower class boundary of each interval. A re-examination of Table 13.17 will reveal that 64 workers were absent for more than 15.5 days, 57 workers were absent for more than 20.5 days and so on. Extending this process to the remaining data results in Table 13.19.

Table 13.19

Number of days absent from work (more than . . .)	Number of workers
15.5	64
20.5	57
25.5	40
30.5	26
35.5	14
40.5	6
45.5	1

The graph of this **more than** cumulative frequency table with lower class boundary along the horizontal axis and cumulative frequency plotted vertically is presented in Figure 13.2.

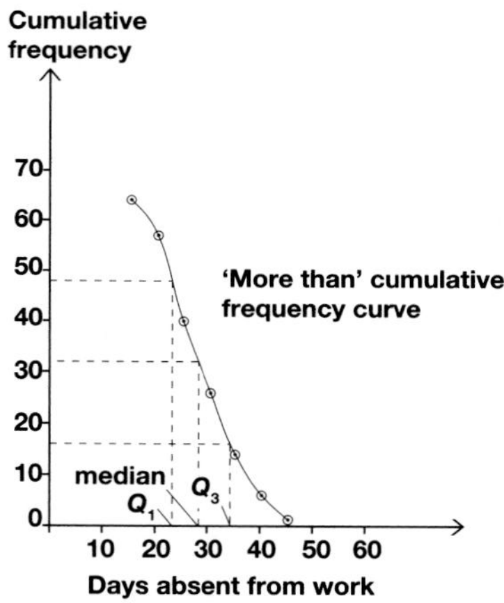

Figure 13.2

Estimating the median from an ogive

An examination of Table 13.18 will reveal that, for example, 63 workers have been absent for less than 45.5 days, 50 workers have been absent for less than 35.5 days and so on. Choosing the point half-way up the cumulative frequency axis (cumulative frequency = 32) in Figure 13.1 and finding the corresponding value on the horizontal axis it is possible to state that half of the workers have been absent for less than 28 days.

Since the median is the middle value when the number of days absent are ranked, and since half of the employees have been absent for less than 28 days in the last year, then 28 days is the median number of days absent. Figure 13.2 shows the same process applied to the 'more than' cumulative frequency curve.

The quartiles and the semi-interquartile range

Three further statistical measures may be estimated from the ogive, namely: the upper quartile, the lower quartile and the semi-interquartile range. The statistical importance of the latter measure will be made clear later in this chapter. **The lower quartile, Q_1, is the value below which 25% of the distribution lies, the upper quartile, Q_3, is the value below which 75% of the distribution lies** and the **semi-interquartile range, SIR**, is given by

$$\text{SIR} = \tfrac{1}{2}(Q_3 - Q_1)$$

As 25% of 64 is 16 and 75% of 64 is 48, the quartiles may be read from Figure 13.1 by locating those points on the horizontal axis corresponding to 16 and 48 on the vertical axis; the construction lines are shown in Figure 13.1.

Since $Q_1 = 23$ and $Q_3 = 34.5$ then SIR = 5.8.

A glance at Figure 13.2 will reveal how similar deductions may be made from a 'more than' ogive.

The median by calculation

In order to avoid a discussion of linear interpolation, the formula for the median is offered without explanation.

It is

$$\text{median} = L_1 + \frac{\left\{\frac{N}{2} - (\Sigma f)_1\right\}c}{f_{\text{med}}}$$

where L_1 is the lower class boundary of the median class, N is the total frequency, $(\Sigma f)_1$ is the sum of frequencies up to (but not including) the median class, f_{med} is the frequency of the median class and c is the size of the median class.

Returning to Table 13.17, an illustration of the formula will be given. The first stage is to establish the median class i.e. the class in which, at some point, the cumulative frequency equals half of the total frequency. Now as half of the total frequency is 32 and the total frequency for the first two classes is 24 while that for the first three classes is 38, the median class is identified as the third class. The lower class boundary for this class is 25.5.

$$\frac{N}{2} = 32,\ (\Sigma f)_1 = 24,\ f_{\text{med}} = 14$$

and $c = 30.5 - 25.5 = 5$

Hence the median $= 25.5 + \left(\frac{32 - 24}{14}\right)5 = 28.4$

EXAMPLE 8

Using the median formula, estimate the median weekly wage for the group of part-time workers given in Table 13.20.

Table 13.20

Weekly wage (to the nearest £)	Numbers of workers earning this wage
0–20	0
21–40	2
41–60	4
61–80	7
81–100	11
101–120	28
121–140	20

Solution 8

Since the total frequency is 72, the sum of the first five frequencies is 24 and the sum of the first six is 52, it follows that the median class is the class 101–120.

$$L_1 = 100.5$$

$$\frac{N}{2} = 36$$

$$(\Sigma f)_1 = 24$$

$$f_{\text{med}} = 28$$

$$c = 120.5 - 100.5 = 20$$

Therefore $\quad \text{median} = 100.5 + 20\frac{(36-24)}{28} = 109$

$\text{median wage} = £109$

EXERCISE 13.3

Estimate the median by calculation for each of the distributions below.

1. **Table 13.21**

Class interval	1–5	6–10	11–15	16–20	21–25	26–30
Frequency	4	12	21	24	13	6

2. **Table 13.22**

Class interval	56–60	61–65	66–70	71–75	76–80
Frequency	8	24	43	19	6

3. **Table 13.23**

Class interval	4–6	7–9	10–12	13–15	16–18	19–21	22–24
Frequency	6	15	24	19	10	4	2

4. *Table 13.24*

Class interval	3–6	7–10	11–14	15–18	19–22	23–26
Frequency	5	10	15	14	9	3

EXERCISE 13.4

1. A number of pupils recorded how long it took them to walk to school on a particular morning A cumulative frequency distribution was formed and is shown below.

Table 13.25

Time taken (min)	< 5	< 10	< 15	< 20	< 25	< 30	< 35	< 40	< 45
Cumulative frequency	25	34	50	65	76	83	88	94	100

Draw a cumulative frequency curve and from it estimate

(i) the median,
(ii) the upper and lower quartiles,
(iii) the semi-interquartile range.

Confirm your median estimate by calculation.

2. The pH value of 50 solutions was tested and recorded as below in the form of a cumulative frequency distribution.

Table 13.26

pH	< 4.6	< 4.8	< 5.0	< 5.2	< 5.4	< 5.6	< 5.8	< 6.0
Cumulative frequency	6	10	19	30	39	44	47	50

Draw the cumulative frequency curve and from it estimate

(i) the median,
(ii) the upper and lower quartiles,
(iii) the semi-interquartile range.

Confirm your median estimate by calculation.

3. The time taken by various motorists to cover a particular journey was recorded and a cumulative frequency distribution formed, as shown below.

Table 13.27

Time (min)	< 80	< 85	< 90	< 95	< 100	< 105	< 110	< 115	< 120
Cumulative frequency	12	24	47	72	90	102	111	115	120

Draw the cumulative frequency curve and use this to estimate

(i) the median,
(ii) the upper and lower quartiles,
(iii) the semi-interquartile range.

Confirm your median estimate by calculation.

4. The heights of various pupils in a year group were recorded as below.

Table 13.28

Height (cm)	< 140	< 145	< 150	< 155	< 160	< 165	< 170
Cumulative frequency	16	24	40	72	156	182	200

Draw the cumulative frequency curve and use this to estimate

(i) the median,
(ii) the upper and lower quartiles,
(iii) the semi-interquartile range.

Confirm your median estimate by calculation.

5. The weights, in kg, of 100 people were distributed as follows.

Table 13.29

Weight (kg)	< 50	< 60	< 70	< 80	< 90	< 100
Cumulative frequency	11	20	60	91	98	100

Draw the cumulative frequency curve and use this to estimate

(i) the median,
(ii) the upper and lower quartiles,
(iii) the semi-interquartile range.

Confirm your median estimate by calculation.

6. The age of people attending a leisure centre was recorded on a particular day and the results are shown below.

Table 13.30

Age	< 10	< 20	< 30	< 40	< 50	< 60	< 70	< 80
Cumulative frequency	22	46	84	160	212	228	236	240

Draw the cumulative frequency curve and hence estimate

(i) the median,
(ii) the upper and lower quartiles,
(iii) the semi-interquartile range.

Confirm your median estimate by calculation.

7. The percentage scores of pupils in a summer test was recorded as follows.

Table 13.31

%	< 20	< 40	< 60	< 80	< 100
Cumulative frequency	38	70	164	178	180

Draw the cumulative frequency curve and hence estimate
(i) the median,
(ii) the upper and lower quartiles,
(iii) the semi-interquartile range.

8. The amount of money spent by each of 1000 late-night shoppers in a store was recorded and the results are shown below.

Table 13.32

Amount of money (£)	> 40	> 35	> 30	> 25	> 20	> 15	> 10	> 5	> 0
Cumulative frequency	20	50	100	170	250	510	790	970	1000

Plot the 'more-than' cumulative frequency curve and hence estimate
(i) the median,
(ii) the upper and lower quartiles,
(iii) the semi-interquartile range.

9. The number of words typed per minute by each of 120 students was recorded as below.

Table 13.33

Number of words	> 160	> 120	> 80	> 40	> 0
Cumulative frequency	13	55	106	116	120

Sketch the 'more-than' cumulative frequency curve and use it to estimate
(i) the median,
(ii) the upper and lower quartiles,
(iii) the semi-interquartile range.

10. The wages of a number of employees in a factory were recorded as follows.

Table 13.34

Wage (£)	> 200	> 160	> 120	> 80	> 40	> 0
Frequency	2	4	8	17	46	50

Draw the 'more-than' cumulative frequency curve and use it to estimate
(i) the median,
(ii) the upper and lower quartiles,
(iii) the semi-interquartile range.

THE MODAL CLASS OF GROUPED DATA

The modal class for grouped data is the class which occurs most frequently.

EXAMPLE 9

The results of 100 pupils in a test are presented in Table 13.35.

Table 13.35

Score	Frequency
60–64	5
65–72	18
73–84	42
85–92	26
93–97	9

Find the modal class.

Solution 9

The class which occurs with the highest frequency is 73–84 and so the modal class is the mark range 73–84.

MEASURES OF DISPERSION

While measures of central tendency, such as the mean, mode and median, help statisticians locate the 'centre' of a distribution, **measures of dispersion permit an assessment to be made of the extent to which the data spreads.** If the data set consists of values close to the mean, for example, the dispersion is likely to be small whereas a data set comprising values which differ greatly from the mean is likely to have a large associated measure of dispersion. The measures of dispersion explored will be the range, the semi-interquartile range, the mean absolute deviation from the mean, the standard deviation and the variance.

MEASURES OF DISPERSION FOR SMALL NUMERICAL DATA SETS

The range

The range of a set of data is simply the largest value minus the smallest value. Therefore, for example, the data set

4 2 7 1 2 15 2 11 4

has range = 15 − 1 = 14

EXERCISE 13.5

Find the range for each of the data sets in questions 1–10 of Exercise 13.1.

The standard deviation

The standard deviation of n numbers is calculated through the formula

$$s = \sqrt{\frac{\Sigma (x - \bar{x})^2}{n}}$$

where x represents a number, and $\bar{x}$ is the mean of the n numbers. The value of the expression $(x - \bar{x})^2$ is large for those x values greatly different from $\bar{x}$ and small for x values close to $\bar{x}$. It follows that the standard deviation will be large for a data set where values are widely spread about the mean and small for a data set where values are clustered around the mean.

EXAMPLE 10

Find the standard deviation of the data below, which represents the number of days absent from school in one year of eight primary school children.

5 7 8 4 2 1 9 13

Solution 10

The standard deviation is given by

$$s = \sqrt{\frac{\Sigma(x - \bar{x})^2}{n}}$$

To facilitate its calculation, Table 13.36 is constructed.

Table 13.36

x	$x - \bar{x}$	$(x - \bar{x})^2$
5	− 1.125	1.265625
7	0.875	0.765625
8	1.875	3.515625
4	− 2.125	4.515625
2	− 4.125	17.015625
1	− 5.125	26.265625
9	2.875	8.265625
13	6.875	47.265625

$$\Sigma(x - \bar{x})^2 = 108.875$$

where $\bar{x} = \frac{1}{8}(5 + 7 + 8 + 4 + 2 + 1 + 9 + 13) = 6.125$

Using the formula above

$$s = \sqrt{\frac{108.875}{8}} = 3.69$$

It can be shown that the formula

$$s = \sqrt{\frac{\Sigma(x - \bar{x})^2}{n}}$$

may be rewritten

$$s = \sqrt{\left(\frac{\Sigma x^2}{n} - \frac{(\Sigma x)^2}{n^2}\right)}$$

This latter formula is much more convenient to use – a glance at the calculation below will confirm this.

Table 13.37

x	x^2
5	25
7	49
8	64
4	16
2	4
1	1
9	81
13	169

$\Sigma x = 49 \qquad \Sigma x^2 = 409$

$$s = \sqrt{\left(\frac{409}{8} - \frac{(49)^2}{64}\right)} = 3.69$$

Mathematics beyond the scope of this book can be used to verify that **if the data set $x_1, x_2 \ldots x_n$ has standard deviation s, then the data set $mx_1 + c, mx_2 + c \ldots, mx_n + c$ has standard deviation ms.**

Finally, a measure of considerable interest in advanced statistics is the variance where

variance = (standard deviation)^2

EXERCISE 13.6

Find the standard deviation and variance for each of the data sets in Exercise 13.1.

EXAMPLE 11

In a class of 30 pupils there are 20 boys and 10 girls. The mean score of the boys in an end of term examination is 60 and the standard deviation of their scores is 10. In the same examination the mean score of the girls is 51 with standard deviation 12. Find the mean score for the complete class and the corresponding standard deviation.

Solution 11

Let x_B represent a boy's score, x_G a girl's score and x_C a pupil's score.

Now $\quad 60 = \dfrac{\Sigma x_B}{20}$ and $51 = \dfrac{\Sigma x_G}{10}$

Therefore $\Sigma x_B = 1200$ and $\Sigma x_G = 510$

Hence Σx_C (for the class) $= 1200 + 510$

or $\quad \bar{x}_C = \dfrac{1200 + 510}{30} = 57$

Furthermore

$$10 = \sqrt{\left(\frac{\Sigma x_B^{\,2}}{20} - \frac{(\Sigma x_B)^2}{400}\right)} \text{ and } \quad 12 = \sqrt{\left(\frac{\Sigma x_G^{\,2}}{10} - \frac{(\Sigma x_G)^2}{100}\right)}$$

But $\frac{\Sigma x_B}{20} = 60$ and $\frac{\Sigma x_G}{10} = 51$

Hence $10 = \sqrt{\left(\frac{\Sigma x_B^{\,2}}{20} - 60^2\right)}$ and $12 = \sqrt{\left(\frac{\Sigma x_G^{\,2}}{10} - 51^2\right)}$

or $100 = \frac{\Sigma x_B^{\,2}}{20} - 3600$ and $144 = \frac{\Sigma x_G^{\,2}}{10} - 2601$

Therefore $\Sigma x_B^{\,2} = 74\,000$ and $\Sigma x_G^{\,2} = 27\,450$

So $\Sigma x_C^{\,2} = 74\,000 + 27\,450 = 101\,450$

It follows that

$$s_c = \sqrt{\left(\frac{\Sigma x_C^{\,2}}{30} - \frac{(\Sigma x_C)^2}{900}\right)}$$

becomes $s_c = \sqrt{\left(\frac{101\,450}{30} - (57)^2\right)}$

$$s_c = 11.52$$

The mean absolute deviation from the mean

The mean absolute deviation from the mean is calculated by the formula

$$MAD = \frac{\Sigma\,|x - \bar{x}|}{n}$$

Where $|x - \bar{x}|$ is the modulus of the difference between a given value, x, and the mean of the data. The effect of the modulus symbol is to ensure that the magnitude of $x - \bar{x}$ only is included in the summation i.e. negative values of $x - \bar{x}$ are treated as being positive. The method is now illustrated through an example.

EXAMPLE 12

One hundred boxes of drawing pins were taken and the number of pins in each box counted. The results are presented below.

Table 13.38

Number of pins	18	19	20	21	22
Number of boxes	5	20	40	30	5

Calculate the mean absolute deviation from the mean.

Solution 12

Table 13.39

x	f	fx
18	5	90
19	20	380
20	40	800
21	30	630
22	5	110
	$\Sigma f = 100$	$\Sigma fx = 2010$

Therefore $\bar{x} = \dfrac{\Sigma fx}{\Sigma f} = \dfrac{2010}{100} = 20.1$

Proceeding now to the calculation of the *MAD*.

Table 13.40

x	f	$\lvert x - \bar{x}\rvert$	$f\lvert x - \bar{x}\rvert$
18	5	2.1	10.5
19	20	1.1	22
20	40	0.1	4
21	30	0.9	27
22	5	1.9	9.5
	$n = \Sigma f = 100$		$\Sigma f\lvert x - \bar{x}\rvert = 73$

$MAD = \frac{73}{100} = 0.73$

The *MAD* clearly qualifies as a measure of dispersion since data comprising x values close to $\bar{x}$ (data which is minimally dispersed) will have a small *MAD*, since the terms in the numerator sum will all be small.

EXERCISE 13.7

Find the mean absolute deviation from the mean for each data set in questions 1–10 of Exercise 13.1.

MEASURES OF DISPERSION FOR GROUPED DATA

The semi-interquartile range

The semi-interquartile range has been discussed earlier. The median and the upper quartile enclose 25% of any distribution, as do the median and lower quartile. It follows that a distribution which is clustered about the median will have both Q_1 and Q_3 close to the median, and, as a consequence, close to one another. Therefore a distribution with small dispersion will have Q_3 close to Q_1 and hence $Q_3 - Q_1$ small. In summary, the semi-interquartile range, $\frac{1}{2}(Q_3 - Q_1)$, qualifies as a measure of dispersion.

Standard deviation

Having demonstrated how the formula

$$s = \sqrt{\left(\frac{\Sigma x^2}{n} - \frac{(\Sigma x)^2}{n^2}\right)}$$

may be used to calculate the standard deviation for small data sets, the use of the grouped-data equivalent

$$s = \sqrt{\left(\frac{\Sigma f x^2}{\Sigma f} - \frac{(\Sigma f x)^2}{(\Sigma f)^2}\right)}$$

will be illustrated in the following examples.

EXAMPLE 13

The time in months (measured to the nearest month) between the onset of a particular illness and its recurrence was recorded for 80 patients. The following frequency table was obtained.

Table 13.41

Time (months)	4–6	7–9	10–12	13–15	16–18	19–21	22–24
Frequency	6	15	24	19	10	4	2

Estimate the standard deviation.

Solution 13

Table 13.42

Time (months)	Frequency, f	x	x^2	fx	fx^2
4–6	6	5	25	30	150
7–9	15	8	64	120	960
10–12	24	11	121	264	2904
13–15	19	14	196	266	3724
16–18	10	17	289	170	2890
19–21	4	20	400	80	1600
22–24	2	23	529	46	1058

$\Sigma f = 80$ $\quad\Sigma fx = 976$ $\quad\Sigma fx^2 = 13\,286$

The standard deviation for grouped data is given by

$$s = \sqrt{\left(\frac{\Sigma f x^2}{\Sigma f} - \frac{(\Sigma f x)^2}{(\Sigma f)^2}\right)}$$

Hence $s = \sqrt{\left(\frac{13\,286}{80} - \frac{(976)^2}{6400}\right)} = 4.15$ months

The standard deviation by coding

The coding method has already been introduced earlier in this chapter. The coding formula for the mean was presented as

$$\bar{x} = A + c\frac{\Sigma fu}{\Sigma f}$$

where A is an 'assumed' mean, c is usually the class size and $u = \frac{(x - A)}{c}$.

The corresponding coding formula for the standard deviation is

$$s = c\sqrt{\left(\frac{\Sigma fu^2}{\Sigma f} - \frac{(\Sigma fu)^2}{(\Sigma f)^2}\right)}$$

The use of this formula is demonstrated in the example below.

EXAMPLE 14

Table 13.43 shows the frequencies with which words of various lengths appear in the first paragraph of a particular book.

Table 13.43

Number of letters	1–3	4–6	7–9	10–12	13–15
Number of words	35	33	34	13	5

Estimate the standard deviation using coding.

Solution 14

Table 13.44

Number of letters	Frequency	x	$x - A$	$u = \frac{x - A}{c}$	fu	fu^2
1–3	35	2	− 6	− 2	− 70	140
4–6	33	5	− 3	− 1	− 33	33
7–9	34	8←A	0	0	0	0
10–12	13	11	3	1	13	13
13–15	5	14	6	2	10	20
	$\Sigma f = 120$			$\Sigma fu = -80$		$\Sigma fu^2 = 206$

A is set equal to a value near the 'middle' of the x column and c is the class size i.e.

$3.5 - 0.5 = 3$ in the case of the first class interval, for example.

Now the coding formula for the standard deviation is

$$s = c\sqrt{\left(\frac{\Sigma fu^2}{\Sigma f} - \frac{(\Sigma fu)^2}{(\Sigma f)^2}\right)}$$

and so $$s = 3\sqrt{\left(\frac{206}{120} - \frac{(-80)^2}{120^2}\right)} = 3.38$$

The standard deviation is hence 3.38 letters.

EXERCISE 13.8

1. Estimate the standard deviation and variance of the history marks of 70 pupils recorded below.

Table 13.45

Mark	30–39	40–49	50–59	60–69	70–79	80–89
Frequency	7	11	12	24	14	2

2. Estimate the standard deviation and variance of the shoe sizes of 40 pupils recorded below.

Table 13.46

Shoe size	1–2	3–4	5–6	7–8	9–10
Frequency	3	8	16	7	6

3. The numbers of hours worked by 100 employees in a factory are shown below. Estimate the standard deviation and variance.

Table 13.47

Number of hours	0–4	5–9	10–14	15–19	20–24	25–29	30–34
Frequency	4	9	12	16	24	29	6

4. The heights of 40 children were recorded to the nearest centimetre as below. Estimate the standard deviation and variance of these heights.

Table 13.48

Height (cm)	140–144	145–149	150–154	155–159	160–164
Frequency	5	10	9	8	8

5. Table 13.49 shows the frequency distribution of weights (to the nearest kg) of 60 women. Estimate the standard deviation and variance.

Table 13.49

Weight (kg)	40–44	45–49	50–54	55–59	60–64	65–69
Frequency	1	3	6	25	21	4

6. The delays (in minutes) for several trains were recorded to the nearest minute as shown below. Estimate the standard deviation and variance.

Table 13.50

Delay	1–3	4–6	7–9	10–12	13–15	16–18	19–21	22–24
Frequency	2	3	3	5	4	1	2	2

7. The numbers of errors made by students in a typing exercise had the following frequency distribution. Estimate the standard deviation and variance.

Table 13.51

Number of errors	0–2	3–5	6–8	9–11	12–14	15–17	18–20
Frequency	16	24	22	21	13	15	9

8. The numbers of days missed from school by a class of 40 pupils were recorded as below. Estimate the standard deviation and variance.

Table 13.52

Days off	0–1	2–3	4–5	6–7	8–9	10–11	12–13	14–15
Frequency	0	3	5	16	12	2	1	1

9. The ages of the workers in a factory had the following frequency distribution. Estimate the standard deviation and variance.

Table 13.53

Age	15–19	20–24	25–29	30–34	35–39	40–44
Frequency	16	19	12	23	26	4

10. The results of 120 pupils in an exam had the following frequency distribution. Estimate the standard deviation and variance.

Table 13.54

% score	0–9	10–19	20–29	30–39	40–49	50–59	60–69	70–79	80–89	90–99
Frequency	4	2	0	3	16	21	18	13	24	19

ADVANTAGES AND DISADVANTAGES OF THE VARIOUS MEASURES OF DISPERSION

The range, calculated as the difference of extreme values, is clearly sensitive to these extreme values or 'outliers'. However, it has the advantage of being easy to calculate. The semi-interquartile range does not suffer from sensitivity to outliers but is insensitive to information from within the upper and lower 25% of data. The mean absolute deviation from the mean and the standard deviation both make use of all the available data. The standard deviation is often cited as ensuring that deviations from the mean are positive by 'natural' mathematical techniques (squaring) as opposed to the 'unnatural' mathematical techniques of the mean absolute deviation (taking the modulus).

MISCELLANEOUS EXAMINATION QUESTIONS

EXAMPLE 15

(a) The number of hours, x, of clear sunshine per day during June and July of 1981 in central Belfast was recorded and tabulated as in Table 13.55.

Table 13.55

Hours of sunshine (x)	Number of days
$0 \le x < 2$	14
$2 \le x < 4$	12
$4 \le x < 6$	5
$6 \le x < 8$	5
$8 \le x < 10$	6
$10 \le x < 12$	19

(i) Draw a frequency polygon to illustrate the above data.

(ii) By visual inspection of Table 13.55 or the frequency polygon, describe briefly what you can conclude regarding the distribution of sunny or dull days.

(iii) Calculate the mean and standard deviation of the number of hours of sunshine. Show clearly the intermediate stages of your working.

(b) A test, given to a class of six men and four women, resulted in a overall mean mark of 16.8. If the six men achieved a mean mark of 16.0, find the mean mark achieved by the four women.

(N. Ireland Additional Mathematics, 1987)

Solution 15

(a) (i) See Figure 13.3.

(ii) This region is prone to extremes of weather; the majority of days are dull or very sunny.

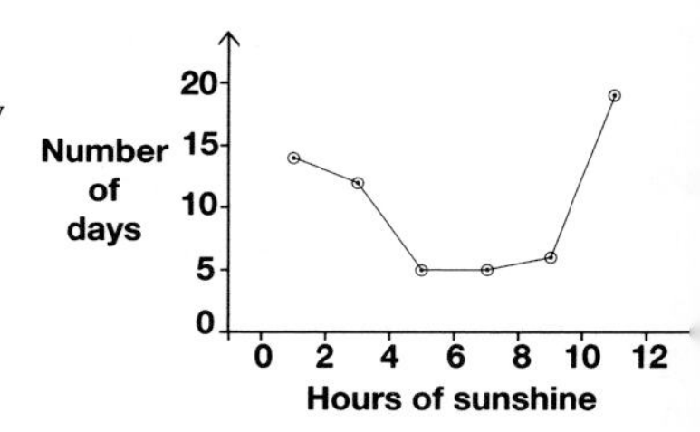

Figure 13.3

(iii) **Table 13.56**

Hours of sunshine	f	x	fx	fx^2
$0 \le x < 2$	14	1	14	14
$2 \le x < 4$	12	3	36	108
$4 \le x < 6$	5	5	25	125
$6 \le x < 8$	5	7	35	245
$8 \le x < 10$	6	9	54	486
$10 \le x < 12$	19	11	209	2299

$\Sigma f = 61$ $\Sigma fx = 373$ $\Sigma fx^2 = 3277$

$$\bar{x} = \frac{\Sigma fx}{\Sigma f} = \frac{373}{61} = 6.1$$

$$s = \sqrt{\left(\frac{\Sigma fx^2}{\Sigma f} - \frac{(\Sigma fx)^2}{(\Sigma f)^2}\right)}$$

$$= \sqrt{\left(\frac{3277}{61} - \frac{(373)^2}{61^2}\right)} = 4.04$$

Therefore the mean hours of sunshine is 6.1 with standard deviation 4.04 hours.

(b) $$\frac{\Sigma x_c}{10} = 16.8 \text{ or } \Sigma x_c = 168$$

where Σx_c denotes the sum of the marks for the complete class.

$$\frac{\Sigma x_m}{6} = 16 \text{ or } \Sigma x_m = 96$$

Hence $\Sigma x_w = 168 - 96 = 72$

Therefore $\bar{x}_w = \frac{\Sigma x_w}{4} = 18.0$

or the mean score of the women is 18.

EXAMPLE 16

(a) A firm employs 200 men. From records in the personnel department each employee's date of birth was obtained, his age calculated and rounded to the nearest year. The frequency distribution in Table 13.57 was obtained.

(i) Write down the boundaries of the modal class.

Construct the corresponding 'less than' cumulative frequency table and hence draw the ogive on graph paper using suitable axes and scales.

From the ogive estimate

(ii) the median of the distribution,
(iii) the interquartile range,
(iv) the minimum age of an employee to whom an offer of early retirement should be made, if the firm wishes to offer early retirement to $7\frac{1}{2}$% of its employees.

Table 13.57

Age (years)	Number of employees
15–19	15
20–24	13
25–29	23
30–34	37
35–39	40
40–44	35
45–49	17
50–54	13
55–59	4
60–64	3

(b) In a village there are 100 families. The number of children in each family was recorded and the results are given in Table 13.58.

Table 13.58

Number of children	0	1	2	3	4	5 or more
Number of families	12	38	25	15	6	4

(i) State why the mean would not be a suitable statistic to use to represent the above data.

(ii) Write down the median and mode of the above distribution.

(N. Ireland Additional Mathematics, 1986)

Solution 16

(a) (i) The modal class is 35–39 and so the boundaries of the modal class are 34.5 and 39.5.

Table 13.59

Age (less than . . .)	Cumulative frequency
19.5	15
24.5	28
29.5	51
34.5	88
39.5	128
44.5	163
49.5	180
54.5	193
59.5	197
64.5	200

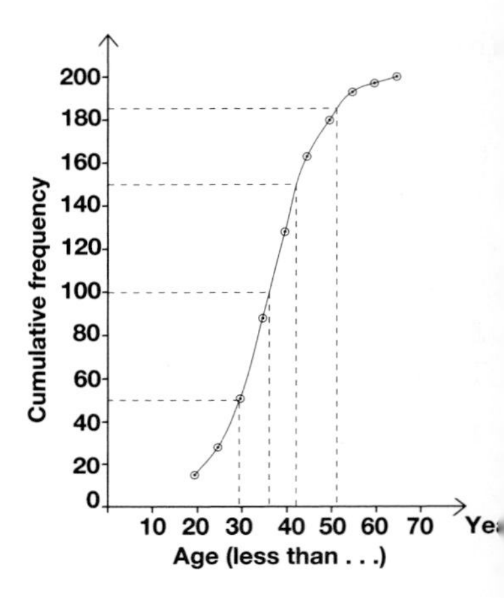

Figure 13.4

(ii) From the ogive in Figure 13.4, the median is approximately 36 years.

(iii) From the ogive

$Q_1 = 29$ approximately,
$Q_3 = 42.5$ approximately,

and so the interquartle range is 13.5 years.

(iv) $7\frac{1}{2}\%$ of $200 = 15$.

Hence the ogive may be used to find a minimum age of 51 years.

(b) (i) The final interval is open since it has no top limit.

(ii) The mode is 1 and the median is calculated by first placing the data in increasing order

$$\underbrace{0\ 0\ 0\ 0\ \ldots\ 0}_{12}\quad \underbrace{1\ 1\ 1\ 1\ \ldots\ 1}_{38}\quad \underbrace{2\ 2\ 2\ldots\ 2\ldots}_{25}$$

As 100 families are involved, the median is the average of the 50th and 51st result.

The median $= \frac{1}{2}(1 + 2) = 1.5$ children

EXAMPLE 17

(a) The lifetimes of 40 electronic diodes have been recorded to the nearest tenth of a year and the frequency distribution in Table 13.60 was obtained.

Construct the corresponding 'less than' cumulative frequency table and hence draw the ogive on graph paper using suitable axes and scales.

From the ogive, estimate

(i) the median,
(ii) the percentage of diodes with recorded lives from 2 years to 3.2 years inclusive,
(iii) the eightieth percentile.

(b) State, briefly, two reasons why the mean is the most widely used measure of central tendency in statistics.

(N. Ireland Additional Mathematics, 1987)

Table 13.60

Life interval (years)	Number of diodes
1.5–1.9	3
2.0–2.4	2
2.5–2.9	4
3.0–3.4	15
3.5–3.9	10
4.0–4.4	6

Solution 17

The cumulative frequency table is given in Table 13.61.

Table 13.61

Life in years (less than . . .)	Cumulative frequency
1.95	3
2.45	5
2.95	9
3.45	24
3.95	34
4.45	40

The ogive corresponding to Table 13.61 is given in Figure 13.5.

(i) From the ogive the median is 3.35 years approximately.

(ii) Since (from Figure 13.5) 3 diodes have lifetimes less than 2 years and 14 have lifetime less than 3.2 years, it follows that 11 diodes have lifetimes between these two values or

$\frac{11}{40} \times 100\% = 27.5\%$ of diodes have lifetimes in this range.

(iii) From the ogive the 80th percentile (the age below which 80% of the distribution lies) is 3.8 years.

(b) The mean is the most widely used measure of central tendency because

(i) it is well understood by many people and
(ii) it makes use of all the data.

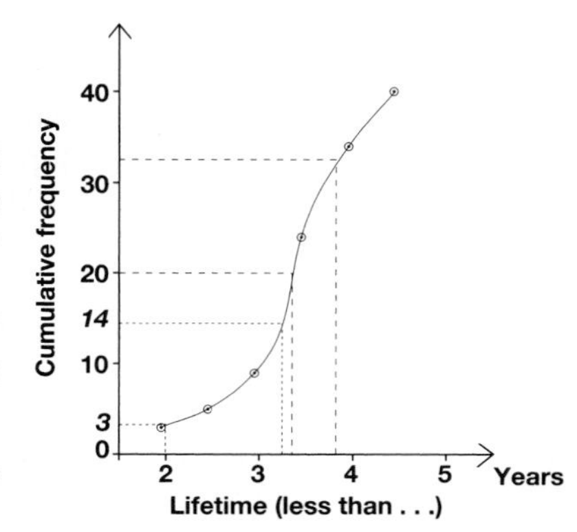

Figure 13.5

EXERCISE 13.9

1. (a) Table 13.62 gives the distribution of marks for 50 candidates who sat an examination in which the maximum possible mark was 150.

Table 13.62

Mark range	0–39	40–59	60–79	80–99	100–139
Number of candidates	5	15	13	10	7

(i) Construct a histogram to show the distribution.

(ii) Calculate the mean and standard deviation of these marks.

(b) Comment, briefly, on the shortcomings of each of the diagrams in Figure 13.6.

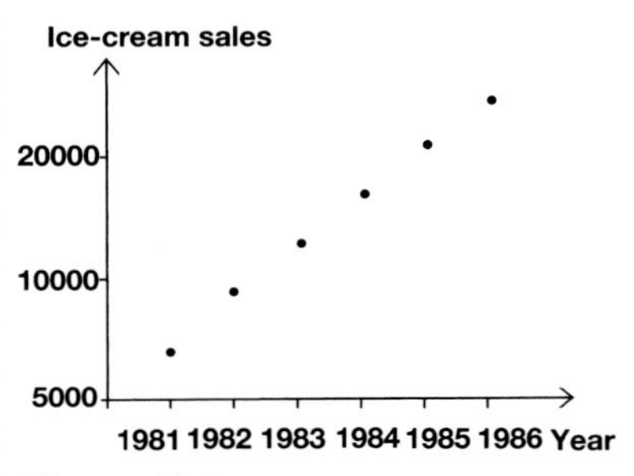

Ice-cream sales

1979 1980 1982 1985 1986 Year

Figure 13.6

(N. Ireland Additional Mathematics, 1988)

2. (a) In a certain town the distribution of female deaths according to age is given as in Table 13.63.

(i) Construct the corresponding 'more than' cumulative frequency table and hence draw the ogive on graph paper using suitable axes and scales.

(ii) Determine the median age at death.

(iii) Use the ogive to estimate the number of new born females who will survive to 65 years of age and hence estimate the proportion of females aged 65 who will survive to age 70.

Table 13.63

Age (x years) at death	Number of families
$0 \leq x < 10$	21
$10 \leq x < 20$	6
$20 \leq x < 30$	9
$30 \leq x < 40$	12
$40 \leq x < 50$	35
$50 \leq x < 60$	103
$60 \leq x < 70$	225
$70 \leq x < 80$	333
$80 \leq x$	256

(b) Describe, briefly, why the range is not often used in practice as a convenient measure of spread.

(N. Ireland Additional Mathematics, 1988)

3. The mean and standard deviation of a set of raw examination marks are 45 and 15 respectively. Teacher A scales the raw marks by subtracting 10 from each mark.

(i) Calculate the mean and standard deviation of the scaled marks.

Teacher B scales the raw marks by dividing each by 5.

(ii) Calculate the mean and standard deviation of the scaled marks.

(N. Ireland Additional Mathematics, 1989)

4. A random sample of 100 students was chosen from a school. Each student's blood pressure was measured to the nearest millimetre. The results are summarised in Table 13.64.

Table 13.64

Blood pressure (mm)	55–59	60–64	65–69	70–74	75–79	80–84	85–89
Number of students	1	3	8	17	30	25	16

Construct the corresponding 'less than' cumulative frequency table and hence draw the ogive on graph paper using suitable axes and scales.

(a) From the ogive estimate

(i) the median blood pressure,
(ii) the interquartile range,
(iii) the percentage of students with blood pressures between 67 mm and 76 mm.

(b) Find, by calculation, the median, giving your answer correct to one place of decimals.

(N. Ireland Additional Mathematics, 1989)

5. (a) In an archery competition the following scores were recorded for an archer: 10, 20, 30, 40 and 50.

Later it was found that the score 40 should have been recorded as 50. State a measure of central tendency which would be

(i) affected by the change,
(ii) unaffected by the change.

(b) The mean of a set of lengths is x. If each of these lengths is halved and then increased by 2, find, in terms of x, the mean of the new set of lengths.

(N. Ireland Additional Mathematics, 1990)

6. The distribution of the number of goals scored per match by a particular football player during last season was as follows.

Table 13.65

Number of goals	0	1	2	3	4 or more
Number of matches	16	18	4	2	0

Determine the mean and standard deviation of the number of goals scored per match.

(N. Ireland Additional Mathematics, 1990)

7. Seven students recorded the marks which they obtained in an intelligence test. Their mean mark was 65 and the standard deviation was 10. An eighth student obtained a mark of 65. Calculate the mean and standard deviation of the marks of all eight students.

(N. Ireland Additional Mathematics, 1991)

8. The increases in monthly salaries of 200 employees in a certain shoe company were recorded to the nearest pound. The following frequency distribution was obtained.

Table 13.66

Salary increase (£)	10–29	30–49	50–69	70–89	90–109	110–129	130–149
Frequency	8	13	21	85	62	9	2

Construct the corresponding 'greater than' cumulative frequency table. Hence draw the 'greater than' ogive on graph paper using suitable axes and scales.

From the ogive determine

(i) the median of the distribution,
(ii) the percentage of employees who received an increase of more than £45 per month,
(iii) the probability of an employee selected at random having received a monthly increase of less than £95.

An employee is selected at random from those who received an increase of less than £95.

(iv) Calculate the probability that this employee received an increase of more than £45.

(N. Ireland Additional Mathematics, 1991)

9. (a) In 1985 the mean length of service of the employees in a factory was 13.6 years with a standard deviation of 4.2 years. Three years later the factory closed and all of these employees were paid a redundancy payment of £200 for every year's service. Calculate

(i) the mean payment paid to these employees,
(ii) the standard deviation of payments made.

(b) Six positive numbers are such that the sum of the squares of the numbers is 390 and the standard deviation is 4. Calculate the mean of the numbers.

(c) In a game, a group of players obtained scores as shown in Table 13.67.

Table 13.67

Score	4	5	6	7	8
Number of players	2	4	9	7	x

Given that the mean score is 6.5, find the value of x.

(d) The lengths of stay, to the nearest minute, of cars in a car park were recorded and grouped as shown in Table 13.68.

Table 13.68

Length of stay (min)	20–39	40–49	50–59	60–79	80–119
Number of cars	18	38	46	50	64

Draw a histogram to represent this information.

(U.C.L.E.S. Additional Mathematics, November 1991)

10. (a) The lengths of time, measured to the nearest second, of 180 telephone calls made by one department of a large store were recorded and grouped as shown in Table 13.69.

Table 13.69

Length of call (to nearest s)	20–39	40–59	60–79	80–119	120–159	160–199
Number of calls	16	24	48	52	28	12

Draw the cumulative frequency curve.

Use the curve to estimate the interquartile range.

The store charges the department according to the following scale.

Calls under 72 s are charged at 8p.
Calls over 135 s are charged at 25p.
All other calls are charged at 15p.

Use the curve to estimate the total charge to the department.

(b) The ages of 160 members of a dancing club are grouped as shown in Table 13.70.

Table 13.70

Age	40–	50–	60–	70–
Number of members	24	48	60	28

Without drawing the cumulative frequency curve, estimate the median age.

(U.C.L.E.S. Additional Mathematics, November 1991, part question)

11. (a) The masses, measured to the nearest kilogram, of 200 girls were recorded and tabulated as shown below.

Table 13.71

Mass (kg)	46–50	51–55	56–60	61–65	66–70	71–75
Number of girls	20	60	56	35	19	10

Construct the cumulative frequency table for this distribution and draw the cumulative frequency curve.

Use your curve to estimate

(i) the interquartile range,
(ii) the percentage of these girls having a mass greater than 58 kg

(b) A group of 125 children raised money for a charity by sponsored activities. The amount raised by each child was recorded. These amounts, taken to the nearest £, are grouped in Table 13.72.

Table 13.72

Amount raised (£)	1–5	6–10	11–15
Number of children	70	36	19

State the smallest possible amount which may have been raised by one child.

Without drawing a cumulative frequency curve estimate the median amount raised.

Also estimate the mean amount raised and explain briefly why this is larger than the median.

(U.C.L.E.S. Additional Mathematics, June 1992)

CHAPTER 14

Probability

The French mathematician **Pierre Simon Laplace** (1749–1827) was born at Beaumont-en-Auge and moved to Paris at age eighteen, with letters of recommendation to Jean Le Rond D'Alembert. D'Alembert all but ignored the young Laplace but reckoned without his tenacity and ambition. Laplace presented D'Alembert with an original paper on the fundamentals of mechanics which opened the necessary doors and eventually secured him a professorship at the Military School of Paris.

Laplace's *Theorie Analytique des Probabilites*, published in 1812, sets him apart as the mathematician who contributed most to probability theory. His five volume *Mecanique Celeste* was preceded by the famous 'memoir' to the Academy of Science, alerting the mathematical world to his great intellect.

This memoir addressed the so-called 'perturbation' in planetary motion which results from apparent gains in speed of Jupiter and Saturn relative to the other planets. Newton's laws predicted dire consequences for the world and Newton himself felt that only the intervention of God could avert disaster and protect mankind from the cumulative effect of these perturbations. With mathematical reasoning deemed 'the most remarkable ever presented to a scientific society', Laplace reassured the scientific community that the perturbations were not cumulative but periodic and self-correcting. However, he made little of the fact that his analysis was for an ideal solar system, free from tidal friction.

To his peers who reviewed his mathematical works, Laplace had two singularly unendearing qualities. He frequently avoided page upon page of complicated analysis through liberal use of the phrase 'it is easy to see'. Laplace often struggled when asked to provide the missing equations. In *Theorie Analytique des Probabilites* he calls upon the reader to use 'common sense' to deal with material which Augustus De Morgan evaluated as 'very much the most difficult mathematical work I have ever met with'. Furthermore, Laplace rarely acknowledged the work of colleagues in his writings, leaving the reader with the mistaken impression that the ideas of others were the fruits of his own mathematical efforts.

INTRODUCTION

Consider an event E which can happen in h ways out of a total of n. For example, the event 'throwing a fair die and scoring 2' can happen in 1 way out of a possible 6.

Textbooks on probability often refer to h as the 'number of favourable outcomes' while n is referred to as 'the number of possible outcomes'. The probability of occurrence of the event E is defined to be $P(E)$ where

$$\mathbf{P}(E) = \frac{h}{n} = \frac{\textbf{the number of favourable outcomes}}{\textbf{the number of possible outcomes}}$$

There are clearly $n-h$ unfavourable outcomes and so the probability of the event not occurring is

$$P(\sim E) = \frac{n-h}{n} = 1 - \frac{h}{n} = 1 - P(E)$$

where the notation $\sim E$ is used to indicate **not *E***.

It follows therefore that

$$\mathbf{P}(\sim E) + \mathbf{P}(E) = \mathbf{1}$$

When the occurrence of the event ***E* is certain**

$$\mathbf{P}(E) = \mathbf{1} \quad \textbf{and} \quad \mathbf{P}(\sim E) = \mathbf{0}$$

And when there is certainty that event ***E* will not occur**

$$\mathbf{P}(\sim E) = \mathbf{1} \qquad \mathbf{P}(E) = \mathbf{0}$$

EXCLUSIVE, INDEPENDENT AND EXHAUSTIVE EVENTS

Mutually exclusive events

Consider the event C: choosing a club from a deck of cards, the event S: choosing a spade from a deck of cards and the event Q: choosing a queen from a deck of cards. $P(CQ)$ will be used to denote the probability that the card chosen is the queen of clubs i.e. that it is a club *and* a queen. Similarly $P(SQ)$ represents the probability that the chosen card is a spade *and* a queen i.e. the probability that the queen of spades is chosen. Since it is impossible to choose a card which is both a club *and* a spade it follows that $P(CS) = 0$. When the occurrence of any one event excludes the occurrence of the other these events are described as mutually exclusive. The events C and S are mutually exclusive events.

Dependent events

Two cards are to be chosen from a deck of cards. Let C_1 denote the event: choosing a club as the first card and let C_2 denote the event: choosing a club as the second card. Now since there are 13 club cards in a pack of 52

$$P(C_1) = \tfrac{13}{52} = \tfrac{1}{4}$$

However, the probability of choosing a club as the second card is influenced by the choice of a club as the first. There are now only 12 club cards remaining in the pack of 51 cards. Hence

$$P(C_2) = \tfrac{12}{51}$$

Note that $P(C_1) \neq P(C_2)$

$P(C_2)$ is more precisely written $P(C_2|C_1)$ where **$P(C_2|C_1)$ translates as the probability of C_2 given that C_1 has occurred**. The probability of choosing clubs on both occasions is computed by multiplying the probability of a club on the first choice by the probability of a club on the second given that a club had been chosen on the first i.e.

$$P(C_1C_2) = P(C_1)P(C_2|C_1)$$

This equation may be rearranged as

$$\mathbf{P(C_2|C_1) = \frac{P(C_1C_2)}{P(C_1)}} \qquad (1)$$

Independent events

Consider two events E_1 and E_2 which are independent and therefore the probability of E_2 occurring is in no way influenced by the occurrence or non-occurrence of E_1. This state of affairs may be restated: the probability of E_2 occurring is equal to the probability of E_2 occurring given that E_1 has already occurred.

Presented mathematically:

$$\mathbf{P(E_2) = P(E_2|E_1)} \qquad (2)$$

Replacing E for C in equation (1)

$$P(E_2|E_1) = \frac{P(E_1E_2)}{P(E_1)}$$

and substituting this expression for $P(E_2|E_1)$ in equation (2)

$$P(E_2) = \frac{P(E_1E_2)}{P(E_1)}$$

or $\quad P(E_1E_2) = P(E_1)P(E_2)$

Accordingly, the probability of events E_1 and E_2 both occurring is therefore equal to the probability of E_1 occurring multiplied by the probability of E_2 occurring provided E_1 and E_2 are independent events.

Exhaustive events

If at least one of the events E_1 and E_2 *must* happen then these events are described as exhaustive.

This is written mathematically as

$\mathbf{P(E_1 + E_2) = 1}$ where $E_1 + E_2$ symbolises 'E_1 or E_2 (or both)'.

Equation (3) below is extremely valuable in tackling probability questions. It is quoted here without proof although its logic can easily be illustrated through elementary set theory.

$$\mathbf{P(E_1 + E_2) = P(E_1) + P(E_2) - P(E_1E_2)} \qquad (3)$$

This translates as 'the probability of E_1 or E_2 or both occurring equals the probability of E_1 occurring plus the probability of E_2 occurring minus the probability of both E_1 and E_2 occurring'. While equation (3) is merely stated here, the interested reader with a basic grasp of the union and intersection of sets (as represented on a Venn diagram) will find a simple justification for equation (3) in any Advanced Level Statistics text.

It is important to note that, in the case of **mutually exclusive events** where $P(E_1E_2) = 0$, equation (3) becomes

$$\mathbf{P(E_1 + E_2) = P(E_1) + P(E_2)}$$

The important relationships presented so far in this chapter may be summarised.

$$P(E) + P(\sim E) = 1$$

$$P(E_1E_2) = 0 \quad \text{for mutually exclusive events } E_1 \text{ and } E_2$$

$$P(E_1E_2) = P(E_1)P(E_2) \quad \text{for independent events } E_1 \text{ and } E_2$$

$$P(E_1 + E_2) = 1 \quad \text{for exhaustive events } E_1 \text{ and } E_2$$

$$P(E_2|E_1) = \frac{P(E_2E_1)}{P(E_1)}$$

$$P(E_1 + E_2) = P(E_1) + P(E_2) - P(E_1E_2)$$

CONTINGENCY TABLES

The contingency table approach serves as an effective alternative to set theory in tackling probability questions. Consider the event C: the choice of a club from a pack of playing cards and Q: the choice of a queen from the pack. Table 14.1 is drawn up with C and not C (represented '$\sim C$'), Q and $\sim Q$ forming the 'margins'. The positions of the labels can be interchanged i.e. C and $\sim C$ could appear in the locations currently occupied by Q and $\sim Q$.

With reference to Table 14.1 the following probabilities are assigned to positions specified by the numbers (1) to (8): in position (1) the probability of choosing a card which is both a club and a queen (i.e. the queen of clubs) is entered; in (2) the probability of a card which is any club except the queen of clubs; in (3) the probability of a card which is a queen but not the queen of clubs and in (4) the probability of choosing a card which is neither a queen nor a club. In position (5) is entered the probability of a club; in (6) the probability of choosing a non-club card; in (7) the probability of a queen and in (8) the probability of choosing a card other than a queen. The 1, placed in the lower right hand position in Table 14.1, acts as a sum of marginal entries since

Table 14.1

	Q	$\sim Q$	
C	(1)	(2)	(5)
$\sim C$	(3)	(4)	(6)
	(7)	(8)	1

$$P(C) + P(\sim C) = 1$$
$$P(Q) + P(\sim Q) = 1$$

Furthermore, it will be demonstrated below that

entry (1) + entry (3) = entry (7)
entry (2) + entry (4) = entry (8)
entry (1) + entry (2) = entry (5)
and entry (3) + entry (4) = entry (6)

Using the symbolism introduced on page 311, Table 14.1 becomes Table 14.2.

Table 14.2

	Q	$\sim Q$	
C	$P(CQ)$	$P(C\sim Q)$	$P(C)$
$\sim C$	$P(\sim CQ)$	$P(\sim C\sim Q)$	$P(\sim C)$
	$P(Q)$	$P(\sim Q)$	1

The probabilities of Table 14.2 may be computed as follows.

$P(CQ)$ = the probability of choosing the queen of clubs $= \frac{1}{52}$

$P(C\sim Q)$ = the probability of choosing a club card which is not a queen $= \frac{12}{52} = \frac{3}{13}$

Similarly $P(\sim CQ) = \frac{3}{52}$; $P(\sim C\sim Q) = \frac{9}{13}$; $P(C) = \frac{1}{4}$; $P(\sim C) = \frac{3}{4}$; $P(Q) = \frac{4}{52}$ and $P(\sim Q) = \frac{12}{13}$.

Inserting the appropriate numerical values transforms Table 14.2 to Table 14.3 where the reader should note (with reference to the entries of Table 14.3) that

Table 14.3

	Q	$\sim Q$	
C	$\frac{1}{52}$	$\frac{3}{13}$	$\frac{1}{4}$
$\sim C$	$\frac{3}{52}$	$\frac{9}{13}$	$\frac{3}{4}$
	$\frac{4}{52}$	$\frac{12}{13}$	1

$$\tfrac{1}{52} + \tfrac{3}{52} = \tfrac{4}{52}$$
$$\tfrac{3}{13} + \tfrac{9}{13} = \tfrac{12}{13}$$
$$\tfrac{1}{52} + \tfrac{3}{13} = \tfrac{1}{4}$$
and $\frac{3}{52} + \frac{9}{13} = \frac{3}{4}$

The application of contingency tables to the solution of problems in probability will be illustrated in the examples which follow.

EXAMPLE 1

The independent events S and T are such that $P(S) = 0.4$ and $P(S + T) = 0.7$. Calculate $P(T)$.

Solution 1

Since the events are independent

$$P(ST) = P(S)P(T) \Rightarrow P(ST) = 0.4P(T)$$

But equation (3) yields

$$P(S + T) = P(S) + P(T) - P(ST)$$

therefore $0.7 = 0.4 + P(T) - 0.4P(T) \Rightarrow 0.3 = 0.6P(T)$

i.e. $P(T) = 0.5$

EXAMPLE 2

The mutually exclusive events S and T are such that $P(S) = 0.4$ and $P(S + T) = 0.7$. Calculate $P(T)$.

Solution 2

Since S and T are mutually exclusive $P(ST) = 0$

or $P(S + T) = P(S) + P(T) - 0$ from equation (3)

therefore $0.7 = 0.4 + P(T) \Rightarrow P(T) = 0.3$.

EXAMPLE 3

Two events L and M are such that $P(L) = 0.2$, $P(M) = 0.3$ and $P(L + M) = 0.4$. Calculate

(i) $P(LM)$, (ii) $P(M|\sim L)$.

Solution 3

(i) Since $P(L + M) = P(L) + P(M) - P(LM)$

then $0.4 = 0.2 + 0.3 - P(LM)$

or $P(LM) = 0.1$

(ii) The contingency Table 14.4 may be partially completed from the information to hand thus far – see Table 14.5.

Using the properties of contingency tables cited earlier in this section, Table 14.5 may be completed as Table 14.6.

Now $P(M|\sim L)$ is required. Using equation (1)

$$P(M|\sim L) = \frac{P(M \sim L)}{P(\sim L)}$$

and from Table 14.6 this becomes

$$P(M|\sim L) = \tfrac{0.2}{0.8} = 0.25$$

Table 14.4

	L	$\sim L$	
M	?	?	
$\sim M$	?	?	
			1

Table 14.5

	L	$\sim L$	
M	0.1	?	0.3
$\sim M$	?	?	?
	0.2	?	1

Table 14.6

	L	$\sim L$	
M	0.1	0.2	0.3
$\sim M$	0.1	0.6	0.7
	0.2	0.8	1

EXAMPLE 4

The events A and B are such that $P(A) = \frac{1}{2}$, $P(\sim A|B) = \frac{1}{3}$, $P(A + B) = \frac{3}{5}$. Determine $P(B|\sim A)$, $P(BA)$ and $P(A|\sim B)$.

The event C is independent of A and $P(AC) = \frac{1}{8}$. State, with a reason in each case, whether

(a) A and B are independent,
(b) A and C are mutually exclusive.

Solution 4

Let $P(B) = x$. Since

$$P(A + B) = P(A) + P(B) - P(AB)$$

then $\quad 0.6 = 0.5 + x - P(AB)$

or $\quad P(AB) = x - 0.1$

The information available so far is summarised in contingency Table 14.7. Completing this table in the usual way produces Table 14.8.

Table 14.7

	A	$\sim A$	
B	$x - 0.1$	?	x
$\sim B$	?	?	?
	0.5	0.5	1

Table 14.8

	A	$\sim A$	
B	$x - 0.1$	0.1	x
$\sim B$	$0.6 - x$	0.4	$1 - x$
	0.5	0.5	1

No use has yet been made of the information: $P(\sim A|B) = \frac{1}{3}$. From equation (1)

$$P(\sim A|B) = \frac{P(\sim AB)}{P(B)}$$

and using Table 14.8, $P(\sim A|B) = \frac{1}{3}$ becomes

$$\frac{0.1}{x} = \frac{1}{3}$$

or $\quad x = 0.3$

Inserting $x = 0.3$ in Table 14.8 above results in Table 14.9.

Table 14.9

	A	$\sim A$	
B	0.2	0.1	0.3
$\sim B$	0.3	0.4	0.7
	0.5	0.5	1

The three probabilities referred to in the first part of the question may now be computed.

$$P(B|\sim A) = \frac{P(B\sim A)}{P(\sim A)} = \frac{0.1}{0.5} = 0.2$$

$$P(BA) = 0.2$$

$$P(A|\sim B) = \frac{P(A\sim B)}{P(\sim B)} = \frac{0.3}{0.7}$$

$$= \frac{3}{7}$$

Turning now to the final part of the question, the independence of A and C implies

$$P(AC) = P(A)P(C)$$
hence $\frac{1}{8} = \frac{1}{2}P(C)$
or $P(C) = \frac{1}{4}$

(a) If A and B are independent then $P(AB) = P(A)P(B)$.
But $P(AB) = 0.2$ and $P(A)P(B) = (0.5)(0.3) = 0.15$
and so A and B are not independent.

(b) If A and C are mutually exclusive then $P(AC) = 0$.
But $P(AC) = \frac{1}{8}$ and so A and C are not mutually exclusive.

EXERCISE 14.1

1. In each case below calculate $P(A + B)$, $P(A|B)$ and $P(B|A)$ using a contingency table.

(i) $P(AB) = 0.5$, $P(B) = 0.8$, $P(\sim A) = 0.4$
(ii) $P(AB) = 0.45$, $P(B) = 0.7$, $P(A) = 0.62$
(iii) $P(B \sim A) = 0.27$, $P(A) = 0.62$, $P(\sim B) = 0.18$
(iv) $P(A \sim B) = 0.12$, $P(B \sim A) = 0.32$, $P(B) = 0.8$
(v) $P(A \sim B) = 0.24$, $P(A) = 0.4$, $P(B) = 0.4$
(vi) $P(B) = 0.8$, $P(A \sim B) = 0.16$, $P(A) = 0.3$
(vii) $P(B) = 0.6$, $P(\sim A \sim B) = 0.21$, $P(\sim A) = 0.3$
(viii) $P(B) = 0.3$, $P(\sim A \sim B) = 0.37$, $P(\sim A) = 0.5$
(ix) $P(AB) = 0.21$, $P(B) = 0.27$, $P(A \sim B) = 0.24$
(x) $P(AB) = 0.2$, $P(B) = 0.32$, $P(A \sim B) = 0.24$

2. In each case below find $P(A + B)$, $P(\sim A + B)$, $P(A| \sim B)$.

(i) $P(AB) = 0.51$, $P(B) = 0.9$, $P(\sim A) = 0.48$
(ii) $P(AB) = 0.43$, $P(B) = 0.68$, $P(A) = 0.6$
(iii) $P(B \sim A) = 0.25$, $P(A) = 0.6$, $P(\sim B) = 0.16$
(iv) $P(B \sim A) = 0.33$, $P(A \sim B) = 0.11$, $P(B) = 0.79$
(v) $P(B) = 0.43$, $P(A \sim B) = 0.27$, $P(A) = 0.43$
(vi) $P(B) = 0.72$, $P(AB) = 0.47$, $P(A) = 0.64$
(vii) $P(B) = 0.47$, $P(A \sim B) = 0.23$, $P(A) = 0.37$
(viii) $P(AB) = 0.22$, $P(B) = 0.34$, $P(A \sim B) = 0.26$
(ix) $P(AB) = 0.17$, $P(B) = 0.23$, $P(A \sim B) = 0.2$
(x) $P(B \sim A) = 0.28$, $P(\sim B) = 0.19$, $P(A) = 0.63$
(xi) $P(A \sim B) = 0.13$, $P(B) = 0.77$, $P(A) = 0.27$
(xii) $P(\sim A \sim B) = 0.35$, $P(B) = 0.28$, $P(\sim A) = 0.48$
(xiii) $P(B) = 0.38$, $P(A) = 0.38$, $P(A \sim B) = 0.22$
(xiv) $P(B \sim A) = 0.3$, $P(B) = 0.78$, $P(A \sim B) = 0.1$
(xv) $P(AB) = 0.46$, $P(B) = 0.71$, $P(A) = 0.63$

TREE DIAGRAMS

In some cases probability questions are best resolved using 'tree diagrams'. The role of such diagrams in probability will be illustrated through a series of examples.

EXAMPLE 5

Ann, Betty and Carol are three secretaries working in an office. Each morning they drink either tea or coffee. Ann and Betty choose independently what to drink. The probability of Ann choosing coffee is $\frac{2}{3}$ and the probability of Betty choosing coffee is $\frac{3}{4}$.

Calculate the probability that

(i) Ann and Betty both have coffee,
(ii) Ann and Betty both have tea,
(iii) Ann and Betty have different drinks.

Carol always waits to see what Ann and Betty are drinking before deciding what she will have. If Ann and Betty both drink coffee or have different drinks then the probability that Carol will have coffee is $\frac{3}{5}$, whereas if Ann and Betty both drink tea then the probability that Carol will have tea is $\frac{4}{5}$.

(iv) Calculate the probability that the three girls will not all be having the same drink.
(v) Show that the probability that Carol will have coffee on any particular morning is $\frac{17}{30}$, and hence,
(vi) Show that in a working year of 240 days almost twice as much coffee as tea is likely to be consumed by the three girls.

(N. Ireland Additional Mathematics, 1986)

Solution 5

Since Ann and Betty choose independently

(i) $P(A_c B_c) = P(A_c)P(B_c) = \frac{2}{3} \times \frac{3}{4} = \frac{1}{2}$

where A_c, for example, denotes the event 'Ann has coffee'.

(ii) $P(A_T B_T) = P(A_T)P(B_T) = (1 - \frac{2}{3})(1 - \frac{3}{4}) = \frac{1}{12}$

(iii) P (Ann and Betty have different drinks)

$= P(A_c B_T) + P(A_T B_c)$

since these two arrangements are mutually exclusive.

Hence the probability that Ann and Betty have different drinks is

$\frac{2}{3} \times \frac{1}{4} + \frac{1}{3} \times \frac{3}{4} = \frac{5}{12}$.

To illustrate the rationale behind tree diagrams, consider Figure 14.1 showing the probability of the various choices made by Ann and Betty.

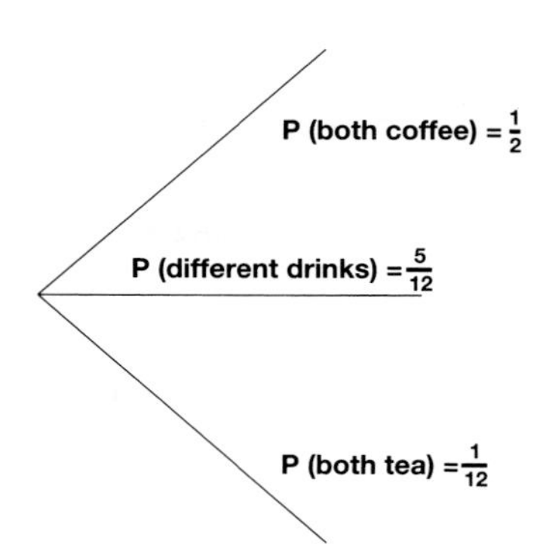

Figure 14.1

The probability that Carol chooses a particular drink is dependent upon the decisions of Ann and Betty. This dependence is captured in the tree diagram of Figure 14.2 where P(diff) represents the probability of Ann and Betty having different drinks.

The reader is reminded that $P(C_c|A_cB_c)$ translates as 'the probability that Carol will have coffee given that Ann and Betty have both had coffee'.

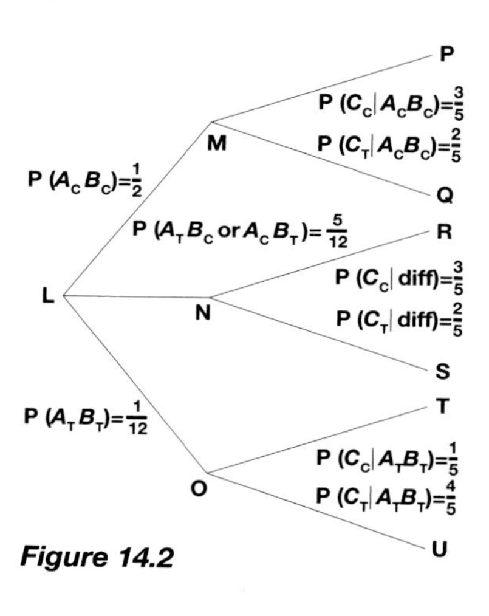

Figure 14.2

The path LMP on the tree of Figure 14.2, for example, corresponds to the situation where Ann and Betty have coffee and then Carol decides to have coffee also, i.e. the event: all three secretaries have coffee.

Now $P(A_cB_cC_c) = P(A_cB_c)P(C_c|A_cB_c)$

and it is therefore clear that the probability of all the secretaries having coffee may be computed by multiplying the probabilities associated with 'branches' LM and MP. Similarly the probability of Ann and Betty having different beverages while Carol has tea is represented by the route LNS through the tree. In probability notation

P (Ann and Betty have different drinks and Carol has tea)
$= P(\text{diff})\ P(C_T|\text{diff})$

and therefore this probability is $\frac{5}{12} \times \frac{2}{5} = \frac{1}{6}$.

Finally, to calculate the probability that Carol has coffee, for example, three routes are relevant; LMP, LNR and LOT since all of these 'end' with Carol choosing coffee. Since these three routes are mutually exclusive then

$$P(C_c) = P(A_cB_c)P(C_c|A_cB_c) + P(\text{diff})P(C_c|\text{diff}) + P(A_TB_T)P(C_c|A_TB_T)$$

In terms of the tree diagram this probability is computed as

P(route LM) × P(route MP)
+ P(route LN) × P(route NR)
+ P(route LO) × P(route OT)

we can now return to the question.

(iv) P (the three girls do not have the same drink)

$$= P(A_cB_c)P(C_T|A_cB_c) + P(\text{diff})P(C_c|\text{diff}) + P(\text{diff})P(C_T|\text{diff}) + P(A_TB_T)P(C_c|A_TB_T)$$
$$= (\tfrac{1}{2})(\tfrac{2}{5}) + (\tfrac{5}{12})(\tfrac{3}{5}) + (\tfrac{5}{12})(\tfrac{2}{5}) + (\tfrac{1}{12})(\tfrac{1}{5}) = \tfrac{19}{30}$$

(v) As detailed above

$$P(C_c) = (\tfrac{1}{2})(\tfrac{3}{5}) + (\tfrac{5}{12})(\tfrac{3}{5}) + (\tfrac{1}{12})(\tfrac{1}{5}) = \tfrac{17}{30}$$

(vi) By following each route through the tree

P (all 3 secretaries have coffee) $= \frac{1}{2} \times \frac{3}{5} = \frac{3}{10}$
P (2 have coffee, 1 has tea) $= \frac{1}{2} \times \frac{2}{5} + \frac{5}{12} \times \frac{3}{5} = \frac{9}{20}$
P (2 have tea, 1 has coffee) $= \frac{5}{12} \times \frac{2}{5} + \frac{1}{12} \times \frac{1}{5} = \frac{11}{60}$
P (all 3 secretaries have tea) $= \frac{1}{12} \times \frac{4}{5} = \frac{1}{15}$

In a working year

on $\frac{3}{10} \times 240$ days i.e. 72 days all 3 have coffee
on $\frac{9}{20} \times 240$ days i.e. 108 days 2 have coffee, 1 has tea
on $\frac{11}{60} \times 240$ days i.e. 44 days 1 has coffee, 2 have tea
and on $\frac{1}{15} \times 240$ days i.e. 16 days all 3 have tea

Hence

$72 \times 3 + 108 \times 2 + 44 \times 1 = 476$ cups of coffee and
$108 \times 1 + 44 \times 2 + 16 \times 3 = 244$ cups of tea are consumed.

Since $476 = 2(244)$ approximately, the required result is confirmed.

EXAMPLE 6

If a day is dry then the probability that the next day will be dry is $\frac{4}{5}$. If a day is wet then the probability that next day will be wet is $\frac{3}{5}$.

(i) If Monday is a dry day find the probability that Wednesday will be wet.

If a day is wet then I will go to work by car. If it is dry then I will go to work either by car or bus, and the probability that I will go by car is $\frac{2}{5}$.

Monday is a dry day. Find the probability that

(ii) I will go to work by bus on Tuesday,
(iii) I will go to work by car on Wednesday.
(iv) Show that the probability that I will use the same means of transport on both Tuesday and Wednesday is approximately $\frac{1}{2}$.

(N. Ireland Additional Mathematics, 1984)

Solution 6

(i) The probabilities quoted at the beginning of the question are entered on the tree diagram for the first three days of the week as shown in Figure 14.3.

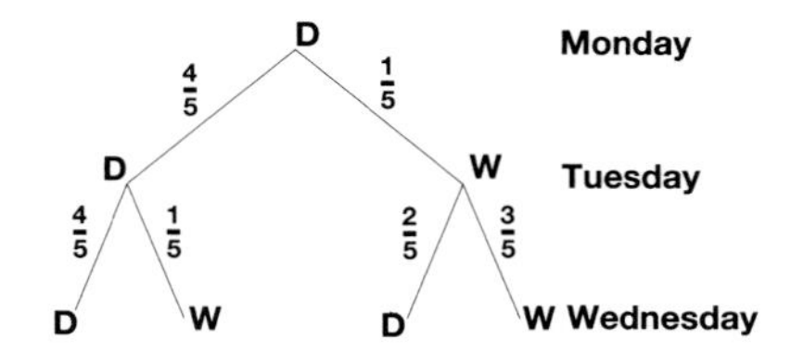

Figure 14.3

By considering routes through the tree diagram which end in Wednesday being wet

$$\text{P (Wednesday wet)} = \tfrac{4}{5} \times \tfrac{1}{5} + \tfrac{1}{5} \times \tfrac{3}{5} = \tfrac{7}{25}$$

(ii) Choice of method of transport may now be incorporated – see Figure 14.4.

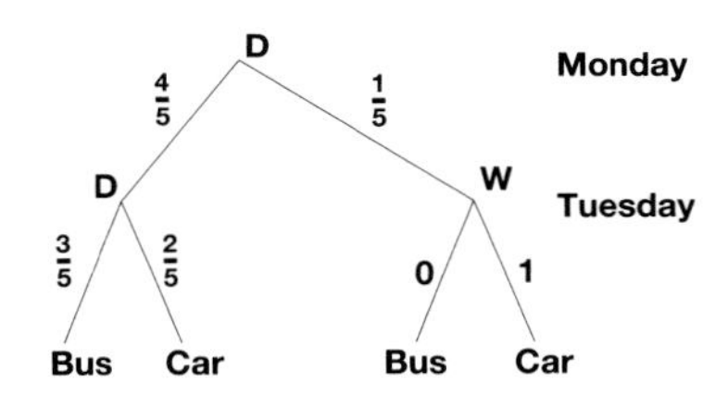

Figure 14.4

It follows that

$$\text{P (I will travel by bus on Tuesday)} = \tfrac{4}{5} \times \tfrac{3}{5} + \tfrac{1}{5} \times 0 = \tfrac{12}{25}$$

(iii) From part (i) the probability that Wednesday is wet is $\tfrac{7}{25}$ and, of course, the probability that it is dry is thus $\tfrac{18}{25}$. The tree diagram of Figure 14.5 may therefore be constructed and so

$$\begin{aligned}\text{P (I will travel by car on Wednesday)} &= \tfrac{7}{25} \times 1 + \tfrac{18}{25} \times \tfrac{2}{5} \\ &= \tfrac{71}{125}.\end{aligned}$$

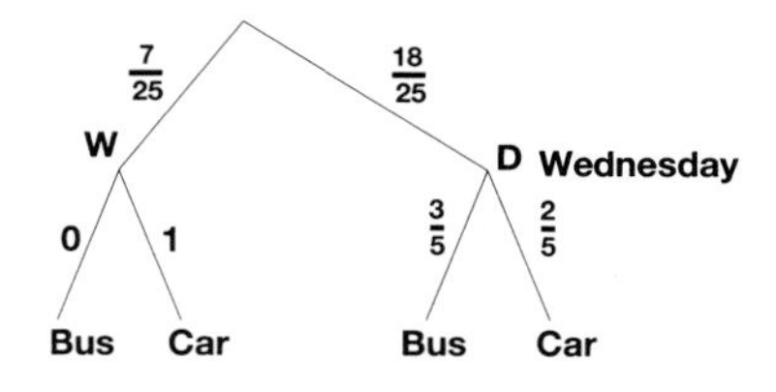

Figure 14.5

(iv) Since the probability of using the bus on Tuesday is $\frac{12}{25}$, it follows that the probability of using the car is $\frac{13}{25}$. Similarly, the probability of travelling by bus on Wednesday is $(1 - \frac{71}{125}) = \frac{54}{125}$.

Using the same means of transport on both days can come about in two ways – using the car on Tuesday and on Wednesday or using the bus on both days. These two events are mutually exclusive, therefore

P (same transport) = P (bus on both days) + P (car on both days).

Hence

$$P\text{ (same transport)} = \frac{12}{25} \times \frac{54}{125} + \frac{13}{25} \times \frac{71}{125}$$
$$= \frac{1571}{3125} = 0.5 \text{ approximately}$$

EXAMPLE 7

When arrested a man who claims to be innocent is given a lie detector test. If he really is guilty the probability that the lie detector will indicate that he is guilty is $\frac{4}{5}$. If he really is innocent the probability that the lie detector will indicate that he is innocent is $\frac{9}{10}$.

Fred is guilty and Bill is innocent, but when arrested both claim to be innocent. Find the probability that the lie detector will indicate that

(i) Bill is guilty,
(ii) both men are guilty,
(iii) exactly one of them is guilty.

If the lie detector indicates that a person is innocent then he is released. If it indicates that he is guilty then he is taken to court. The probability that the court will convict him is $\frac{7}{10}$ if he really is guilty and $\frac{1}{10}$ if he really is innocent.

Dave is guilty and Jack is innocent, but when arrested both claim to be innocent. Find the probability that

(iv) Dave will be convicted,
(v) Jack will be released,
(vi) Dave will be released and Jack will be convicted.

(N. Ireland Additional Mathematics, 1983)

Solution 7

(i) Since Bill is innocent the lie detector will find him guilty with probability $(1 - \frac{9}{10}) = \frac{1}{10}$.

(ii) Since the probability that Fred is found guilty is $\frac{4}{5}$ and their tests are independent then the probability that both men are found guilty is $\frac{1}{10} \times \frac{4}{5} = \frac{2}{25}$.

(iii) There are two possibilities here: Fred is found guilty and Bill innocent or vice-versa.

Now $P(\text{Bill found innocent}) = \frac{9}{10}$ and
$P(\text{Fred found innocent}) = \frac{1}{5}$

and so, once again appealing to the independence of decisions associated with the two men:

$$\begin{aligned} &P(\text{exactly 1 man found guilty}) \\ &= P(\text{Fred guilty})\, P(\text{Bill innocent}) \\ &\quad + P(\text{Bill guilty})\, P(\text{Fred innocent}) \\ &= (\tfrac{4}{5})(\tfrac{9}{10}) + (\tfrac{1}{10})(\tfrac{1}{5}) = \tfrac{37}{50}. \end{aligned}$$

(iv) The information on Dave and Jack may be summarised in the tree diagrams of Figure 14.6.

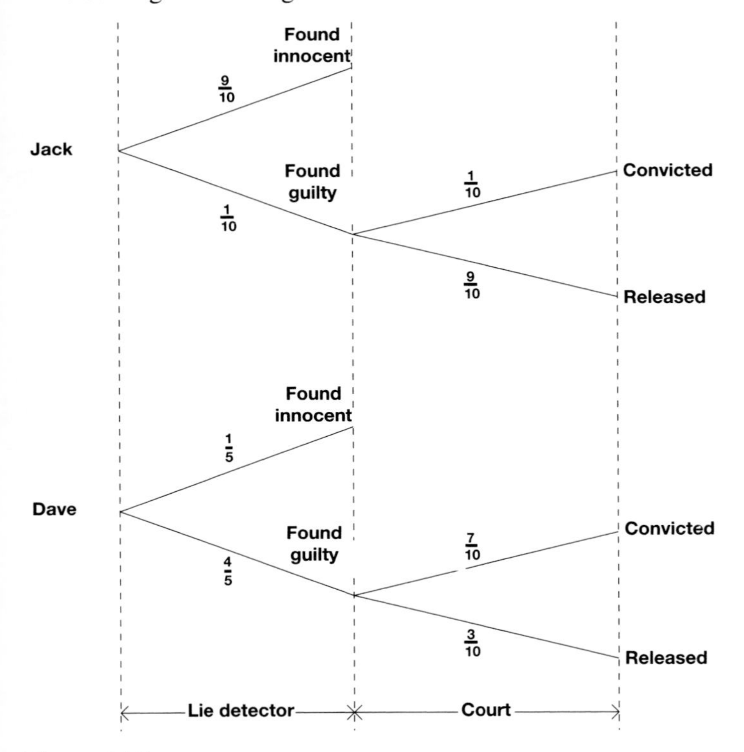

Figure 14.6

With reference to Dave's tree diagram (see Figure 14.6)

$$P(\text{Dave convicted}) = \tfrac{4}{5} \times \tfrac{7}{10} = \tfrac{14}{25}$$

(v) Now Jack can be released in two ways; either he passes the lie detector test or he fails it and is subsequently set free by the court. Hence, from the appropriate routes of Jack's tree diagram (Figure 14.6)

$$P(\text{Jack released}) = \tfrac{9}{10} + \tfrac{1}{10} \times \tfrac{9}{10} = 0.99$$

(vi) Furthermore

$$\text{P(Dave released)} = \tfrac{1}{5} + \tfrac{4}{5} \times \tfrac{3}{10} = \tfrac{11}{25}$$
$$\text{P(Jack convicted)} = \tfrac{1}{10} \times \tfrac{1}{10} = \tfrac{1}{100}.$$

Therefore the probability that Dave will be released and Jack convicted is

$$\tfrac{11}{25} \times \tfrac{1}{100} = \tfrac{11}{2500}$$

EXAMPLE 8

A team believe that they are more likely to win, with probability $\frac{2}{3}$, if it rains, and that if it is dry, then the probability that they will win is $\frac{1}{4}$. The probability that it will rain is $\frac{1}{5}$.

(a) Draw and complete a tree diagram.
(b) Calculate the probability that they will not win.
(c) Given that they won, calculate the probability that it rained.

Solution 8

(a) The information is summarised as shown in the tree diagram of Figure 14.7.

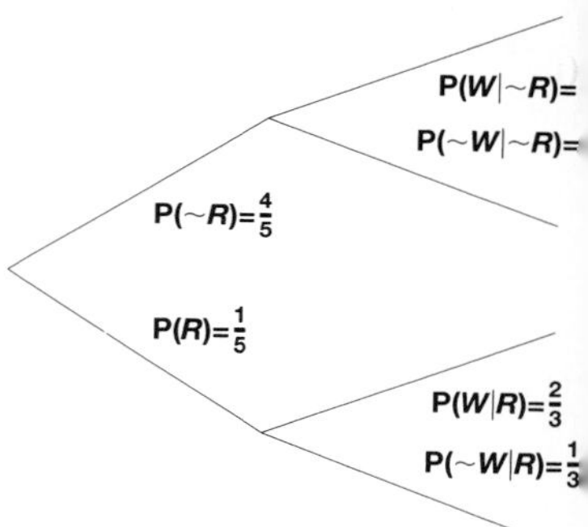

Figure 14.7

(b) The probability that the team will not win is given by

$$\begin{aligned} P(\sim W) &= P(\sim R)P(\sim W|\sim R) + P(R)P(\sim W|R) \\ &= (\tfrac{4}{5})(\tfrac{3}{4}) + (\tfrac{1}{5})(\tfrac{1}{3}) = \tfrac{2}{3} \end{aligned}$$

(c) Now $P(W) = 1 - P(\sim W) = \frac{1}{3}$.

In probability notation the probability that it rained, given that the team won may be written $P(R|W)$.

Since $P(R|W) = \dfrac{P(RW)}{P(W)}$

and, from Figure 14.7

$$P(R|W) = (\tfrac{1}{5})(\tfrac{2}{3})$$

then $P(R|W) = \dfrac{(\frac{1}{5})(\frac{2}{3})}{\frac{1}{3}} = \frac{2}{5}$

EXAMPLE 9

On average I get up early three days out of ten and get up late one day in ten. I forgot something on two out of every five days on which I am late and on one-third of the days when I am early. I do not forget anything on the days when I get up on time. Draw a tree diagram showing all this information.

(i) State the probability that I am

(a) early and forgetful,
(b) late and forgetful,
(c) punctual and forgetful.

(ii) Hence calculate
   (a) the probability of my not forgetting anything,
   (b) the probability that, if I have forgotten something, I got up late.

(iii) Nevertheless confirm that for 80 per cent of the time I am neither forgetful nor late.

Solution 9

Summarising the information given using a common-sense notation

$$\mathrm{P}(E) = \tfrac{3}{10} \qquad \mathrm{P}(L) = \tfrac{1}{10} \qquad \mathrm{P}(P) = \tfrac{6}{10}$$
$$\mathrm{P}(F|L) = \tfrac{2}{5} \qquad \mathrm{P}(F|E) = \tfrac{1}{3} \qquad \mathrm{P}(F|P) = 0$$

facilitates the construction of the tree diagram of Figure 14.8.

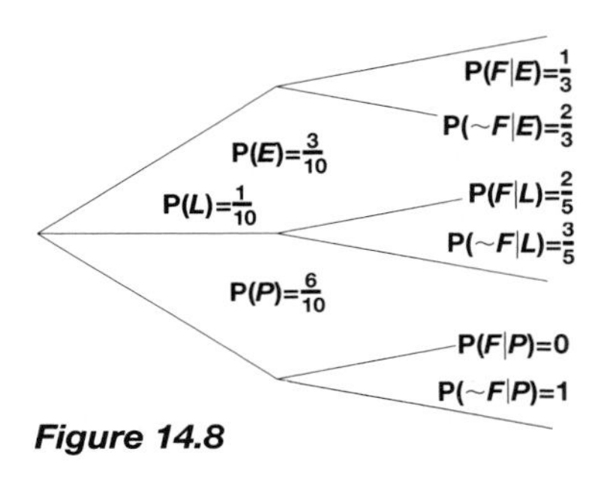

Figure 14.8

(i) Using the notation introduced above
   (a) $\mathrm{P}(EF) = \mathrm{P}(E)\mathrm{P}(F|E) = \frac{3}{10} \times \frac{1}{3} = \frac{1}{10}$
   (b) $\mathrm{P}(LF) = \mathrm{P}(L)\mathrm{P}(F|L) = \frac{1}{10} \times \frac{2}{5} = \frac{1}{25}$
   (c) $\mathrm{P}(PF) = \mathrm{P}(P)\mathrm{P}(F|P) = 0$

(ii) (a)
$$\mathrm{P}(\sim F) = \mathrm{P}(L)\mathrm{P}(\sim F|L) + \mathrm{P}(E)\mathrm{P}(\sim F|E) + \mathrm{P}(P)\mathrm{P}(\sim F|P)$$
or $\mathrm{P}(\sim F) = \frac{1}{10} \times \frac{3}{5} + \frac{3}{10} \times \frac{2}{3} + \frac{6}{10} \times 1 = \frac{43}{50}$

   (b) Since (ii) (a) yields $\mathrm{P}(\sim F) = \frac{43}{50}$ then $\mathrm{P}(F) = \frac{7}{50}$ and

$$\mathrm{P}(L|F) = \frac{\mathrm{P}(LF)}{\mathrm{P}(F)}$$

   leads to

$$\mathrm{P}(L|F) = \frac{\frac{1}{10} \times \frac{2}{5}}{\frac{7}{50}} = \frac{2}{7}$$

   where the product in the numerator is calculated using the tree diagram of Figure 14.8.

(iii) Since the probability of not being late is equal to the probability of either being early or punctual then

$$\mathrm{P}(\sim F \sim L) = \mathrm{P}(\sim FE) + \mathrm{P}(\sim FP)$$
$$= \tfrac{3}{10} \times \tfrac{2}{3} + \tfrac{6}{10} = 0.80$$

EXERCISE 14.2

1. Of the employees of a large factory, $\frac{1}{8}$ travel to work by bus, $\frac{3}{8}$ by train, and the remainder by car. Those travelling by bus have a probability of $\frac{1}{3}$ of being late, those by train will be late with probability $\frac{2}{5}$, and those by car will be late with probability $\frac{3}{4}$.

Draw and complete a tree diagram and calculate the probability that an employee chosen at random will be late.

If an employee is late, calculate the probability that he travelled by car.

(SMP Additional Mathematics, 1982)

2. (a) On a certain route the probability of a bus being on time is $\frac{2}{3}$. Bob finds that if a bus is on time the probability of getting on board is $\frac{1}{2}$, but if the bus is late then the probability of getting on board is reduced to $\frac{1}{4}$.

Find the probability that Bob will

(i) get on the first bus,
(ii) still be waiting after the second bus has passed.

(b) In a small school there are 5 boys and 1 girl in P1 and 4 boys and 2 girls in P2. A pupil is chosen at random from P1 and transferred to P2.

One of the classes is then chosen at random and 2 pupils from that class are randomly selected.

Calculate the probability that the 2 pupils are of opposite sex.

(N. Ireland Additional Mathematics, 1980)

3. It is known that for all first class letters the probability of delivery on the following day is $\frac{4}{5}$. The probability of delivery two days after posting is $\frac{3}{20}$ and the probability of delivery three days after posting is $\frac{1}{20}$.

Similarly, for all second class letters the probability of delivery on the following day is $\frac{1}{10}$, the probability of delivery on the second day is $\frac{3}{5}$, the probability of delivery on the third day is $\frac{1}{5}$ and the probability of delivery on the fourth day is $\frac{1}{10}$.

Two letters are posted on Monday, one by first class post and the other by second class post.

What is the probability that

(i) both will be delivered on Tuesday,
(ii) neither will be delivered before Thursday,
(iii) the letter posted second class will be delivered before the letter posted first class?

On Monday Mr Smith posts a letter to Mr Jones. As soon as he receives it Mr Jones posts a reply. Show that the probability that Mr Smith receives the reply on or before Thursday is almost 7 times greater if first class post is used throughout rather than second class post.

(N. Ireland Additional Mathematics, 1981)

4. A competition in a school produced 10 winners, 8 girls and 2 boys. Two of the winners are to be chosen at random to represent the school in a national competition.

Calculate the probability that

(i) no boys will be chosen,
(ii) one boy will be chosen,
(iii) two boys will be chosen.

The headmaster wishes to ensure that the probability of at least one boy being chosen is a maximum.

Determine which of the following options he should select.

(iv) Choose two pupils at random from the 10 winners.
(v) Divide the winners into two groups of 5 pupils, with one boy in each group, and then choose one pupil at random from each group.

(vi) Divide the winners into two groups of 5 pupils, with the two boys in one of the groups, and then choose one pupil at random from each group.

(N. Ireland Additional Mathematics, 1982)

5. In a provincial town there are two newspapers, the *Gazette* and the *Reporter*. The *Gazette* is read by $\frac{2}{3}$ of the population of the town. Of those who read the *Gazette* $\frac{4}{5}$ read the *Reporter* and of those who do not read the *Gazette* $\frac{3}{5}$ read the *Reporter*.

Two residents of the town, Jack and Jill, are chosen at random. Calculate the probability that

(i) Jack reads both the *Gazette* and the *Reporter*,
(ii) Jill reads the *Reporter*,
(iii) Jack reads just one newspaper,
(iv) Jack and Jill have the same reading habits regarding the number and names of newspapers (if any) read.

(N. Ireland Additional Mathematics, 1985)

6. (a) The two electronic systems C_1, C_2 of a communications satellite operate independently and have probabilities of 0.1 and 0.05 respectively of failing. Find the probability that

(i) neither circuit fails,
(ii) at least one circuit fails,
(iii) exactly one circuit fails.

(b) In a certain boxing competition all fights are either won or lost; draws are not permitted.

If a boxer wins a fight then the probability that he wins his next fight is $\frac{3}{4}$, if he loses a fight the probability of him losing the next fight is $\frac{2}{3}$.

Assuming that he won his last fight, use a tree diagram, or otherwise, to calculate the probability that of his next three fights

(i) he wins exactly two fights,
(ii) he wins at most two fights.

State the most likely and least likely sequence of results for these three fights.

(N. Ireland Additional Mathematics, 1987)

7. In a university degree course, the senior tutor has asked the students for nominations for two vacancies on a committee. Eight women and four men put their names forward for nomination.

If a random selection of two students is made from the 12, calculate the probability that

(i) no women will be chosen,
(ii) exactly one woman will be chosen,
(iii) two women will be chosen.

For each of the following procedures, determine the probability of two women being selected.

(iv) The 12 students are divided into two groups of six with two men in each group. One student is then selected at random from each group.

(v) Two students are selected at random from the 12.

(vi) The 12 students are divided into two groups of six with three men in one group and one in the other. One student is then selected at random from each group.

If the course tutor wishes to maximise the probability of at least one man being selected for the committee state which of the above three procedures she should favour.

Give a reason for your answer.

(N. Ireland Additional Mathematics, 1988)

8. Michael drives his mother's car three times each week, his mother drives it six times each week and no other person drives the car.

On each occasion when Michael is driving, the probability that the car will *not* break down is 0.99. On each occasion when his mother is driving the probability that the car will break down is 0.002. Using a tree diagram, or otherwise, calculate the probability that the car breaks down.

(N. Ireland Additional Mathematics, 1989)

9. Ann, Barbara and Carol are three investment consultants who, at the beginning of each month either advise clients to invest in UK bonds or International bonds. As Ann and Barbara have independent investment strategies, the probability of Ann advising UK bonds is $\frac{3}{4}$, whereas the probability of Barbara advising UK bonds is $\frac{1}{3}$.

Find the probability that

(i) Ann and Barbara both recommend International bonds,
(ii) Ann and Barbara both recommend UK bonds,
(iii) Ann and Barbara recommend different bonds.

The investment consultant, Carol, is influenced by the recommendations of Ann and Barbara. If Ann and Barbara both recommend International bonds then Carol will recommend International bonds to her clients with probability $\frac{2}{3}$. If Ann and Barbara recommend different investment strategies or both recommend UK bonds, Carol will recommend UK bonds with probability $\frac{4}{5}$.

The tree diagram of Figure 14.9 represents the recommendations for Ann, Barbara and Carol.

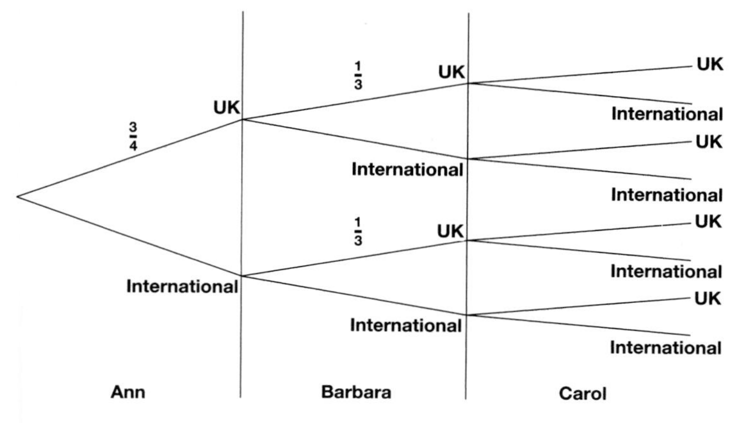

Figure 14.9

Copy and complete the tree diagram by inserting the appropriate missing probabilities.

Calculate the probability that

(iv) the three investment consultants will not all advise the same type of investment,
(v) Carol will recommend International bonds at the beginning of any particular month.

(N. Ireland Additional Mathematics, 1990)

10. In a multiple choice test each question has five given answers, only one of which is correct. A student thinks she knows the correct answer to three-quarters of the questions but, in fact, only 95% of these answers are correct.

For the remaining quarter of the questions she selects the answers at random.

Using a tree diagram, or otherwise, calculate to two decimal places the probability that a question chosen at random was correctly answered by the student.

(N. Ireland Additional Mathematics, 1991)

11. If I roll a red die followed by a green die, what is the probability that the red die will have an even number on it?

(i) (a) Give the probability that my total score will be 9.
(b) If the red die does come up with an even number, show, by enumerating all the possible solutions, that I am no less likely to get a final total of 9.
(c) If my total score turns out to be 9, find the probability that the red die came up with an even number.

(ii) Answer the same three questions as for part (i), but for a total of 8 instead of 9.

Hence confirm that scoring 9 is independent of whether the red score is even or odd (i.e. the probability of scoring 9 is unaltered), but that scoring 8 is not.

(SMP Additional Mathematics, 1976)

12. In a probability model, $P(A) = 0.6$, $P(B) = 0.5$ and $P(A \text{ or } B) = 0.7$.

(i) With the aid of a Venn diagram, or otherwise, calculate

(a) $P(A \text{ and } B)$,
(b) $P(A \text{ and } \sim B)$.

(ii) The tree diagram of Figure 14.10 represents the same model. From your answers to (i) calculate the probabilities x and y.

(SMP Additional Mathematics, 1977)

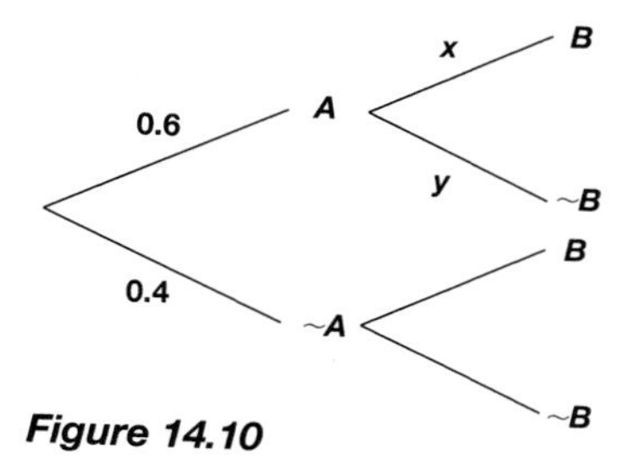

Figure 14.10

13. There are 50 books on my shelf, some paperbacks and the rest hardback, and I pick one a random to take on holiday. The probability that it is a paperback is 0.4, the probability that it is a novel is 0.3, and the probability that it is neither is 0.4.

(i) By using a Venn diagram, or otherwise, find

(a) the probability that it is a paperback novel,
(b) the probability that it is a hardback novel.

(ii) If it is a paperback, find

(a) the probability that it is a novel,
(b) the probability that it is a hardback.

(SMP Additional Mathematics, 1978)

14. $P(A) = 0.5$, $P(B) = 0.7$, $P(A \text{ or } B) = 0.85$; where $P(X)$ means the probability of outcome X.

(a) Calculate $P(A \text{ and not } B)$ and verify that A and B are independent, i.e. tha $P(A \text{ and } B) = P(A) \times P(B)$.

It is also given that $P(C) = 0.6$ and that A and C are independent.

(b) Write down $P(A \text{ and } C)$ and hence find $P(A \text{ or } C)$.

Finally it is given that $P(B \text{ or } C) = 0.84$.

(c) State, with reasons, whether or not B and C are independent.

(d) Show that there is some probability of neither A nor B nor C.

(SMP Additional Mathematics, 1979)

15. In one session in a particular habitat, the probability of seeing a redbreast is 0.8 in winter bu only 0.1 in summer; the probability of observing a bluetit is 0.3 in winter but 0.6 in summer Assume that all these events are independent.

(i) In summer what is the probability of spotting

(a) a redbreast and a bluetit,
(b) a bluetit but no redbreast?

(ii) In winter what is the probability of spotting

(a) a redbreast and a bluetit,
(b) a redbreast or a bluetit?

(iii) If both birds are seen, what is the probability that it is summer?

(iv) If only one species is seen in a summer session, what is the probability that it is a bluetit?

(SMP Additional Mathematics, 1980)

16. (a) In a particular examination, all students in set 1 and set 2 obtained grade A, B or C, where A is the highest grade and C is the lowest. Table 14.10 shows the proportion of students in each set obtaining these grades.

Table 14.10

	Grade A	Grade B	Grade C
Set 1	0.6	0.2	0.2
Set 2	0.2	0.3	0.5

One student is selected at random from each set. Find the probability that

(i) only one of these two students obtained grade A,
(ii) both students obtained the same grade,
(iii) the student from set 1 obtained a higher grade than the student from set 2.

(b) Two events A and B are independent and are such that the probability of A happening is p and the probability of B happening is $2p$. Given that the probability that both A and B happen is 0.08, find

(i) the value of p,
(ii) the probability that neither event happens,
(iii) the probability that only one of A and B happens.

(U.C.L.E.S. Additional Mathematics, November 1991)

17. (a) Seeds from a particular variety of plant produce new plants that have either red, blue or pink flowers and these occur in the ratio 1:6:3 respectively. Three seeds chosen at random are planted in a bowl. Assuming that all three produce flowering plants, find the probability that

(i) all the flowers are the same colour,
(ii) there are exactly two plants with blue flowers.

(b) Three men are visiting a town and agree to meet at the Crown Hotel. There are, however, three hotels with this name in the town and each man picks one of these hotels at random. Find the probability that

(i) each of the three men goes to the same hotel,
(ii) each of the three men goes to a different hotel.

(c) A and B are two events such that the probability of A happening is 0.45 and the probability of B happening is 0.40. Find the probability of neither A nor B happening in the case where

(i) A and B are mutually exclusive,
(ii) A and B are independent.

(U.C.L.E.S. Additional Mathematics, June 1991)

CHAPTER **15**

The normal distribution

The normal distribution curve was developed from a model originally propounded by the German mathematician **Carl Friedrich Gauss** (1777–1855) for dealing with measurement error. Gauss was born in Brunswick, the son of a labourer, and his *Disquisitiones arithmeticae* gave birth to that branch of mathematics now known as 'Number theory'. Despite his immense contribution to mathematical thinking (his diaries reveal that he published only a fraction of his mathematical discoveries), Gauss also found time to calculate the orbit of Ceres, the first 'planetoid', and to invent the electric telegraph in 1833. In writing what is now referred to as 'Pure Mathematics', W. W. R. Ball, in his 1907 *History of Mathematics* (p. 463) contrasts the mathematical *style* of Gauss with those of his two great contemporaries.

The great masters of modern analysis are Lagrange, Laplace and Gauss, who were contemporaries. It is interesting to note the marked contrast in their styles.

Lagrange is perfect both in form and matter, he is careful to explain his procedure, and though his arguments are general they are easy to follow.

Laplace on the other hand explains nothing, is indifferent to style, and, if satisfied that his results are correct, is content to leave them either with no proof or with a faulty one. Gauss is as exact and elegant as Lagrange, but even more difficult to follow than Laplace, for he removes every trace of the analysis by which he reached his results, and studies to give a proof which while rigorous shall be as concise and synthetical as possible.

INTRODUCTION

Many variables have frequency distributions which are 'bell' shaped or 'Gaussian'. Figure 15.1 shows the marks of pupils on a mathematics test.

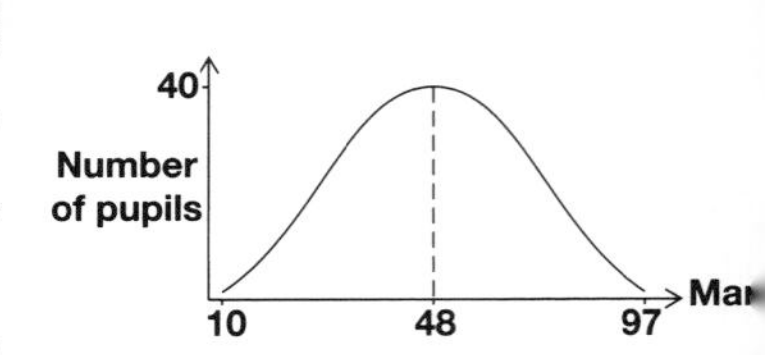

Figure 15.1

Very few pupils have very low or very high scores and the majority of pupils score around 48. If pupil weight were plotted horizontally a similar shape would result. The graph of the heights of pupils in a school would also be Gaussian since very few pupils would be extremely tall, very few extremely short, and the majority of pupils would be of average height. The frequent occurrence of this 'distribution' curve in social and natural science make it worthy of particular study. It is known as the **normal distribution curve**.

PROPERTIES OF THE NORMAL CURVE

If a variable is normally distributed with mean μ and standard deviation σ it can be shown that the normal distribution curve is as shown in Figure 15.2.

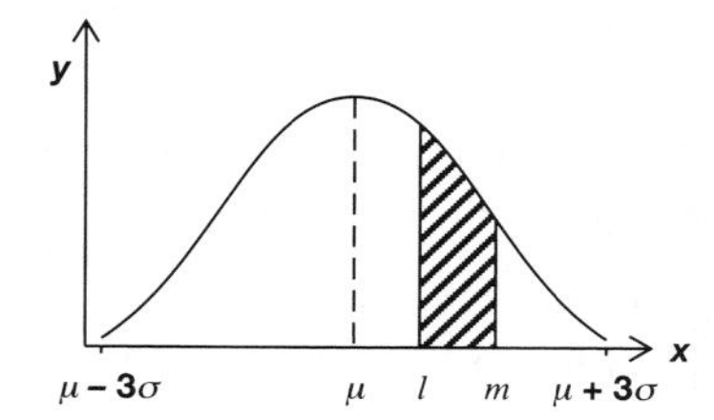

Figure 15.2

The curve is symmetrical about $x = \mu$ (the mean) and approaches very near to the x axis at the points $(\mu - 3\sigma,0)$ and $(\mu + 3\sigma,0)$. The curve never touches the axis; it is said to be 'asymptotic' to the axis. When the y axis represents probability, the area under the complete curve is 1 and the probability that x lies between l and m is given by the area contained by the curve, the x axis and the lines $x = l$ and $x = m$ i.e. the shaded area in Figure 15.2.

CONVERTING FROM RAW VARIABLES TO STANDARD VARIABLES

A raw variable, x, is converted to a standard variable, z, through the equation

$$z = \frac{x - \mu}{\sigma} \qquad (1)$$

The standard variable z represents the 'distance' of x from the mean in standard deviations. To illustrate this suppose $x = \mu + t\sigma$ i.e. x is t standard deviations more than the mean. Using equation (1) the corresponding value of z is

$$z = \frac{\mu + t\sigma - \mu}{\sigma} = t$$

In order to establish the appearance of Figure 15.2 when probability is plotted against z instead of x, $x = \mu - 3\sigma$, $x = \mu$ and $x = \mu + 3\sigma$ are inserted in turn into equation (1). It follows that

$$\begin{array}{lll} & x = \mu - 3\sigma & \text{corresponds to} \quad z = -3 \\ & x = \mu & \text{corresponds to} \quad z = 0 \\ \text{and} & x = \mu + 3\sigma & \text{corresponds to} \quad z = +3 \end{array}$$

Figure 15.2 therefore becomes Figure 15.3. The reader should note that the transformation from Figure 15.2 to Figure 15.3 can also be effected by simply setting $\mu = 0$ and $\sigma = 1$.

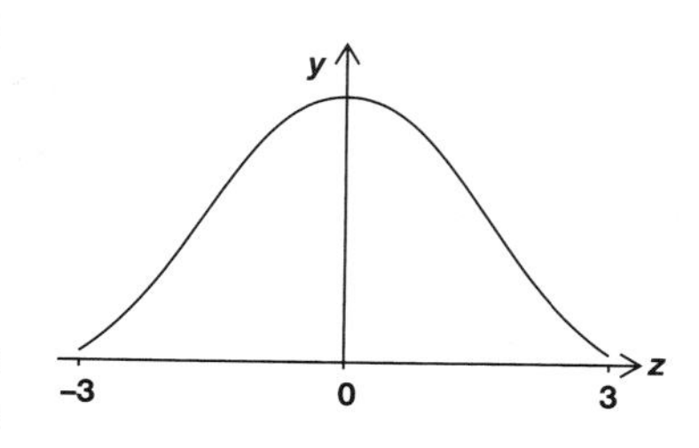

Figure 15.3

It follows that the conversion process from raw variable to standard variable allows the normal distribution curve to be considered as one with mean zero and standard deviation one. This is of great value since areas under the normal curve are tabulated for the curve with mean zero and standard deviation one.

USING 'NORMAL TABLES' TO CALCULATE THE AREA UNDER THE STANDARDISED NORMAL CURVE

Suppose the standardised normal curve is to be used to calculate the probability that z is less than 2, i.e. $P(z < 2)$ is to be found.

This probability may be represented diagrammatically by the shaded area in Figure 15.4.

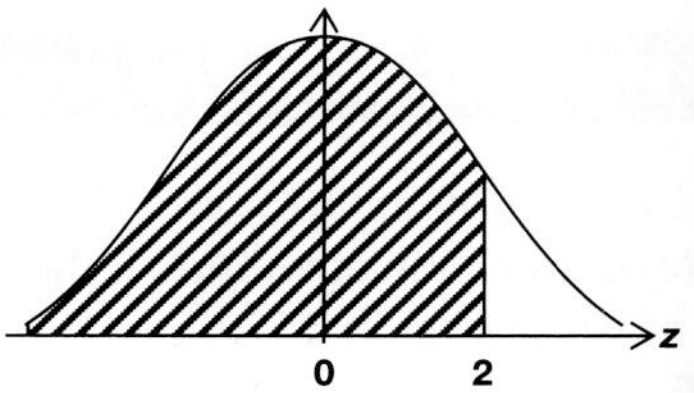

Figure 15.4

Now three figure normal tables consist of tabulated areas, $\mathbf{A}(z)$, for various z values based upon the diagram in Figure 15.5.

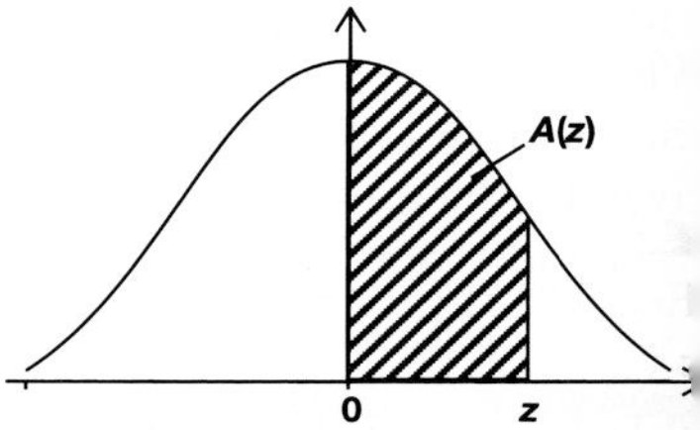

Figure 15.5

To use the tables to compute $P(z < 2)$, the first step is to note that the shaded area in Figure 15.4 is given by

$$A(2) + 0.5$$

(the area to the left of the line $z = 0$ is 0.5 since the total area beneath the curve is 1).

Hence $P(z < 2) = 0.477 + 0.5 = 0.977$

EXAMPLES

EXAMPLE 1

Find the area under the standardised normal curve to the left of $z = 2.14$.

Solution 1

The required area is shown in Figure 15.6.

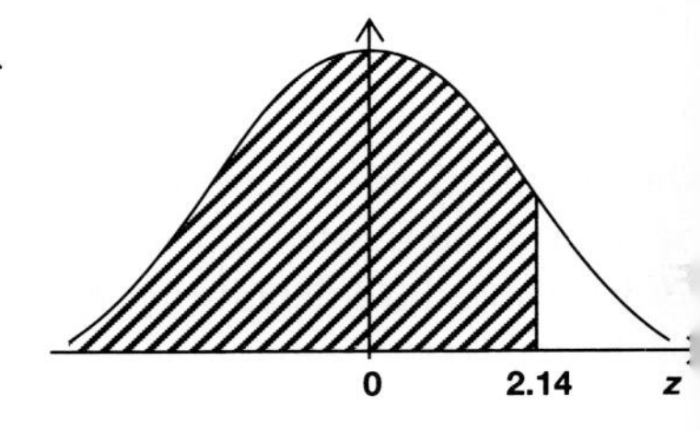

Figure 15.6

$$\text{area} = 0.5 + A(2.14)$$

Hence
$$\text{area} = 0.5 + 0.484$$
$$\text{area} = 0.984$$

EXAMPLE 2

Find the area under the standardised normal curve to the right of $z = 2.37$.

Solution 2

The required area is shown in Figure 15.7.

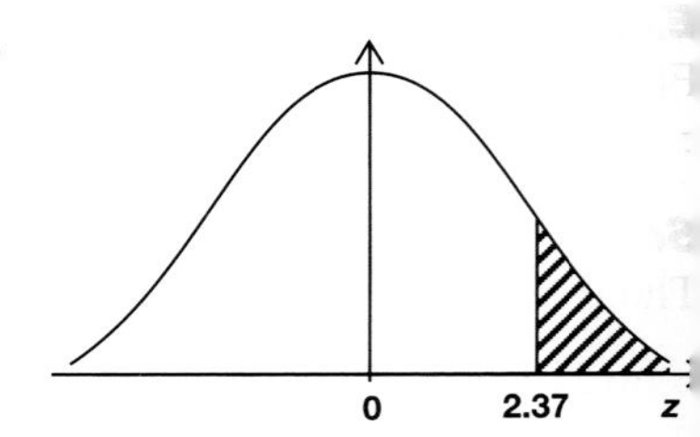

Figure 15.7

$$\begin{aligned}\text{area} &= 0.5 - A(2.37)\\ &= 0.5 - 0.491\\ &= 0.009\end{aligned}$$

EXAMPLE 3

Find the area under the standardised normal curve to the left of $z = -1.25$.

Solution 3

The required area is shown in Figure 15.8.

By symmetry, an equivalent area is shown in Figure 15.9.

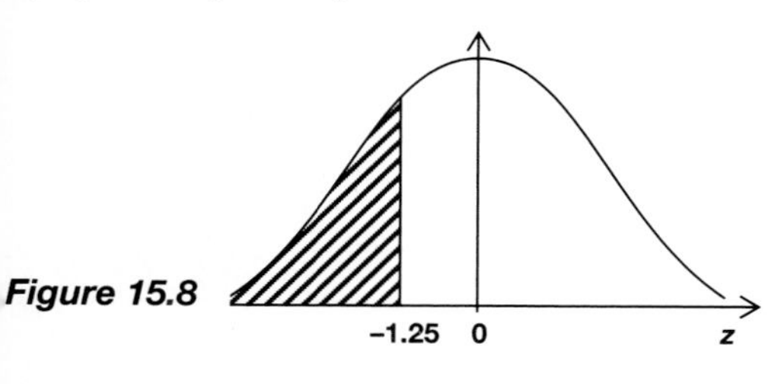

Figure 15.8

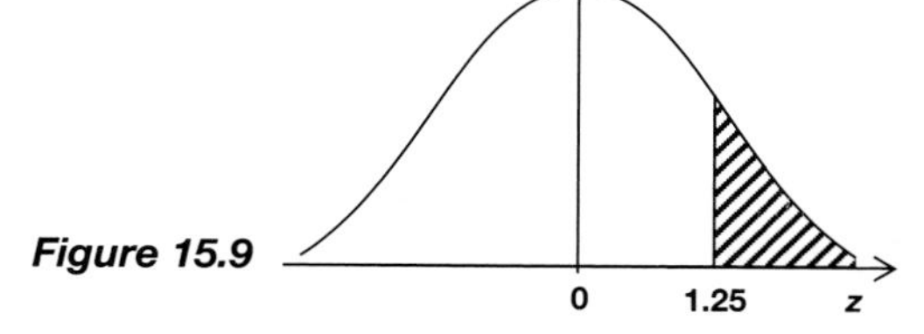

Figure 15.9

$$\begin{aligned} \text{Therefore}\quad \text{area} &= 0.5 - A(1.25) \\ &= 0.5 - 0.394 \\ &= 0.106 \end{aligned}$$

EXAMPLE 4

Find the area under the standardised normal curve to the right of $z = -2$.

Solution 4

The required area is shown in Figure 15.10.

By symmetry, an equivalent area is shown in Figure 15.11.

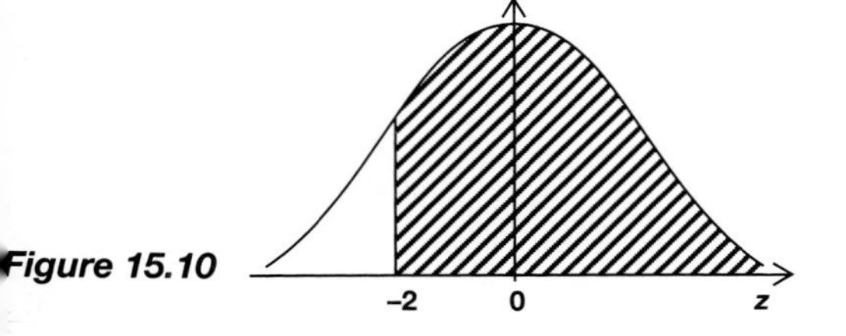

Figure 15.10

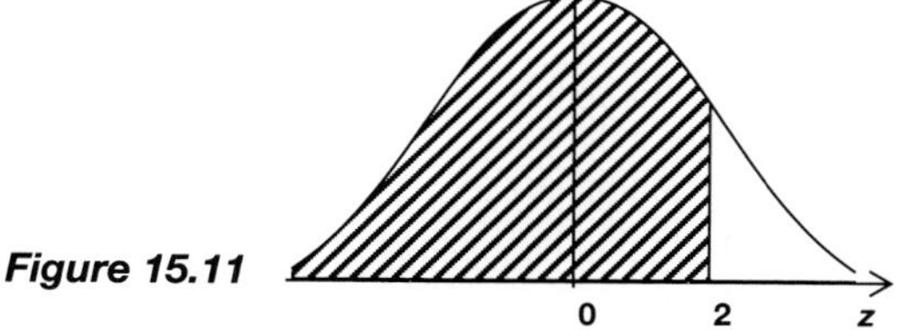

Figure 15.11

$$\begin{aligned} \text{Hence}\quad \text{area} &= 0.5 + A(2) \\ &= 0.5 + 0.477 \\ &= 0.977 \end{aligned}$$

EXAMPLE 5

Find the area under the standardised normal curve and between $z = 0.2$ and $z = 2.81$.

Solution 5

The required area is shown in Figure 15.12.

$$\begin{aligned} \text{area} &= A(2.81) - A(0.2) \\ &= 0.498 - 0.079 \\ &= 0.419 \end{aligned}$$

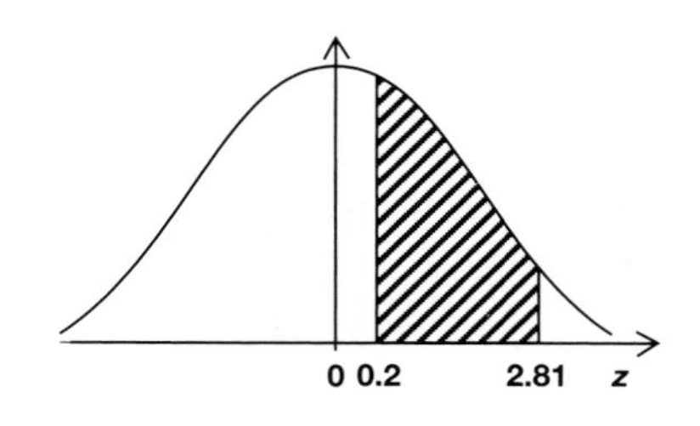

Figure 15.12

EXAMPLE 6

Find the area under the standardised normal curve and between $z = -2.4$ and $z = -0.8$.

Solution 6

The required area is shown in Figure 15.13.

By symmetry, an equivalent area is shown in Figure 15.14.

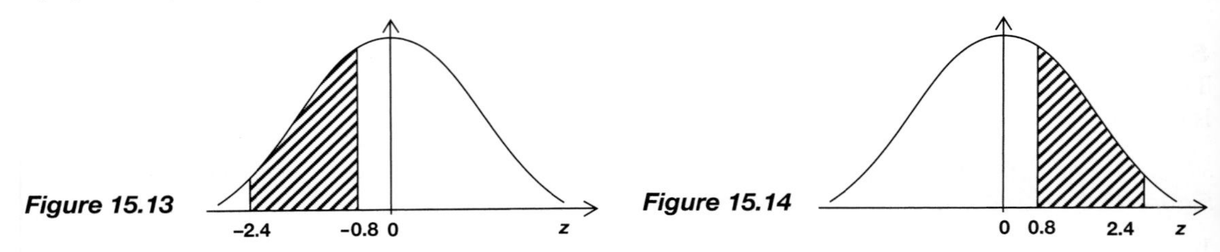

Figure 15.13

Figure 15.14

$$
\begin{aligned}
\text{area} &= A(2.4) - A(0.8) \\
&= 0.492 - 0.288 \\
&= 0.204
\end{aligned}
$$

EXAMPLE 7

Find the area under the standardised normal curve and between $z = -1.27$ and $z = 3$.

Solution 7

The required area is shown in Figure 15.15.

The area is split as in Figure 15.16.

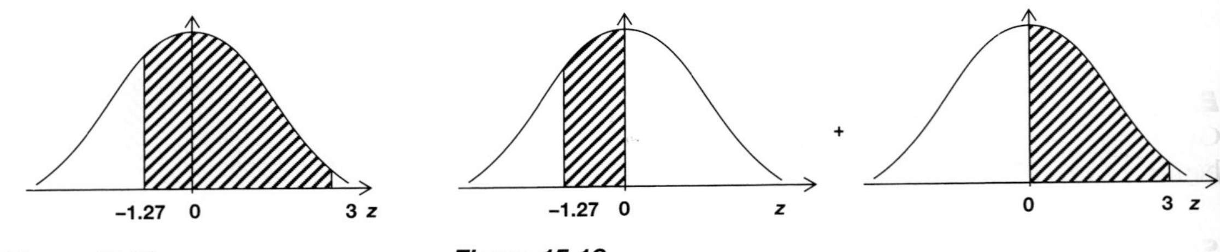

Figure 15.15

Figure 15.16

The curve on the left in Figure 15.16 may be replaced through considerations of symmetry to produce Figure 15.17.

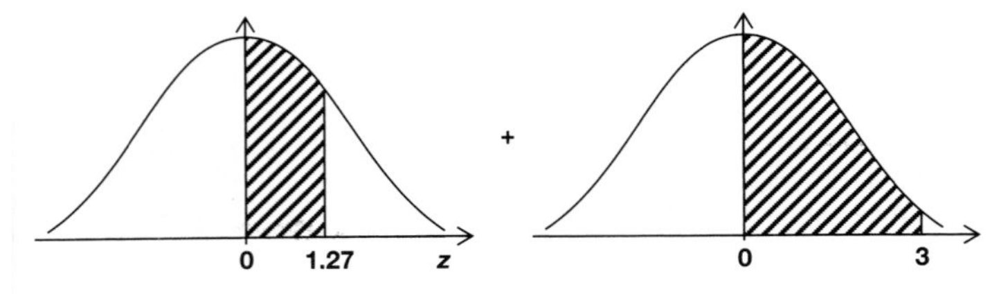

Figure 15.17

Therefore area $= A(1.27) + A(3)$
$= 0.398 + 0.499$
$= 0.897$

EXAMPLE 8

Given that the area under the standardised normal curve to the left of $z = \alpha$ is 0.911. Find α.

Solution 8

The given area is shown in Figure 15.18. Since the total shaded area is 0.911 and the area under the normal curve is 1, it follows that

$A(\alpha) = 0.911 - 0.5$
$A(\alpha) = 0.411$

Using the normal tables 'in reverse'

$\alpha = 1.35$

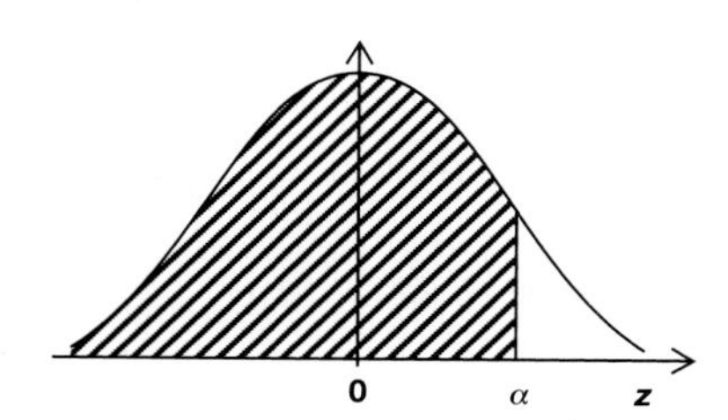

Figure 15.18

EXAMPLE 9

Given that the area under the standardised normal curve and between $z = 0.92$ and $z = \alpha(\alpha > 0.92)$ is 0.03, find α.

Solution 9

The given area is shown in Figure 15.19 since $\alpha > 0.92$.

Now area $= A(\alpha) - A(0.92)$
Hence $0.03 = A(\alpha) - 0.321$
$A(\alpha) = 0.351$
$\alpha = 1.04$ using the tables in reverse.

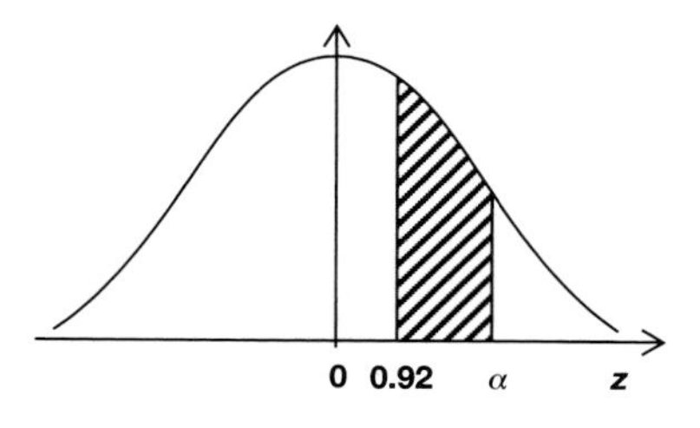

Figure 15.19

EXAMPLE 10

Given that the area under the standardised normal curve and between $z = \alpha$ and $z = 2.17$ is 0.826, find α.

Solution 10

Consider the two possibilities shown in Figure 15.20.

α to the left of $z = 2.17$ and α to the right of $z = 2.17$

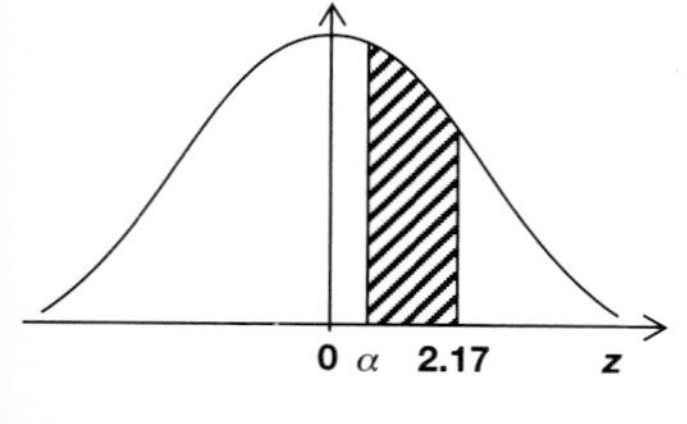

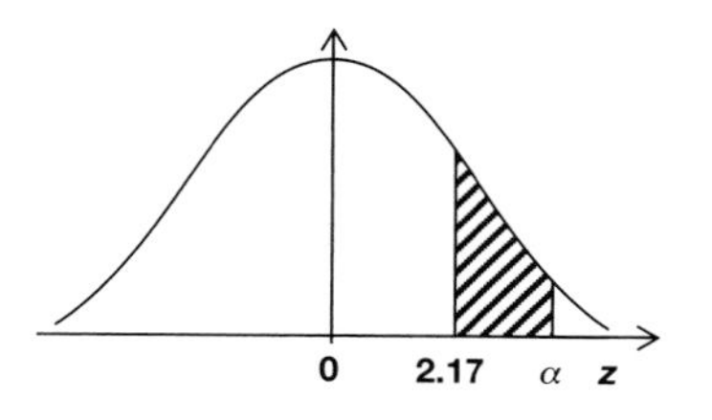

Figure 15.20

The diagram on the right is clearly out of the question since, no matter what the value of α, the shaded area is clearly less than 0.5 and cannot possibly equal 0.826. The diagram on the left is chosen. Now for the shaded area to be 0.826, α a must be negative and a more realistic representation might be that in Figure 15.21.

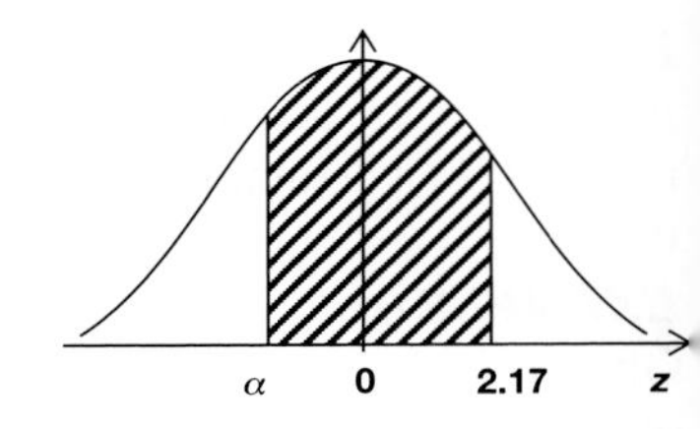

Figure 15.21

The area in Figure 15.21 may be split as in Figure 15.22.

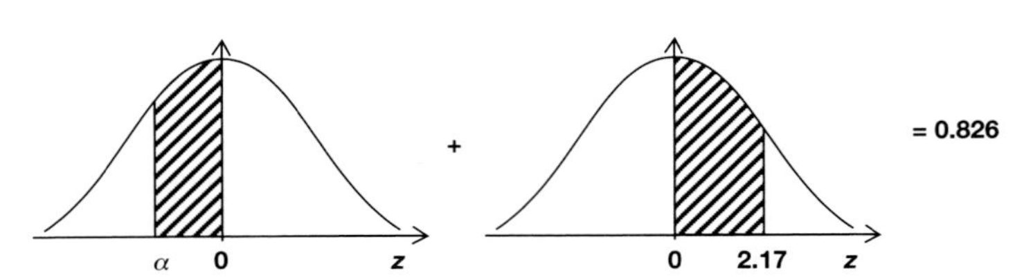

Figure 15.22

Finally, Figure 15.22 may be converted to Figure 15.23.

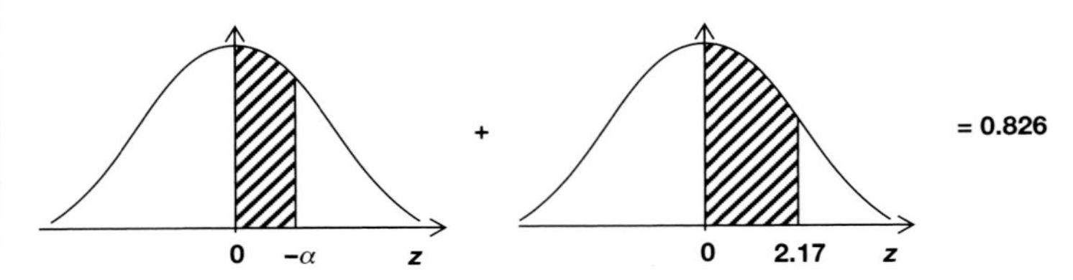

Figure 15.23

$$\begin{aligned} \text{Hence} \quad A(-\alpha) + A(2.17) &= 0.826 \\ A(-\alpha) + 0.485 &= 0.826 \\ A(-\alpha) &= 0.341 \\ -\alpha &= 1 \\ \alpha &= -1 \end{aligned}$$

EXAMPLE 11

On a final examination the mean was 74 and the standard deviation 12. Determine the standard score of a pupil who was awarded a score of 92.

Solution 11

Equation (1) is used to convert raw scores to standard scores.

$$\begin{aligned} \text{Here} \quad \mu &= 74 \\ \sigma &= 12 \\ x &= 92 \end{aligned}$$

$$\text{So} \quad z = \frac{(92 - 74)}{12} = 1.5$$

i.e. 92 is 1.5 standard deviations above the mean:

$92 = 74 + (1.5)(12)$

EXAMPLE 12

A pupil is awarded a standard score of -0.5 in the examination referred to in Example 11. What mark was the pupil awarded on the paper?

Solution 12

$$z = \frac{x - \mu}{\sigma}$$

becomes $\quad -0.5 = \dfrac{x - 74}{12}$

or
$$\begin{aligned} 12(-0.5) &= x - 74 \\ -6 &= x - 74 \\ x &= 68 \end{aligned}$$

Alternatively, this pupil's score is said to be half of one standard deviation below the mean.

i.e.
$$\begin{aligned} x &= 74 - (0.5)(12) \\ x &= 68 \end{aligned}$$

EXAMPLE 13

Two pupils scoring 29 and 52 in their summer English paper were informed by their English teacher that their standard scores on the paper were -3.8 and 0.8 respectively. Find the mean and standard deviation of the English scores.

Solution 13

Using $\quad z = \dfrac{x - \mu}{\sigma} \quad$ for these two pupils

$$-3.8 = \frac{29 - \mu}{\sigma}$$

$$0.8 = \frac{52 - \mu}{\sigma}$$

or
$$\begin{aligned} 29 - \mu &= -3.8\sigma \\ 52 - \mu &= 0.8\sigma \end{aligned}$$

Subtracting
$$\begin{aligned} 29 - 52 &= -3.8\sigma - 0.8\sigma \\ -23 &= -4.6\sigma \\ \sigma &= 5 \end{aligned}$$

But from above $\quad 29 - \mu = -3.8\sigma \quad$ gives
$$\begin{aligned} 29 - \mu &= -3.8(5) \\ \Rightarrow \quad 29 - \mu &= -19 \\ \mu &= 29 + 19 \\ \mu &= 48 \end{aligned}$$

So the mean score on the paper was 48 and the standard deviation was 5.

EXAMPLE 14

(a) An airline offers a continental breakfast on its early morning flights. The percentages of passengers on each flight who accept breakfast are normally distributed with a mean of 75% and a standard deviation of 5%.

(i) On a plane carrying 125 passengers find the probability that fewer than 90 will want breakfast.

(ii) On a plane carrying 150 passengers determine the minimum number of breakfasts which should be carried if the probability that there will not be enough breakfasts on board is to be less than $\frac{1}{100}$.

(b) A rival airline serves hot breakfasts on its flights. On $\frac{2}{5}$ of its flights less than 78% want breakfast while on $\frac{1}{5}$ of its flights more than 87% want breakfast.

Assuming a normal distribution, calculate the mean and standard deviation for the percentage of passengers on each flight who want the hot breakfast.

(N. Ireland Additional Mathematics, 1985)

Solution 14

(a) (i) 90 passengers is $\frac{90}{125} \times 100\% = 72\%$ of passengers

Compute 72 as a standard score.

$$z_1 = \frac{72 - 75}{5} = -0.6$$

Hence, using Figure 15.24, the area between the z_1 ordinate and the $z = 0$ ordinate is

$$A(0.6) = 0.226$$

The shaded area is therefore

$$0.5 - 0.226 = 0.274$$

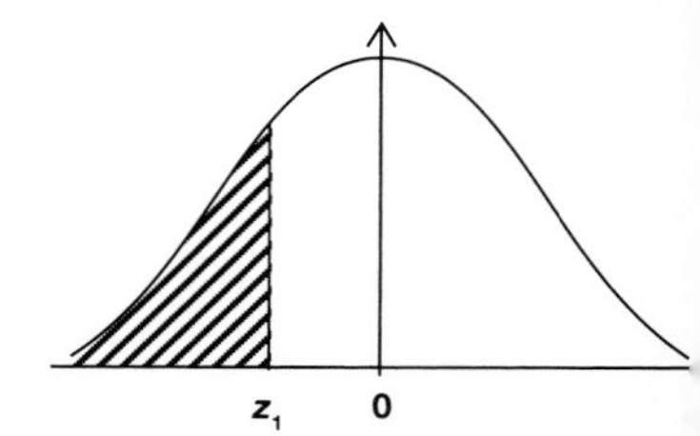

Figure 15.24

And so the required probability is 0.274.

(ii) At the limit i.e. when the probability of there being insufficient breakfasts on board is 0.01, the shaded area in Figure 15.25 equals 0.01.

So $A(z) = 0.49$ i.e. $z = 2.32$

To convert z to a percentage of passengers taking breakfasts, let x be the number of breakfasts corresponding to a standard score of 2.32.

$$\Rightarrow \quad 2.32 = \frac{x - 75}{5} \quad \text{or } x = 86.6\%$$

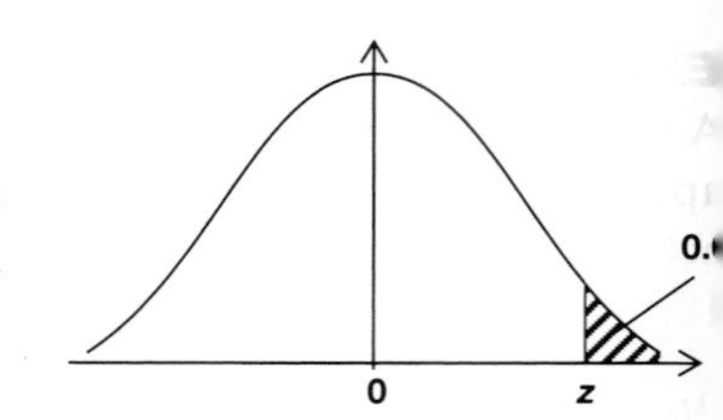

Figure 15.25

Therefore the number of breakfasts is

86.6% of 150 = 129.9

The minimum number of breakfasts to be carried is 130.

(b) The information given concerning the rival aircraft may be summarised in the two diagrams of Figure 15.26.

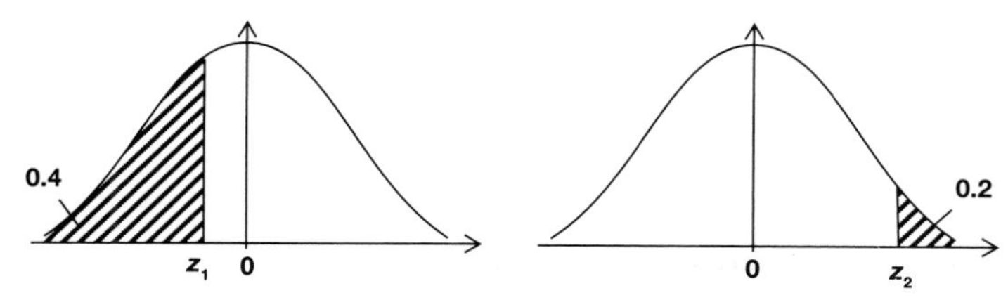

Figure 15.26

z_1 is the standard score corresponding to 78% and z_2 that corresponding to 87%.

$$\Rightarrow \quad z_1 = \frac{78 - \mu}{\sigma} \text{ and } z_2 = \frac{87 - \mu}{\sigma}$$

But the curve on the left of Figure 15.26 gives

$$A(-z_1) = 0.1$$
i.e. $\quad -z_1 = +0.25$
$$z_1 = -0.25$$

The curve on the right gives

$$A(z_2) = 0.3$$
$$z_2 = 0.84$$

Substituting in the appropriate equations yields

$$-0.25 = \frac{78 - \mu}{\sigma} \qquad 0.84 = \frac{87 - \mu}{\sigma}$$

$$\Rightarrow \qquad 78 - \mu = -0.25\sigma$$
$$87 - \mu = 0.84\sigma$$

$$78 - 87 = -0.25\sigma - 0.84\sigma$$
$$-9 = -1.09\sigma \qquad \Rightarrow \quad \sigma = 8.26$$
Hence $\quad 78 - \mu = -0.25(8.26) \quad \Rightarrow \quad \mu = 80$

EXAMPLE 15

A firm uses a machine to produce washers with inside diameters of approximately 10 mm. Over a long period it was found that the actual diameters were normally distributed with a mean of 10.2 mm and a standard deviation of 0.25 mm.

(i) Calculate an estimate of the percentage of washers which have inside diameters of less than 10.0 mm.

(ii) In a sample of 500 washers estimate the number which have inside diameters between 9.5 mm and 10.5 mm.

Each washer costs 2p to produce. Only those washers with inside diameters between 9.5 mm and 10.5 mm are usable.

(iii) Calculate an estimate of the cost of producing 1000 usable washers.

A new machine is available with which each washer costs 2.1p to produce. With this machine 0.3% of washers have inside diameters of less than 9.5 mm and 50% of washers have inside diameters of greater than 10.1 mm.

(iv) Determine whether or not the firm could produce usable washers more cheaply using the new machine.

(N. Ireland Additional Mathematics, 1986)

Solution 15

(i) The shaded area in Figure 15.27, where z is the standard score corresponding to 10, is to be found.

$$z = \frac{10 - 10.2}{0.25} = -0.8$$

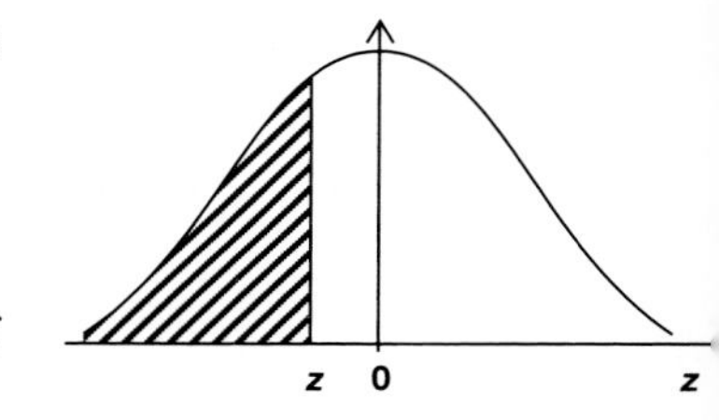

Figure 15.27

The area contained between the z ordinate and the axis of symmetry is

$$A(0.8) = 0.288$$

The shaded area is therefore

$$0.5 - 0.288 = 0.212$$

The percentage of washers with diameter less than 10 mm is 21.2%.

(ii) Given z_1 to be the standard score corresponding to 9.5 mm and z_2 that corresponding to 10.5 mm then the probability of a washer having diameter between 9.5 mm and 10.5 mm is represented by the shaded area in Figure 15.28.

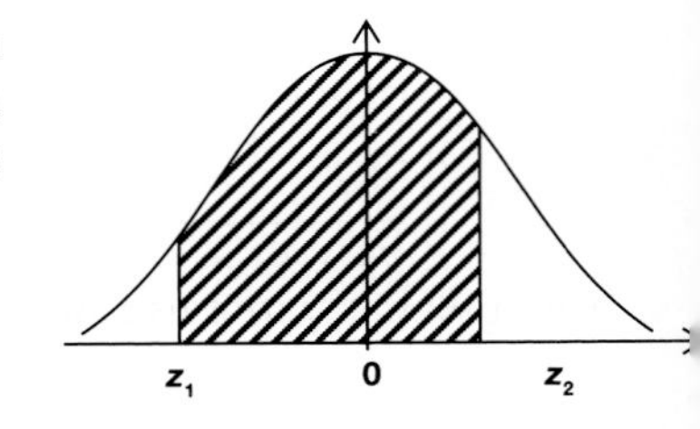

Figure 15.28

$$z_1 = \frac{9.5 - 10.2}{0.25} = -2.8$$

$$z_2 = \frac{10.5 - 10.2}{0.25} = 1.2$$

The probability represented by the shaded area in Figure 15.28 is

$$\begin{aligned} A(-z_1) + A(z_2) &= A(2.8) + A(1.2) \\ &= 0.497 + 0.385 \\ &= 0.882 \end{aligned}$$

As there are 500 washers, the expected number is $0.882 \times 500 = 441$ washers.

(iii) Since some washers will be unusable it will be necessary to produce in excess of 1000 in order to ensure that 1000 are usable. Assume n are produced.

$$\text{Now} \quad (0.882)n = 1000$$
$$\text{Therefore} \quad n = 1134$$

If follows that the cost will be

$$1134 \times £0.02 = £22.68$$

(iv) Since 50% of washers have inside diameter greater than 10.1 mm, it follows that $\mu = 10.1$.

Let z be the standardised value corresponding to 9.5 mm.

As the shaded area in Figure 15.29 represents 0.3% then

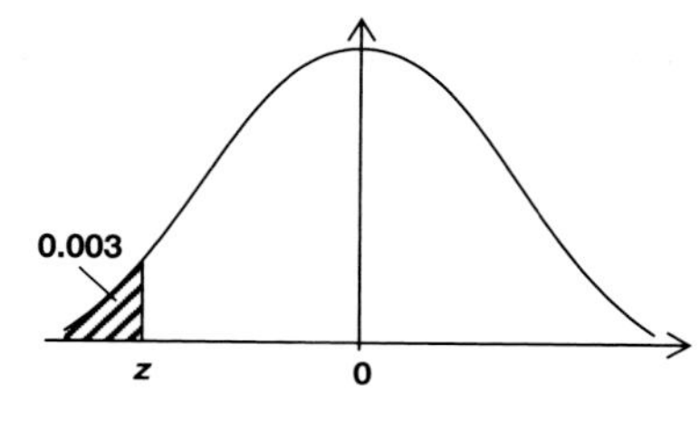

Figure 15.29

$$A(-z) = 0.497$$
$$-z = +2.75$$
$$z = -2.75$$

But -2.75 is the standard score corresponding to 9.5 and, since $\mu = 10.1$, it follows that

$$-2.75 = \frac{9.5 - 10.1}{\sigma}$$

So $\sigma = 0.218$

The probability of a washer being usable is represented by the shaded area in Figure 15.30 when z_2 is the dimension 10.5 mm in standard form i.e.

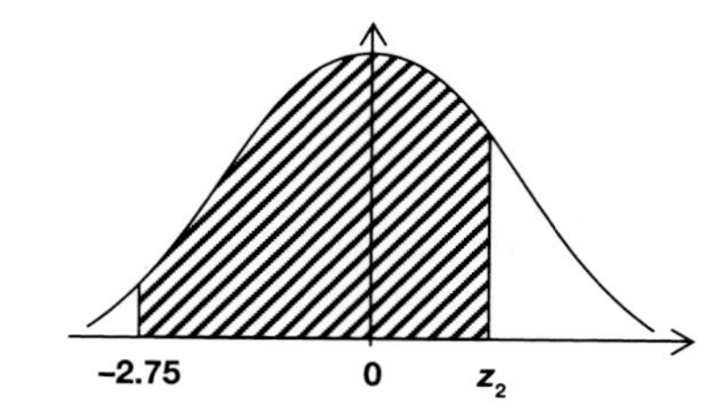

Figure 15.30

$$z_2 = \frac{10.5 - 10.1}{0.218} = 1.83$$

So the probability of a usable washer

$$= A(2.75) + A(1.83)$$
$$= 0.497 + 0.466 = 0.963$$

Suppose that, with this new machine, m washers require to be manufactured in order to produce 1000 usable washers.

$$(0.963)m = 1000$$
$$m = 1039$$
$$\text{cost} = 1039 \times £0.021 = £21.82$$

It would hence be cheaper to manufacture the washers on the new machine.

EXAMPLE 16

(a) The average height of a certain breed of dog is 30 cm with a standard deviation of 4 cm. The heights may be considered to follow a normal distribution.

(i) What percentage of such dogs exceeds 35 cm in height assuming that the heights can be measured with complete accuracy?
(ii) If the heights are all measured to the nearest centimetre what percentage of such dogs exceed 35 cm in height?

Account for the difference in your answers to (i) and (ii).

(b) For another breed of dog it is known that 12.5% have heights less than 35 cm and 20% have heights less than 42 cm. Find the mean and standard deviation of the heights of this breed of dog. Assume the heights can be measured with complete accuracy.

(N. Ireland Additional Mathematics, 1987)

Solution 16

(a) (i) The area in the sketch in Figure 15.31 represents the probability of a dog exceeding 35 cm in height where

$$z = \frac{35 - 30}{4} = 1.25$$

This probability $= 0.5 - A(1.25)$
$= 0.5 - 0.394$
$= 0.106$

The percentage of dogs exceeding 35 cm in height is 10.6%.

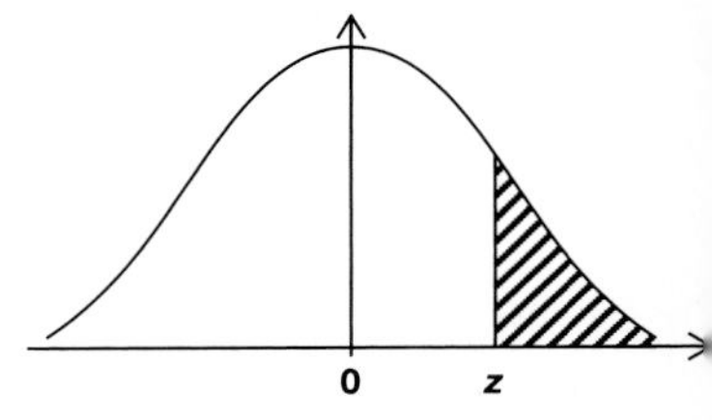

Figure 15.31

(ii) Figure 15.31 still applies except that in this case z is the standardised variable corresponding to 35.5.

$$\Rightarrow \qquad z = \frac{35.5 - 30}{4} = 1.375$$

The probability $= 0.5 - A(1.375)$
$= 0.5 - 0.4155$
$= 0.0845$

The percentage in this case is 8.45%.

No measure can be known exactly. Part (i) of this question is purely theoretical. The difference (2.15%) is accounted for by those dogs with heights in the range 35–35.5 cm which, due to the inexact nature of measurement, are recorded as being 35 cm high.

(b) The information given is summarised in Figure 15.32.

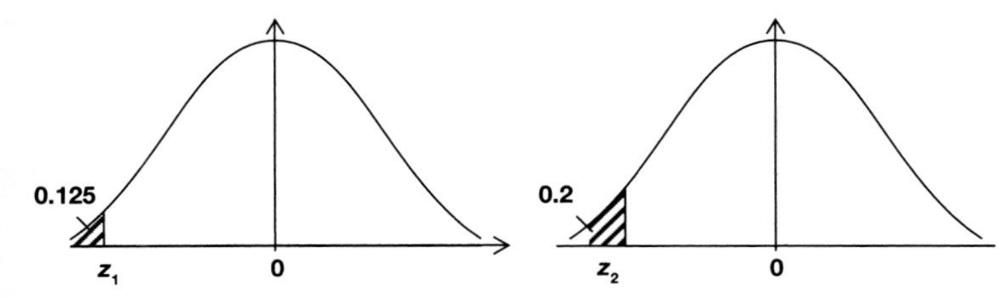

Figure 15.32

z_1 is the standardised height corresponding to 35 cm and z_2 that corresponding to 42 cm.

$$\Rightarrow \quad z_1 = \frac{35 - \mu}{\sigma}$$

$$z_2 = \frac{42 - \mu}{\sigma}$$

But $A(-z_1) = 0.5 - 0.125$ $\quad$ $A(-z_2) = 0.5 - 0.2$

$A(-z_1) = 0.375$ $\quad$ $A(-z_2) = 0.3$

$-z_1 = 1.15$ $\quad$ $-z_2 = 0.84$

$z_1 = -1.15$ $\quad$ $z_2 = -0.84$

$$z_1 = \frac{35 - \mu}{\sigma} \quad \Rightarrow \quad 35 - \mu = -1.15\sigma$$

$$z_2 = \frac{42 - \mu}{\sigma} \quad \Rightarrow \quad 42 - \mu = -0.84\sigma$$

Subtracting $35 - 42 = -1.15\sigma + 0.84\sigma$

$-7 = -0.31\sigma$

$\sigma = 22.6\,\text{cm}$

hence $35 - \mu = -1.15(22.6)$

$\Rightarrow$ $\mu = 61\,\text{cm}$

EXERCISE 15.1

Three figure normal tables should be used throughout this exercise.

1. Find the area under the standardised normal curve to the left of the following values of z.

(i) 1.5 (ii) 1.72 (iii) 1.89 (iv) 2.14 (v) 3.14 (vi) 3.02 (vii) 1.37 (viii) 0.14 (ix) 0.52 (x) 0.17

2. Find the area under the standardised normal curve to the right of the following values of z.

(i) 3.12 (ii) 1.52 (iii) 3.02 (iv) 1.72 (v) 0.24 (vi) 1.13 (vii) 0.27 (viii) 0.02 (ix) 1.34 (x) 2.07

3. Find the area under the standardised normal curve to the left of the following values of z.

(i) -3.12 (ii) -2.14 (iii) -1.42 (iv) -1.37 (v) -1.8

4. Find the area under the standardised normal curve to the right of the following values of z.

(i) -3.06 (ii) -1.24 (iii) -2.17 (iv) -1.19 (v) -1.8

5. Find the area under the standardised normal curve and between the following values of z.

(i) 1.42, 3.16 (ii) 1.72, 2.15 (iii) 1.04, 2.18 (iv) 0.07, 1.41 (v) 0.14, 2.1
(vi) 0.21, 1.33 (vii) 0.14, 2.41 (viii) 0.21, 3.01 (ix) 1.21, 2.21 (x) 1.37, 2.1

6. Find the area under the standardised normal curve and between the following values of z.

(i) -1.24, -0.24 (ii) -1.37, -0.28 (iii) -1.42, -0.21 (iv) -2.14, -1.1
(v) -3.12, -1.02 (vi) -3.04, -0.42 (vii) -2.82, -0.11 (viii) -3.14, -2
(ix) -2.11, -1.11 (x) -2.17, -0.72

7. Find the area under the standardised normal curve and between the following values of z.

(i) -1.24, 2.18 (ii) -1.37, 2.21 (iii) -1.42, 3.01 (iv) -2.14, 3.01
(v) -0.41, 0.47 (vi) -0.28, 0.24 (vii) -0.14, 0.26 (viii) -0.2, 0.24
(ix) -0.3, 0.6 (x) -1.42, 2.14

8. The area to the left of z is given. Find z.

(i) 0.957 (ii) 0.972 (iii) 0.984 (iv) 0.915 (v) 0.567

9. The area to the right of z is given. Find z.

(i) 0.064 (ii) 0.043 (iii) 0.129 (iv) 0.492 (v) 0.019

10. The area to the left of z is given. Find z.

(i) 0.016 (ii) 0.078 (iii) 0.085 (iv) 0.035

11. The area to the right of z is given. Find z.

(i) 0.893 (ii) 0.967 (iii) 0.883 (iv) 0.967

12. (i) The area between z and 1.42 ($z > 1.42$) is 0.007. Find z.
(ii) The area between z and 1.72 ($z > 1.72$) is 0.027. Find z.
(iii) The area between z and 1.04 ($z > 1.04$) is 0.134. Find z.
(iv) The area between z and 0.07 ($z > 0.07$) is 0.393. Find z.
(v) The area between z and 0.14 ($z > 0.14$) is 0.429. Find z.

13. (i) The area between z and -0.24 $(z < -0.24)$ is 0.298. Find z.

(ii) The area between z and -0.28 $(z < -0.28)$ is 0.305. Find z.

(iii) The area between z and -0.21 $(z < -0.21)$ is 0.339. Find z.

(iv) The area between z and -1.11 $(z < -1.11)$ is 0.117. Find z.

(v) The area between z and 2.18 is 0.878. Find z.

(vi) The area between z and 2.21 is 0.901. Find z.

(vii) The area between z and 3.01 is 0.921. Find z.

14. In each of the following the mean and standard deviation of a set of test scores is given. The student's raw score is supplied. Compute the corresponding standardised score in each case to one decimal place.

	Mean	Standard deviation	Student's score
(i)	24	5	27
(ii)	27.20	2.70	20.72
(iii)	25.2	1.2	26.64
(iv)	27	2.4	19.8
(v)	92.00	4.70	102.34
(vi)	250.00	21.20	296.64
(vii)	127	42	22
(viii)	10.40	2.70	7.16
(ix)	2.40	0.20	2.94
(x)	5.70	0.40	5.42

15. In each case below the percentage of variates below a given raw score, and the percentage above another raw score are given. Compute the mean and standard deviation in each case assuming a normal distribution.

(i) 33% above 31.2; 23% below 28.0

(ii) 6.2% above 105.86; 11.5% below 81.20

(iii) 73.2% above 15.82; 96.7% below 37.96

(iv) 1.8% above 149.1; 2.4% below 63.63

(v) 23.6% above 227.8; 17.4% below 202.9

16. The masses of engine components produced by a machine are normally distributed with mean μ and standard deviation σ. Given 5.7% of components have mass less than 13.16 g and 2.5% have mass less than 10.12 g, find μ and σ.

17. The leaves of a bush have lengths which are normally distributed with mean μ and standard deviaton σ. Given 14.7% of leaves have lengths above 85 mm and 23.6% have lengths greater than 81.04 mm, find μ and σ.

18. A particular type of integrated circuit has a mean lifetime of 2.0 years with a standard deviatio of 0.5 years. The lifetimes may be considered to follow a normal distribution. Find th probability that

(i) a circuit will function for more than 2.5 years,
(ii) a circuit has a lifetime between 1.75 and 2.50 years,
(iii) the lifetime of the circuit lies within 2 standard deviations of the mean.

It is decided to grade the circuits with respect to lifetime as A, B, C where A denotes the 15% o circuits which have the longest lifetime, B represents the next 50% and C the remainder.

Determine the lifetime of a circuit which

(iv) just manages to be graded A,
(v) just manages to be graded B.

(N.Ireland Additional Mathematics, 1988

19. (a) The resistances, in ohms, of resistors produced by a certain machine are distribute normally with a mean of 100 and a standard deviation of 4.

(i) If resistors with resistances, in ohms, outside the range 94 to 106 are to be rejected estimate the number rejected in each batch of 100.

(ii) To reduce the number of rejected resistors the machine is adjusted so that the standar deviation is reduced from 4. If 10% of the resistors now have resistances, in ohms outside the range 94 to 106, determine the value of the new standard deviation assuming the mean remains constant.

(b) In a batch of 1000 resistors produced by another machine, 67 resistors had resistances o more than 78 ohms and 23 resistors had resistances of less than 29 ohms. Assuming th distribution to be normal, estimate the mean resistance and the standard deviation.

(N. Ireland Additional Mathematics, 1989

20. A hi-tech company manufactures two types of long-life batteries. The standard type of batter has a lifetime which is normally distributed with a mean of 150 hours and a standard deviation o 10 hours.

(i) Determine the percentage of standard batteries which have lifetimes of less than 155 hours

(ii) In a mainframe computer which utilises 25 standard batteries, calculate the expecte number of batteries which have lifetimes of between 135 hours and 170 hours.

The de luxe type of battery has a lifetime which is normally distributed with a standard deviatio of 8 hours.

A microcomputer has two de luxe batteries and operates when at least one of the batteries i working. It has been found that the probability that the microcomputer will operate for mor than 168 hours is 0.16.

(iii) If x equals the probability that a de luxe battery works properly for less than 168 hours show that $x = 0.9165$.

(iv) Calculate the mean lifetime of a de luxe battery.

(N. Ireland Additional Mathematics, 1990

21. The lifetimes, in hours, of a certain type of electric light bulb, are distributed normally with a mean of 1500 and a standard deviation of 150.

(i) Find the probability that a bulb selected at random will burn for more than 1700 hours.

(ii) If two bulbs are selected at random, calculate the probability that both will burn for more than 1450 hours but less than 1600 hours.

(N. Ireland Additional Mathematics, 1991)

22. When cars are serviced, the garage records the number of miles travelled by a car since the last service. Given that this mileage is normally distributed with a mean of 5200 miles and a standard deviation of 1200 miles, find the probability that a car selected at random will have travelled

(i) less than 7000 miles since the last service,
(ii) between 4600 miles and 6100 miles since the last service.

The garage carries out three types of service –

'Low Service' for cars with low mileage,
'Super Service' for cars with high mileage,
'Normal Service' for all other cars.

Given that 15% of all services are 'Low' and 20% are 'Super', estimate, to the nearest 50 miles,

(iii) the greatest mileage for cars having a 'Low Service',
(iv) the lowest mileage for cars having a 'Super Service'.

On average a 'Low Service' takes a mechanic $1\frac{1}{2}$ hours, a 'Normal Service' takes 2 hours and a 'Super Service' takes $2\frac{1}{2}$ hours. On the assumption that the garage will service 200 cars a week and that a mechanic works for 40 hours a week, estimate the number of mechanics that the garage needs for service work.

(U.C.L.E.S. Additional Mathematics, November 1991)

23. (a) The volumes of wine in bottles are normally distributed with a mean of 760 ml and a standard deviation of 12 ml.

(i) Find the probability that a bottle of wine picked at random from this distribution contains more than 769 ml.

(ii) Find the interquartile range of the distribution.

A wine merchant orders 400 bottles of wine.

(iii) Estimate the number of these bottles containing less than 750 ml.

The machine that fills the bottles is readjusted so that the mean amount of wine in a bottle is altered but the standard deviation remains the same.

(iv) Find the value of the new mean if only 5% of the bottles are to contain less than 750 ml.

(U.C.L.E.S. Additional Mathematics, June 1991, part question)

CHAPTER 16

The binomial distribution

The life of French mathematician **Abraham de Moivre** (1667–1754) was not a happy one. He fled France for Britain to escape religious persecution only to find himself disadvantaged in securing a university teaching post because he wasn't a British citizen. He was forced to work as a tutor, augmenting his meagre income by advising gambling syndicates and insurance companies. His work received scant recognition in his lifetime and he tragically lost his sight. He was a close friend of Sir Isaac Newton and played a significant role in the debate between Newton and Leibniz over 'ownership' of the calculus. While his friendship with Newton played an important part in his election to the Royal Society, he won the recognition of the Academies of Paris and Berlin on his own terms.

His famous *Doctrine of Chances*, published in 1718, discussed probability in contexts that are still familiar today; choosing balls from bags and throwing dice are much in evidence. The aim of the treatise was to present an 'algebra' of probability although the work also comprises a section 'Annuities upon Lives', setting out the principles which underpin modern actuarial thinking.

The *Doctrine of Chances* is presented as a series of probability 'problems' including the problem which bears de Moivre's name: i.e. the probability of throwing any given number with n dice each having m faces. In his *History of Mathematics* (London, 1991) W. W. R. Ball recounts:

The manner of de Moivre's death has a certain interest for psychologists. Shortly before it, he declared that it was necessary for him to sleep some ten minutes or a quarter of an hour longer each day than the preceding one: the day before he had thus reached a total of something over twenty-three hours he slept up to the limit of twenty-four hours, and then died in his sleep.

INTRODUCTION

Consider the expansion of $(a + b)^n$ for successive values of n.

$$(a + b)^0 = 1$$
$$(a + b)^1 = a + b$$
$$(a + b)^2 = a^2 + 2ab + b^2$$
$$(a + b)^3 = a^3 + 3a^2b + 3ab^2 + b^3$$
etc.

Pascal's triangle can be constructed from the coefficients on the right-hand sides of these equations as follows.

1
1 1
1 2 1
1 3 3 1
etc.

Close study of the triangle below reveals how each new row may be generated from the one before by addition. For example, the 6 in the fifth row is the sum of the two 3s in the row immediately above it and each 5 in the sixth row is the sum of the 1 and 4 in the row above.

1
1 1
1 2 1
1 3 3 1
1 4 6 4 1
1 5 10 10 5 1
etc.

The expansion below, for instance, may be generated from the numbers in the fifth row of Pascal's triangle.

$$(a + b)^4 = a^4 + 4a^3b + 6a^2b^2 + 4ab^3 + b^4$$

and those in the sixth row suggest

$$(a + b)^5 = a^5 + 5a^4b + 10a^3b^2 + 10a^2b^3 + 5ab^4 + b^5$$

and so on.

EXAMPLE 1

Use Pascal's triangle to expand $(a + b)^7$.

Solution 1

Since it has been established that the expansion of $(a + b)^4$ may be constructed from the fifth row of Pascal's triangle and $(a+b)^5$ from the sixth, it is necessary to focus on the eighth row in order to determine the expansion above.

1
1 1
1 2 1
1 3 3 1
1 4 6 4 1
1 5 10 10 5 1
1 6 15 20 15 6 1
1 7 21 35 35 21 7 1

and so

$$(a+b)^7 = a^7 + 7a^6b + 21a^5b^2 + 35a^4b^3 + 35a^3b^4 + 21a^2b^5 + 7ab^6 + b^7$$

EXAMPLE 2

Find the third term in the expansion of $(\frac{2}{7}+\frac{5}{7})^6$.

Solution 2

From Pascal's triangle (seventh row)

$$(\tfrac{2}{7}+\tfrac{5}{7})^6 = (\tfrac{2}{7})^6 + 6(\tfrac{2}{7})^5(\tfrac{5}{7}) + 15(\tfrac{2}{7})^4(\tfrac{5}{7})^2 + 20(\tfrac{2}{7})^3(\tfrac{5}{7})^3 + 15(\tfrac{2}{7})^2(\tfrac{5}{7})^4 + 6(\tfrac{2}{7})(\tfrac{5}{7})^5 + (\tfrac{5}{7})^6$$

The third term is

$$15(\tfrac{2}{7})^4(\tfrac{5}{7})^2 = \frac{6000}{117649} = 0.0510$$

EXERCISE 16.1

1. Use Pascal's triangle to expand the following.

(a) $(a+b)^2$ (b) $(p+q)^5$ (c) $(q+p)^3$ (d) $(x+y)^6$ (e) $(b+a)^7$ (f) $(m+n)^4$
(g) $(x+\frac{1}{3})^2$ (h) $(\frac{1}{2}+b)^5$ (i) $(a+\frac{1}{4})^4$ (j) $(\frac{1}{2}+\frac{1}{2})^3$ (k) $(\frac{3}{5}+\frac{2}{5})^6$ (l) $(\frac{1}{4}+\frac{3}{4})^4$

2. (a) What is the second term in the expansion of $(q+p)^4$?
(b) What is the sixth term in the expansion of $(a+b)^6$?
(c) What is the fourth term in the expansion of $(p+\frac{1}{3})^6$?
(d) What is the first term in the expansion of $(\frac{1}{3}+\frac{2}{3})^3$?
(e) What is the third term in the expansion of $(\frac{1}{4}+\frac{3}{4})^4$?
(f) What is the seventh term in the expansion of $(\frac{1}{2}+\frac{1}{2})^7$?

THE BINOMIAL DISTRIBUTION

Consider an experimental trial in which p is the probability of success and q $(=1-p)$ the probability of failure. A binomial distribution arises when n experimental trials are performed providing **the probability of success is constant and equal to p and that the success of any trial is not dependent upon the success or otherwise of previous trials.**

In a binomial distribution the expansions of the previous section can be used to compute the probability of s successes (say), represented $\mathrm{P}(s)$, in n trials according to the equation

$$(p+q)^n = \mathrm{P}(n) + \mathrm{P}(n-1)... + \mathrm{P}(1) + \mathrm{P}(0) \qquad (1)$$

Here

$\mathrm{P}(n)$ is equal to the first term in the expansion of the left-hand side of equation (1).

$\mathrm{P}(n-1)$ is equal to the second term in the expansion of the left-hand side of equation (1), and so on until

$\mathrm{P}(1)$ is the second last term in the expansion of the left-hand side of equation (1) and

$\mathrm{P}(0)$ is the last term in the expansion of the left-hand side of equation (1).

EXAMPLE 3

The probability that a young person on a training scheme will secure permanent employment at the end of the scheme is 0.4.

Determine the probability that, from a sample of five young people on the scheme,

(i) 2 (ii) 3 (iii) all (iv) none

(v) more than 3 (vi) fewer than 2 (vii) at least one

will secure permanent employment at the end of the scheme.

Solution 3

Since the sample size (n) is 5, the following expansion, constructed from Pascal's triangle, is of interest.

$$(p+q)^5 = p^5 + 5p^4q + 10p^3q^2 + 10p^2q^3 + 5pq^4 + q^5$$

Let $p = 0.4$ the probability of success
and $q = 0.6$ the probability of failure.

Using the equation (1) above

$$(0.4+0.6)^5 = \mathrm{P}(5) + \mathrm{P}(4) + \mathrm{P}(3) + \mathrm{P}(2) + \mathrm{P}(1) + \mathrm{P}(0)$$

$$\Rightarrow \quad \begin{aligned} P(5) &= 0.4^5 \\ P(4) &= 5(0.4)^4(0.6) \\ P(3) &= 10(0.4)^3(0.6)^2 \\ P(2) &= 10(0.4)^2(0.6)^3 \\ P(1) &= 5(0.4)(0.6)^4 \\ P(0) &= (0.6)^5 \end{aligned}$$

The probabilities required in Example 3 may now be compiled.

(i) $P(2) = 10(0.4)^2(0.6)^3 = 0.3456$

(ii) $P(3) = 10(0.4)^3(0.6)^2 = 0.2304$

(iii) $P(5) = 0.4^5 = 0.01024$

(iv) $P(0) = (0.6)^5 = 0.07776$

(v) $P(>3) = P(4) + P(5) = 5(0.4)^4(0.6)^1 + (0.4)^5 = 0.08704$

(vi) $P(<2) = P(0) + P(1) = (0.6)^5 + 5(0.4)(0.6)^4 = 0.3370$

(vii) $P(\text{at least } 1) = P(1) + P(2) + P(3) + P(4) + P(5)$

Note, however, that

$$(0.4 + 0.6)^5 = P(5) + P(4) + P(3) + P(2) + P(1) + P(0)$$

and so $\quad 1 - P(0) = P(5) + P(4) + P(3) + P(2) + P(1)$

The right-hand side of this equation represents the probability of at least one young person securing employment while the left-hand side represents 1 minus the probability of no young person securing employment. Statisticians are quick to take advantage of this result. It eliminates much tedious computation since the necessity to calculate and sum five probabilities (P(1) to P(5)) is replaced by the less onerous task of subtracting a single probability, P(0), from unity. This result is often stated

probability of at least one success = 1 − probability of no successes

For this particular example

$$\mathbf{P(\text{at least } 1) = 1 - P(0)}$$
$$= 1 - (0.6)^5$$
$$= 0.9222$$

EXERCISE 16.2

1. In a survey 60% of people said they owned a cassette recorder. What is the probability that three people in a group of four own a cassette recorder?

2. In a certain school, $\frac{1}{3}$ of pupils walk to school. What is the probability that among a group of six pupils, exactly two walk to school?

3. In a class of 20 pupils, 16 study French and four study German. What is the probability that among a group of three friends, all study French?

4. If two out of three houses have a telephone what is the probability that, in a group of five houses, none has a telephone?

5. If a survey reveals that 40% of people smoke, what is the probability that among a group of four friends exactly two smoke?

6. If three out of four people are right handed, what is the probability that in a family of five exactly one person is left handed?

7. In a collection of schoolbooks, 70% are maths books and the rest are English books. Calculate the probability of obtaining three maths books in a group of five school books.

8. If six fair coins are tossed, what is the probability of obtaining exactly one head?

9. In a test on light bulbs it is found that 25% are defective. What is the probability of finding exactly two defective bulbs in a batch of four?

10. The pass rate in an examination is 85%. What is the probability that among a group of seven friends, exactly one of them will fail?

EXERCISE 16.3

1. In a certain year the probability of a graduate obtaining employment is 0.8. What is the probability that among four students at least three will gain employment?

2. Exactly $\frac{1}{3}$ of a workforce travel to work by train. What is the probability that among six workers at least three travel by train?

3. In an examination the probability of getting a top grade is 55%. What is the probability of more than one person receiving the top grade in a group of five?

4. In a factory, 15% of items are defective. What is the probability of more than three items in a batch of six being defective?

5. The probability of a dartsman hitting the bullseye is 0.3. What is the probability that he will hit it at least once with three darts to throw?

6. One in five of the population wear glasses. What is the probability that among five people at least two wear glasses?

7. Two out of three cars have a sunroof. What is the probability that in a group of six cars at least five have a sunroof?

8. 30% of people travel abroad for holidays. What is the probability that among four people more than one of them travels abroad?

9. What is the probability of obtaining more than one six when a fair die is thrown six times?

10. 20% of the population rent their TVs. What is the probability that in a survey of six houses, at least four owned their TV?

EXERCISE 16.4

1. If the probability of hitting the bullseye is $\frac{1}{10}$, what is the probability of a dartsman hitting the bullseye at least once if he has three darts?

2. The probability of an egg being cracked during transport is 0.4. What is the probability that in a box containing half a dozen eggs at least one will be cracked?

3. The probability that a pupil will be absent on a particular day is 0.35. What is the probability of at least one absentee in 10 pupils?

4. 85% of road traffic accidents are caused by excess speed. What is the probability that among eight accidents reported, at least one will have been caused by excess speed?

5. The probability of having no rain on a particular day is 0.36. What is the probability of having at least one dry day in a particular week?

MORE ADVANCED EXAMPLES

EXAMPLE 4

Write down the binomial expansion of $(p + q)^4$.

In a small town there are two primary schools. Three-quarters of the primary school children of the town attend the East Primary and the remainder attend the West Primary.

The East Primary School ran a trip to the city and one-third of its pupils went on the trip. If four pupils are chosen at random from the school, calculate the probability that

(i) exactly two of them went on the trip,
(ii) at least three of them went on the trip.

The West Primary School also ran a trip to the city and $\frac{3}{5}$ of its pupils went on this trip.

(iii) If a primary school child from the town is chosen at random, calculate the probability that this child went on a school trip to the city.
(iv) Calculate the probability that if four primary school children from the town are chosen at random then at least one of them went on a school trip to the city.

(N. Ireland Additional Mathematics, 1985)

Solution 4

The fifth row of Pascal's triangle yields

$$(p + q)^4 = p^4 + 4p^3q + 6p^2q^2 + 4pq^3 + q^4$$

One-third of East Primary pupils went on the trip to the city. Let p represent the probability of a pupil going on the trip and q the probability of not going. Clearly $p = \frac{1}{3}$ and $q = \frac{2}{3}$. An examination of the terms of the above expansion reveals

(i) probability of two pupils going on the trip $= 6(\frac{1}{3})^2(\frac{2}{3})^2 = \frac{8}{27}$

(ii) probability of at least three pupils going on the trip $= 4(\frac{1}{3})^3(\frac{2}{3}) + (\frac{1}{3})^4 = \frac{1}{9}$

(iii) A tree diagram may be used to summarise this information.

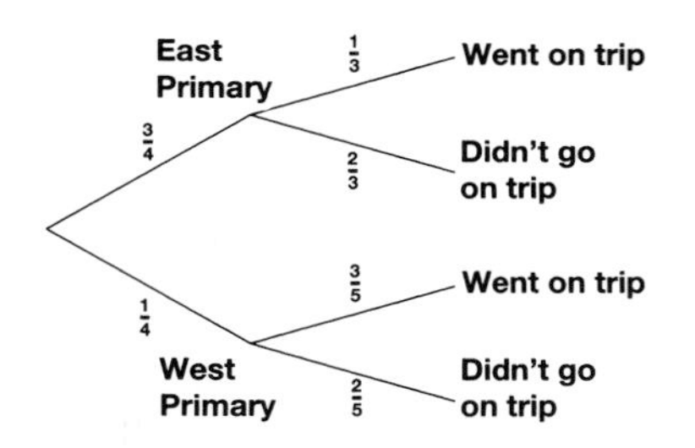

Figure 16.1

Using Figure 16.1, the probability of the child going on the trip is given by

$$P(E)P(T|E) + P(W)P(T|W) = \tfrac{3}{4} \times \tfrac{1}{3} + \tfrac{1}{4} \times \tfrac{3}{5} = \tfrac{2}{5}$$

where $P(E)$ represents the probability of choosing a child from East Primary and $P(T|E)$ represents the probability that a child went on the trip given that he or she attends East Primary. $P(W)$ and $P(T|W)$ may be interpreted similarly for children attending West Primary.

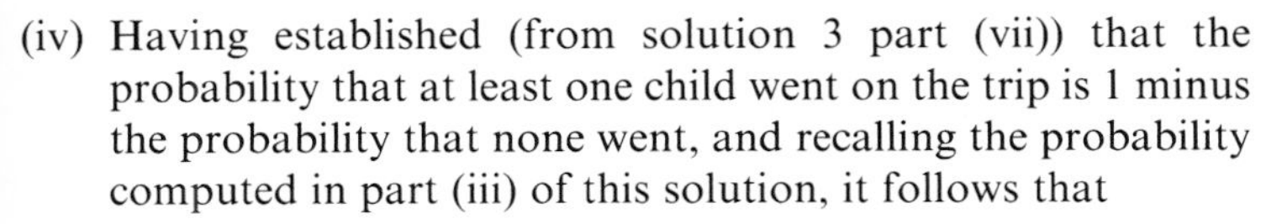

(iv) Having established (from solution 3 part (vii)) that the probability that at least one child went on the trip is 1 minus the probability that none went, and recalling the probability computed in part (iii) of this solution, it follows that

probability that at least one pupil went on the trip
$= 1 - (\tfrac{3}{5})^4 = \tfrac{544}{625}$

EXAMPLE 5

State, briefly, two assumptions one makes when applying a binomial distribution.

(a) In the UK the probability of recovering from a certain blood disorder is $\frac{1}{3}$.

Three patients suffering from this disorder are selected at random.

Calculate the probability that

(i) all three patients in the sample recover from the disorder,
(ii) two or more of these patients recover.
(iii) A treatment is found which increases the probability of recovery to a value, x. Determine, correct to two decimal places, the value of x if the probability of all three patients in a sample recovering is now $\frac{2}{27}$.

(b) (i) Write down an expression for the probability that in a sample of n untreated patients at least one patient recovers.

(ii) If the probability that at least one patient recovers in a sample of n untreated patients selected at random is greater than 0.8 determine the least value of n.

(N. Ireland Additional Mathematics, 1987)

Solution 5

One assumes that the probability of success in a trial is constant for every trial and that the success of any trial is independent of the success or otherwise of previous trials.

(a) (i) the probability of all three patients recovering is

$$(\tfrac{1}{3})^3 = \tfrac{1}{27}$$

(ii) the probability that two or more patients recover is given by

$$3(\tfrac{1}{3})^2(\tfrac{2}{3}) + (\tfrac{1}{3})^3 = \tfrac{7}{27}$$

(iii) The probability of all three patients recovering $= x^3$

Hence $x^3 = \frac{2}{27}$
or $x = 0.42$

(b) (i) $1 - (\tfrac{2}{3})^n$

(ii) $1 - (\tfrac{2}{3})^n > 0.8$
$0.2 > (\tfrac{2}{3})^n$
$\log(0.2) > n\log(\tfrac{2}{3})$
$-0.699 > -0.1761n$
$0.699 < 0.1761n$
$3.969 < n$
$n = 4$

EXAMPLE 6

Black and White were the only candidates standing in an election.

A week before the election 50% of all voters in a certain area of the constituency said that they would vote for Black, 40% said that they would vote for White and 10% were undecided.

(i) Find the probability that a voter did not say that he would vote for White.

Three voters were chosen at random.

(ii) Calculate the probability that exactly two of them said that they would vote for White.

On the day of the election 20% of those who said they would vote for Black actually voted for White and 10% of those who said they would vote for White actually voted for Black. Of those who were undecided 80% voted for White and 20% voted for Black.

(iii) Calculate the probability that a person chosen at random voted for Black in the election.

In the election the voting pattern in this area of the constituency reflected exactly the voting pattern throughout the constituency.

Three voters were chosen at random after the election.

(iv) Calculate the probability that at least one of them voted for the winning candidate.

(N. Ireland Additional Mathematics, 1984)

Solution 6

Let p = the probability that a voter said he/she would vote for White.

Let q = the probability that a voter said he/she would not vote for White.

So $p = \frac{2}{5}$, $q = \frac{3}{5}$.

(i) It follows immediately that the probability that a voter did not say he/she would vote for White is $\frac{3}{5}$.

(ii) Since

$$\left(\tfrac{2}{5} + \tfrac{3}{5}\right)^3 = \left(\tfrac{2}{5}\right)^3 + 3\left(\tfrac{2}{5}\right)^2\left(\tfrac{3}{5}\right) + 3\left(\tfrac{2}{5}\right)\left(\tfrac{3}{5}\right)^2 + \left(\tfrac{3}{5}\right)^3$$

then the probability that exactly two said they would vote for White is

$$3\left(\tfrac{2}{5}\right)^2\left(\tfrac{3}{5}\right) = \tfrac{36}{125}$$

(iii) The tree diagram in Figure 16.2 may be useful in summarising this information.

$$\text{Probability of a vote for Black} = \tfrac{1}{2} \times \tfrac{4}{5} + \tfrac{2}{5} \times \tfrac{1}{10} + \tfrac{1}{10} \times \tfrac{1}{5} = \tfrac{23}{50}$$

Hence White wins.

(iv) The following probabilities result from the probability computed in part (iii).

$$\text{P(vote for winner)} = \tfrac{27}{50}; \text{ P(vote for loser)} = \tfrac{23}{50}$$

Therefore the probability that at least one voter chose the winning candidate is

$$\left(\tfrac{27}{50}\right)^3 + 3\left(\tfrac{27}{50}\right)^2\left(\tfrac{23}{50}\right) + 3\left(\tfrac{27}{50}\right)\left(\tfrac{23}{50}\right)^2 = 0.9027$$

or simply $1 - \left(\frac{23}{50}\right)^3$

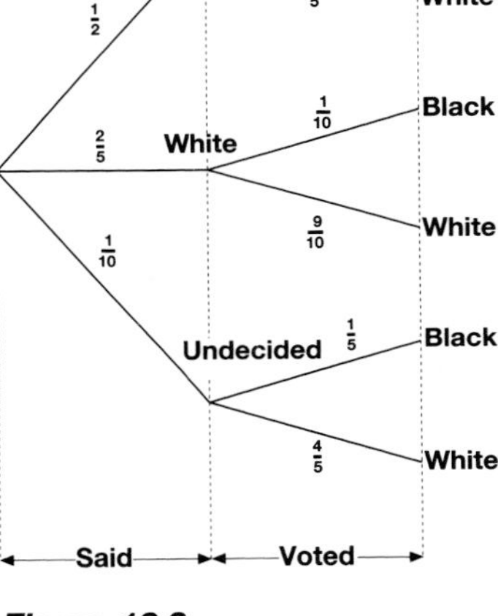

Figure 16.2

EXERCISE 16.5

1. (a) Write down the binomial expansion of $(p + q)^4$.

(b) In Form V of a certain school the probability of a student being successful in the mathematics examination is $\frac{2}{3}$.

Four students are selected at random form Form V. Calculate the probability that

(i) exactly three of these students will be successful,
(ii) at least two of these students will be successful.

To increase the probability of passing the mathematics examination to $\frac{3}{4}$, the school provides extra classes in mathematics. These are attended by half of the students in Form V.

(iii) Calculate the probability that a student selected at random is successful in passing the mathematics examination.

(iv) If four students are now chosen at random, determine the probability that at most three students will pass the examination.

(N. Ireland Additional Mathematics, 1988)

2. The probability that a certain type of seed will germinate is $\frac{3}{4}$. Six of these seeds are sown. Determine the probability that

(i) three seeds will germinate,
(ii) at least two seeds will germinate.

(N. Ireland Additional Mathematics, 1989)

3. A firm receives a large consignment of bolts from its supplier. A random sample of six bolts is taken from the consignment. If 25% of the bolts in the consignment are defective, calculate the probability of

(i) finding no defective bolts in the sample,
(ii) finding two or more defective bolts in the sample.

(N. Ireland Additional Mathematics, 1990)

4. (a) State, briefly, two assumptions made when applying a binomial distribution.

(b) In an archery competition each of ten competitors has a probability of $\frac{1}{4}$ of scoring a gold with one shot.

If four competitors are chosen at random find the probability that, with one shot each,

(i) all four competitors score a gold,
(ii) at least two competitors miss scoring a gold.

Six of the competitors are women. They are given extra lessons in archery with the result that the probability of any woman scoring a gold with one shot is increased to $\frac{3}{4}$.

If two of the ten competitors are now chosen at random, calculate

(iii) the probability that both are of the same sex and that at least one of them scores a gold.

(N. Ireland Additional Mathematics, 1991)

5. Write down the binomial expansions of each of the following.

$(p+q)^3$ and $(p+q)^5$

The probability of a marksman hitting a target with any shot is $\frac{1}{3}$.

(i) If he takes three shots what is the probability that he will hit the target three times?

(ii) If he takes five shots what is the probability that he will hit the target exactly three times?

(iii) If he takes five shots what is the probability that he will hit the target at least three times?

The target is repositioned and the probability of the marksman hitting it with any shot is reduced to $\frac{1}{4}$.

(iv) If he takes three shots what is the probability that he will not hit the target at all?

(v) Determine the least number of shots he must take in order that the probability of hitting the target at least once is greater than $\frac{3}{5}$.

6. Write down the binomial expansion of $(p+q)^4$.

Fred buys an old car with four cylinders, each of which has one sparking plug. In each cylinder the probability that the sparking plug will work is p and the probability that it will fail is q.

Using the binomial expansion verify that the probability that exactly two plugs will work is

$$6p^2(1-2p+p^2)$$

If $p=\frac{9}{10}$ calculate the probability that

(i) only one plug will work,
(ii) less than three plugs will work.

The engine will start if at least three plugs work. What is the probability that the engine will start?

Fred cleans two of the plugs so that the probability that each of these plugs will work is increased to $\frac{19}{20}$. What is the probability that the engine will start?

7. The probability that a lady solves a crossword puzzle on her way to work is 0.65. Over a period of six days, she attempts the puzzle every day. Find the probability that

(i) she solves the puzzle on exactly four days,
(ii) she solves the puzzle more times than she fails to solve it,
(iii) she fails to solve the puzzle on two consecutive days, but solves it on the other four days.

(U.C.L.E.S. Additional Mathematics, November 1991, part question)

CHAPTER 17

Time series

This chapter is concerned with the method for establishing trends in fluctuating measures such as industrial output, unemployment data, and sales, in order to make reliable forecasts of future behaviour. No treatment of economic forecasting would be complete without reference to **John Maynard Keynes** (1883–1946). Keynes was born in Cambridge and educated at King's College Cambridge where he was a member of the 'Apostles', a secret society boasting the finest minds in the University.

Upon leaving Cambridge he took up a position in the Civil Service at the India Office. In 1909 he was elected to Fellowship at King's but moved to the Treasury during the period 1915–1919. Although Keynes wrote a *Treatise on Probability* which was much praised by Bertrand Russell, it is for his two great works in economics for which he will be remembered.

In his *Treatise on Money* he stood economic logic on its head by recommending that Governments spend their way out of recession: '... in times of depression and unemployment it is desirable to encourage spending and lavishness'.

In the later *General Theory of Employment, Interest and Money*, he departed from the prevailing orthodoxy and suggested that high wages do not necessarily result in increased unemployment rates, through workers 'pricing themselves out of jobs'. Keynes showed early promise as an Eton schoolboy and his famous 'women are more fitted to rule than men' speech alerted his teachers to his great skills as an orator. It is therefore ironical that Keynesian principles fell from favour with the election of Britain's first woman Prime Minister, Margaret Thatcher. Lady Thatcher, a monetarist and free marketeer, believed Keynesian economics to be the engine of socialism, an ideology for which Keynes had as little stomach as Lady Thatcher.

INTRODUCTION

Everyday life abounds with examples of data reported at regular intervals of time; the Government provides unemployment figures monthly, reports from public bodies appear annually, a weather report follows the six o'clock news each evening and so on. Furthermore, such reports frequently point to the future using phrases such as 'seasonally adjusted statistics' and 'the underlying trend'. In some instances there is little difficulty predicting future figures from currently available data. Consider Figure 17.1. The number of people unemployed (in millions in the UK) is given for years 1990 to 1992.

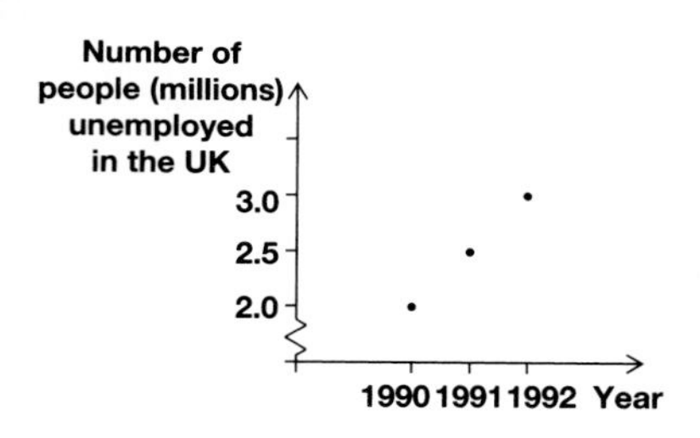

Figure 17.1

It would be a simple exercise to draw a line through the three points in the figure (a 'trend' line) and predict the 1993 unemployment figure to be 3.5 million by extrapolating this line. However, real-life data rarely presents itself in linear form.

The electricity bill, to the nearest £, for a pensioner living in a small flat is given in Table 17.1.

Table 17.1

Year	1990				1991				1992			
Quarter	1	2	3	4	1	2	3	4	1	2	3	4
Bill (£)	86	70	74	98	94	78	74	106	102	86	90	114

The first quarter of the year comprises January, February and March, the second April, May and June, and so on. It is clear that the pensioner's electricity bill is high during the cold months of the year i.e. in quarters one and four. Figure 17.2 illustrates this data and, while it is clear that the trend is upwards, the means of construction of the trend line is not so obvious – a single straight line cannot be drawn through all of these points. A possible

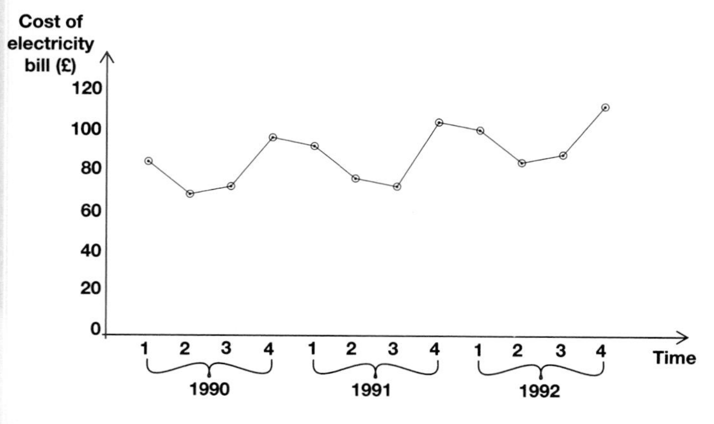

Figure 17.2

solution might be to construct a line passing as close as possible to all the points and use this for prediction purposes. The actual data would then fluctuate about the trend line.

This example typifies a situation frequently encountered by statisticians and this chapter details how time series methods are employed to 'smooth out' these fluctuations, thereby revealing the trend of the data.

CLASSIFYING FLUCTUATIONS

Four principal variations or fluctuations will be considered in this chapter.

Secular variation (or trend) concerns itself with the general direction of the data over a long time period. A common example of secular variation is the Olympic high jump record; one is assured that this will increase in successive games. A downward moving secular variation might be the winning time in the 100 m Olympic sprint.

Seasonal variation refers to the patterns which occur in data during corresponding months in successive years. In the example of the pensioner's bill it is clear that it is least in quarters two and three each year and maximum in quarter four.

Cyclical variations are long (several years) oscillations about the trend line. Periods of prosperity and recession are examples of such variations in business life.

Irregular or random variations are impossible to 'smooth out' using simple time series methods. These variations may be caused by sudden unpredictable events such as a stock market 'crash' or a political crisis.

CONSTRUCTING THE TREND LINE - MOVING AVERAGES

This brief section details how 'moving averages' are calculated so that the predictive power of the trend line drawn through these averages may be demonstrated in the next section.

Returning to the example of the pensioner's electricity bill, it is worth noting that the data is tabulated in three groups of four; a glance at Figure 17.2 reveals three similarly shaped 'periods' of four observations each.

The skilled statistician must be able to examine data such as that displayed in Figure 17.2 and detect these 'repeats'. The pensioner's data will be analysed using a four-point moving average. Having decided upon four-point moving averages, the averages are calculated as follows: the first is the average of observations 1, 2, 3 and 4 in the data, the second moving average is the average of observations 2, 3, 4 and 5 and so on. The complete calculation appears in Table 17.2.

Table 17.2

Year	Quarter	Bill (£)	Moving average
1990	1	86	
	2	70	
			82
	3	74	
			84
	4	98	
			86
1991	1	94	
			86
	2	78	
			88
	3	74	
			90
	4	106	
			92
1992	1	102	
			96
	2	86	
			98
	3	90	
	4	114	

where

$$82 = \tfrac{1}{4}(86 + 70 + 74 + 98)$$
$$84 = \tfrac{1}{4}(70 + 74 + 98 + 94)$$

and so on.

The location of the moving average in Table 17.2 is very important. The moving average is positioned at the 'mid-point' of the data from which it is computed. For example, the first moving average is computed from observations taken at quarters 1, 2, 3 and 4 of 1990. The 'mid-point' is midway between quarters 2 and 3 and so the first moving average is positioned half-way between quarters 2 and 3 of 1990 – see the fourth column of Table 17.2. Similarly the second moving average is computed from observations taken at quarters 2, 3 and 4 of 1990 together with quarter 1 of 1991. The 'mid-point' in this case is midway between quarters 3 and 4 of 1990 and the reader should confirm that the position of the second moving average in column four reflects this consideration.

Care in positioning the moving averages against the original data is rewarded when the averages are graphed. The contents of Table 17.2 are graphed in Figure 17.3 with the moving averages plotted (using crosses) in a manner which reflects their positioning in

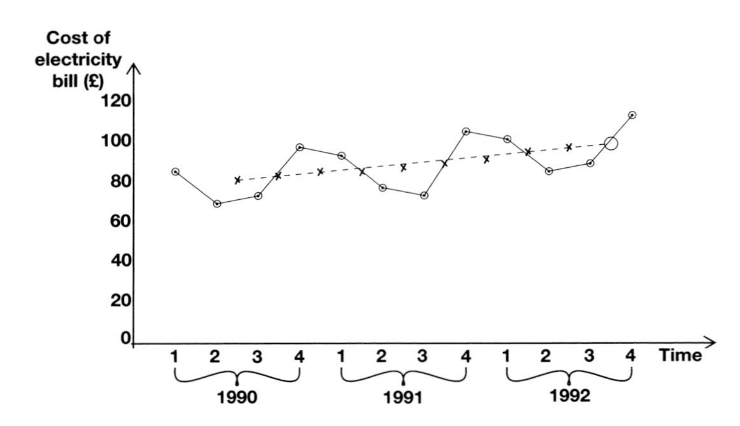

Figure 17.3

column four of the table. The trend line is drawn (by eye) as close as possible to the moving averages; it is shown dotted and has been extended beyond the last moving average.

MAKING PREDICTIONS USING MOVING AVERAGES

By extrapolating the trend line, an estimate of the electricity bill for the first quarter of 1993 may be made. The moving average located at the 'mid-point' of quarters 3 and 4 of 1992 may be estimated to be £100 (see point A on Figure 17.3). This moving average would normally be calculated from the average of bills for the second quarter of 1992, the third quarter of 1992, the fourth quarter of 1992 and the first quarter of 1993.

Hence

$\frac{1}{4}(86 + 90 + 114 + \text{electricity bill for first quarter of 1993}) = 100$

$\Rightarrow \quad 290 + \text{bill for first quarter of 1993} = 400$

i.e. the estimated bill for the first quarter of 1993 is £110.

EXAMPLE 1

(a) With which characteristic movement of a time series would you mainly associate each of the following?

- (i) the increase in sales of TV sets caused by a pre-budget spending rush
- (ii) the total monthly bookings recorded by a travel agency
- (iii) the cost of a medium-sized family saloon car since 1960

(b) Table 17.3 gives the quarterly revenue (in millions of pounds) of an airline on a particular route over a three year period.

Table 17.3

Year	1st quarter	2nd quarter	3rd quarter	4th quarter
1977	14	17	28	13
1978	18	21	30	15
1979	24	23	34	19

(i) Plot these data on graph paper using suitable axes and scales.

(ii) Calculate the four-quarterly moving averages and plot these on the graph obtained in (i).

(iii) Draw the trend line and use it to estimate the income expected in the first quarter of 1980.

(iv) The cost of an air ticket in the first quarter of 1977 was £100. Assuming that the same number of passengers use the air route during the first quarter of each year, estimate the cost of a ticket in the first quarter of 1980.

(N. Ireland Additional Mathematics, 1980)

Solution 1

(a) (i) Irregular or seasonal.

(ii) Seasonal.

(iii) Secular.

(b) (i) See Figure 17.4.

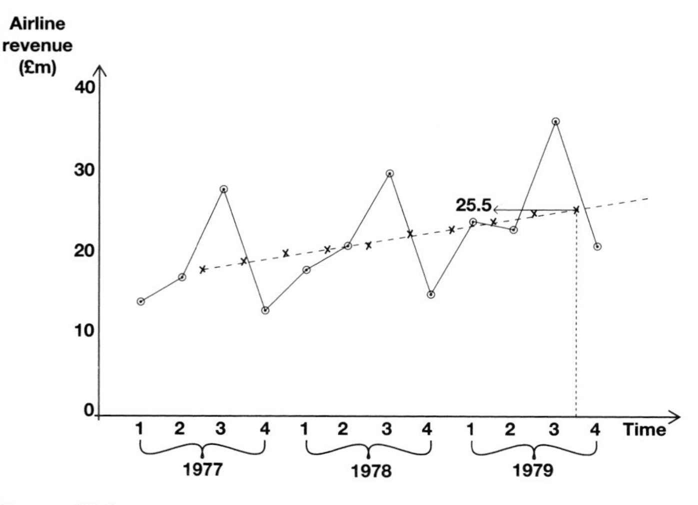

Figure 17.4

(ii) ***Table 17.4***

Year	Quarter	Revenue (£m)	Moving average
1977	1	14	
	2	17	18
	3	28	19
	4	13	20
1978	1	18	20.5
	2	21	21
	3	30	22.5
	4	15	23
1979	1	24	24
	2	23	25
	3	34	
	4	19	

The moving averages of column 4 appear in Figure 17.4 together with the appropriate trend line.

(iii) The first task is to estimate the revenue for the first quarter of 1980. The moving average located midway between quarters 3 and 4 of 1979 can be estimated from the extrapolated trend line to be £25.5m. Now this average is based upon the second, third and fourth quarters of 1979 together with the first quarter of 1980.

Hence

$$\tfrac{1}{4}(23 + 34 + 19 + \text{revenue for first quarter of 1980}) = 25.5$$

$\Rightarrow$ revenue for first quarter of 1980 is £26m.

(iv) Now if the cost of a ticket in the first quarter of 1977 was £100 then 140 000 people used the airline in the first quarter of 1977.

Furthermore, if the same number of people use the air route in the first quarter of 1980 (for which the estimated revenue is £26m) then

140 000 times the cost of a ticket = £26 000 000

or the ticket cost is £186 approximately.

EXAMPLE 2

Following the discovery of an exploding star, its magnitude (measure of brightness) was observed over a period of 16 days and the results in Table 17.5 were obtained.

(i) Plot these data on graph paper using suitable axes and scales.

(ii) State which of the 3, 4 or 5-point moving averages is the most appropriate to smooth the time series.

Using your choice, calculate the moving averages and plot them on your graph obtained in (i).

(iii) Draw the trend line and use it to estimate the magnitude of the star on the 17th day.

(N. Ireland Additional Mathematics, 1981)

Table 17.5

Day	Magnitude
1	3.3
2	4.2
3	5.0
4	6.0
5	5.5
6	5.3
7	6.2
8	7.0
9	7.5
10	6.5
11	6.3
12	6.7
13	7.5
14	8.0
15	7.0
16	6.8

Solution 2

(i)

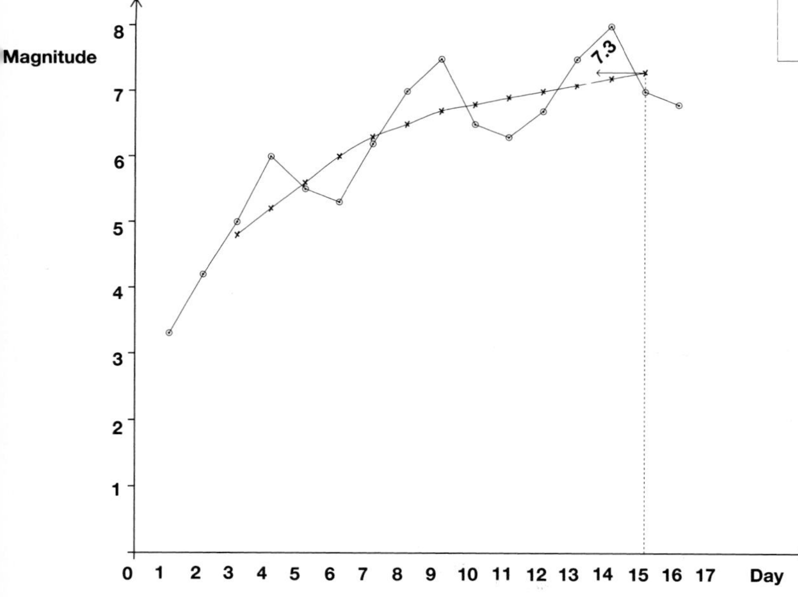

Figure 17.5

(ii) An examination of the data graphed in (i) reveals that cycles of five observations are repeated three times. A 5-point moving average seems advisable.

The results are displayed in Table 17.6 and on the graph. The reader is asked to note the location of the moving averages both in the table and graph. The principle of locating averages at the mid-point of the interval is adhered to.

Table 17.6

Day	Magnitude	Moving average
1	3.3	
2	4.2	
3	5	4.8
4	6	5.2
5	5.5	5.6
6	5.3	6
7	6.2	6.3
8	7	6.5
9	7.5	6.7
10	6.5	6.8
11	6.3	6.9
12	6.7	7
13	7.5	7.1
14	8	7.2
15	7	
16	6.8	

(iii) The moving average centred on day 15 may be estimated by extrapolating the trend line. The result is that this moving average has magnitude 7.3.

Hence

$$7.3 = \tfrac{1}{5}(7.5 + 8 + 7 + 6.8 + \text{magnitude on day 17})$$

$\Rightarrow$ $36.5 = 29.3 + \text{magnitude on day 17}$

i.e. the magnitude on day 17 = 7.2

EXAMPLE 3

(a) With which characteristic movement of a time series would you associate each of the following?

(i) the average price of a gallon of petrol measured annually
(ii) the monthly sales of lettuce in the UK

(b) State, briefly, two reasons why one applies the technique of moving averages to data.

(c) The number of optical devices sold by a company during each month of 1986 was as follows.

Table 17.7

Month	J	F	M	A	M	J	J	A	S	O	N	D
Number sold	25	20	22	27	22	26	30	24	28	31	27	29

(i) Plot the data on graph paper using suitable axes and scales.
(ii) Calculate appropriate moving averages to smooth the data and plot these on the graph obtained in (i).

(iii) Draw the trend line and use it to estimate the number of optical devices which the company will expect to sell in January of 1987.

(N. Ireland Additional Mathematics, 1987)

Solution 3

(a) (i) Secular.

(ii) Seasonal.

(b) (i) To smooth out variations and establish whether trend is upwards, downwards, or flat.

(ii) To predict behaviour at time points beyond those for which the given data is available.

(c) (i) See Figure 17.6.

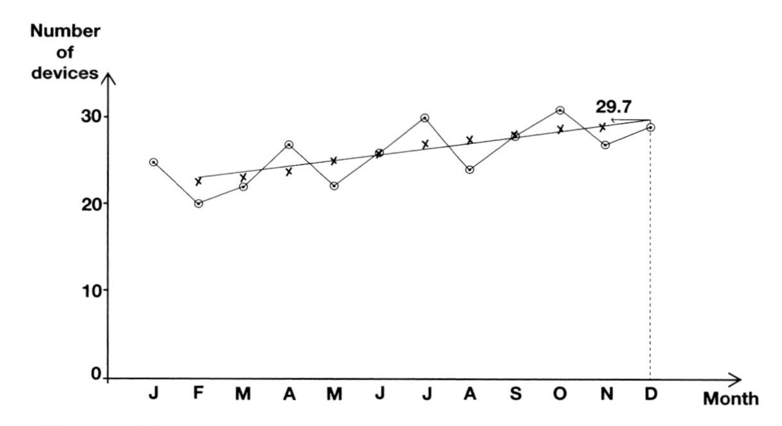

Figure 17.6

(ii) An examination of the data graphed in Figure 17.6 reveals that cycles of three observations are repeated four times and therefore 3-point averages are adopted in Table 17.8.

The moving averages in column three are graphed with the original data.

(iii) The moving average centred on December may be estimated from the extrapolated trend line to be 29.7 devices – see Figure 17.6.

Table 17.8

Month	Devices sold	Moving averages
J	25	
F	20	22.3
M	22	23.0
A	27	23.7
M	22	25.0
J	26	26.0
J	30	26.7
A	24	27.3
S	28	27.7
O	31	28.7
N	27	29.0
D	29	

It follows therefore that

$$29.7 = \tfrac{1}{3}(27 + 29 + \text{expected number of devices sold in January 1987})$$

$$\Rightarrow \quad 89.1 = 56 + \text{expected number of devices sold in January 1987}$$

Hence the estimated number of devices which the company might expect to sell in January 1987 is 33.

EXERCISE 17.1

1. (a) With which characteristic movement of a time series would you associate each of the following?

(i) the quarterly sales of caravans in Northern Ireland
(ii) the total population of the world measured annually

(b) A manufacturer of toothpaste organises a special advertising campaign on television every three months.

The numbers of tubes of toothpaste (in thousands) sold by the manufacturer between January 1983 and February 1984 were as shown in Table 17.9.

(i) Plot these data on graph paper using suitable axes and scales.

(ii) Calculate the appropriate moving averages to smooth the time series and plot these on the graph obtained in (i).

(iii) Draw the trend line and use it to estimate the number of tubes of toothpaste that the manufacturer can expect to sell in March 1984.

(iv) In the first quarter of 1983 the manufacturer sold a tube of toothpaste for £0.50. When all production costs were deducted from the total income for this quarter the net profit was £64 100.

Table 17.9

Number of tubes (thousands)	Month	Year
460	J	1983
418	F	
404	M	
480	A	
439	M	
429	J	
500	J	
459	A	
451	S	
521	O	
478	N	
473	D	
542	J	1984
500	F	

In the first quarter of 1984 the manufacturer sold toothpaste at the same price, but the cost of producing each tube of toothpaste was 5% higher than in the first quarter of 1983. Calculate the net profit which the manufacturer can expect to make in the first quarter of 1984.

(N. Ireland Additional Mathematics, 1984)

2. (a) Explain briefly what is meant by the terms given below, each of which describes a characteristic movement of a time series. Give two examples to illustrate each movement.

(i) secular
(ii) seasonal

(b) The amount of fruit (in thousands of tonnes) handled by a fruit importer over a three year period was as shown in Table 17.10.

Table 17.10

Year	Jan.–Mar.	Apr.–Jun.	Jul.–Sep.	Oct.–Dec.
1982	20	13	8	15
1983	28	20	15	23
1984	38	32	27	37

(i) Plot these data on graph paper using suitable axes and scales.

(ii) Calculate appropriate moving averages to smooth the data and plot these on the graph obtained in (i).

(iii) Draw the trend line and use it to estimate the amount of fruit which the importer will handle between January 1985 and March 1985.

(N. Ireland Additional Mathematics, 1985)

3. (a) State which characteristic movement of a time series would be mainly associated with each of the following.

(i) the monthly output of goods from a firm for which a strike delayed production for several weeks

(ii) the amount of rates paid annually by a householder

(iii) the value of sales in the toy department of a large store, measured monthly

(b) State the reason for using moving averages.

(c) Table 17.11 gives the numbers of patients who attended a health centre on each weekday for the first three weeks in November.

Table 17.11

	Mon.	Tues.	Wed.	Thurs.	Fri.
Week 1	100	92	87	89	97
Week 2	108	99	94	97	104
Week 3	115	106	103	104	112

(i) Plot these data on graph paper using suitable axes and scales.

(ii) Calculate appropriate moving averages and plot these on the graph obtained in (i).

(iii) Draw the trend line and use it to estimate the number of patients expected on Monday of week 4.

(N. Ireland Additional Mathematics, 1986)

4. A car sales firm sells the following numbers of new cars in each quarter of the years 1983–85.

Table 17.12

	1st quarter	2nd quarter	3rd quarter	4th quarter
1983	24	45	36	23
1984	29	52	39	28
1985	36	57	44	35

(i) Plot this information on a graph.

(ii) Calculate 4-point 'moving averages' for this data, and plot the 'moving averages' on the same graph.

(iii) Using your graph, or otherwise, estimate the sales for the first quarter of 1986, on the assumption that established trends will persist.

(N. Ireland Additional Mathematics, 1989 *specimen paper*)

5. (a) With which characteristic movement of a time series is each of the following associated?

(i) the amount spent annually on food by a householder

(ii) the number of floods in Northern Ireland per year

(b) The sales (in thousands of pounds) of a city boutique during the past 15 months were as shown in Table 17.13.

Table 17.13

Month	M	A	M	J	J	A	S	O	N	D	J	F	M	A	M
Sales	38	47	55	65	60	58	67	75	80	70	68	72	80	85	75

(i) Plot the data on graph paper using suitable axes and scales.

(ii) Calculate appropriate moving averages to smooth the data and plot these on the graph obtained in (i).

(iii) Draw the trend line and use it to estimate the sales (in thousands of pounds) for the next month.

(iv) By visual inspection of the trend line, state what type of trend the sales are following in recent months.

(N. Ireland Additional Mathematics, 1988)

6. (a) Explain briefly the difference between cyclical and seasonal movements of a time series.

(b) The numbers of lawnmowers sold by a garden centre in each quarter of the years 1986, 1987 and 1988 are given in Table 17.14.

Table 17.14

Year	Quarter 1	Quarter 2	Quarter 3	Quarter 4
1986	234	926	653	431
1987	275	978	704	472
1988	296	1003	728	498

(i) Plot these data on graph paper using suitable axes and scales.

(ii) Calculate appropriate moving averages and plot these on the graph obtained in (i).

(iii) Draw the trend line and use it to estimate the expected number of lawnmowers sold in the first quarter of 1989.

(N. Ireland Additional Mathematics, 1989)

7. (a) Give a reason why one applies the technique of moving averages to data.

(b) A finance director recorded the profit (in thousands of £) of his company for each month of 1989 as in Table 17.15.

(i) Plot the data on graph paper using suitable scales and axes.

(ii) Calculate appropriate moving averages to smooth the data and plot these averages on the graph obtained in (i).

(iii) Draw the trend line and use it to estimate the profit for January 1990.

Table 17.15

Month	Profit (£'000)
Jan.	1.6
Feb.	1.8
Mar.	2.2
Apr.	2.0
May	2.1
June	2.5
July	2.3
Aug.	2.5
Sept.	2.9
Oct.	2.7
Nov.	2.9
Dec.	3.2

(N. Ireland Additional Mathematics, 1990)

8. (a) With what characteristic movement of a time series would you associate each of the following?

(i) the quarterly sales of iced lollipops in Northern Ireland

(ii) the total number of cars in France recorded annually

(iii) the total number of earthquakes in California recorded every ten years

(b) A department store sold the following numbers of video-recorders over 15 consecutive weekdays.

Table 17.16

	Week 1					Week 2					Week 3				
Day	M	T	W	T	F	M	T	W	T	F	M	T	W	T	F
Number of recorders	8	9	11	13	10	9	10	12	14	12	11	13	15	17	16

(i) Plot these data on graph paper using suitable axes and scales.

(ii) Calculate appropriate moving averages to smooth the time series and plot these on the graph obtained in (i).

(iii) Draw the trend line and use it to estimate the number of video-recorders which the department store can expect to sell on the Monday of week 4.

(N. Ireland Additional Mathematics, 1991)

Answers

EXERCISE 1.1

1. $\dfrac{8x+1}{(x+2)(x-3)}$

2. $\dfrac{x(14x+3)}{4x^2-9}$

3. $\dfrac{2x^2}{x-y}$

4. $\dfrac{x(13x+5)}{4(x+5)}$

5. $\dfrac{x(9x-1)}{x^2-1}$

6. $\dfrac{7x^2-5y^2}{(x+y)^2(x-y)}$

7. $\dfrac{x^3+x^2+xy^2-y^2}{x(x-y)(x+y)}$

8. $\dfrac{9x^2+7xy+10y^2}{6(x-y)(x+y)}$

9. $\dfrac{6x-10}{(x-7)(x^2-4)}$

10. $\dfrac{x(3x^2-1)}{(x-1)(x+1)}$

EXERCISE 1.2

1. $\frac{8}{5}$

2. $\dfrac{3(x+3)}{2x(x-1)}$

3. 2

4. $\dfrac{8z}{3y(x+2y)}$

5. $\dfrac{x+y}{x}$

6. $3x(x-3)$

7. $\dfrac{3(x+y)^3}{2(x-y)^2}$

8. $\dfrac{x(y-x)^2}{3}$

9. $\dfrac{x-4}{x+5}$

EXERCISE 1.3

1. $x = -2$, $y = 4$ or $x = 3$, $y = 9$

2. $x = -\frac{7}{3}$, $y = \frac{98}{3}$ or $x = \frac{5}{2}$, $y = \frac{75}{2}$

3. $x = 5$, $y = -30$ or $x = 6$, $y = -30$

4. $x = -3$, $y = -12$

5. $x = 2.615$, $y = 0.77$ or $x = 2$, $y = 2$

6. $x = 6.894$, $y = -2.671$ or $x = 0.249$, $y = 2.313$

7. $x = -7.772$, $y = 3.193$ or $x = 2.216$, $y = 0.696$

8. $x = 1.943$, $y = -0.886$ or $x = 0.057$, $y = 2.886$

9. $x = -7.2$, $y = -0.4$ or $x = 6$, $y = 4$

10. $x = 0.374$, $y = -1.951$ or $x = 7.126$, $y = 6.151$

11. $x = 0.747$ or $x = -1.784$
$y = 2.807$ $y = -1.622$

12. $x = 2.437$ or $x = -1.437$
$y = 2.291$ $y = -0.291$

13. $x = -0.827$ or $x = 3.736$
$y = 2.942$ $y = 1.421$

14. $x = 7.695$ or $x = -0.195$
$y = 2.065$ $y = -0.565$

15. (i) $(1,2), (-\frac{1}{2}, -2\frac{1}{2})$ (ii) $(\frac{1}{4}, -\frac{1}{4})$

16. $(3,1), (2, -1)$

17. $(2,2), (-0.4, -2.8)$

EXERCISE 1.4

1.	$x = 2$	$y = 3$	$z = 4$
2.	$x = -1$	$y = 5$	$z = 2$
3.	$x = 0$	$y = 2$	$z = -2$
4.	$x = -5$	$y = 3$	$z = -1$
5.	$x = 3.5$	$y = 6$	$z = -4.5$
6.	$x = 5$	$y = 5$	$z = 5$
7.	$x = 7$	$y = -1$	$z = 2$
8.	$x = \frac{5}{13}$	$y = \frac{21}{13}$	$z = \frac{58}{13}$
9.	$x = 6$	$y = 2$	$z = -3$
10.	$x = 2$	$y = 0$	$z = 1$
11.	$x = -4$	$y = 7$	$z = 0$
12.	$x = 4$	$y = -2$	$z = -1$
13.	$x = 1$	$y = -1$	$z = 5$
14.	$x = -3$	$y = 3$	$z = 2$
15.	$x = -\frac{5}{6}$	$y = \frac{8}{3}$	$z = -\frac{23}{6}$

16. (a) The equations are inconsistent

(b) (i) elec = 2p; oil = 10p; gas = 15p
(ii) £1.52

17. (i) $f = 10$; $m = 0.25$; $h = 12$ (ii) £83

18. $x = 3$; $y = -1$; $z = 7$

19. (i) $x = 7$; $y = 5$; $z = 4$ (ii) 8

20. $x = 20$; $y = 15$; $z = 10$; 65p

21. (i) 300 (ii) 90, 60, 150; 50, 100, 150
125, 125, 50 (iii) 3; 5; 6

EXERCISE 2.1

1.	1.779	**2.**	-13.661
3.	1.477	**4.**	-4.695
5.	-2.635	**6.**	-0.412
7.	-7.184	**8.**	1.348
9.	1.498	**10.**	1.477

EXERCISE 2.2

1. (i) 52°, 128° (ii) $-36.25°$, $-143.75°$
(iii) 66.9°, 113.1°

2. (i) 36°, $-36°$ (ii) 72°, $-72°$
(iii) 147°, $-147°$

3. (i) 158°, $-22°$ (ii) 25°, $-155°$
(iii) $-69°$, 111°

4. (i) 44°, 136° (ii) 52°, 128°
(iii) $-78°$, $-102°$

5. (i) 83°, $-83°$ (ii) 21°, $-21°$
(iii) 165.9°, $-165.9°$

6. (i) 15°, $-165°$ (ii) 151°, $-29°$
(iii) 146°, $-34°$

EXERCISE 2.3

1. $-48.6°$, $-131.4°$, $-41.8°$, $-138.2°$

2. $\pm 67.98°$, $\pm 113.6°$

3. $-56.4°$, $-123.6°$, 41.8°, 138.2°

4. 41.8°, 138.2°, 48.6°, 131.4°

5. $\pm 113.6°$, $\pm 138.6°$

6. 30°, 150°, $-19.5°$, $-160.5°$

7. $-90°$

EXERCISE 2.4

1. $\pm 75.52°, \pm 60°$
2. $-19.47°, -160.53°$
3. $45°, 63.43°, -135°, -116.57°$
4. $26.57°, 135°, -153.4°, -45°$
5. $\pm 60°$
6. $30°, 41.81°, 138.19°, 150°$

EXERCISE 2.5

1.	$-\frac{24}{25}, \frac{7}{24}$	2.	$\frac{4}{5}, -\frac{3}{4}$
3.	$\frac{5}{4}, -\frac{3}{5}$	4.	$-\frac{5}{13}, -\frac{5}{12}$
5.	$-\frac{25}{7}, \frac{401}{600}$	6.	$-\frac{8}{17}, -\frac{8}{15}$
7.	$-\frac{15}{17}, -\frac{17}{8}$	8.	$-\frac{7}{25}, -\frac{24}{7}$
9.	$\frac{4}{5}, -\frac{3}{4}$	10.	$\frac{21}{29}, \frac{21}{20}$
11.	$-\frac{25}{7}, \frac{24}{25}$	12.	$\frac{15}{17}, \frac{15}{8}$
13.	$-\frac{15}{17}, \frac{17}{8}$	14.	$\frac{15}{17}, -\frac{8}{15}$
15.	$-\frac{25}{24}, -\frac{7}{25}$	16.	$\frac{3}{5}, \frac{3}{4}$
17.	$-\frac{13}{12}, \frac{5}{13}$	18.	$\frac{8}{17}, \frac{8}{15}$
19.	$-\frac{21}{29}, -\frac{29}{20}$		

EXERCISE 2.6

1.	$C = 62°$	$a = 6.82$	$b = 9.89$
2.	$B = 59°$	$a = 17.84$	$c = 28.07$
3.	$A = 44°$	$b = 9.94$	$c = 10.08$
4.	$E = 63.4°$	$d = 61.30$	$f = 125.12$
5.	$D = 76.3°$	$e = 27.53$	$f = 32.08$
6.	$F = 67.4°$	$d = 6.76$	$e = 6.76$
7.	$Q = 63.4°$	$p = 4.05$	$r = 2.28$
8.	$P = 69.88°$	$q = 69.16$	$r = 57.14$
9.	$Q = 73.4°$	$p = 16.92$	$r = 28.88$
10.	$R = 78°$	$p = 8.36$	$q = 9.10$

EXERCISE 2.7

1.	$A = 28.96°$	$B = 46.57°$	$C = 104.48°$
2.	$A = 18.19°$	$B = 128.68°$	$C = 33.12°$
3.	$A = 55.85°$	$B = 33.48°$	$C = 90.67°$
4.	$A = 39.33°$	$B = 105.31°$	$C = 35.36°$
5.	$A = 114.93°$	$B = 34.82°$	$C = 30.25°$
6.	$a = 10.06$	$B = 47.15°$	$C = 77.82°$
7.	$b = 22.34$	$A = 42.30°$	$C = 67.72°$
8.	$c = 38.32$	$A = 55.79°$	$B = 42.21°$
9.	$b = 35.34$	$A = 30.35°$	$C = 39.65°$
10.	$a = 42.92$	$B = 13.58°$	$C = 32.45°$

EXERCISE 2.8

1.	$C = 80°$	$a = 5.87$	$b = 7.91$
2.	$c = 11.55$	$A = 32.45°$	$B = 85.25°$
3.	$A = 29.54°$	$B = 38.05°$	$C = 112.41°$
4.	$B = 73.50°$	$C = 33°$	$c = 3.98$
5.	$A = 92°$	$b = 5.29$	$c = 7.10$
6.	$A = 18.72°$	$B = 25.33°$	$C = 135.95°$
7.	$a = 7.94$	$B = 40.89$	$C = 79.08°$
8.	$A = 34.77°$	$B = 58.81°$	$C = 86.42°$
9.	$C = 36°$	$a = 15.03$	$b = 8.41$
10.	$A = 15.37°$	$B = 25.89°$	$C = 138.74°$
11.	$A = 43.16°$	$C = 69.84°$	$c = 29.78$
12.	$B = 65°$	$a = 5.22$	$c = 5.90$
13.	$b = 5.56$	$A = 32.49°$	$C = 91.45°$
14.	$a = 9.06$	$B = 18.98°$	$C = 35.97°$
15.	$A = 26.94°$	$B = 43.53°$	$C = 109.52°$
16.	$C = 40.39°$	$A = 76.61°$	$a = 12.01$
17.	$B = 53.46°$	$A = 86.54°$	$a = 6.21$
18.	$c = 7.33$	$A = 25.86°$	$B = 101.08°$
19.	$A = 43°$	$b = 5.61$	$c = 6.09$
20.	$B = 83.5°$	$a = 14.77$	$c = 30.22$

EXERCISE 2.9

1.	$A = 52.04°$	$C = 85.66°$	$c = 6.65$
	$A = 127.96°$	$C = 9.74°$	$c = 1.13$
2.	$A = 63.48°$	$B = 84.32°$	$b = 10.01$
	$A = 116.52°$	$B = 31.28°$	$b = 5.22$
3.	$B = 73.57°$	$C = 50.13°$	$c = 111.63$
	$B = 106.43°$	$C = 17.27°$	$c = 43.18$
4.	$C = 83.48°$	$B = 85.22°$	$b = 122.06$
	$C = 96.52°$	$B = 72.18°$	$b = 116.61$
5.	$B = 56.18°$	$A = 74.52°$	$a = 21.23$
	$B = 123.82°$	$A = 6.88°$	$a = 2.64$
6.	$B = 47.62°$	$C = 102.88°$	$c = 19.80$
	$B = 132.38°$	$C = 18.12°$	$c = 6.32$
7.	$B = 72.89°$	$C = 43.11°$	$c = 107.97$
	$B = 107.11°$	$C = 8.89°$	$c = 24.42$
8.	$C = 71.19°$	$A = 71.71°$	$a = 10.23$
	$C = 108.81°$	$A = 34.09°$	$a = 6.04$
9.	$C = 50.58°$	$A = 98.42°$	$a = 11.52$
	$C = 129.42°$	$A = 19.58°$	$a = 3.9$
10.	$A = 58.53°$	$C = 85.67°$	$c = 8.52$
	$A = 121.47°$	$C = 22.73°$	$c = 3.30$

EXERCISE 2.10

1.	14.31	**2.**	4.69	**3.**	13.38
4.	30.68	**5.**	2.31	**6.**	7.08
7.	55.03	**8.**	2709		

EXERCISE 2.11

1.	122.93; 20.78	**2.**	55.61; 25.63
3.	18.14; 7.13	**4.**	21.04; 8.44
5.	7571; 1334	**6.**	1455; 1390
7.	26.71; 9.05	**8.**	18.17; 7.04
9.	6478; 2506	**10.**	8.19; 2.94

EXERCISE 2.12

1.

(i)	0.26	(ii)	0.59	(iii)	5.10
(iv)	0.82	(v)	6.32	(vi)	3.37
(vii)	0.98	(viii)	5.48	(ix)	0.07
(x)	3.68	(xi)	0.65	(xii)	0.35
(xiii)	5.34	(xiv)	1.94	(xv)	0.97

2.

(i)	314°	(ii)	362°	(iii)	184°
(iv)	42°	(v)	36.2°	(vi)	53.7°
(vii)	148.4°	(viii)	311°	(ix)	227.3°
(x)	721.9°	(xi)	286.5°	(xii)	559.2°
(xiii)	48.7°	(xiv)	407.4°	(xv)	120.3°

3.

	Arc length	Sector area
(i)	4.95 m	6.81 m^2
(ii)	7.68 m	31.50 m^2
(iii)	5.12 m	10.58 m^2
(iv)	10.5 m	13.13 m^2
(v)	2.26 m	10.60 m^2
(vi)	17.93 m	48.86 m^2
(vii)	31.28 m	212.70 m^2
(viii)	2.40 m	9.00 m^2
(ix)	15.40 m	72.78 m^2
(x)	3.94 m	4.65 m^2
(xi)	8.67 m	39.89 m^2
(xii)	0.22 m	0.36 m^2
(xiii)	5.87 m	34.05 m^2
(xiv)	2.50 m	5.52 m^2
(xv)	3.83 m	7.51 m^2
(xvi)	6.12 m	17.20 m^2
(xvii)	9.69 m	36.32 m^2
(xviii)	17.05 m	102.29 m^2
(xix)	1.57 m	2.28 m^2
(xx)	3.46 m	23.19 m^2

4.

	Sector	Triangle	Segment
(i)	$32.29\,cm^2$	$30.09\,cm^2$	$2.2\,cm^2$
(ii)	$102.1\,cm^2$	$88.65\,cm^2$	$13.45\,cm^2$
(iii)	$37.63\,cm^2$	$24.49\,cm^2$	$13.14\,cm^2$
(iv)	$97.34\,cm^2$	$77.19\,cm^2$	$20.15\,cm^2$
(v)	$34.85\,cm^2$	$32.95\,cm^2$	$1.9\,cm^2$
(vi)	$19.67\,cm^2$	$17.62\,cm^2$	$2.05\,cm^2$
(vii)	$126.22\,cm^2$	$120.15\,cm^2$	$6.07\,cm^2$
(viii)	$11.18\,cm^2$	$8.32\,cm^2$	$2.86\,cm^2$
(ix)	$13.4\,cm^2$	$13.02\,cm^2$	$0.38\,cm^2$
(x)	$244.97\,cm^2$	$205.23\,cm^2$	$39.74\,cm^2$

5. (i) 4.44 m (ii) $4.57\,m^2$

6. (i) 0.644 rad (ii) 7.22 cm (iii) $2.04\,cm^2$

7. (i) $\frac{1}{2}$ (ii) $8\,m^2$

EXERCISE 2.13

1. (a) 109.47° or − 109.47°
   (b) (ii) 5.736 cm
   (iii) $11.746\,cm^2$
   (iv) $3.526\,cm^2$

2. (i) 330.61 m
   (ii) 72.84°; 6.06°
   (iii) 464.76 m
   (iv) 177.98 m
   (v) 137.12 m; 226.94 m

3. (i) 29.68 m
   (ii) 15.15 m
   (iii) 29.12°
   (iv) 4.04 m

4. (i) 134.56 m (ii) 48°
   (iii) 122.09 m (iv) 6.99°
   (v) 125.01° (vi) 26.56 m

EXERCISE 2.14

1. (a) 8 cm (b) 14.49 cm (c) 20.19° (d) 24.44°
2. (a) 5.45 cm (b) 4.84 cm (c) 62.70°
3. (a) 4.87 cm (b) 22.09° (c) 21.28° (d) 54.31°
4. (a) 9.43 cm (b) 9.90 cm (c) 32.01° (d) 17.65°
5. (a) 16.97 cm (b) 4.53 cm (c) 12.73 cm (d) 9 cm
6. (a) 71.94° (b) 15.33 cm (c) 58.39° (d) 54.34°
7. $66.69\,m^2$; 18.11°
8. 123.6 m; 224.7 m
9. 296.2 m
10. (a) 130.3 m (b) 83.21 m (c) 35.49°
11. (a) 86.96 m (b) 54.34 m (c) 32.59°
12. 90.03 m
13. (a) 41.85° (b) 4.67 cm (c) 32.88°
14. (i) 1.368 m (ii) 1.286 m (iii) 21.87° (iv) 20.90°
15. (i) 67.38° (ii) 1.393 m (iii) 0.9231 m (iv) 41.5°
16. (i) 14.12 m (ii) 17.65°
17. 82.68 cm; 76.6 cm; 82.94 cm
18. (i) 1.376 cm (ii) 1.701 cm (iii) $6.88\,cm^2$ (iv) 79.3° (v) 81.31°
19. (i) $a = 9$, $b = 6$ (iii) 23.41°
20. (i) 24.78° (ii) 4.615 cm (iii) 69.91° (iv) 52.43°
21. (i) 10 (ii) 72.08° (iii) 8.617 (iv) 7.762 (v) 75.05°

EXERCISE 3.1

1. (i) 1.724 (ii) 1.996 (iii) 1.623
(iv) 1.924 (v) -1 (vi) -0.155
(vii) 0.431 (viii) 0.628 (ix) 2.004
(x) 2.678 (xi) 2.449 (xii) 2.960
(xiii) 3.242 (xiv) 3.010 (xv) 5.000

2. (i) 1 (ii) 2 (iii) 3 (iv) 4
(v) -1 (vi) -2 (vii) $\frac{1}{2}$ (viii) $\frac{1}{3}$
(ix) $\frac{3}{2}$ (x) $\frac{2}{3}$ (xi) $\frac{3}{2}$ (xii) $\frac{5}{2}$
(xiii) $-\frac{1}{2}$ (xiv) $-\frac{3}{2}$ (xv) $\frac{2}{3}$ (xvi) $\frac{5}{3}$

3. (i) $\log a + \log b + \log c$
(ii) $\log a + 2\log b + 3\log c$
(iii) $\log a + 2(\log b + \log c)$
(iv) $\log a + \log b - \log c$
(v) $\log a - (\log b + \log c)$
(vi) $3\log a - 2\log b - \log c$
(vii) $3(\log a + \log b) - 2\log c$
(viii) $\log a + \frac{1}{2}\log b$
(ix) $\log a + \frac{1}{2}(\log b + \log c)$
(x) $\frac{1}{2}\log a + \frac{1}{2}\log b + \frac{1}{2}\log c$
(xi) $\log a + \log b - \frac{1}{2}\log c$
(xii) $\frac{1}{2}\log a + \frac{1}{2}\log b - \frac{1}{2}\log c$
(xiii) $\frac{1}{2}\log a - \log b - \log c$
(xiv) $\log a + \frac{1}{2}\log b - \frac{1}{2}\log c$
(xv) $\frac{3}{2}\log a + \log b - \frac{1}{2}\log c$
(xvi) $1 + \log a + \log b + \log c$
(xvii) $1 + 2\log a + \log b + 3\log c$
(xviii) $2 + 2\log a$
(xix) $\frac{1}{2} + \frac{1}{2}\log a - \log b - \log c$
(xx) $\frac{1}{2} + \log a - \log b - \frac{1}{2}\log c$

4. (i) 0.24 (ii) 1.2 (iii) 1.68
(iv) 1.44 (v) 0.3 (vi) 0.42
(vii) 0.72 (viii) 0.54 (ix) 0.12

EXERCISE 3.2

1. (i) 1000 (ii) 3.16×10^{-5}
(iii) 1.585×10^{7} (iv) 251.189
(v) 1.0×10^{9} (vi) 5.012×10^{-9}
(vii) 2.512×10^{5} (viii) 5.012
(ix) 2.512 (x) 0.0398
(xi) 5.62×10^{-6} (xii) 31.623
(xiii) 1.995 (xiv) 338.844
(xv) 208.93 (xvi) 3.802

2. (i) 3.170 (ii) 2.044
(iii) 2.104 (iv) 3.841
(v) 1.077 (vi) 0.889
(vii) 4.551 (viii) 3.872
(ix) 1.657 (x) 4.590
(xi) 0.471 (xii) -0.313
(xiii) -2.710 (xiv) -2.819
(xv) 1.066 (xvi) -1.950
(xvii) -0.212 (xviii) 0.042
(xix) -0.050 (xx) 0.202

3. (i) 1, 1.585 (ii) 0, 1.893
(iii) 1.585, 2 (iv) 0.431, 0.683
(v) 0, 1.953 (vi) 1.585
(vii) 0.631, 2.524 (viii) 1, 3.907
(ix) 1.631, 1.771 (x) 1, 2
(xi) 1.585 (xii) 0, 1.893
(xiii) -1, 1 (xiv) -0.569, 0
(xv) -0.369, 0.631

4. (a) $\frac{1}{4}$, 2 (b) 0, 1.89

EXERCISE 3.3

1.
$\log T = n\log l + \log k$
$\log T = \frac{1}{2}\log l + 0.3$
$\Rightarrow \quad n = \frac{1}{2}$
$\log k = 0.3$
$\Rightarrow \quad k = 2$
Answer: $T = 2l^{\frac{1}{2}}$

2.
$\log D = n\log T + \log k$
$\log D = 2\log T + 0.69$
$\Rightarrow \quad n = 2$
$\log k = 0.69$
$\Rightarrow \quad k = 5$
Answer: $D = 5T^2$

3.
$\log I = n\log d + \log k$
$\log I = -2\log d + 1.65$
$\Rightarrow \quad n = -2$
$\log k = 1.65$
$\Rightarrow \quad k = 45$
Answer: $I = 45d^{-2}$

4.
$\log P = n\log s + \log k$
$\log P = 3\log s - 3.55$
$\Rightarrow \quad n = 3$
$\log k = -3.55$
$\Rightarrow \quad k = 0.000\,28$
Answer: $P = (0.000\,28)s^3$

5.
$\log R = n\log s + \log k$
$\log R = \frac{1}{2}\log s + 2.857$
$\Rightarrow \quad n = \frac{1}{2}$
$\log k = 2.857$
$\Rightarrow \quad k = 720$
Answer: $R = 720s^{\frac{1}{2}}$

6.
$\log F = n\log l + \log k$
$\log F = -\log l + 4.22$
$\Rightarrow \quad n = -1$
$\log k = 4.22$
$\Rightarrow \quad k = 16\,500$
Answer: $F = 16\,500\,l^{-1}$

7.
$\log s = n\log R + \log k$
$\log s = \frac{1}{2}\log R + 0.3$
$\Rightarrow \quad n = \frac{1}{2}$
$\log k = 0.3$
$\Rightarrow \quad k = 2$
Answer: $s = 2R^{\frac{1}{2}}$

8.
$\log F = n\log \lambda + \log k$
$\log F = -\log \lambda + 5.47$
$\Rightarrow \quad n = -1$
$\log k = 5.47$
$\Rightarrow \quad k = 300\,000$
Answer: $F = 300\,000\,\lambda^{-1}$

9.
$\log e = n\log T + \log k$
$\log e = \log T - 0.477$
$\Rightarrow \quad n = 1$
$\log k = -0.477$
$\Rightarrow \quad k = \frac{1}{3}$
Answer: $e = \frac{1}{3}T$

10.
$\log P = n\log V + \log k$
$\log P = -\log V + 3.07$
$\Rightarrow \quad n = -1$
$\log k = 3.07$
$\Rightarrow \quad k = 1200$
Answer: $P = 1200\,V^{-1}$

11. $k = 0.2$, $n = 1.5$
(i) 688 days (ii) 108.2×10^6 km

12. $a = 0.92$, $k = -0.29$
(i) 218 000 K (ii) 8.18×10^6 km

13. $k = 0.045$, $n = 2$
(i) 55.1 N (ii) 66.67 m/s

EXERCISE 4.1

1. 2 **2.** 5
3. 12 **4.** $2x + 3$
5. $4x + 4$ **6.** $-2x - 2$
7. $3x^2$ **8.** $3x^2 + 4$
9. $3x^2 + 4x + 7$ **10.** $6x^2 + 10x - 9$

EXERCISE 4.2

1. $4x^3$ **2.** $7x^6$
3. $9x^8$ **4.** $6x^2$
5. $32x^7$ **6.** $28x^3$
7. $\dfrac{3x^2}{4}$ **8.** $3x^5$

9. $\frac{9x^5}{2}$
10. $-15x^4$
11. $-36x^3$
12. $-15x^2$
13. $3mx^2$
14. $7px^6$
15. $-4nx^3$
16. 2
17. 5
18. 9
19. 0
20. 0
21. 0
22. $2x + 7$
23. $4x^3 - 4x$
24. $5x^4 - 6x + 7$
25. $12x^2 + 14x + 2$
26. $4x + 5$
27. $27x^2 + 4x + 7$
28. $\frac{3x^2}{2} + \frac{x}{2} + \frac{1}{2}$
29. $2ax + b$
30. $2x + 14$
31. $2x - 2$
32. $8x + 20$
33. $2x + 1$
34. $2x + 4$
35. $4x + 15$
36. $28x + 67$
37. $12x + 22$
38. $30x + 16$
39. $24x - 29$
40. $4x^3 + 4x$
41. $6x^5 + 30x^2$
42. $8x^7 - 40x^3$
43. $3x^2 + 42x + 147$
44. $4x^3 + 24x^2 + 48x + 32$
45. $3x^2 - 2x - 24$
46. $6x^5 - 36x^3 + 54x$
47. $3x^2 - 8x - 3$
48. $4x^3 + 30x^2 + 70x + 50$

EXERCISE 4.3

1. $\frac{5x^{\frac{3}{2}}}{2}$
2. $\frac{2}{5\sqrt[5]{x^3}}$
3. $\frac{1}{3\sqrt[3]{x^2}}$
4. $\frac{4\sqrt[3]{x}}{3}$
5. $\frac{5}{6\sqrt[6]{x}}$
6. $-\frac{3}{x^4}$
7. $-\frac{1}{x^2}$
8. $-\frac{2}{x^3}$
9. $-\frac{7}{x^8}$
10. $-\frac{5}{x^6}$
11. $-\frac{4}{x^3}$
12. $-\frac{9}{x^4}$
13. $-\frac{8}{x^5}$
14. $-\frac{10}{x^3}$
15. $-\frac{6}{x^3}$
16. $-\frac{3}{2x^4}$
17. $-\frac{1}{x^4}$
18. $-\frac{2}{3x^3}$
19. $-\frac{1}{2x^3}$
20. $-\frac{3}{5x^4}$

EXERCISE 4.4

1. $\frac{1}{3x^{\frac{2}{3}}}$
2. $\frac{1}{2\sqrt{x}}$
3. $\frac{2}{3\sqrt[3]{x}}$
4. $\frac{3}{4\sqrt[4]{x}}$
5. $\frac{3\sqrt{x}}{2}$
6. $-\frac{4}{x^5}$
7. $-\frac{2}{x^3}$
8. $-\frac{3}{x^4}$
9. $-\frac{5}{x^6}$
10. $-\frac{6}{x^7}$
11. $-\frac{2}{x^2}$
12. $-\frac{1}{x^2}$
13. $-\frac{3}{x^2}$
14. $-\frac{4}{x^3}$
15. $-\frac{6}{x^4}$
16. $-\frac{1}{2x^{\frac{3}{2}}}$
17. $-\frac{1}{3x^{\frac{4}{3}}}$
18. $-\frac{2}{x^4}$
19. $-\frac{8}{5x^3}$
20. $-\frac{4}{3x^5}$
21. $-\frac{1}{4x^{\frac{3}{2}}}$
22. $-\frac{1}{9x^{\frac{4}{3}}}$
23. $-\frac{1}{4x^{\frac{3}{2}}}$
24. $-\frac{1}{2x^2}$
25. $-\frac{10}{x^6}$

EXERCISE 4.5

1. $1 - \frac{2}{x^3}$
2. $\frac{1}{2\sqrt{x}} - \frac{1}{x^2}$
3. $3 + 3x^{-2}$
4. $2x + 2x^{-3}$
5. $6x + \frac{1}{2x^3}$
6. $\frac{1}{3} - 3x^{-2}$

7. $\frac{2x}{7} - \frac{14}{x^3}$

8. $12x^3 - \frac{3}{2x^4}$

9. $\frac{7}{2\sqrt{x}} - \frac{16}{3x^{\frac{5}{3}}}$

10. $2x + \frac{5x^{\frac{3}{2}}}{2}$

11. $\frac{1}{3x^{\frac{2}{3}}} - \frac{2}{3x^{\frac{1}{3}}}$

12. $3x^2 + 3 - 3x^{-2} - 3x^{-4}$

13. $6x^5 - 6x^{-7}$

14. $-4x^{-2}$

15. $3x^{-2}$

16. $1 + 6x^{-2}$

17. $4x - 14x^{-3}$

18. $\frac{1}{2\sqrt{x}} + 2x^{-\frac{3}{2}}$

19. $-\frac{3}{2x^{\frac{3}{2}}} - \frac{1}{2\sqrt{x}}$

20. $-8x^{-3} - 27x^{-4}$

21. $\frac{1}{8\sqrt{x}} - \frac{1}{4x^{\frac{3}{2}}}$

22. $-2x^{-3} - 3x^{-2} + 1$

23. $-6x^{-2} + 1$

24. $\frac{3\sqrt{x}}{2} - \frac{5}{2\sqrt{x}}$

25. $-\frac{1}{x^2} - \frac{2}{3x^3} - \frac{2}{x^4} - \frac{20}{3x^5}$

26. $8x - \frac{8}{\sqrt{x}} - \frac{16}{x^2}$

27. $2\frac{1}{2}x^{\frac{3}{2}} + 9\sqrt{x} + \frac{6}{\sqrt{x}} - \frac{4}{x^{\frac{3}{2}}}$

28. $2x - \frac{1}{\sqrt{x}} - \frac{1}{x^2}$

29. $\frac{5x^{\frac{2}{3}}}{3} + \frac{5}{3x^{\frac{1}{3}}} - \frac{7}{6x^{\frac{4}{3}}}$

30. $\frac{5x^{\frac{3}{2}}}{4} + \frac{9x^{\frac{1}{2}}}{4} + \frac{3}{4x^{\frac{1}{2}}} - \frac{1}{4x^{\frac{3}{2}}}$

EXERCISE 4.6

1. $12;\ y = 12x - 16$
2. $3;\ y = 3x$
3. $\frac{1}{6};\ y = \frac{x}{6} + \frac{3}{2}$
4. $48;\ y = 48x + 64$
5. $-1;\ y = -x + 2$
6. $\frac{1}{27};\ y = \frac{x}{27} + 2$

EXERCISE 4.7

1. $y = -6x - 9$
2. $y = 3x - 2$
3. $y = 4x$
4. $y = -6x + 10$
5. $y = 20x - 18$
6. $y = -96x - 144$
7. $y = \frac{x}{12} + \frac{4}{3}$
8. $y = -\frac{2x}{9} + 1$
9. $y = \frac{5x}{8} + 10$
10. $y = 16x - 22$

EXERCISE 4.8

1. $y = 12x - 8$
2. $(2,22)$
3. $y = 12x - 27;\ (0, -27);\ (\frac{9}{4},0)$
4. $y = 16x - 24;\ (\frac{3}{4}, -12)$
5. $(\frac{1}{2}, -\frac{3}{2});\ (-\frac{1}{2},\frac{3}{2})$
6. $y = -3x + 2$
7. $(3,0)(-1, -\frac{8}{3});\ y = 6x - 18;\ y = 6x + \frac{10}{3}$
8. $y = 3x - 9$
9. $M = (\frac{5}{2},0),\ N = (0, -5);\ 5.59$
10. $y = 3x - \frac{3}{2};\ y = -3x - 3;\ (-\frac{1}{4}, -2\frac{1}{4})$
11. $y = 2x + 1;\ y = -4x - 2$
12. $y = -12x - 9;\ y = 12x - 9;\ (0, -9)$

13. (a) $\dfrac{9\sqrt{x}}{2}+\dfrac{2}{3x^{\frac{5}{3}}}$ (b) $y=4x+2$

14. (a) $-\dfrac{5}{4x^{\frac{5}{4}}}$

(b) (i) $y=-3x+6$
(ii) 6 square units

15. (i) $6x+3x^{-\frac{5}{2}}$
(ii) $y=12x-16$

16. (i) 5,6 (ii) $2y=7x+12$

17. (2,4), -6

EXERCISE 4.9

1. (0,0) min.

2. (2, $-$4) min.

3. (0,6) max.

4. (0,0) min.

5. (1, $-$2) min., ($-$1,2) max.

6. (1, $-$4) min., ($-$1,4) max.

7. $(\frac{1}{2},-\frac{25}{4})$ min.

8. $(-\frac{5}{4},-\frac{1}{8})$ min.

9. $(1,\frac{4}{3})$ max., (3,0) min.

10. (2.215, $-$4.225) min., (0.451, 1.262) max.

EXERCISE 4.10

1. Intersects x axis: (0,0)(6,0)
Intersects y axis: (0,0)
Turning point: (3, $-$9) min.

2. Intersects x axis: ($-$3,0)(2,0)
Intersects y axis: (0, $-$6)
Turning point: ($-$0.5, $-$6.25) min.

3. Intersects x axis: ($-$3,0)(3,0)
Intersects y axis: (0,9)
Turning point: (0,9) max.

4. Intersects x axis: (2,0)
Intersects y axis: (0,4)
Turning point: (2,0) min.

5. Intersects y axis: (0,4)
Turning point: ($-$0.5,3.75) min.

6. Intersects y axis: (0, $-$9)
Turning point: (1, $-$8) max.

7. Intersects x axis: (0,0)(2,0)
Intersects y axis: (0, 0)
Turning points: (2,0) min., $(\frac{2}{3},\frac{32}{27})$ max.

8. Intersects x axis: ($-$2,0)(0,0)(2,0)
Intersects y axis: (0,0)
Turning points: $\left(\dfrac{2}{\sqrt{3}},\dfrac{16}{3\sqrt{3}}\right)$ max.,
$\left(-\dfrac{2}{\sqrt{3}},-\dfrac{16}{3\sqrt{3}}\right)$ min

9. Intersects x axis: (0,0)(4,0)
Intersects y axis: (0,0)
Turning points: (0,0) min.,
$\left(\dfrac{8}{3},\dfrac{256}{27}\right)$ max.

10. Intersects x axis: $(-\sqrt{3},0)(-\sqrt{2},0)$
$(\sqrt{2},0)(\sqrt{3},0)$
Intersects y axis: (0,6)
Turning points: $(-\sqrt{2.5},-0.25)$ min
$(\sqrt{2.5},-0.25)$ min.,
(0, 6) max.

EXERCISE 4.11

1. $800\,m^2$ **2.** $128\,cm^3$

3. $x^2+\dfrac{16}{x}$, 2 m

4. 4 cm × 12 cm × 6 cm

5. $6\frac{2}{3}$ cm × 10 cm × 4 cm

6. (i) $S=2\pi rl+4\pi r^2$

(iii) $S=\dfrac{\pi}{3r}+\dfrac{4\pi r^2}{3}$

(iv) $\pi\,m^2$

7. (i) $V=4x^3-10x^2+6x$

(ii) $A=6-4x^2$

(iii) $P=400x^3-950x^2+600x-85$

(iv) 29.16

8. (i) $l = (\pi + 2)r + 2x$

(ii) $\frac{\pi r^2}{2} + 2rx$

(iv) £15.12, $r = 1.06$ m, $x = 1.06$ m

9. $A = 2\pi rh + \pi r^2$; $h = \frac{64}{r^2}$; $r = 4$ cm; $h = 4$ cm; $A = 48\pi$ cm^2

10. 1151 cm^2

11. (i) (a) 10 cm = CP; 6 cm = CD

(b) angle DCP = NQC; angle CPD = NCQ; angle CDP = CNQ; scale factor = 3

(c) QC = 30 cm

12. $y = \frac{27 - 4x^3}{\pi x^2}$ (i) 1.5 (ii) 54, 1.91, min.

13. (a) $(\frac{1}{2}, 8)$ min. (b) $\frac{60x - 5x^2}{12}$, 15 cm^2

14. (a) $(10x + 2y)$ cm; $225x^2 - 5x^3$; (30,75) max.

EXERCISE 5.1

1. (a) (i) 4 (ii) 3

(b) g, h, i (c) b, e, h, k

(d) (i) b (ii) d (iii) i

(e) (i) first row, first column

(ii) second row, second column

(iii) third row, second column

(iv) first row, third column

(v) fourth row, first column

(vi) fourth row, third column

2. (a) 2,1; -6 (b) 2,2; 7 (c) 2,3; s

(d) 3,2; 0 (e) 2,4; 5 (f) 3,3; 4

4. (a) 3×4 (b) 3×3 (c) 3×2

(d) 1×3 (e) 4×1 (f) 2×4

6. **A** = **K**, **B** = **M**, **C** = **L**

7. 2×3, 2×2, 1×3, 1×3, 3×1, 2×1, 2×3, 2×2, 1×3, 2×3, 1×3, 2×2

8. (a) $x = -4, y = 1$ (b) $x = 2, y = 2$

(c) $x = 3, y = 1$ (d) $x = 4, y = -3$

EXERCISE 5.2

1. $\begin{pmatrix} 10 \\ 10 \end{pmatrix}$ **2.** $\begin{pmatrix} 5 \\ 1 \end{pmatrix}$ **3.** $\begin{pmatrix} -2x \\ 3y \end{pmatrix}$

4. $\begin{pmatrix} 2a+3 \\ b+4 \end{pmatrix}$ **5.** $\begin{pmatrix} a+2c \\ b+d \end{pmatrix}$ **6.** $\begin{pmatrix} 0 \\ 0 \end{pmatrix}$

7. $(4 \quad 6)$ **8.** $(5 \quad -5)$

9. $\begin{pmatrix} 6 & 2 \\ 3 & 9 \end{pmatrix}$ **10.** $\begin{pmatrix} 5 & 2 \\ 2 & 9 \end{pmatrix}$

11. $\begin{pmatrix} 1 & 2 \\ 2 & -1 \end{pmatrix}$ **12.** $\begin{pmatrix} 4 & 0 \\ 0 & 1 \end{pmatrix}$

13. $\begin{pmatrix} 15 & 3 & 13 \\ 3 & 3 & 0 \end{pmatrix}$ **14.** $\begin{pmatrix} 4x & 3y \\ x & 0 \end{pmatrix}$

15. $\begin{pmatrix} 4a & 4b \\ -2a & 2b \end{pmatrix}$

16. (a) (i) $\begin{pmatrix} 6 & 8 \\ -1 & 3 \end{pmatrix}$ (ii) $\begin{pmatrix} 5 & 2 \\ -2 & -5 \end{pmatrix}$

(iii) $\begin{pmatrix} 9 & 9 \\ 0 & -2 \end{pmatrix}$ (iv) $\begin{pmatrix} 9 & 9 \\ 0 & -2 \end{pmatrix}$

(b) Yes. Associative law of addition holds for matrix addition.

17. (a) $\begin{pmatrix} 4 \\ -6 \end{pmatrix}$ (b) $\begin{pmatrix} -3 \\ -2 \\ -1 \end{pmatrix}$

(c) $\begin{pmatrix} -8 & 5 \\ 3 & -2 \end{pmatrix}$ (d) $(-2 \quad 3)$

(e) $\begin{pmatrix} -4 & -2 & 1 \\ 3 & -8 & -2 \end{pmatrix}$

18. $\mathbf{A} = \begin{pmatrix} -1 & -3 \\ -5 & 7 \\ 0 & -9 \end{pmatrix}$

19. (a) $\begin{pmatrix} 2 \\ 3 \end{pmatrix}$ (b) $\begin{pmatrix} 7 \\ 0 \end{pmatrix}$

(c) $\begin{pmatrix} 0 \\ 9 \end{pmatrix}$ (d) $\begin{pmatrix} -3p \\ 4q \end{pmatrix}$

(e) $\begin{pmatrix} a-2 \\ 2-3y \end{pmatrix}$ (f) $\begin{pmatrix} r-t \\ s-2u \end{pmatrix}$

20. (a) $\begin{pmatrix} 3 & 5 \\ 2 & 0 \end{pmatrix}$ (b) $\begin{pmatrix} 3 & 0 \\ 10 & 5 \end{pmatrix}$

(c) $\begin{pmatrix} -3 & 7 \\ -2 & -2 \end{pmatrix}$ (d) $\begin{pmatrix} a & 4 \\ 3 & 5b \end{pmatrix}$

21. (a) $\begin{pmatrix} 5 & 2 \\ -3 & 5 \end{pmatrix}$ (b) $\begin{pmatrix} 5 & 9 \\ -1 & 1 \end{pmatrix}$

(c) $\begin{pmatrix} 7 & 6 \\ -2 & 5 \end{pmatrix}$ (d) $\begin{pmatrix} 0 & -7 \\ -2 & 4 \end{pmatrix}$

(e) $\begin{pmatrix} 1 & 1 \\ -3 & 1 \end{pmatrix}$ (f) $\begin{pmatrix} 0 & 7 \\ 2 & -4 \end{pmatrix}$

(g) $\begin{pmatrix} 10 & 11 \\ -4 & 6 \end{pmatrix}$ (h) $\begin{pmatrix} 1 & 1 \\ -3 & 1 \end{pmatrix}$

22. $\mathbf{X} = \begin{pmatrix} -8 \\ 5 \end{pmatrix}$

23. $a = 5, \quad b = 2, \quad c = 2$

24. (a) $\begin{pmatrix} -1 & 3 \\ 9 & 5 \end{pmatrix}$ (b) $\begin{pmatrix} -3 & -8 \\ 3 & -1 \end{pmatrix}$

(c) $\begin{pmatrix} 1 & 2 \\ 2 & 13 \end{pmatrix}$ (d) $\begin{pmatrix} 5 & -3 \\ 0 & 2 \end{pmatrix}$

25. (a) $a = 4, \quad b = 0, \quad c = 2, \quad d = 4$

(b) $a = 3, \quad b = 5, \quad c = -13, \quad d = 3$

26. (a) $\begin{pmatrix} 4 & 2 \\ 0 & 4 \\ 3 & 9 \end{pmatrix}$ (b) $\begin{pmatrix} -1 & 2 \\ 1 & 1 \\ -1 & 1 \end{pmatrix}$

(c) $\begin{pmatrix} 1 & 1 \\ 2 & 1 \\ 2 & 5 \end{pmatrix}$ (d) $\begin{pmatrix} 1 & 1 \\ 2 & 1 \\ 2 & 5 \end{pmatrix}$

EXERCISE 5.3

1. (a) $\begin{pmatrix} 10 & 6 \\ 0 & 26 \end{pmatrix}$ (b) $\begin{pmatrix} 6 \\ 15 \end{pmatrix}$

(c) $\begin{pmatrix} 2 & -4 & 0 \\ 6 & -10 & 4 \end{pmatrix}$

(d) $\begin{pmatrix} 27 & 15 \\ 6 & 24 \end{pmatrix}$ (e) $\begin{pmatrix} -3 \\ 2 \\ 0 \end{pmatrix}$

(f) $\begin{pmatrix} -4 & -8 & -12 \end{pmatrix}$

(g) $\begin{pmatrix} 6 & 3 & -9 \\ 15 & 12 & 0 \end{pmatrix}$

(h) $\begin{pmatrix} -2 & 6 & -4 \\ 4 & -6 & -8 \end{pmatrix}$

(i) $\begin{pmatrix} 4a & -4b & 12c \\ 8a & -28b & 20c \end{pmatrix}$

2. (a) $\begin{pmatrix} 5 & 0 & -2 \\ 6 & 4 & -3 \end{pmatrix}$

(b) $\begin{pmatrix} 15 & 0 & -6 \\ 18 & 12 & -9 \end{pmatrix}$

(c) $\begin{pmatrix} 6 & -3 & 0 \\ 15 & 0 & -9 \end{pmatrix}$

(d) $\begin{pmatrix} 9 & 3 & -6 \\ 3 & 12 & 0 \end{pmatrix}$

(e) $\begin{pmatrix} 15 & 0 & -6 \\ 18 & 12 & -9 \end{pmatrix}$

(f) $\begin{pmatrix} 12 & -6 & 0 \\ 30 & 0 & -18 \end{pmatrix}$

(g) $\begin{pmatrix} 12 & -6 & 0 \\ 30 & 0 & -18 \end{pmatrix}$

(h) $\begin{pmatrix} 27 & 9 & -18 \\ 9 & 36 & 0 \end{pmatrix}$

(i) $\begin{pmatrix} 27 & 9 & -18 \\ 9 & 36 & 0 \end{pmatrix}$

3. (a) $\begin{pmatrix} 2 & 3 \\ 4 & 1 \end{pmatrix}$ (b) $\begin{pmatrix} 3 & -1 \\ -4 & 0 \end{pmatrix}$

(c) $\begin{pmatrix} 1 & -2 & 0 \\ -3 & 0 & 2 \end{pmatrix}$

(d) $\begin{pmatrix} 4 & -4 & 0 \\ 12 & 2 & 8 \end{pmatrix}$

4. (a) $\begin{pmatrix} 6 & 2 \\ -8 & 4 \end{pmatrix}$ (b) $\begin{pmatrix} -15 & 30 \\ 24 & 27 \end{pmatrix}$

(c) $\begin{pmatrix} 12 & 4 \\ -16 & 8 \end{pmatrix}$ (d) $\begin{pmatrix} -25 & 50 \\ 40 & 45 \end{pmatrix}$

(e) $\begin{pmatrix} -2 & 11 \\ 4 & 11 \end{pmatrix}$ (f) $\begin{pmatrix} -6 & 33 \\ 12 & 33 \end{pmatrix}$

(g) $\begin{pmatrix} 8 & -9 \\ -12 & -7 \end{pmatrix}$

(h) $\begin{pmatrix} 16 & -18 \\ -24 & -14 \end{pmatrix}$

(i) $\begin{pmatrix} -6 & 33 \\ 12 & 33 \end{pmatrix}$ (j) $\begin{pmatrix} 16 & -18 \\ -24 & -14 \end{pmatrix}$

(k) $\begin{pmatrix} 5 & 25 \\ -4 & 28 \end{pmatrix}$ (l) $\begin{pmatrix} 23 & -4 \\ -32 & 3 \end{pmatrix}$

5. (a) $\begin{pmatrix} 15 & 13 & 5 \\ -2 & -17 & 15 \end{pmatrix}$

(b) $\begin{pmatrix} 6 & -4 & 0 \\ 14 & 8 & 10 \end{pmatrix}$

6. (a) $\begin{pmatrix} 2 & 3 \\ 4 & 1 \end{pmatrix}$ (b) $\begin{pmatrix} 3 & -4 \\ 2 & -1 \end{pmatrix}$

(c) $\begin{pmatrix} 3 & 4 \\ 2 & 2 \end{pmatrix}$ (d) $\begin{pmatrix} -3 & 2 \\ 6 & 2 \end{pmatrix}$

7. (a) $\begin{pmatrix} -2 \\ 21 \\ -5 \end{pmatrix}$ (b) $\begin{pmatrix} 1 & -7 & 1 \\ 2 & 11 & -6 \end{pmatrix}$

8. $a = 4$, $b = 7$, $c = 5$, $d = -4$

9. (a) (13) (b) (62) (c) (23) (d) (32)

(e) (-5) (f) $(ad + be + cf)$

10. (a) 1 (b) 4 (c) ± 2 (d) ± 5

11. (a) $\begin{pmatrix} 17 \\ 39 \end{pmatrix}$ (b) $\begin{pmatrix} 49 \\ 39 \end{pmatrix}$ (c) $\begin{pmatrix} -40 \\ 54 \end{pmatrix}$

(d) $\begin{pmatrix} -28 \\ 8 \end{pmatrix}$ (e) $\begin{pmatrix} 48 \\ 0 \end{pmatrix}$ (f) $\begin{pmatrix} 21 \\ 55 \end{pmatrix}$

(g) $\begin{pmatrix} 7 \\ 7 \end{pmatrix}$ (h) $\begin{pmatrix} -2a \\ -2a \end{pmatrix}$ (i) $\begin{pmatrix} -7a \\ -6a \end{pmatrix}$

12. **XY**, **YZ**, **ZY**, **ZX**

$\mathbf{XY} = \begin{pmatrix} 4 \\ -10 \end{pmatrix}$ $\mathbf{YZ} = \begin{pmatrix} -10 & -14 \\ 5 & 7 \end{pmatrix}$

$\mathbf{ZY} = (-3)$ $\mathbf{ZX} = (16 \quad -18)$

13. (a) $\begin{pmatrix} 7 \\ 5 \end{pmatrix}$ (b) Not possible

(c) (29) (d) $\begin{pmatrix} 14 \\ 32 \\ 50 \end{pmatrix}$ (e) Not possible

(f) $\begin{pmatrix} 13 \\ 1 \\ 6 \end{pmatrix}$ (g) Not possible

(h) $(27 \quad 7)$

14. (a) $x - y = 5$, $x + y = 11$; $x = 8$, $y = 3$

(b) $x - y = 0$, $x + y = 8$; $x = 4$, $y = 4$

(c) $3x - 4y = 18$, $5x + y = 7$; $x = 2$, $y = -3$

(d) $5x - 3y = 9$, $7x - 6y = 9$; $x = 3$, $y = 2$

(e) $2x - y = 10$, $5x - 6y = -45$; $x = 15$, $y = 20$

(f) $2x + y = 11$, $2y - x = 2$; $x = 4$, $y = 3$

15. (a) $\begin{pmatrix} 29 & 22 \\ 18 & 16 \end{pmatrix}$ (b) $\begin{pmatrix} 19 & 33 \\ 11 & 17 \end{pmatrix}$

(c) $\begin{pmatrix} 31 & 25 \\ 17 & 13 \end{pmatrix}$ (d) $\begin{pmatrix} 10 & 7 \\ 22 & 15 \end{pmatrix}$

(e) $\begin{pmatrix} 4 & 10 & 6 \\ 14 & 3 & -7 \end{pmatrix}$

(f) $\begin{pmatrix} 7 & 3 & -5 & 2 \\ -7 & -3 & 5 & -2 \end{pmatrix}$

16. (a) $\begin{pmatrix} 2 & -3 \\ 5 & 6 \end{pmatrix}$ (b) $\begin{pmatrix} 7 & -2 \\ 1 & 5 \end{pmatrix}$

(c) $\begin{pmatrix} p & q \\ r & s \end{pmatrix}$ (d) $\begin{pmatrix} p & q \\ r & s \end{pmatrix}$

(e) $\begin{pmatrix} 1 & 0 \\ 0 & 1 \end{pmatrix}$ (f) $\begin{pmatrix} 1 & 1 \\ 1 & 1 \end{pmatrix}$

17. $\mathbf{X}^2 = \begin{pmatrix} 4 & -5 \\ 0 & 9 \end{pmatrix}$ $\mathbf{X}^3 = \begin{pmatrix} 8 & -19 \\ 0 & 27 \end{pmatrix}$

18. $p = 3, \quad q = -5, \quad r = -1, \quad s = 2$

19. $w = 5, \quad x = -3, \quad y = -3, \quad z = 2$

20. (a) $\begin{pmatrix} -7 & 8 \\ -1 & 2 \end{pmatrix}$ (b) $\begin{pmatrix} -1 & -2 \\ -5 & -4 \end{pmatrix}$

(c) $\begin{pmatrix} 7 & -8 \\ -1 & -10 \end{pmatrix}$ (d) $\begin{pmatrix} -11 & 10 \\ -1 & 8 \end{pmatrix}$

(e) $\begin{pmatrix} 19 & 8 \\ 3 & 4 \end{pmatrix}$ (f) $\begin{pmatrix} 13 & 6 \\ 13 & 10 \end{pmatrix}$

21. (a) $\begin{pmatrix} 6 & 10 \\ 17 & 24 \\ 11 & 14 \end{pmatrix}$ (b) $\begin{pmatrix} 8 & 12 \\ 6 & 10 \\ 21 & 32 \end{pmatrix}$

(c) $\begin{pmatrix} 3 & 5 \\ 6 & 5 \\ 9 & 5 \end{pmatrix}$

22. (a) $\begin{pmatrix} 4 & -2 \\ 0 & 2 \end{pmatrix}$ (b) $\begin{pmatrix} -2 & 6 \\ -4 & 4 \end{pmatrix}$

(c) $\begin{pmatrix} 0 & 16 \\ -8 & 8 \end{pmatrix}$ (d) $\begin{pmatrix} -3 & 8 \\ -8 & 5 \end{pmatrix}$

(e) $\begin{pmatrix} 1 & -8 \\ 4 & -7 \end{pmatrix}$

No

23. (a) $\begin{pmatrix} 6 & 1 \\ 9 & 0 \end{pmatrix}$ (b) $\begin{pmatrix} 45 & 6 \\ 54 & 9 \end{pmatrix}$

(c) $\begin{pmatrix} 7 & 2 \\ -4 & 23 \end{pmatrix}$ (d) $\begin{pmatrix} 4 & 2 \\ -58 & -42 \end{pmatrix}$

(e) $\begin{pmatrix} 23 & 16 \\ 56 & 39 \end{pmatrix}$

No

26. $a = 5, \quad b = -3$

27. $\mathbf{AB} = \begin{pmatrix} -1 & -2 & 7 \\ -3 & -2 & 11 \\ -5 & -2 & 15 \end{pmatrix}$

$\mathbf{BA} = \begin{pmatrix} 8 & 6 \\ 4 & 4 \end{pmatrix}$

28. $\mathbf{AB} = \begin{pmatrix} 12 & 9 & 7 \\ 2 & -1 & 3 \\ 11 & 11 & 3 \end{pmatrix}$

$\mathbf{BA} = \begin{pmatrix} -4 & -3 & 1 \\ 3 & 9 & 12 \\ 5 & 10 & 9 \end{pmatrix}$

EXERCISE 5.4

7. $\begin{pmatrix} 2 & -5 \\ -1 & 3 \end{pmatrix}$ **8.** $\begin{pmatrix} 5 & -4 \\ -6 & 5 \end{pmatrix}$

9. $\begin{pmatrix} 3 & -4 \\ -5 & 7 \end{pmatrix}$ **10.** $\begin{pmatrix} 1 & 2 \\ 2 & 5 \end{pmatrix}$

11. $\begin{pmatrix} 2 & 9 \\ 1 & 5 \end{pmatrix}$ **12.** $\begin{pmatrix} 1 & -3 \\ -3 & 10 \end{pmatrix}$

13. $\begin{pmatrix} 0 & 1 \\ -1 & -1 \end{pmatrix}$ **14.** $\begin{pmatrix} 12 & 11 \\ -11 & -10 \end{pmatrix}$

16. $\frac{1}{4}\begin{pmatrix} 1 & -3 \\ -1 & 7 \end{pmatrix}$ **17.** $\frac{1}{2}\begin{pmatrix} 3 & -2 \\ -2 & 2 \end{pmatrix}$

18. $-\frac{1}{5}\begin{pmatrix} 2 & 3 \\ 7 & 8 \end{pmatrix}$ **19.** None

20. $\begin{pmatrix} 7 & -3 \\ -9 & 4 \end{pmatrix}$ **21.** $\begin{pmatrix} 2 & -1 \\ -1 & 1 \end{pmatrix}$

22. None

23. $-\frac{1}{11}\begin{pmatrix} 9 & -5 \\ -4 & 1 \end{pmatrix}$

24. $-1\begin{pmatrix} 1 & -1 \\ -1 & 0 \end{pmatrix}$

25. $\frac{1}{32}\begin{pmatrix} 8 & 4 \\ -2 & 3 \end{pmatrix}$

26. $-\frac{1}{11}\begin{pmatrix} 7 & -9 \\ -2 & 1 \end{pmatrix}$

27. $\frac{1}{20}\begin{pmatrix} 2 & -3 \\ 2 & 7 \end{pmatrix}$

28. $-1\begin{pmatrix} 2 & 3 \\ 3 & 4 \end{pmatrix}$

29. $\frac{1}{6}\begin{pmatrix} 1 & 5 \\ 2 & 4 \end{pmatrix}$

30. $-\frac{1}{2}\begin{pmatrix} 1 & 0 \\ -3 & -2 \end{pmatrix}$

31. (a) $\begin{pmatrix} 12 & 29 \\ 14 & 34 \end{pmatrix}$ (b) $\begin{pmatrix} 20 & 37 \\ 14 & 26 \end{pmatrix}$

(c) $\frac{1}{2}\begin{pmatrix} 4 & -3 \\ -2 & 2 \end{pmatrix}$ (d) $\begin{pmatrix} 5 & -7 \\ -2 & 3 \end{pmatrix}$

(e) $\frac{1}{2}\begin{pmatrix} 34 & -29 \\ -14 & 12 \end{pmatrix}$

(f) $\frac{1}{2}\begin{pmatrix} 26 & -37 \\ -14 & 20 \end{pmatrix}$

(g) $\frac{1}{2}\begin{pmatrix} 34 & -29 \\ -14 & 12 \end{pmatrix}$

(h) $\frac{1}{2}\begin{pmatrix} 26 & -37 \\ -14 & 20 \end{pmatrix}$

$(\mathbf{AB})^{-1} = \mathbf{B}^{-1}\mathbf{A}^{-1}$, $\mathbf{A}^{-1}\mathbf{B}^{-1} = (\mathbf{BA})^{-1}$

33. $\mathbf{X} = \begin{pmatrix} 0 & 4 \\ 1 & -1 \end{pmatrix}$

EXERCISE 5.5

1. $x = 8, y = 2$ **2.** $x = 3, y = 3$

3. $x = 7, y = 2$ **4.** $x = 1, y = 6$

5. $x = -2, y = 4$ **6.** $x = -3, y = 2$

7. $x = 4, y = 1$ **8.** $x = 3, y = 1$

9. $x = 3, y = 2$ **10.** $x = 2, y = 1$

11. $x = 12, y = 5$ **12.** $x = 2, y = -3$

13. (a) $x = 7, y = -2$

(b) $x = 1, y = 2$

(c) $x = 1, y = -2$

14. $\begin{pmatrix} -4 & -12 \\ 6 & 19 \end{pmatrix}$ **15.** $\begin{pmatrix} 47 & 23 \\ 33 & 16 \end{pmatrix}$

16. $\frac{1}{6}\begin{pmatrix} 4 & -2 \\ 2 & -4 \end{pmatrix}$

17. $-\frac{1}{9}\begin{pmatrix} 18 & 21 \\ 27 & 29 \end{pmatrix}$

EXERCISE 5.6

1. (a) 8 (b) 7 (c) 10 (d) 7

(e) 2 (f) 12 (g) 7 (h) 8

(i) 20 (j) 16 (k) 2 (l) 7

(m) 1 (n) 14 (o) 7 (p) 6

(q) -4 (r) 17 (s) -1 (t) 14

(u) 0

2. (a) 6 (b) $12\,\text{cm}^2$

3. (a) 3 (b) $12\,\text{cm}^2$

4. $24\,\text{cm}^2$ **5.** $5\,\text{cm}^2$ **6.** $4\,\text{cm}^2$

7. $2\,\text{cm}^2$ **8.** $4\,\text{cm}^2$ **9.** $5\,\text{cm}^2$

10. $12\,\text{cm}^2$ **11.** $\frac{9}{2}\,\text{cm}^2$

12. (a) The image has vertices: (1,1), (1,5), (3,5), (3, 1)

(b) Area of image = 8 square units; area scale factor = $\frac{2}{3}$

$\Rightarrow$ area of object = 12 square units

13. (a) The image has vertices: (9,9), (0,9), (0,18), (9,18)

(b) Area of image = 81 square units; area scale factor = 9

$\Rightarrow$ area of object = 9 square units

EXERCISE 5.7

1. (a) (d) (e) (f) (i) (k) (l)

2. (a) Vertices: (1,3), (4,12), (5,15), (8,24)

(b) Area of parallelogram = 1 square unit; area of image = 0 square units; area scale factor = 0

(c) Determinant = 0 (d) $y = 3x$

3. (a) Vertices: (4,8), (6,12), (10,20)

(b) Area of triangle = $\frac{1}{2}$ square unit; area of image = 0 square units; area scale factor = 0

(c) Determinant = 0 (d) $y = 2x$

4. (b) Area scale factor = 0; determinant = 0

(c) $y = \frac{1}{2}x$

5. (b) Area scale factor = 0; determinant = 0

(c) $y = -3x$

EXERCISE 5.8

1. (0,0), (6, − 1), (18, − 5)

(0,0), (3,8), (3,4)

$M_1M_2 \neq M_2M_1$

Area of 0AB = 6 square units. The area is preserved under both transformations

2. (a) (14,4) (− 26, − 12)

(b) $-\frac{1}{4}\begin{pmatrix}-2 & 3\\-2 & 5\end{pmatrix} = \begin{pmatrix}\frac{1}{2} & -\frac{3}{4}\\\frac{1}{2} & -\frac{5}{4}\end{pmatrix}$

(c) (0, − 4) (d) $\begin{pmatrix}19 & -9\\6 & -2\end{pmatrix}$

3. (a) A reflection in the line $y = x$

(b) A reflection in the line $y = x - 2$

4. (a) A reflection in the line $y = x$

(b) A reflection in the line $y = x - 3$

5. (a) A′(1, − 2) B′(3, − 3) C′(2, − 5)

(c) Rotation of triangle clockwise about the origin through 90°

6. (a) $\begin{pmatrix}-1 & 0\\0 & -1\end{pmatrix}$

(e) R and S are identical

7. (a) $\begin{pmatrix}1 & 0\\0 & -1\end{pmatrix}$ (c) $\begin{pmatrix}0 & -1\\1 & 0\end{pmatrix}$

(d) Reflection in the line $y = x$; $\begin{pmatrix}0 & 1\\1 & 0\end{pmatrix}$

8. (a) $\begin{pmatrix}0 & 1\\1 & 0\end{pmatrix}$

(d) Quarter turn clockwise about the origin

$$\begin{pmatrix}0 & 1\\-1 & 0\end{pmatrix}\begin{pmatrix}1\\2\end{pmatrix} = \begin{pmatrix}2\\-1\end{pmatrix}$$

$$\begin{pmatrix}0 & 1\\-1 & 0\end{pmatrix}\begin{pmatrix}1\\4\end{pmatrix} = \begin{pmatrix}4\\-1\end{pmatrix}$$

$$\begin{pmatrix}0 & 1\\-1 & 0\end{pmatrix}\begin{pmatrix}2\\4\end{pmatrix} = \begin{pmatrix}4\\-2\end{pmatrix}$$

(f) Quarter turn anticlockwise about the origin;

$$\begin{pmatrix}0 & -1\\1 & 0\end{pmatrix}\begin{pmatrix}1\\2\end{pmatrix} = \begin{pmatrix}-2\\1\end{pmatrix}$$

$$\begin{pmatrix}0 & -1\\1 & 0\end{pmatrix}\begin{pmatrix}1\\4\end{pmatrix} = \begin{pmatrix}-4\\1\end{pmatrix}$$

$$\begin{pmatrix}0 & -1\\1 & 0\end{pmatrix}\begin{pmatrix}2\\4\end{pmatrix} = \begin{pmatrix}-4\\2\end{pmatrix}$$

9. (a) $\begin{pmatrix}0 & -1\\-1 & 0\end{pmatrix}$

(d) Quarter turn clockwise about the origin

$$\begin{pmatrix} 0 & 1 \\ -1 & 0 \end{pmatrix}\begin{pmatrix} -2 \\ 4 \end{pmatrix} = \begin{pmatrix} 4 \\ 2 \end{pmatrix}$$

$$\begin{pmatrix} 0 & 1 \\ -1 & 0 \end{pmatrix}\begin{pmatrix} -3 \\ 7 \end{pmatrix} = \begin{pmatrix} 7 \\ 3 \end{pmatrix}$$

$$\begin{pmatrix} 0 & 1 \\ -1 & 0 \end{pmatrix}\begin{pmatrix} -1 \\ 7 \end{pmatrix} = \begin{pmatrix} 7 \\ 1 \end{pmatrix}$$

10. (a) $\frac{1}{2}\begin{pmatrix} 3 & -1 \\ -4 & 2 \end{pmatrix}$; $x = 3$, $y = -2$

(b) $a = -5$, $b = 2$, $c = 3$, $d = 7$

11. (i) $\frac{1}{6}\begin{pmatrix} 2 & 1 \\ 0 & 3 \end{pmatrix}$ (ii) $\begin{pmatrix} 2 & 1 \\ -1 & 0 \end{pmatrix}$

(iii) $\begin{pmatrix} 1 & 0 \\ 0 & -1 \end{pmatrix}$

Reflection in the x-axis

12. (i) $S(3\sqrt{2},0)$ $T(2\sqrt{2},\sqrt{2})$

(ii) 3 square units

(iii) $\frac{1}{4}\begin{pmatrix} -2 & 0 \\ 0 & -2 \end{pmatrix}$

(iv) $\frac{1}{2\sqrt{2}}\begin{pmatrix} -1 & -1 \\ 1 & -1 \end{pmatrix}$

(v) $\frac{3}{4}$

13. (i) $\frac{1}{6}\begin{pmatrix} 3 & 0 \\ 1 & -2 \end{pmatrix}$

(ii) $\begin{pmatrix} -1 & 0 \\ 0 & -1 \end{pmatrix}$

(iii) 180° rotation about the origin

14. (i) $\mathbf{M} = \begin{pmatrix} 3 & 1 \\ 1 & 2 \end{pmatrix}$

(ii) 20 square units

(iv) $\mathbf{Z} = \frac{1}{5}\begin{pmatrix} -1 & -2 \\ 3 & 1 \end{pmatrix}$; $Q(-3,4)$

EXERCISE 6.1

1. (a) $x + c$ (b) $\frac{x^2}{2} + c$ (c) $\frac{x^3}{3} + c$

(d) $\frac{x^4}{4} + c$ (e) $\frac{x^7}{7} + c$

2. (a) $2x + c$ (b) $2x^2 + c$ (c) $2x^3 + c$

(d) $-2x^4 + c$ (e) $x^8 + c$

3. (a) $-\frac{1}{x} + c$ (b) $-\frac{1}{2x^2} + c$

(c) $-\frac{1}{3x^3} + c$ (d) $-\frac{1}{4x^4} + c$

(e) $\frac{1}{x^5} + c$

4. (a) $\frac{2x^{\frac{3}{2}}}{3} + c$ (b) $\frac{3x^{\frac{4}{3}}}{4} + c$ (c) $\frac{3x^{\frac{5}{3}}}{5} + c$

(d) $2\sqrt{x} + c$ (e) $\sqrt{x} + c$

5. $3x + c$

6. $\frac{x^2}{2} + c$

7. $\frac{x^2}{2} + 3x + c$

8. $\frac{x^3}{3} - 2x + c$

9. $2x - 2x^4 + c$

10. $3x^3 + 3x^2 + 7x + c$

11. $4x^3 - 5x + c$

12. $3x^5 + 3x^4 + c$

13. $\frac{x^3}{3} - 9x + c$

14. $\frac{x^3}{3} - 2x^2 + 4x + c$

15. $9x - 18x^2 + 12x^3 + c$

16. $\frac{x^4}{2} + \frac{x^3}{3} - \frac{3x^2}{2} + c$

17. $x - \frac{1}{x} + c$

18. $9x + x^2 + c$

19. $\frac{x^2}{2} + \frac{x^4}{2} + \frac{x^6}{6} + c$

20. $\frac{x^3}{3} + 2x - \frac{1}{x} + c$

21. $\frac{2x^{\frac{5}{2}}}{5} - \frac{3x^{\frac{5}{3}}}{5} + c$

22. $2x^{\frac{5}{2}} + 6x^{\frac{3}{2}} + c$

23. $-\frac{1}{x} - \frac{2}{\sqrt{x}} + c$

24. $2x^3 + \frac{2}{x} + c$

25. $\frac{x^3}{3} - \frac{2}{x} + c$

26. $\frac{x^3}{3} + \frac{1}{x} + c$

27. $\frac{x^3}{3} + \frac{x^2}{2} - \frac{2}{\sqrt{x}} + c$

28. $2\sqrt{x} - 6x^{\frac{3}{2}} + c$

29. $\frac{x^3}{3} - 2x - \frac{1}{x} + c$

30. $-\frac{1}{x} + \frac{2}{x^2} - \frac{4}{3x^3} + c$

31. $\frac{2x^{\frac{5}{2}}}{5} + 2x^{\frac{1}{2}} + c$

32. $\frac{x^2}{2} + 8\sqrt{x} - \frac{4}{x} + c$

EXERCISE 6.2

1. $y = 3x^2 - 7$

2. $y = x^2 - 3x + 2$

3. $y = x^3 - 1$

4. $y = 2x^2 + \frac{2}{x^3}$

5. $y = 2x + 4\sqrt{x} + 1$

6. $y = \frac{3x^2}{2} - \frac{x^3}{3} - \frac{1}{3}$

7. (a) $y = 3x^3 + 3x^2 + 7x - 11$

   (b) $y = \frac{x^3}{3} + 2x - \frac{1}{x} + 18\frac{2}{3}$

   (c) $y = \frac{2x^{\frac{5}{2}}}{5} - \frac{3x^{\frac{5}{3}}}{5} + \frac{1}{5}$

   (d) $y = -\frac{1}{x} - \frac{2}{\sqrt{x}} + 8$

   (e) $y = \frac{x^2}{2} + 8\sqrt{x} - \frac{4}{x} + 1$

8. $y = 5x^2 + 7x + 2$

9. $y = 4x^3 + 2x^2 + 7x + 2$

10. $y = 2x^3 - 4x^2 + 2x - 3$

11. $y = 4x^4 + 2x^3 - 3x^2 + 17x - 2$

12. $y = 2 - \frac{1}{\sqrt{x}}$

13. (i) $a = 6, b = 9$

    (ii) $y = x^3 + 3x^2 - 9x + 5$

EXERCISE 6.3

1. 6
2. 6
3. 2
4. $\frac{2}{5}$
5. 4
6. 20
7. -2
8. $3\frac{5}{6}$
9. 42
10. 10
11. $56\frac{1}{3}$
12. 8
13. 4
14. 0
15. 2
16. -4
17. 12
18. 13
19. -4
20. -39
21. (a) $\frac{2x^{\frac{3}{2}}}{3} - 3x^{\frac{1}{3}} + c$ (b) $x^2 + 2$; $2\frac{1}{3}$

EXERCISE 6.4

1. (i) 12 (ii) 21 (iii) $37\frac{1}{2}$ (iv) 8

2. (i) 72 (ii) $10\frac{2}{3}$ (iii) $5\frac{1}{3}$ (iv) 36
(v) $42\frac{2}{3}$ (vi) 64 (vii) $409\frac{3}{5}$

3. (a) $10\frac{2}{3}$ (b) $85\frac{1}{3}$ (c) 108

4. $21\frac{1}{3}$

5. 72 and 36

6. 0

7. $(-3,0)$ $(0,0)$ $(3,0)$; area $= 40\frac{1}{2}$

8. $(-2,0)$ $(0,0)$ $(2,0)$; area $= 8$

9. $(-1,0)$ $(0,0)$ $(4,0)$; area $= 32\frac{3}{4}$

10. A(0,5) B(1,0) C(5,0); area = 13

11. (a) $\dfrac{3x^4}{4} - \dfrac{2x^{\frac{3}{2}}}{3} + c$ (b) $1\frac{1}{3}$

12. $\frac{4}{3}$ units^2

EXERCISE 6.5

1. $10\frac{2}{3}$ **2.** $1\frac{1}{3}$ **3.** $4\frac{1}{2}$ **4.** $4\frac{1}{2}$

5. $20\frac{5}{6}$ **6.** $1\frac{1}{8}$ **7.** $\frac{1}{3}$ **8.** $\frac{1}{24}$

9. $41\frac{2}{3}$ **10.** $13\frac{1}{2}$

11. (i) $-x^{-2} - 4x^{\frac{7}{4}} + c$ (ii) $\frac{2}{3}$

EXERCISE 7.1

1.
(i) 2 m/s 1.8 m/s $1.5\,\text{m/s}^2$
$-0.775\,\text{m/s}^2$ 123.2 m

(ii) 14 m/s 14 m/s $4.143\,\text{m/s}^2$
$-11.6\,\text{m/s}^2$ 1.753 km

(iii) 12 m/s 30 m/s $2.5\,\text{m/s}^2$
$-1.091\,\text{m/s}^2$ 804 m

(iv) 20 km/h 0 km/h $1.25\,\text{km/h}^2$
$-4.167\,\text{km/h}^2$ 515 km

(v) 0 km/h 10 km/h $2.4\,\text{km/h}^2$
$-2.8\,\text{km/h}^2$ 445 km

2.
(i) 48 m/s $-1.647\,\text{m/s}^2$ 20 m/s
(ii) 52 m/s $-3.357\,\text{m/s}^2$ 5 m/s
(iii) 155 m/s $-1.701\,\text{m/s}^2$ 7 m/s
(iv) 100 m/s $-1\frac{2}{3}\,\text{m/s}^2$ 40 m/s
(v) 50 m/s $-0.7143\,\text{m/s}^2$ 40 m/s

3. $2\frac{2}{3}\,\text{m/s}^2$

4. $1\,\text{m/s}^2$; $2\,\text{m/s}^2$

5. 3 s

6. 12 m

7. (ii) 90 m

8. (i) 21 km (ii) 8.5 min

EXERCISE 7.2

9. $6\,\text{m/s}^2$; 3 s

10. 30 m/s; 8.1 km

11. $983\frac{1}{3}$ m; $28\frac{1}{3}$ s

12. (i) $1.12\,\text{m/s}^2$ (ii) 31.25 s

13. (i) 45 m/s (ii) 2.175 km

14. 200 s

15. (i) 20 (ii) 45.75 s; $\pm 1.5\,\text{m/s}^2$

16. (a) (i) $0.25\,\text{m/s}^2$ (ii) 350 m (iii) 400 m
(iv) 300 m (b) 14 s

17. (a) (i) 4 s (ii) $1.25\,\text{m/s}^2$ (iii) 8 s (iv) 40 m
(b) (i) 180 m (ii) 54 m (iii) 33 (iv) 65.4 m

EXERCISE 7.3

1. 322.5 m; 567.5 m

2. 1.960 km; 40 s

3. 10.17 s; 181.63 m

4. (i) 31.25 m (ii) 2.5 s
(iii) 5 m/s downwards (iv) 30 m
(v) 1 and 4 s

5. (i) 30 m (ii) 7 s

6. (a) (i) $t = 0.6$; $v = 30$ (ii) 3 s
(b) (i) 96.8 m (ii) 44 m/s (iii) 5 s; 135 m

7. (i) 30 m/s (ii) 45 m (iii) 31.25 m
(iv) 20 m; $(45 - 5t^2)$; $(20 + 15t - 5t^2)$; $t = \frac{5}{3}$

EXERCISE 8.1

1. (i) $A = 30$ N
(ii) $B = 50$ N
(iii) $A = 20$ N, $B = 30$ N
(iv) $A = 90$ N
(v) $B = 30$ N
(vi) $A = 60$ N, $B = 10$ N
(vii) $A = 90$ N, $B = 30$ N
(viii) $A = 90$ N, $B = 20$ N
(ix) $A = 20$ N, $B = 60$ N
(x) $A = 30$ N, $B = 50$ N

2. (i) $A = 50$ N
(ii) $A = 20$ N, $B = 10$ N
(iii) $A = 40$ N, $B = 50$ N
(iv) $A = 40$ N, $B = 40$ N
(v) $A = 80$ N, $B = 90$ N
(vi) $A = 50$ N, $B = 40$ N
(vii) $A = 90$ N, $B = 60$ N
(viii) $A = 30$ N, $B = 110$ N
(ix) $A = 30$ N, $B = 90$ N
(x) $A = 80$ N, $B = 20$ N

3. (i) $A = 106$ N
(ii) $B = 152$ N
(iii) $A = 110$ N
(iv) $A = 156$ N
(v) $A = 70$ N, $B = 10$ N
(vi) $A = 150$ N, $B = 55$ N
(vii) $A = 20$ N, $B = 100$ N
(viii) $A = 20$ N, $B = 60$ N
(ix) $A = 155$ N
(x) $A = 65$ N, $B = 60$ N

EXERCISE 8.2

1. 6 m/s^2 **2.** 4 m/s^2 **3.** 6.5 m/s
4. 24 N **5.** 120 N **6.** 84 N
7. 6 kg **8.** 8 kg **9.** 7.5 kg
10. 40 m/s^2 **11.** 750 N **12.** 800 N
13. 3 m/s^2 **14.** 2 m/s^2
15. (a) 1600 N (b) 2100 N

EXERCISE 8.3

1. (i) 0.102 (ii) 0.5 (iii) 0.05
(iv) 0.6 (v) 0.169 (vi) 0.18

2. $\frac{2}{7}$

3. 10 N; no

4. 23.52 N; 0.123 m/s^2

5. 0.3675 N; 6.18 m/s^2

6. 0.08

7. 19.8 N

8. 70.8 N

9. 0.05

10. 12.76 m

11. 0.61 m

12. (i) $\frac{5}{6}$ m/s^2 (ii) 40 N (iii) $\frac{1}{3}$

13. (ii) 6.05 m/s^2

14. (i) $50 - 0.5T$ (ii) $\frac{3}{\sqrt{2}}T$ (iii) 12.6 N
(iv) 2.46 m/s^2

15. (i) $P\cos 30° + 2g$ (ii) $\frac{1}{2}P$ (iii) 0.5

EXERCISE 8.4

i) 25.46 N; 3.1 m/s^2

ii) 33.95 N; 2.6 m/s^2

iii) 15.05 N; 2.973 m/s^2

iv) 41.55 N; 7.593 m/s^2

v) 35.53N; 4.642 m/s^2

vi) 19.74N; 4.01 m/s^2

vii) 15.05 N; 15.57 m/s^2

viii) 19.74 N; 13.81 m/s^2

ix) 16.41 N; 1.819 m/s^2

x) 21.3 N; 3.763 m/s^2

xi) 16.41 N; 14.42 m/s^2

xii) 21.3 N; 13.56 m/s^2

EXERCISE 8.5

1. (i) 24.13N; remains at rest

(ii) 18.77 N; accelerates

(iii) 23.66 N; remains at rest

(iv) 15.75 N; accelerates

(v) 12.06 N; remains at rest

(vi) 11.83 N; accelerates

(vii) 10.61 N; accelerates

(viii) 12.13 N; remains at rest

2. 3.18 N; the body will slide

3. 0.04

4. 0.505

5. 3.19 m/s

6. 0.2

7. 25.66 N

8. 22.39 N

9. 0.226

10. 22.19 N; 18.08 N

11. (i) 0.25 m/s^2 (ii) 13.75 kN (iii) 7.5 kN

12. (ii) 0.577 (iii) 100 N (iv) 5 N

13. (a) (ii) 164 N (iii) 0.7 (b) 187.5 N

EXERCISE 8.6

1. (i) 5 m/s^2; 5 N; 25 N; 65 N

(ii) 5 m/s^2; 30 N; 70 N

(iii) 4 m/s^2; 16 N; 24 N

(iv) 20 m/s^2; 160 N; 400 N; 480 N

(v) 3 m/s^2; 36 N; 78 N; 132 N; 204 N

2. (i) 4.2 m/s^2; 28 N

(ii) 5.88 m/s^2; 31.36 N

(iii) 0.98 m/s^2; 97.02 N

(iv) 7.35 m/s^2; 17.15 N

(v) 7 m/s^2; 33.6 N

3. (i) 7.84 m/s^2; 15.68 N

(ii) 6.533 m/s^2; 39.2 N; 32.67 N

(iii) 2.191 m/s^2; 23.98 N

(iv) 3.23 m/s^2; 20.4 N

(v) 5.765 m/s^2; 22.19 N; 10.67 N

4. (i) 1.429 m/s^2 (ii) 6.857 N

5. (i) 1.667 m/s^2 (ii) 6.325 m/s

(iii) 2 m (iv) 1.265 s

6. (a) (i) 37.5 N (ii) 3

(b) (i) 5 m/s^2 (ii) 6.4 m (iii) 2.68 m/s^2

7. (a) (i) 50 N (ii) 90 N

(b) (i) 7 m/s^2, 0.6 N (ii) 2.8 m/s; 4.2 m/s

8. (a) (i) 25 (ii) 1.5 m/s^2

(b) (i) 3 m/s^2 (ii) 3.46 m/s, 2.8 m

EXERCISE 9.1

1. (i)

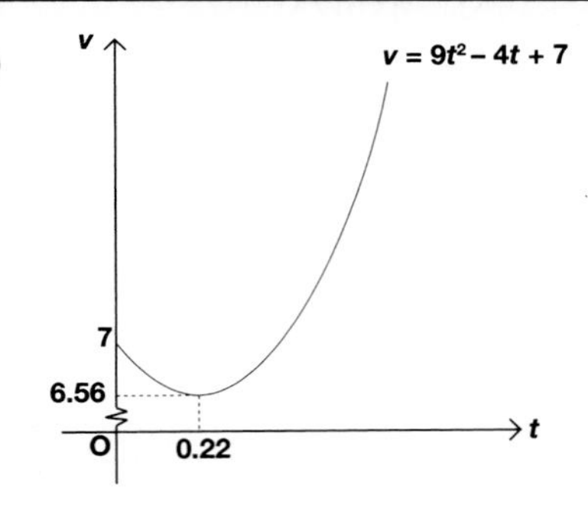

Figure A.1

(ii) Never (iii) 12 m/s

(iv) 68 m/s^2

2. (i)

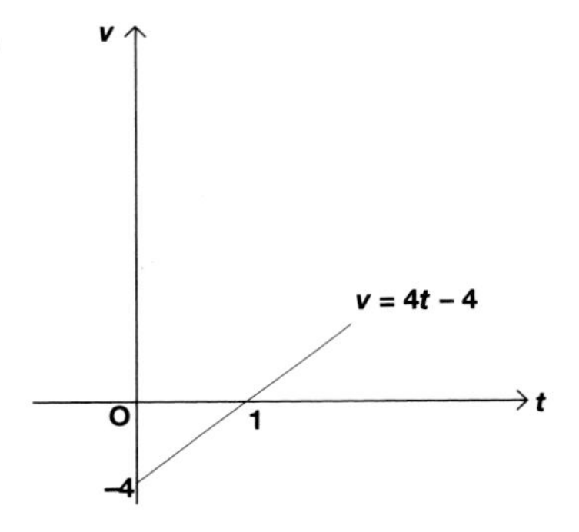

Figure A.2

(ii) $t = 1$ (iii) 12 m/s

(iv) 4 m/s^2

3. (i)

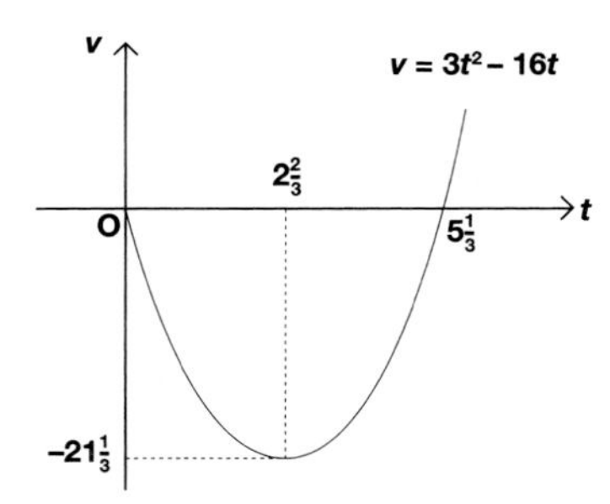

Figure A.3

(ii) $t = 0$ and $t = 5\frac{1}{3}$

(iii) -20 m/s (iv) -10 m/s^2

4. (i)

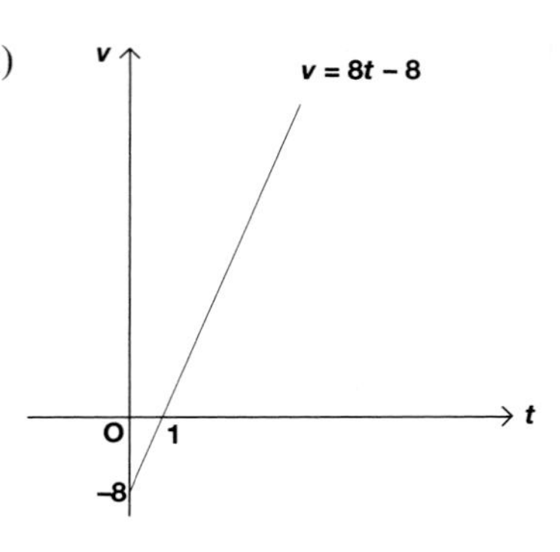

Figure A.4

(ii) $t = 1$ (iii) 88 m/s

(iv) 8 m/s^2

5. (i)

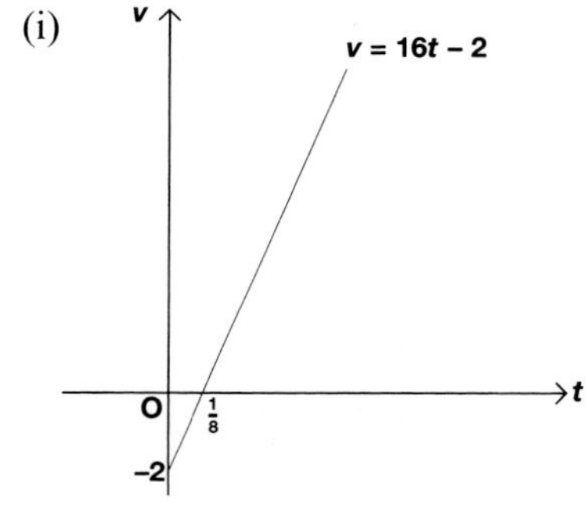

Figure A.5

(ii) $t = \frac{1}{8}$ (iii) 30 m/s

(iv) 16 m/s^2

6. (i)

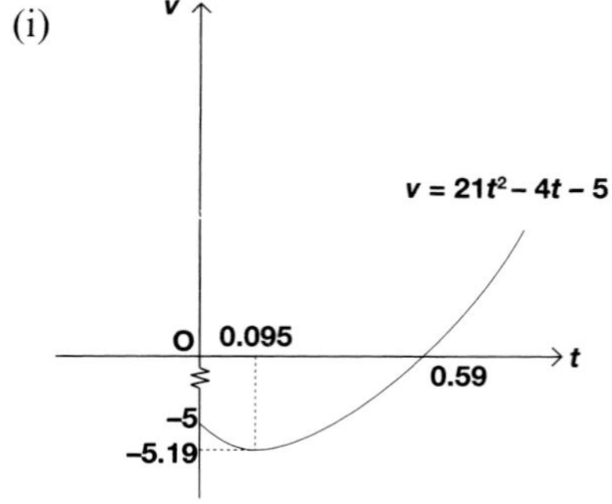

Figure A.6

(ii) $t = -0.4$ and $t = 0.58$
($t = -0.4$ is unphysical)

(iii) 315 m/s (iv) 206 m/s^2

7. (i)

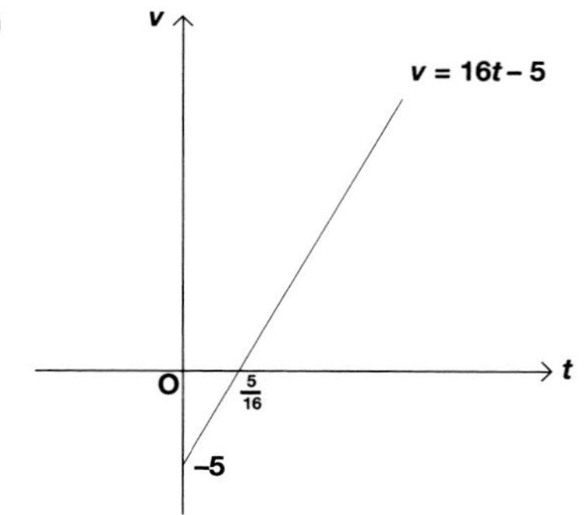

Figure A.7

(ii) $t = \frac{5}{16}$ (iii) 27 m/s

(iv) 16 m/s^2

8. (i)

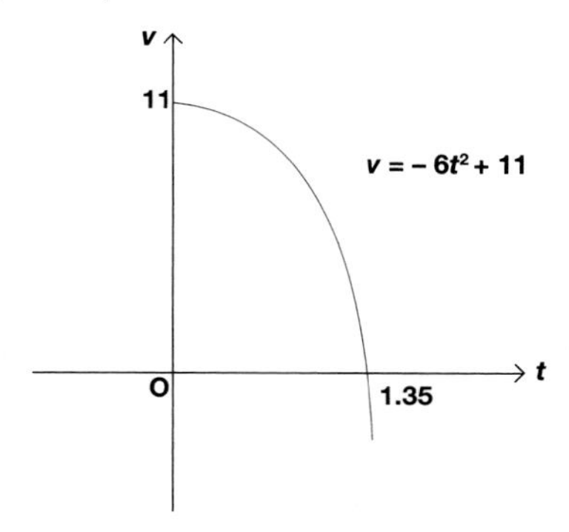

Figure A.8

(ii) $t = \pm 1.35$ ($t = -1.35$ is unphysical)

(iii) 5 m/s (iv) -48 m/s^2

9. (i)

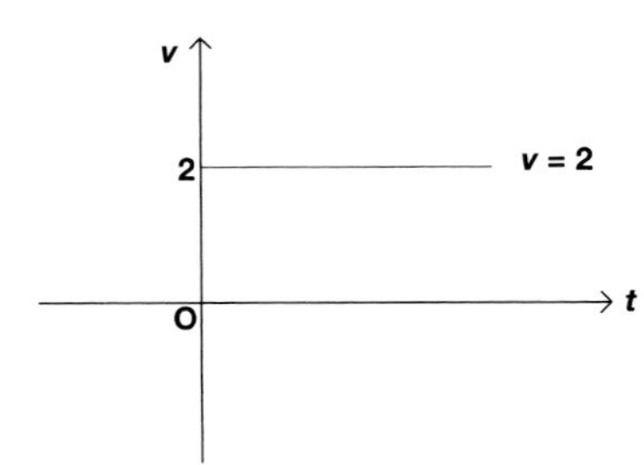

Figure A.9

(ii) Never (iii) 2 m/s

(iv) 0 m/s^2

10. (i)

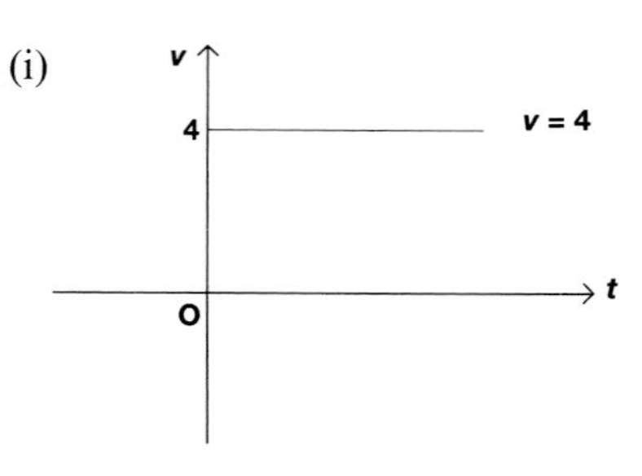

Figure A.10

(ii) Never (iii) 4 m/s (iv) 0 m/s^2

EXERCISE 9.2

1. Left: $0 \le t < 1$; right: $1 < t < 4$; left: $t > 4$
2. Right: $0 \le t < 2$; left: $2 < t < 4$; right: $t > 4$
3. Left: $0 \le t < 4$; right: $4 < t < 8$; left: $t > 8$
4. Right: $0 \le t < 5$; left: $5 < t < 9$; right: $t > 9$
5. Right: $0 \le t < 3$; left: $3 < t < 4$; right: $t > 4$; $36\frac{2}{3}$
6. Left: $0 \le t < 1$; right: $1 < t < 2$; left: $t > 2$; $7\frac{1}{3}$
7. Left: $0 \le t < \frac{1}{3}$; right: $\frac{1}{3} < t < 1$; left: $t > 1$; $2\frac{8}{27}$
8. Right: $0 \le t < 2$; left: $2 < t < 5$; right: $t > 5$; 15
9. Left: $0 \le t < 1$; right: $1 < t < 3$; left: $t > 3$; $9\frac{1}{3}$

EXERCISE 9.3

1. (i) $t = 2, 5$ (ii) 60 m/s
   (iii) -18 m/s^2
2. (i) $3t^2 - 15t + 12$ (ii) $t_1 = 1$; $t_2 = 4$
   (iii) $5\frac{1}{2}, -8$ (iv) 45 m
   (v) 2.314, 5.186 (vi) 30

3. (i) $3t^2 - kt + b$ (ii) $t^3 - \frac{1}{2}kt^2 + bt$
(iii) $k = 4, b = 1$ (iv) $\frac{4}{27}$ m

4. (i) 13, 15 m/s (ii) 6 s
(iii) 15.6 m/s (iv) 71.3 m

5. (i) $t_1 = 2; t_2 = 3$ (ii) $4\frac{2}{3}, 4\frac{1}{2}$ m
(iii) $5\frac{2}{3}$ m (iv) $2\frac{1}{2}$, 0 s

6. (i) $3t^2 - 15t + 12; t^3 - \frac{15}{2}t^2 + 12t + 9$
(ii) $t_1 = 1; t_2 = 4$ (iii) $14\frac{1}{2}$, 1 m
(iv) 30 m/s; $6\frac{3}{4}$ m/s
(v)

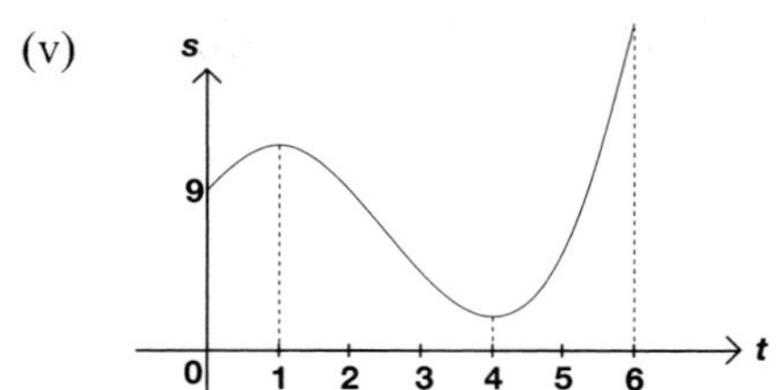

Figure A.11

7. (i) 4 m/s (ii) $\frac{2}{3}$ (iii) $\frac{32}{27}$ m (iv) 0
(v)

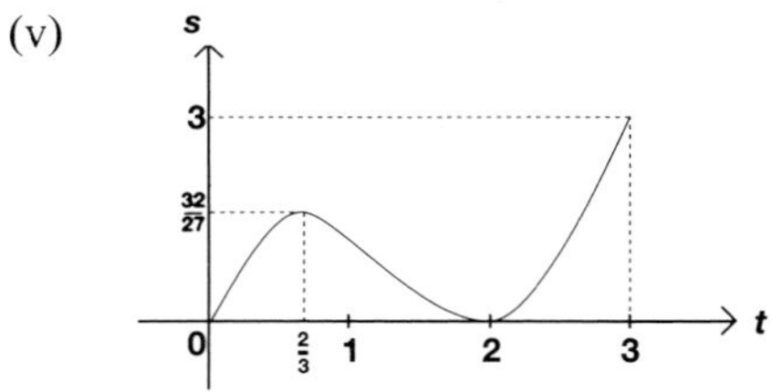

Figure A.12

8. (i) 2 m/s, 8 m/s (ii) 6 m

9. (i) $t = 1, t = 2$ (ii) 0.423, 1.577
(iii) $t = 1$ (iv) $0, \frac{1}{4}, 0$ (v) $2\frac{3}{4}$ m

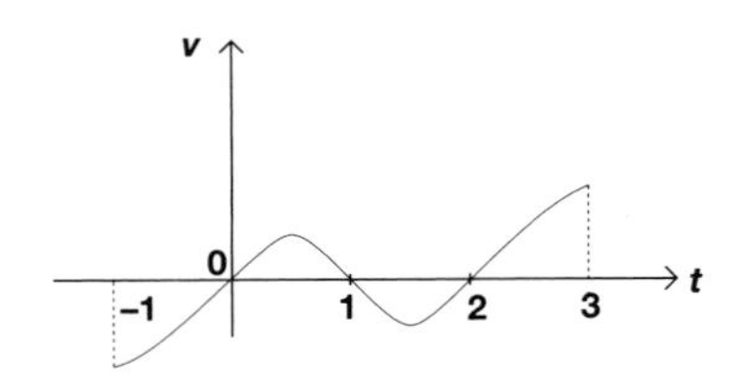

Figure A.13

10. $v = 3(t - 6)(t - 2)$; 2; $- 12$ m/s^2
(i) 6 (ii) 64 m

11. $4 - 10t - 3t^2$; $- 10 - 6t$; 4 m/s, $- 10$ m/s^2; $2\frac{1}{3}$ s; same direction

EXERCISE 10.1

1. (i) $3\mathbf{i} + 5\mathbf{j}$ (ii) $\sqrt{34}$ (iii) $\frac{3\mathbf{i} + 5\mathbf{j}}{\sqrt{34}}$

(i) $2\mathbf{i} + 4\mathbf{j}$ (ii) $\sqrt{20}$ (iii) $\frac{2\mathbf{i} + 4\mathbf{j}}{\sqrt{20}}$

(i) $\mathbf{i} + 7\mathbf{j}$ (ii) $\sqrt{50}$ (iii) $\frac{\mathbf{i} + 7\mathbf{j}}{\sqrt{50}}$

(i) $2\mathbf{i} + 5\mathbf{j}$ (ii) $\sqrt{29}$ (iii) $\frac{2\mathbf{i} + 5\mathbf{j}}{\sqrt{29}}$

(i) $\mathbf{i} + 10\mathbf{j}$ (ii) $\sqrt{101}$ (iii) $\frac{\mathbf{i} + 10\mathbf{j}}{\sqrt{101}}$

(i) $-\mathbf{i} + 3\mathbf{j}$ (ii) $\sqrt{10}$ (iii) $\frac{-\mathbf{i} + 3\mathbf{j}}{\sqrt{10}}$

(i) $\mathbf{i} - 4\mathbf{j}$ (ii) $\sqrt{17}$ (iii) $\frac{\mathbf{i} - 4\mathbf{j}}{\sqrt{17}}$

(i) $2\mathbf{i} - 7\mathbf{j}$ (ii) $\sqrt{53}$ (iii) $\frac{2\mathbf{i} - 7\mathbf{j}}{\sqrt{53}}$

(i) $4\mathbf{i} - 9\mathbf{j}$ (ii) $\sqrt{97}$ (iii) $\frac{4\mathbf{i} - 9\mathbf{j}}{\sqrt{97}}$

(i) $2\mathbf{i}$ (ii) 2 (iii) $\mathbf{i}$

2. (i) 26 (ii) 30 (iii) 37
(iv) 52 (v) $- 13$ (vi) 71
(vii) 8 (viii) $- 24$ (ix) 4
(x) 6

3. (i) $-\mathbf{i} - \mathbf{j}$ (ii) $-2\mathbf{i} + 2\mathbf{j}$
(iii) $-\mathbf{i} + 3\mathbf{j}$ (iv) $\mathbf{j}$
(v) $-2\mathbf{i} + 5\mathbf{j}$ (vi) $2\mathbf{i} - 7\mathbf{j}$
(vii) $\mathbf{i} - 3\mathbf{j}$ (viii) $-2\mathbf{i} + 9\mathbf{j}$
(ix) $-3\mathbf{i} + 3\mathbf{j}$ (x) $-\mathbf{i} - 5\mathbf{j}$

4. (i) -2 (ii) 18 (iii) 30
(iv) -53 (v) -7 (vi) 33
(vii) -198 (viii) 16 (ix) 11
(x) 23

5. (i) 7 (ii) 10 (iii) 0
(iv) 4 (v) 3 (vi) 1
(vii) 4 (viii) 2 (ix) -4
(x) -3
Vector pair (iii) are perpendicular

6. (i) 3, 4 (ii) $0.6\mathbf{i} + 0.8\mathbf{j}$ (iii) 70

7. (i) $5\mathbf{i} - \mathbf{j}$, $11\mathbf{j}$

8. (i) $-9\mathbf{i} + 9\mathbf{j}$ (ii) 13 (iii) $-0.6\mathbf{i} + 0.8\mathbf{j}$
(iv) -24

9. (i) -2, 3 (ii) $0.6\mathbf{i} + 0.8\mathbf{j}$

EXERCISE 10.2

1. (i) (a) $6\mathbf{i} + 7\mathbf{j}$ (b) $12\mathbf{i} + 9\mathbf{j}$
(ii) (a) $8\mathbf{i} + 13\mathbf{j}$ (b) $12\mathbf{i} + 18\mathbf{j}$
(iii) (a) $11\mathbf{i} + 11\mathbf{j}$ (b) $31\mathbf{i} + 24\mathbf{j}$
(iv) (a) $9\mathbf{i} + 13\mathbf{j}$ (b) $38\mathbf{i} + 46\mathbf{j}$
(v) (a) $-5\mathbf{i} + 10\mathbf{j}$ (b) $-7.5\mathbf{i} + 24.5\mathbf{j}$
(vi) (a) $8.5\mathbf{i} + 11\mathbf{j}$ (b) $26.75\mathbf{i} + 28\mathbf{j}$
(vii) (a) $-7\mathbf{i} - 10\mathbf{j}$ (b) $-9\mathbf{i} - 16\mathbf{j}$
(viii) (a) $12\mathbf{i} + 19\mathbf{j}$ (b) $38\mathbf{i} + 59\mathbf{j}$
(ix) (a) $3\mathbf{i} - 8\mathbf{j}$ (b) $6\mathbf{i} - 8\mathbf{j}$
(x) (a) $11\mathbf{i} - 17\mathbf{j}$ (b) $22\mathbf{i} - 37\mathbf{j}$

2. (i) $-3\mathbf{i} + 2\mathbf{j}$ (ii) $-12\mathbf{i} + 12\mathbf{j}$

3. (i) 2, 3 (ii) $-\frac{4}{3}\mathbf{i} + 3\mathbf{j}$

4. (i) 1, 9 (ii) $2.5\mathbf{i} + 6.25\mathbf{j}$

5. $-0.4\mathbf{i} + 2.6\mathbf{j}$

EXERCISE 11.1

2. (a) 29.41°, 22.1 N
(b) 53.3°, 65.75 N
(c) 38.37°, 62.07 N
(d) 32.86°, 75.96 N

3. (a) 36.87°, 1.633 kg
(b) 18.03°, 4.075 kg
(c) 37.57°, 6.632 kg
(d) 48.13°, 6.402 kg

4. (a) 28.37 N, 17.59 N
(b) 46.62 N, 15.96 N
(c) 65.69 N, 18.73 N
(d) 101.5 N, 27.96 N

EXERCISE 11.2

1. 35.35°, 0.975 N
2. 26.44° to the upward vertical, 0.668 N
93.56° to the upward vertical, 1.497 N
3. 79.92°, 68.87°
4. 28.14°, 19.69°
5. 38.4 N, 11.2 N
6. 36.87°, 53.13°
7. 320 N, 600 N

EXERCISE 11.3

1. 49.7 N, 58.1 N
2. 39.2 N, 58.8 N
3. 87.11 N, 10.89 N
4. (a) 54.13 N, 2.707 N (b) 142.1 N
5. 980 N, 588 N
6. 100 kg
7. 500.6 N
8. (a) 0.33 m (b) 0.48 m

9. 428.8 N, 208.3 N

10. 0.6875l from that end of the pipe which supports the 20 kg mass

11. (a) 73.5 N, 0 N, 73.5 N
(b) 147 N, 127.3 N, 73.5 N
(c) 147 N, 127.3 N, 73.5 N
(d) 73.5 N, 0 N, 73.5 N
(e) 51.97 N, 36.75 N, 110.25 N
(f) 51.97 N, 36.75 N, 110.25 N

12. (a) 98 N, 98 N, 60°
(b) 69.3 N, 133.9 N, 30°
(c) 24.5 N, 88.3 N, 13.9°
(d) 28.3 N, 102 N, 16°
(e) 63 N, 37.6 N, 16.9°
(f) 42.4 N, 64.8 N, 19°

13. (a) 14.7 N, 12.7 N, 51.45 N, 0.25
(b) 36 N, 34.8 N, 49.5 N, 0.7
(c) 50.9 N, 25.5 N, 14.7 N, 1.73

14. (a) 147 N, 73.5 N, 73.5 N
(b) 147 N, 61.7 N, 61.7 N
(c) 147 N, 42.4 N, 42.4 N

15. (a) 147 N, 147 N, 294 N, 0.5
(b) 84.9 N, 84.9 N, 294 N, 0.3
(c) 25.9 N, 25.9 N, 294 N, 0.09

16. (a) 29.44 N, 98 N, 29.44 N
(b) 15.84 N, 78.4 N, 15.84 N
(c) 9.525 N, 98 N, 9.525 N
(d) 87.8 N, 137.2 N, 87.8 N
(e) 62.72 N, 98 N, 62.72 N
(f) 20.25 N, 117.6 N, 20.25 N
(g) 49.58 N, 147 N, 49.58 N
(h) 77.54 N, 186.2 N, 77.54 N
(i) 23.72 N, 107.8 N, 23.72 N
(j) 44.29 N, 176.4 N, 44.29 N

17. (a) 52.84 N, 74.98 N, 42.9°
(b) 44.45 N, 90.14 N, 28.84°
(c) 453.1 N, 490.4 N, 63.69°
(d) 75.4 N, 62.7 N, 76.65°
(e) 112.8 N, 121.9 N, 27.55°
(f) 93.59 N, 196.1 N, 24.41°

18. (a) 80 cm
(b) (i) 60° (ii) 42.4 N
(iii) 64.8 N, 19.1° to vertical

19. (i) $60\sqrt{3}$ N (ii) 60 N (iii) 87.2 N
(iv) 53.4°, 156.6°

20. (i) 1182 N
(ii) 1066 N at 58° to the vertical; 85 kg

21. (i) 245 N
(ii) 2581 N at 85.28° to the horizontal

22. (i) 28.28 N (ii) 118 N; 35.23 N; 55.41°

23. (ii) 1346 N
(iii) 1346 N; 21.8° to horizontal
(iv) 242.8 N

24. (a) 33 N, 67 N
(b) (i) $20 = M(2 - x)$; $10 = M(x - 0.5)$
$M = 20$; $x = 1$
(ii) 1 m

25. (ii) 278 N (iii) 1000 N (iv) 1.64 m

26. (i) $T_C = 1.2$ N, $T_D = 0.8$ N (ii) 24 N

27. (ii) 111.8 N

28. (a) 50 N, 150 N
(b) (i) 0.75 m (ii) 4.2 m

29. (a) 0.25, 16 (b) 5

30. (a) 7.41 N, 61.1°
(b) $P = 9.66$, $Q = 13.7$, $R = 4$

31. (a) 19.1°
(b) (i) $P = 5.93$, $R = 1.27$
(c) (i) 7.17 N (ii) 5.86 N

EXERCISE 12.1

1.

Class	Frequency
15–17	2
18–20	5
21–23	6
24–26	7
27–29	3
30–32	5
33–35	2
36–38	2

2.

Class	Frequency
1–2	7
3–4	7
5–6	8
7–8	4
9–10	4

3.

Class	Frequency
140–144	7
145–149	9
150–154	8
155–159	10
160–164	14
165–169	2

4.

Class	Frequency
15–19	7
20–24	5
25–29	9
30–34	7
35–39	6
40–44	3
45–49	3

5.

Class	Frequency
0–9	2
10–19	4
20–29	7
30–39	5
40–49	2
50–59	3
60–69	4
70–79	3
80–89	2

6.

Class	Frequency
30–33	6
33–36	2
36–39	6
39–42	9
42–45	4
45–48	4
48–51	1

7.

Class	Frequency
140–145	4
145–150	6
150–155	5
155–160	5
160–165	9
165–170	1

8.

Class	Frequency
45–50	2
50–55	1
55–60	3
60–65	6
65–70	5
70–75	5
75–80	4
80–85	6

9.

Class	Frequency
100–110	4
110–120	4
120–130	7
130–140	4
140–150	6
150–160	1
160–170	4

10.

Class	Frequency
10–20	7
20–30	10
30–40	8
40–50	3
50–60	1
60–70	3

11.

Class	Frequency
0–	4
5–	5
10–	3
15–	4
20–	6
25–	1
30–	4
35–	2
40–45	1

12.

Class	Frequency
10–	8
20–	13
30–	10
40–50	4

13.

Class	Frequency
140–	3
145–	6
150–	4
155–	4
160–	7
165–170	1

14.

Class	Frequency
30–	3
34–	4
38–	7
42–	3
46–	5
50–54	2

15.

Class	Frequency
10–	1
12–	2
14–	1
16–	3
18–	9
20–	4
22–	2
24–	6
26–	2
28–	0
30–	3
32–34	2

EXERCISE 12.2

1. (i) 6 (ii) 6.5 (iii) 9.5 (iv) 2 (v) 0.2

2. (i) 145 (ii) 164.5 (iii) 152 (iv) 5 (v) 0.12

3. (i) 55 (ii) 54.5 (iii) 67 (iv) 5 (v) 0.32

4. (i) 9 (ii) 89.5 (iii) 34.5 (iv) 10 (v) 0.25

5. (i) 9 (ii) 14.5 (iii) 16 (iv) 3 (v) 0.2

6. (i) 15 (ii) 14.5 (iii) 12 (iv) 5 (v) 0.167

7. (i) 699 (ii) 699.5 (iii) 674.5 (iv) 50 (v) 0.3

8. (i) 8 (ii) 27.5 (iii) 1.5 (iv) 4 (v) 0.15

9. (i) 16 (ii) 35.5 (iii) 21.5 (iv) 4 (v) 0.24

10. (i) 64 (ii) 24.5 (iii) 19.5 (iv) 10 (v) 0.1

EXERCISE 12.3

1.

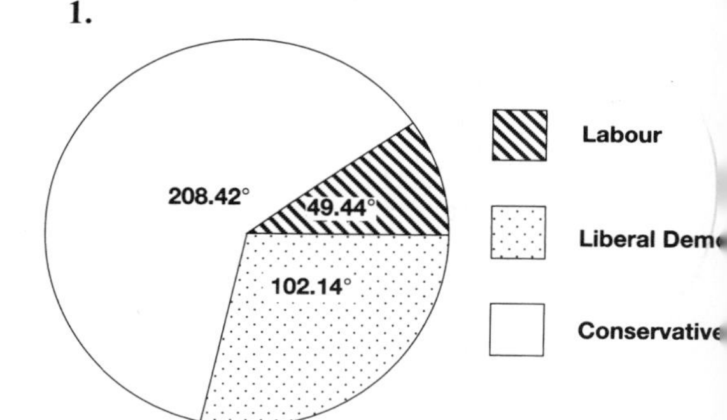

Figure A.14

2.

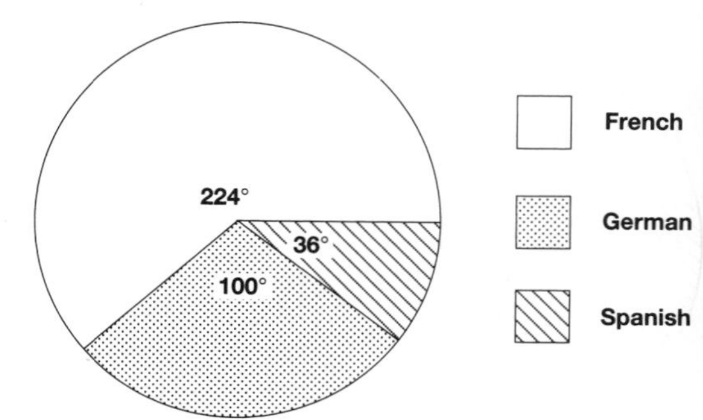

Figure A.15

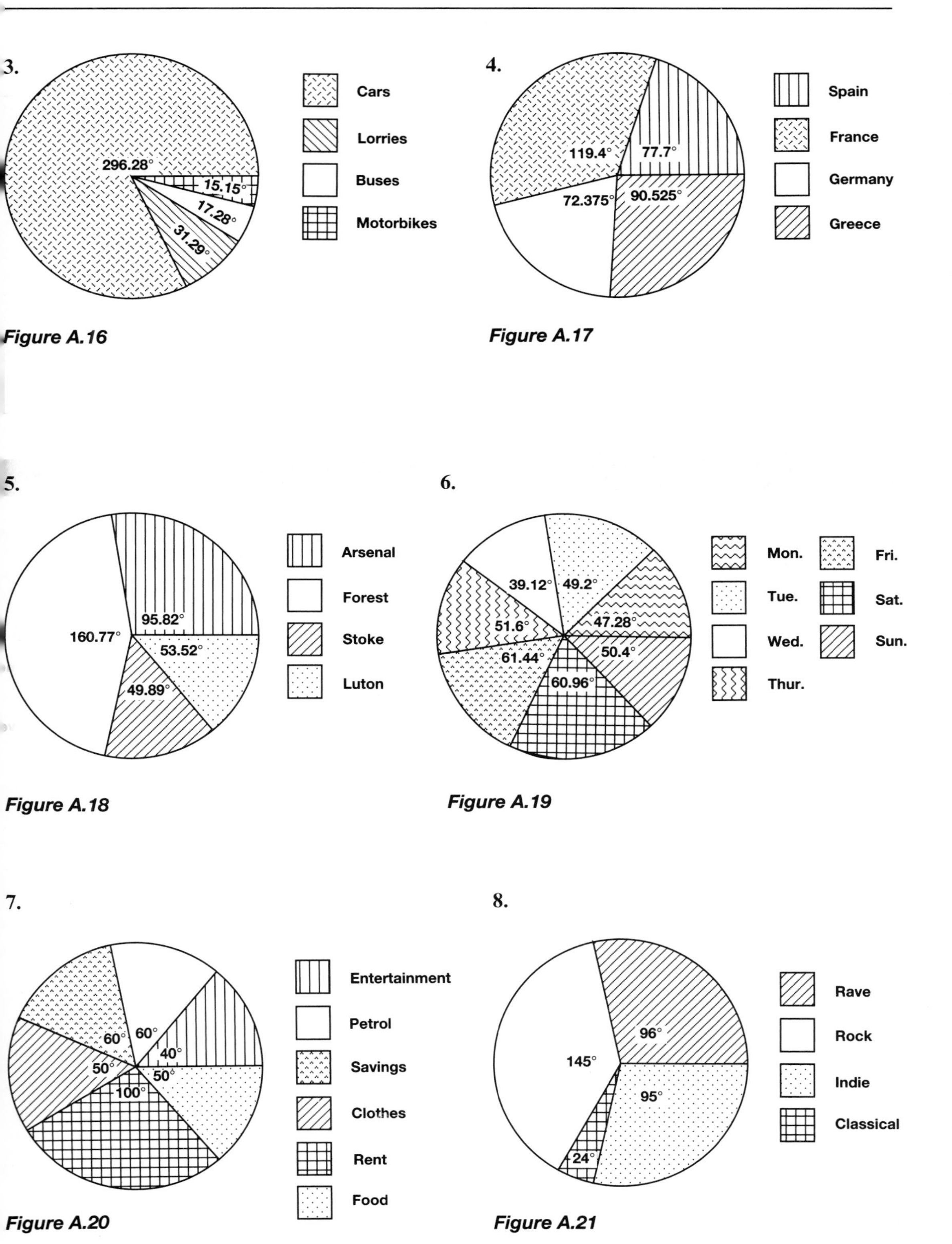

Figure A.16

Figure A.17

Figure A.18

Figure A.19

Figure A.20

Figure A.21

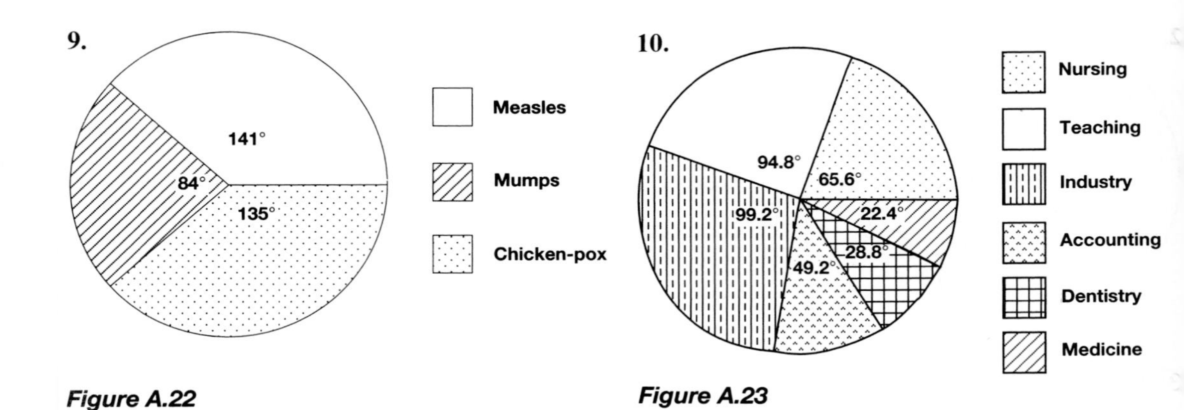

Figure A.22

Figure A.23

EXERCISE 12.4

1.

Interval	f	Boundaries	Size	Frequency density
1–4	2	0.5–4.5	4	0.5
5–8	4	4.5–8.5	4	1
9–15	7	8.5–15.5	7	1
16–18	10	15.5–18.5	3	3.33
19–22	18	18.5–22.5	4	4.5
23–27	12	22.5–27.5	5	2.4
28–37	6	27.5–37.5	10	0.6
38–40	3	37.5–40.5	3	1

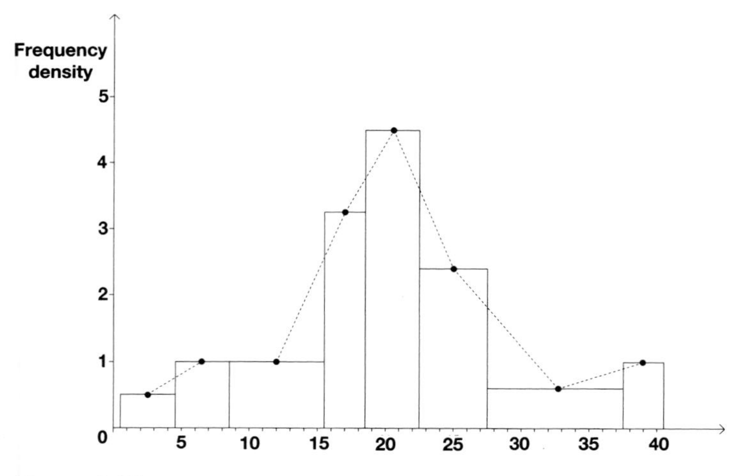

Figure A.24

2.

Interval	f	Boundaries	Size	Frequency density
1–5	3	0.5–5.5	5	0.6
6–10	5	5.5–10.5	5	1
11–19	8	10.5–19.5	9	0.89
20–26	10	19.5–26.5	7	1.43
27–39	14	26.5–39.5	13	1.08
40–52	17	39.5–52.5	13	1.31
53–59	12	52.5–59.5	7	1.71
60–64	4	59.5–64.5	5	0.8

3.

Interval	f	Boundaries	Size	Frequency density
3–8	7	2.5–8.5	6	1.17
9–19	12	8.5–19.5	11	1.09
20–25	18	19.5–25.5	6	3
26–31	19	25.5–31.5	6	3.17
32–38	30	31.5–38.5	7	4.29
39–49	21	38.5–49.5	11	1.91
50–52	20	49.5–52.5	3	6.67
53–59	12	52.5–59.5	7	1.71
60–61	8	59.5–61.5	2	4

4.

Interval	f	Boundaries	Size	Frequency density
4–12	3	3.5–12.5	9	0.33
13–21	6	12.5–21.5	9	0.67
22–30	7	21.5–30.5	9	0.78
31–35	14	30.5–35.5	5	2.8
36–38	20	35.5–38.5	3	6.67
39–40	27	38.5–40.5	2	13.5
41–57	21	40.5–57.5	17	1.24
58–61	15	57.5–61.5	4	3.75
62–70	10	61.5–70.5	9	1.11
71–79	5	70.5–79.5	9	0.56
80–86	4	79.5–86.5	7	0.57

5.

Interval	f	Boundaries	Size	Frequency density
3–6	2	2.5–6.5	4	0.5
7–12	4	6.5–12.5	6	0.67
13–15	3	12.5–15.5	3	1
16–20	9	15.5–20.5	5	1.8
21–23	6	20.5–23.5	3	2
24–26	6	23.5–26.5	3	2
27–29	3	26.5–29.5	3	1
30–35	5	29.5–35.5	6	0.83

6.

Interval	f	Boundaries	Size	Frequency density
1–6	12	0.5–6.5	6	2
7–10	13	6.5–10.5	4	3.25
11–13	12	10.5–13.5	3	4
14–17	12	13.5–17.5	4	3
18–20	10	17.5–20.5	3	3.33
21–23	9	20.5–23.5	3	3
24–25	7	23.5–25.5	2	3.5

7.

Interval	f	Boundaries	Size	Frequency density
4–9	3	3.5–9.5	6	0.5
10–15	12	9.5–15.5	6	2
16–20	15	15.5–20.5	5	3
21–23	4	20.5–23.5	3	1.33
24–25	11	23.5–25.5	2	5.5
26–31	10	25.5–31.5	6	1.67
32–35	10	31.5–35.5	4	2.5
36–40	11	35.5–40.5	5	2.2

8.

Interval	f	Boundaries	Size	Frequency density
10–12	9	9.5–12.5	3	3
13–18	12	12.5–18.5	6	2
19–21	3	18.5–21.5	3	1
22–24	4	21.5–24.5	3	1.33
25–27	15	24.5–27.5	3	5
28–32	16	27.5–32.5	5	3.2
33–35	15	32.5–35.5	3	5
36–40	10	35.5–40.5	5	2

9.

Interval	f	Boundaries	Size	Frequency density
1–5	16	0.5–5.5	5	3.2
6–10	20	5.5–10.5	5	4
11–15	22	10.5–15.5	5	4.4
16–20	13	15.5–20.5	5	2.6
21–30	14	20.5–30.5	10	1.4
31–40	24	30.5–40.5	10	2.4
41–45	16	40.5–45.5	5	3.2
46–50	8	45.5–50.5	5	1.6

EXERCISE 13.1

1. Mean = 5
   Mode = 6
   Median = 5
2. Mean = 3.1
   Mode = 2.4
   Median = 3.0
3. Mean = 5
   Mode = 5 or 6
   Median = 5
4. Mean = 154.5 cm
   Mode = 153 cm or 160 cm
   Median = 156 cm
5. Mean = 62
   Mode = 63
   Median = 62.5
6. Mean = 2.325
   Mode = 1
   Median = 2
7. Mean = 60.75 kg
   Mode = 60 kg or 64 kg
   Median = 61 kg
8. Mean = 74.14
   Mode = 62
   Median = 75
9. Mean = 4
   Mode = 4
   Median = 4
10. Mean = 186.83
    Mode = 184 or 193
    Median = 184
11. 10
12. 6.1
13. 13
14. 122.4

EXERCISE 13.2

1. 25.9
2. 38.1 years
3. 158.7 cm
4. £20.30
5. 14.5 errors
6. 8.3 days
7. 35.27 h
8. 69.3 kg
9. 24.3 years
10. 40 min

EXERCISE 13.3

1. 16.1
2. 67.6
3. 11.9
4. 14.0

EXERCISE 13.4

1. Median = 15.0
   Lower quartile = 5.0
   Upper quartile = 24.25
   Semi-interquartile range = 9.625
2. Median = 5.11
   Lower quartile = 4.86
   Upper quartile = 5.365
   Semi-interquartile range = 0.25
3. Median = 92.25
   Lower quartile = 86.5
   Upper quartile = 100.0
   Semi-interquartile range = 6.75
4. Median = 156.5
   Lower quartile = 152
   Upper quartile = 159.75
   Semi-interquartile range = 3.875
5. Median = 68.0
   Lower quartile = 61.5
   Upper quartile = 73.5
   Semi-interquartile range = 6
6. Median = 35.0
   Lower quartile = 24.0
   Upper quartile = 42.5
   Semi-interquartile range = 9.25
7. Median = 44.5
   Lower quartile = 25
   Upper quartile = 53
   Semi-interquartile range = 14
8. Median = 15.25
   Upper quartile = 20
   Lower quartile = 10.875
   Semi-interquartile range = 4.5625

9. Median = 116
Upper quartile = 136
Lower quartile = 96
Semi-interquartile range = 20

10. Median = 68
Upper quartile = 95
Lower quartile = 56
Semi-interquartile range = 19.5

EXERCISE 13.5

1.	8	**2.**	3.1	**3.**	9	**4.**	22
5.	61	**6.**	6	**7.**	27	**8.**	31
9.	8	**10.**	123				

EXERCISE 13.6

1.	2.12	4.50	**2.**	0.93	0.87
3.	2.42	5.87	**4.**	6.63	43.92
5.	17.46	304.9	**6.**	1.52	2.32
7.	7.08	50.06	**8.**	10.35	107.1
9.	2.05	4.20	**10.**	27.37	749.32
11.	4.2	18.00	**12.**	0.93	0.87
13.	4.84	23.48	**14.**	31.05	936.6

EXERCISE 13.7

1.	1.67	**2.**	0.84	**3.**	1.87
4.	5.5	**5.**	14.53	**6.**	1.239
7.	5.625	**8.**	8.694	**9.**	1.533
10.	21.16				

EXERCISE 13.8

1. Standard deviation = 13.17
Variance = 173.49

2. Standard deviation = 2.24
Variance = 5.04

3. Standard deviation = 7.82
Variance = 61.09

4. Standard deviation = 6.60
Variance = 43.5

5. Standard deviation = 5.03
Variance = 25.31

6. Standard deviation = 6.07
Variance = 36.8

7. Standard deviation = 5.46
Variance = 29.76

8. Standard deviation = 2.42
Variance = 5.84

9. Standard deviation = 7.63
Variance = 58.26

10. Standard deviation = 22.02
Variance = 484.97

EXERCISE 13.9

1. (a) (ii) mean = 69.5; s.d. = 28.28

(b) The vertical scale on the left-hand graph represents increases of 5000 and 10000 by the same increment on the axis similar comments apply to the horizontal scale on the right-hand graph. The vertical axis in this graph is without a scale

2. (ii) 74 years (iii) 730; 80.7%

(b) The range is prone to distortion through its sensitivity to outliers

3. (i) mean = 35; s.d. = 15

(ii) mean = 9; s.d. = 3

4. (a) (i) 78 mm (ii) 8.5 mm (iii) 31%
(b) 78.0 mm

5. (a) (i) mean (ii) median (b) $\frac{1}{2}x + 2$

6. mean = 0.8; s.d. = 0.81

7. mean = 65; s.d. = 9.4

8. (i) £85.50 (ii) 91% (iii) 0.81 (iv) 0.8

9. (a) (i) £3320 (ii) £840 (b) 7 (c) 8

10. (a) 53; £24.90 (b) 61.3

11. (a) (i) 9 kg (ii) 44%

(b) 50p, £5, £5.96

EXERCISE 14.1

1.

(i)	0.9	0.625	0.8333
(ii)	0.87	0.6430	0.7258
(iii)	0.89	0.6707	0.8871
(iv)	0.92	0.6	0.8
(v)	0.64	0.4	0.4
(vi)	0.96	0.175	0.4667
(vii)	0.79	0.85	0.7286
(viii)	0.63	0.5667	0.34
(ix)	0.51	0.7778	0.4667
(x)	0.56	0.625	0.4545

2.

(i)	0.91	0.99	0.1
(ii)	0.85	0.83	0.5313
(iii)	0.85	0.99	0.0625
(iv)	0.9	0.89	0.5238
(v)	0.7	0.73	0.4737
(vi)	0.89	0.83	0.6071
(vii)	0.7	0.77	0.434
(viii)	0.6	0.74	0.3939
(ix)	0.43	0.8	0.2597
(x)	0.91	0.9	0.5263
(xi)	0.9	0.87	0.5652
(xii)	0.65	0.63	0.5139
(xiii)	0.6	0.78	0.3548
(xiv)	0.88	0.9	0.4545
(xv)	0.88	0.83	0.5862

EXERCISE 14.2

1. $\frac{17}{30}$; $\frac{45}{68}$

2. (a) (i) $\frac{5}{12}$ (ii) $\frac{49}{144}$ (b) $\frac{26}{63}$

3. (i) $\frac{2}{25}$ (ii) $\frac{3}{200}$ (iii) $\frac{1}{20}$

4. (i) $\frac{28}{45}$ (ii) $\frac{16}{45}$ (iii) $\frac{1}{45}$
(iv) $\frac{17}{45}$ (v) $\frac{9}{25}$ (vi) 0.4
Choose (vi)

5. (i) $\frac{8}{15}$ (ii) $\frac{11}{15}$ (iii) $\frac{1}{3}$
(iv) 0.36

6. (a) (i) 0.855 (ii) 0.145 (iii) 0.14
(b) (i) $\frac{17}{64}$ (ii) $\frac{37}{64}$
Least likely: LWL; most likely: WWW

7. (i) $\frac{1}{11}$ (ii) $\frac{16}{33}$ (iii) $\frac{14}{33}$
(iv) $\frac{4}{9}$ (v) $\frac{14}{33}$ (vi) $\frac{5}{12}$
Choose (vi)

8. 0.004 67

9. (i) $\frac{1}{6}$ (ii) $\frac{1}{4}$ (iii) $\frac{7}{12}$
(iv) $\frac{31}{45}$ (v) $\frac{5}{18}$

10. 0.76

11. $\frac{1}{2}$
(i) (a) $\frac{1}{9}$ (c) $\frac{1}{2}$
(ii) (a) $\frac{5}{36}$ (b) $\frac{1}{6}$ (c) $\frac{3}{5}$

12. (i) (a) 0.4 (b) 0.2 (ii) $\frac{2}{3}$; $\frac{1}{3}$

13. (i) (a) 0.1 (b) 0.2
(ii) (a) 0.25 (b) 0

14. (a) 0.15 (b) 0.3; 0.8
(c) $P(BC) = 0.46 \neq P(B)P(C)$
and therefore not independent

15. (i) (a) 0.06 (b) 0.54
(ii) (a) 0.24 (b) 0.86
(iii) 0.2 (iv) $\frac{27}{29}$

16. (a) (i) 0.56 (ii) 0.28 (iii) 0.58
(b) (i) 0.2 (ii) 0.48 (iii) 0.44

17. (a) (i) 0.244 (ii) 0.432
(b) (i) $\frac{1}{9}$ (ii) $\frac{2}{9}$
(c) (i) 0.15 (ii) 0.33

EXERCISE 15.1

1. (i) 0.933 (ii) 0.957 (iii) 0.971 (iv) 0.984 (v) 0.999 (vi) 0.999 (vii) 0.915 (viii) 0.556 (ix) 0.698 (x) 0.567

2. (i) 0.001 (ii) 0.064 (iii) 0.001 (iv) 0.043 (v) 0.405 (vi) 0.129 (vii) 0.394 (viii) 0.492 (ix) 0.09 (x) 0.019

3. (i) 0.001 (ii) 0.016 (iii) 0.078 (iv) 0.085 (v) 0.035

4. (i) 0.999 (ii) 0.893 (iii) 0.985 (iv) 0.883 (v) 0.967

5. (i) 0.077 (ii) 0.027 (iii) 0.134 (iv) 0.393 (v) 0.429 (vi) 0.325 (vii) 0.436 (viii) 0.416 (ix) 0.099 (x) 0.07

6. (i) 0.298 (ii) 0.305 (iii) 0.339 (iv) 0.117 (v) 0.153 (vi) 0.336 (vii) 0.454 (viii) 0.022 (ix) 0.116 (x) 0.221

7. (i) 0.878 (ii) 0.901 (iii) 0.921 (iv) 0.983 (v) 0.34 (vi) 0.205 (vii) 0.159 (viii) 0.174 (ix) 0.344 (x) 0.906

8. (i) 1.72 (ii) 1.91 (iii) 2.14 (iv) 1.37 (v) 0.17

9. (i) 1.52 (ii) 1.72 (iii) 1.13 (iv) 0.02 (v) 2.07

10. (i) − 2.14 (ii) − 1.42 (iii) − 1.37 (iv) − 1.81

11. (i) − 1.24 (ii) − 1.84 (iii) − 1.19 (iv) − 1.84

12. (i) 1.47 (ii) 2.14 (iii) 2.16 (iv) 1.41 (v) 2.16

13. (i) − 1.24 (ii) − 1.37 (iii) − 1.42 (iv) − 2.14 (v) − 1.24 (vi) − 1.37 (vii) − 1.42

14. (i) 0.6 (ii) − 2.4 (iii) 1.2 (iv) − 3 (v) 2.2 (vi) 2.2 (vii) − 2.5 (viii) − 1.2 (ix) 2.7 (x) − 0.7

15. (i) 30; 2.7 (ii) 92; 9 (iii) 21.4; 9 (iv) 105; 21 (v) 217; 15

16. 25.8 g; 8 g

17. 72.4 mm; 12 mm

18. (i) 0.159 (ii) 0.533 (iii) 0.955 (iv) 2.51 (v) 1.81

19. (a) (i) 13.4 (ii) 3.64 (b) 57, 14

20. (i) 69.1% (ii) 22 (iv) 157 hours

21. (i) 0.092 (ii) 0.143

22. (i) 0.933 (ii) 0.465 (iii) 3950 or 3960 (iv) 6200 or 6210; 11

23. (a) (i) 0.227 (ii) 16.2 ml (iii) 81 (iv) 770 ml

EXERCISE 16.1

1. (a) $a^2 + 2ab + b^2$

(b) $(p + q)^5 = p^5 + 5p^4q + 10p^3q^2 + 10p^2q^3 + 5pq^4 + q^5$

(c) $q^3 + 3q^2p + 3qp^2 + p^3$

(d) $x^6 + 6x^5y + 15x^4y^2 + 20x^3y^3 + 15x^2y^4 + 6xy^5 + y^6$

(e) $b^7 + 7b^6a + 21b^5a^2 + 35b^4a^3 + 35b^3a^4 + 21b^2a^5 + 7ba^6 + a^7$

(f) $m^4 + 4m^3n + 6m^2n^2 + 4mn^3 + n^4$

(g) $x^2 + \frac{2}{3}x + \frac{1}{9}$

(h) $\frac{1}{32} + \frac{5}{16}b + \frac{5}{4}b^2 + \frac{5}{2}b^3 + \frac{5}{2}b^4 + b^5$

(i) $a^4 + a^3 + \frac{3}{8}a^2 + \frac{1}{16}a + \frac{1}{256}$

(j) $\frac{1}{8} + \frac{3}{8} + \frac{3}{8} + \frac{1}{8}$

(k) $\frac{729}{15625} + \frac{2916}{15625} + \frac{4860}{15625} + \frac{4320}{15625} + \frac{2160}{15625} + \frac{576}{15625} + \frac{64}{15625}$

(l) $\frac{1}{256} + \frac{12}{256} + \frac{54}{256} + \frac{108}{256} + \frac{81}{256}$

. (a) $4q^3p$ (b) $6ab^5$ (c) $\frac{20}{27}p^3$

(d) $\frac{1}{27}$ (e) $\frac{27}{128}$ (f) $\frac{7}{128}$

EXERCISE 16.2

1. 0.3456	**2.** $\frac{80}{243}$	**3.** $\frac{64}{125}$
4. $\frac{1}{243}$	**5.** 0.3456	**6.** $\frac{405}{1024}$
7. 0.3087	**8.** $\frac{3}{32}$	**9.** 0.2109
0. 0.3960		

EXERCISE 16.3

1. 0.8192	**2.** $\frac{233}{729}$	**3.** 0.8688
4. 0.0059	**5.** 0.6570	**6.** $\frac{821}{3125}$
7. $\frac{256}{729}$	**8.** 0.3483	**9.** 0.2632
0. 0.9011		

EXERCISE 16.4

. 0.2710	**2.** 0.9533	**3.** 0.9865
. 0.9999	**5.** 0.9560	

EXERCISE 16.5

1. (a) $p^4 + 4p^3q + 6p^2q^2 + 4pq^3 + q^4$

(b) (i) 0.395 (ii) 0.889 (iii) 0.708 (iv) 0.749

2. (i) 0.132 (ii) 0.995

3. (i) 0.18 (ii) 0.47

4. (a) Probability fixed and events independent

(b) (i) 0.0039 (ii) 0.949 (iii) 0.3708

5. $p^3 + 3p^2q + 3pq^2 + q^3$

$p^5 + 5p^4q + 10p^3q^2 + 10p^2q^3 + 5pq^4 + q^5$

(i) $\frac{1}{27}$ (ii) $\frac{40}{243}$ (iii) $\frac{17}{81}$ (iv) $\frac{27}{64}$

(v) 4 shots

6. $p^4 + 4p^3q + 6p^2q^2 + 4pq^3 + q^4$

(i) 0.0036 (ii) 0.0523; 0.9477, 0.9704

7. (i) 0.328 (ii) 0.647 (iii) 0.109

EXERCISE 17.1

1. (a) (i) Seasonal

(ii) Secular

(b) (ii) 3-point moving averages: 427.3; 434; 441; 449.3; 456; 462.7; 470; 477; 483.3; 490.7; 497.7; 505

(iii) 497 000

(iv) £42 322.50

2. (a) (i) (ii) See pages 363–364.

(b) (ii) 4-point moving averages: 14; 16; 17.75; 19.5; 21.5; 24; 27; 30; 33.5

(iii) 52 000 tonnes

3. (a) (i) Irregular

(ii) Secular

(iii) Seasonal

(b) Moving averages may be used to smooth variations thereby exposing the trend in the data more vividly. In addition, moving averages may be used to make predictions of future behaviour based on current data

(c) (ii) 5-point moving averages: 93; 94.6; 96; 97.4; 99; 100.4; 101.8; 103.2; 105; 106.4; 108

(iii) 121 patients

4. (ii) 32; 33.25; 35; 35.75; 37; 38.75; 40; 41.25; 43

(iii) 40

5. (a) (i) Secular

(ii) Irregular

(b) (ii) 5-point moving averages: 53; 57; 61; 65; 68; 70; 72; 73; 74; 75; 76

(iii) £73 000

(iv) Linear

6. (a) See pages 363–364.

(b) (ii) 4-point moving averages: 561; 571; 584; 597; 607; 613; 619; 625; 631

(iii) 331

7. (a) In order to ‘smooth out’ variations and expose the trend of the data and to make predictions for the future behaviour of that data

(b) (ii) 3-point moving averages: 1.9; 2.0; 2.1; 2.2; 2.3; 2.4; 2.6; 2.7; 2.8; 2.9

(iii) £2900

8. (a) (i) Seasonal

(ii) Secular

(iii) Irregular

(b) (ii) 5-point moving averages: 10.2; 10.4; 10.6; 10.8; 11.0; 11.4; 11.8; 12.4; 13.0; 13.6; 14.4

(iii) 15 video recorders

Index

Index

I

K

L

M